DESIGN, DEPLOYMENT AND PERFORMANCE OF 4G-LTE NETWORKS

DESIGN, DEPLOYMENT AND PERFORMANCE OF 4G-LTE NETWORKS

A PRACTICAL APPROACH

Ayman Elnashar
Emirates Integrated Telecomms Co., UAE

Mohamed A. El-saidny
QUALCOMM Technologies, Inc., USA

Mahmoud R. Sherif
Emirates Integrated Telecomms Co., UAE

Library of Congress Cataloging-in-Publication Data

Elnashar, Ayman.
   Design, deployment and performance of 4G-LTE networks : A Practical Approach / Dr Ayman Elnashar, Mr Mohamed A. El-saidny, Dr Mahmoud Sherif.
      pages cm
   Includes bibliographical references and index.
   ISBN 978-1-118-68321-7 (hardback)
1. Wireless communication systems. 2.   Mobile communication systems.   I. Title.
   TK5103.2.E48 2014
   621.3845′6–dc23

                                                                                    2013037384

A catalogue record for this book is available from the British Library.

ISBN: 978-1-118-68321-7

Typeset in 10/12pt TimesLTStd by Laserwords Private Limited, Chennai, India

1   2014

To my beloved kids Noursin, Amira, and Yousef. You're the inspiration!

This book is dedicated to the memory of my father (God bless his soul) and also my mother, who's been a rock of stability throughout my life. This book is also dedicated to my beloved wife whose consistent support and patience sustain me still.

My sincerest appreciations for a lifetime career that has surpassed anything my imagination could have conceived.

Ayman Elnashar

To my Family for all their continuous support. To my elder brother for his guidance and motivation throughout the years. To my inspirational, intelligent, and beautiful daughter, Hana.

Your work is going to fill a large part of your life, and the only way to be truly satisfied is to do what you believe is great work. And the only way to do great work is to love what you do. If you haven't found it yet, keep looking. Don't settle. As with all matters of the heart, you'll know when you find it. – Steve Jobs

Mohamed A. El-saidny

This work would not have been possible without the consistent and full support of my beloved family. To my beloved wife, Meram, to my intelligent, motivating, and beautiful kids, Moustafa, Tasneem, and Omar. You are my inspiration.

To my Dad, my Mom (God bless her soul), my brother, and my entire family. Thank you for all your support and encouragement.

There is no elevator to success. You have to take the stairs. – Unknown Author

Those who think they have found this elevator will end up falling down the elevator shaft

Mahmoud R. Sherif

Contents

Authors' Biographies

Ayman Elnashar was born in Egypt in 1972. He received the B.S. degree in electrical engineering from Alexandria University, Alexandria, Egypt, in 1995 and the M.Sc. and Ph.D. degrees in electrical communications engineering from Mansoura University, Mansoura, Egypt, in 1999 and 2005, respectively. He obtained his M.Sc. and Ph.D. degrees while working fulltime. He has more than 17 years of experience in telecoms industry including GSM, GPRS/EDGE, UMTS/HSPA+/LTE, WiMax, WiFi, and transport/backhauling technologies. He was part of three major start-up telecom operators in MENA region (Mobinil/Egypt, Mobily/KSA, and du/UAE) and held key leadership positions. Currently, he is Sr. Director of Wireless Broadband, Terminals, and Performance with the Emirates Integrated Telecommunications Co. "du", UAE. He is in charge of mobile and fixed wireless broadband networks. He is responsible for strategy and innovation, design and planning, performance and optimization, and rollout/implementation of mobile and wireless broadband networks. He is the founder of the Terminals department and also the terminals lab for end-to-end testing, validation, and benchmarking of mobile terminals. He managed and directed the evolution, evaluation, and introduction of du mobile broadband HSPA+/LTE networks. Prior to this, he was with Mobily, Saudi Arabia, from June 2005 to Jan 2008 and with Mobinil (orange), Egypt, from March 2000 to June 2005. He played key role in contributing to the success of the mobile broadband network of Mobily/KSA.

He managed several large-scale networks, and mega projects with more than 1.5 billion USD budgets including start-ups (LTE 1800 MHz, UMTS, HSPA+, and WiMAX16e), networks expansions (GSM, UMTS/HSPA+, WiFi, and transport/backhauling) and swap projects (GSM, UMTS, MW, and transport network) from major infrastructure vendors. He obtained his PhD degree in multiuser interference cancellation and smart antennas for cellular systems. He published 20+ papers in wireless communications arena in highly ranked journals such as *IEEE Transactions on Antenna and Propagation, IEEE Transactions Vehicular technology, and IEEE Transactions Circuits and Systems I, IEEE Vehicular technology Magazine, IET Signal Processing*, and international conferences. His research interests include practical performance analysis of cellular systems (CDMA-based & OFDM-based), 3G/4G mobile networks planning, design, and Optimization, digital signal processing for wireless communications, multiuser detection, smart antennas, MIMO, and robust adaptive detection and beamforming. He is currently working on LTE-Advanced and beyond including eICIC, HetNet, UL/DL CoMP, 3D Beamforming, Combined LTE/HSPA+, Combined LTE/WiFi: simultaneous reception, etc …

Mohamed A. El-saidny is a technical expert with 10+ years of international technical and leadership experience in wireless communication systems for mobile phones, modem chipsets, and networks operators. He received the B.Sc. degree in Computer Engineering and the M.Sc.

degree in Electrical Engineering from the University of Alabama in Huntsville, USA in 2002 and 2004, respectively. From 2004 to 2008, he worked in Qualcomm CDMA Technology, Inc. (QCT), San Diego, California, USA. He was responsible for performance evaluation and analysis of the Qualcomm UMTS system and software solutions used in user equipment. As part of his assignments, he developed and implemented system studies to optimize the performance of various UMTS algorithms. The enhancements utilize Cell re-selection, Handover, Cell Search and Paging. He worked on several IOT and field trials to evaluate and improve the performance of 3G systems. Since 2008, he has been working in Qualcomm Corporate Engineering Services division in Dubai, UAE. He has been working on expanding the 3G/4G technologies footprints with operators, with an additional focus on user equipment and network performance as well as technical roadmaps related to the industry. Mohamed is currently supporting operators in Middle East and North Africa in addition to worldwide network operators and groups in LTE commercial efforts. His responsibilities are to ensure the device and network performance are within expectations. He led a key role in different first time features evaluations such as CSFB, C-DRX, IRAT, and load balance techniques in LTE. As part of this role, he is focused on aligning network operators to the device and chipset roadmaps and products in both 3G and 4G. Mohamed is the author of several international IEEE journal papers and contributions to 3GPP, and an inventor of numerous patents.

Mahmoud R. Sherif is a leading technical expert with more than 18 years of international experience in the design, development and implementation of fourth generation mobile broadband technologies and networks. He received his Ph.D. degree in Electrical Engineering from the City University of New York, USA in February 2000. His Ph.D. degree was preceded by the B.Sc. degree in Computer Engineering and the M.Sc. degree in Electrical Engineering from the University of Ain Shams in Cairo, Egypt in 1992, and 1996, respectively. From 1997 to 2008, he was working in the Wireless Business Unit at Lucent Technologies (which became Alcatel-Lucent in 2007), in Whippany, New Jersey, USA. He led the Voice and Data Quality and Performance Analysis team responsible for the end-to-end performance analysis of the different wireless/mobile technologies. In November 2008, he moved to Dubai in the United Arab Emirates to join the Emirates Integrated Telecommunications Co. "du" where he is now the Head of the Mobile Access Planning within du (Senior Director Mobile Access Planning) managing the Radio Planning, Site Acquisition and Capacity and Feature Management Departments. He is responsible for managing the planning of the mobile access network nationwide, Mobile Sites' Acquisition, Strategic Planning on Mobile Access Network Capacity Management, all Feature testing and rollout across 2G, 3G and LTE, defining and managing the financial resources efficiently and with alignment with company's financial targets (CAPEX & OPEX). He is also responsible for the mobile access network technology strategy in coordination with the commercial and marketing teams. He is considered a company expert resource in the various mobile broadband technologies, including HSPA+, LTE, VoLTE and LTE-A. He has published several related papers in various technical journals as well as multiple international conferences. He has multiple contributions to the 3GPP and other telecommunications standards. He also has multiple granted patents in the USA.

Preface

Cellular mobile networks have been evolving for many years. Several cellular systems and networks have been developed and deployed worldwide to provide the end user with quality and reliable communication over the air. Mobile technologies from the first to third generation have been quickly evolving to meet the need of services for voice, video, and data.

Today, the transition to smartphones has steered the user's interest toward a more mobile-based range of applications and services, increasing the demand for more network capacity and bandwidth. Meanwhile, this transition presents a significant revenue opportunity for network operators and service providers, as there is substantially higher average revenue per user (ARPU) from smartphone sales and relevant services. While the rollout of more advanced radio networks is proceeding rapidly, smartphone penetration is also increasing exponentially. Therefore, network operators need to ensure that the subscribers' experience stays the same as, or is even better than, with the older existing systems.

With the growing demand for data services, it is becoming increasingly challenging to meet the required data capacity and cell-edge spectrum efficiency. This adds more demand on the network operators, vendors and device providers to apply methods and features that stabilize the system's capacity and consequently improves the end-user experience. 4G systems and relevant advanced features have the capabilities to keep up with today's widespread use of mobile-communication devices, providing a range of mobile services and quality communications.

This book describes the long term evolution (LTE) technology for mobile systems; a transition from third to fourth generation. LTE has been developed in the 3GPP (Third Generation Partnership Project), starting from the first version in Release 8 and through to the continuing evolution to Release 10, the latest version of LTE, also known as LTE-Advanced. The analysis in this book is based on the LTE of 3GPP Release 8 together with Release 9 and Release 10 roadmaps, with a focus on the LTE-FDD (frequency division duplex) mode . Unlike other books, the authors have bridged the gap between theory and practice, thanks to hands on experience in the design, deployment, and performance of commercial 4G-LTE networks and terminals.

The book is a practical guide for 4G networks designers, planners, and optimizers, as well as other readers with different levels of expertise. The book brings extensive and broad practical hands-on experience to the readers. Practical scenarios and case studies are provided, including performance aspects, link budgets, end-to-end architecture, end-to-end QoS (quality of service) topology, dimensioning exercises, field measurement results, applicable business case studies, and roadmaps.

Chapters 1 and 2 describe the LTE system architecture, interfaces, and protocols. They also introduce the LTE air interface and layers, in addition to downlink and uplink channels and procedures.

Chapters 3 to 8 constitute the main part of the book. They provide a deeper insight into the LTE system features, performance, design aspects, deployment scenarios, planning exercises, VoLTE (voice over long term evolution) implementation, and the evolution and roadmap to LTE-Advanced. Further material supporting this book can be found in www.ltehetnet.com.

Acknowledgments

We would like to express our deep gratitude to our colleagues in Qualcomm and du for assisting in reviewing and providing excellent feedback on this work. We are indebted to Huawei team in the UAE for their great support and review of Chapters 5 and 6, and also for providing the necessary supporting materials. Special thanks go to the wireless broadband and terminals team at du for their valuable support. We acknowledge the support of Harri Holma from NSN, for reviewing and providing valuable comments on Chapters 5 and 6. We wish to express our appreciation to every reviewer who reviewed the book proposal and provided very positive feedback and insightful comments. Thanks for their valuable comments and suggestions. Our thanks go to our families for their patience, understanding, and constant encouragement, which provided the necessary enthusiasm to accomplish this book. Also, our deep and sincere appreciations go to our professors who supervised and guided us through our academic career. Finally, we would like to thank the publishing team at John Wiley & Sons for their competence, extensive support and encouragement throughout the project to bring this work to completion.

Abbreviations and Acronyms

16-QAM	16-Quadrature amplitude modulation
64-QAM	64-Quadrature amplitude modulation
1G, 2G, 3G or 4G	1st, 2nd, 3rd, 4th generation
3GPP	Third generation partnership project
3GPP2	Third generation partnership project 2
AAA	Authentication, authorization and accounting
ACK	Acknowledgment
AES	Advanced encryption standard
AF	Application Function
AIPN	All-IP network
AMBR	Aggregate maximum bit rate
AMC	Adaptive modulation and coding
AMD	Acknowledged mode data
AN	Access network
APN	Access point name
ARP	Allocation and retention priority
ARQ	Automatic repeat request
AS	Access stratum
BC	Business Case
BCCH	Broadcast control channel
BCH	Broadcast channel
BI	Backoff indicator
BLER	Block error rate
BP	Bandwidth part
BSR	Buffer status report
BW	Bandwidth
CAPEX	Capital Expenditure
CCCH	Common control channel
CCE	Control channel elements
CDD	Cyclic delay diversity
CDM	Code Division Multiplexed
CDMA	Code division multiple access

CDS	Channel dependent scheduling
CFI	Control format indicator
CN	Core network
COGS	Cost of Goods Sold
CP	Control plane
	Cyclic prefix
CQI	Channel quality indicator
CRC	Cyclic redundancy check
CRF	Charging Rules Function
C-RNTI	Cell radio network temporary identifier
CS	Circuit switched
CSG	Closed subscriber group
CSI	Channel signal information
CW	Code word
DAS	Distributed Antenna System
DCCH	Dedicated control channel
DCI	Downlink control information
DFT	Discrete Fourier transform
DFTS-OFDM	Discrete Fourier transform spread orthogonal frequency division multiplexing
DL	Downlink
DL-SCH	Downlink shared channel
DM	Demodulation
DM-RS	Demodulation reference signal
DNS	Domain Name System
DRX	Discontinuous transmission
DS	Data services
DTCH	Dedicated traffic channel
E-AGCH	Enhanced absolute granting channel
EBITDA	Earnings Before Interest, Taxes, Depreciation, and Amortization
E-DCH	Enhanced dedicated channel
E-DPCCH	Enhanced dedicated physical control channel
E-DPDCH	Enhanced dedicated physical data channel
E-HICH	Enhanced hybrid indicator channel
EEA	EPS encryption algorithm
EIA	EPS integrity algorithm
EIR	Equipment Identity register
EMM	EPS mobility management
eNB	Evolved node B
EPC	Evolved packet core
EPLMN	Equivalent PLMN
EPRE	Energy per resource element
EPS	Evolved packet system
E-RGCH	Enhanced relative granting channel
ESM	EPS session management
ESP	Encapsulated security protocol

ETWS	Earthquake and tsunami warning system
E-UTRA	Evolved UMTS terrestrial radio access; PHY aspects
E-UTRAN	Evolved UMTS terrestrial radio access network; MAC/L2/L3 aspects
FD	Full-duplex
FDD	Frequency division duplex
FDM	Frequency division multiplexing
FDMA	Frequency division multiple access
FFT	Fast Fourier transform
FH	Frequency hopping
FI	Framing information
FL	Forward link
FMS	First missing sequence
FS	Frame structure
FSTD	Frequency shift time diversity
GBR	Guaranteed bit rate
GERAN	GSM/EDGE radio access network
GGSN	GPRS gateway support node
GPRS	General packet radio service
GSM	Global system for mobiles (European standard)
GTP-U	GPRS tunneling protocol – user
GUMMEI	Globally unique MME identity
GUTI	Globally unique temporary identifier
GW	Gateway
HA	Home agent
HAP ID	HARQ process ID
HARQ	Hybrid ARQ
HD	Half-duplex
HFN	Hyper frame number
HI	Hybrid ARQ indicator
HLD	High Level Design
HLR	Home location register
HNBID	Home evolved node B identifier
HO	Handover
HPLMN	Home public land mobile network
HRPD	High rate packet data
HS	High speed
HSDPA	High speed downlink packet access
HS-DPCCH	High speed dedicated control channel
HSPA	High speed packet access
HSPA+	High speed packet access evolved or enhanced
HSS	Home subscriber service
HSUPA	High speed uplink packet access
IDFT	Inverse discrete Fourier transform
IETF	Internet Engineering Task Force
IFFT	Inverse fast Fourier transform
IMS	IP Multimedia subsystem

IMSI	International Mobile Subscriber Identity
IP	Internet protocol
IP-CAN	IP connectivity access network
ISI	Inter-symbol interference
ISR	Idle signaling load reduction
IRR	Internal Rate of Return
L1, L2, L3	Layer 1, 2, 3
LA	Location area
LAC	Location area code
LAI	Location area identifier
LAU	Location area updating
LCG	Logical channel group
LDAP	Lightweight Directory Access
LFDM	Localized frequency division multiplexing
LI	Lawful Interception
LI	Length indicators
LTE	Long term evolution
LTI	Linear time invariant
MAC	Medium access control
MAC-I	Message authentication code for integrity
MBMS	Multimedia broadcast multicast service
MBR	Maximum bit rate
MBSFN	Multimedia broadcast over a single frequency network
MCCH	Multicast control channel
MCH	Multicast channel
MCS	Modulation and coding schemes
MCW	Multiple code word
ME	Mobile equipment
MIB	Master information block
MIMO	Multiple-input–multiple-output
MME	Mobility management entity
MMEC	MME code
MMEGI	MME group ID
MSISDN	Mobile Subscriber Integrated Services Digital Network-Number
MOS	Mean Opinion Score
MTCH	Multicast traffic channel
MU-MIMO	Multi-user multiple-input–multiple-output
NAK	Negative acknowledgment
NAS	Non-access stratum
NDI	New data indicator
NID	Network ID
NPV	Net Present Value
OCS	Online Charging System
OFCS	Offline Charging System
OFDM	Orthogonal frequency division multiplexing
OFDMA	Orthogonal frequency division multiple access

OS	Operating system
PAPR	Peak-to-average power ratio
PAR	Peak to average ratio
PBCH	Physical broadcast channel
PCC	Policy charging and control
PCCH	Paging control channel
PCFICH	Physical control format indicator channel
PCH	Paging channel
PCRF	Policy and charging rules function
PDCCH	Physical downlink control channel
PDCP	Packet data convergence protocol
PDG	Packet data gateway
PDN	Packet data network
PDSCH	Physical downlink shared channel
PDSN	Packet data serving node
PDU	Protocol data unit
PELR	Packet error loss rate
P-GW	Packet data network gateway
PHICH	Physical hybrid automatic repeat request indicator channel
PHR	Power headroom report
PHY	Physical layer
PIM	Passive Intermodulation
PLMN	Public land mobile network
PMCH	Physical multicast channel
PMI	Precoding matrix indicator
PMIP	Proxy mobile IP
PoC	Push-to-talk over cellular
PRACH	Physical random access channel
PRB	Physical resource block
PS	Packet switched
PSC	Primary synchronization code
P-SCH	Primary synchronization channel
PSS	Primary synchronization signal
PSTN	Packet switched telephone network
PSVT	Packet switched video telephony
PTT	Push-to-talk
PUCCH	Physical uplink control channel
PUSCH	Physical uplink shared channel
QAM	Quadrature amplitude modulation
QCI	QoS class identifier
QoS	Quality of service
QPSK	Quadrature phase shift keying
RA	Routing area
RAC	Routing area code
RACH	Random access channel
RAN	Radio access network

RAPID	Random access preamble identifier
RAR	Random access response
RAU	Routing area updating
RB	Resource block
RBG	Resource block group
RDS	RMS delay spread
RE	Resource element
REG	Resource element group
RI	Rank indicator
RIV	Resource indication value
RL	Reverse link
RLC	Radio link control
RLF	Radio link failure
RMS	Root-mean-square
RN	Relay Node
RNC	Radio network controller
RNL	Radio network layer
RNTI	Radio network temporary identifier
ROHC	Robust header compression
ROI	Return On Investment
RPLMN	Registered PLMN
RRC	Radio resource control
RRM	Radio resource management
RS	Reference signal
RV	Redundancy version
SAE	System architecture evolution
SAW	Stop-and-wait
SC-FDM	Single-carrier frequency division multiplexing
SC-FDMA	Single-carrier frequency division multiple access
SCH	Supplemental channel (CDMA2000)
	Synchronization channel (WCDMA)
SCTP	Stream control transmission protocol
SCW	Single code word
SDF	Service data low
SDM	Spatial division multiplexing
SDMA	Spatial division multiple access
SDU	Service data unit
SFBC	Space frequency block code
SFN	System frame number
SGSN	Serving GPRS support node
S-GW	Serving gateway
SI	System information message
SIB	System information block
SINR	Signal to interference noise ratio
SM	Session management
	Spatial multiplexing

SNR	Signal to noise ratio
SOAP	Simple Object Access Protocol
SPOF	Single Point of Failure
SPS	Semi-persistent scheduling
SR	Scheduling request
SRS	Sounding reference signals
SSC	Secondary synchronization code
S-SCH	Secondary synchronization channel
SSS	Secondary synchronization signal
SU-MIMO	Single-user multiple-input–multiple-output
TA	Tracking area
	Timing advance/alignment
TAC	Tracking area code
TAI (_List)	Tracking area identifier (_List)
TAU	Tracking area update
TDD	Time division duplex
TDM	Time division multiplexing
TDMA	Time division multiple access
TFT	Traffic flow template
TPC	Transmit power control
TTI	Transmission time interval
Tx	Transmit
UCI	Uplink control information
UE	User equipment
UL	Uplink
UL-SCH	Uplink shared channel
UMTS	Universal mobile telecommunications system
UP	User plane
UTRA	UMTS terrestrial radio access
UTRAN	UMTS terrestrial radio access network
VAF	Voice Activity Factor
VoIP	Voice over Internet protocol
VoLTE	Voice over LTE
VRB	Virtual resource block
VT	Video telephony
WACC	Weighted Average Cost of Capital
WCDMA	Wideband code division multiple access
WiMAX	Worldwide interoperability for microwave access
X2	The interface between eNodeBs
ZC	Zadoff–Chu

1

LTE Network Architecture and Protocols

Ayman Elnashar and Mohamed A. El-saidny

Cellular mobile networks have been evolving for many years. The initial networks are referred to as First Generation, or 1G systems. The 1G mobile system was designed to utilize analog. It included the AMPS (advanced mobile phone system). The Second Generation, 2G mobile systems, were introduced utilizing digital multiple access technology; TDMA (time division multiple access) and CDMA (code division multiple access). The main 2G networks were GSM (global system for mobile communications) and CDMA, also known as cdmaOne or IS-95 (Interim Standard 95). The GSM system still has worldwide support and is available for deployment on several frequency bands, such as 900, 1800, 850, and 1900 MHz. CDMA systems in 2G networks use a spread spectrum technique and utilize a mixture of codes and timing to identify cells and channels. In addition to being digital, as well as improving capacity and security, the 2G systems also offer enhanced services, such as SMS (short message service) and circuit switched (CS) data. Different variations of the 2G technology evolved later to extend the support of efficient packet data services, and to increase the data rates. GPRS (general packet radio system) and EDGE (enhanced data rates for global evolution) systems have been the evolution path of GSM. The theoretical data rate of 473.6 kbps enabled the operators to offer multimedia services efficiently. Since it does not comply with all the features of a 3G system, EDGE is usually categorized as 2.75G.

3G (Third Generation) systems are defined by IMT2000 (International Mobile Telecommunications). IMT2000 defines that a 3G system should provide higher transmission rates in the range of 2 Mbps for stationary use and 348 kbps in mobile conditions. The main 3G technologies are:

- **WCDMA (wideband code division multiple access)** – This was developed by the 3GPP (Third Generation Partnership Project). WCDMA is the air interface of the 3G UMTS (universal mobile telecommunications system). The UMTS system has been deployed based on the existing GSM communication core network (CN) but with a totally new radio access technology (RAT) in the form of WCDMA. Its radio access is based on FDD (frequency division duplex). Current deployments are mainly at 2.1 GHz bands. Deployments at lower frequencies are also possible, such as UMTS900. UMTS supports voice and multimedia services.

- **TD-CDMA (time division multiple access)** – This is typically referred to as UMTS TDD (time division duplex) and is part of the UMTS specifications. The system utilizes a combination of CDMA and TDMA to enable efficient allocation of resources.
- **TD-SCDMA (time division synchronous code division multiple access)** – This has links to the UMTS specifications and is often identified as UMTS-TDD low chip rate. Like TD-CDMA, it is also best suited to low mobility scenarios in microcells or picocells.
- **CDMA2000** – This is a multi-carrier technology standard which uses CDMA. It is part of the 3GPP2 standardization body. CDMA2000 is a set of standards including CDMA2000 EV-DO (evolution-data optimized) which has various revisions. It is backward compatible with cdmaOne.
- **WiMAX (worldwide interoperability for microwave access)** – This is another wireless technology which satisfies IMT2000 3G requirements. The air interface is part of the IEEE (Institute of Electrical and Electronics Engineers) 802.16 standard which originally defined PTP (point-to-point) and PTM (point-to-multipoint) systems. This was later enhanced to provide greater mobility. WiMAX Forum is the organization formed to promote interoperability between vendors.

4G (Fourth Generation) cellular wireless systems have been introduced as the latest version of mobile technologies. 4G is defined to meet the requirements set by the ITU (International Telecommunication Union) as part of IMT Advanced.

The main drivers for the network architecture evolution in 4G systems are: all-IP (Internet protocol) -based, reduced network cost, reduced data latencies and signaling load, interworking mobility among other access networks in 3GPP and non-3GPP, always-on user experience with flexible quality of service (QoS) support, and worldwide roaming capability. 4G systems include different access technologies:

- **LTE and LTE-Advanced (long term evolution)** – This is part of 3GPP. LTE as it stands now does not meet all IMT Advanced features. However, LTE-Advanced is part of a later 3GPP release and has been designed specifically to meet 4G requirements.
- **WiMAX 802.16m** – The IEEE and the WiMAX Forum have identified 802.16m as their offering for a 4G system.
- **UMB (ultra mobile broadband)** – This is identified as EV-DO Rev C. It is part of 3GPP2. Most vendors and network operators have decided to promote LTE instead.

1.1 Evolution of 3GPP Standards

The specifications of GSM, GPRS, EDGE, UMTS, and LTE have been developed in stages, known as 3GPP releases. Operators, network, and device vendors use these releases as part of their development roadmap. All 3GPP releases are backward compatible. This means that a device supporting one of the earlier releases of 3GPP technologies can still work on a newer release deployed in the network.

The availability of devices on a more advanced 3GPP release makes a great contribution to the choice of evolution by the operator. Collaboration between network operators, network vendors, and chipset providers is an important step in defining the roadmap and evolution of 3GPP features and releases. This has been the case in many markets.

1.1.1   3GPP Release 99

3GPP Release 99 has introduced UMTS, as well as the EDGE enhancement to GPRS. UMTS contains all features needed to meet the IMT-2000 requirements as defined by the ITU. It is able to support CS voice and video services, as well as PS (packet switched) data services over common and dedicated channels. The theoretical data rate of UMTS in this release is 2 Mbps. The practical uplink and downlink data rates for UMTS in deployed networks have been 64, 128, and 384 kbps.

1.1.2   3GPP Release 4

Release 4 includes enhancements to the CN. The concept of all-IP networks has been introduced in this release. There has not been any significant change added to the user equipment (UE) or air interface in this release.

1.1.3   3GPP Release 5

Release 5 is the first major addition to the UMTS air interface. It adds HSDPA (high speed downlink packet access) to improve capacity and spectral efficiency. The goal of HSDPA in the 3GPP roadmap was to improve the end-user experience and to keep up with the evolution taking place in non-3GPP technologies. During the time when HSDPA was being developed, the increasing interest in mobile-based services demanded a significant improvement in the air interface of the UMTS system.

HSDPA improves the downlink speeds from 384 kbps to a maximum theoretical 14.4 Mbps. The typical rates in the Release 5 networks and devices are 3.6 and 7.2 Mbps. The uplink in Release 5 has preserved the capabilities of Release 99.

HSDPA provides the following main features which hold as the fundamentals of all subsequent 3GPP evolutions:

- **Adaptive modulation** – In addition to the original UMTS modulation scheme, QPSK (quadrature phase shift keying), Release 5 also includes support for 16-QAM (quadrature amplitude modulation).
- **Flexible coding** – Based on fast feedback from the mobile in the form of a CQI (channel quality indicator), the UMTS base station (known as NodeB) is able to modify the effective coding rate and thus increase system efficiency. In Release 99, such adaptive data rate scheduling took place at the RNC (radio network controller) which impacted the cell capacity and edge of cell data rates.
- **Fast scheduling** – HSDPA includes a shorter TTI (time transmission interval) of 2 ms, which enables the NodeB scheduler to quickly and efficiently allocate resources to mobiles. In Release 99 the minimum TTI was 10 ms, adding more latency to the packets being transmitted over the air.

- **HARQ (hybrid automatic repeat request)** – If a packet does not get through to the UE successfully, the system employs HARQ. This improves the retransmission timing, thus requiring less reliance on the RNC. In Release 99, the packet re-transmission was mainly controlled by the physical (PHY) layer as well as the RNC's ARQ (automatic repeat request) algorithm, which was slower in adapting to the radio conditions.

1.1.4   3GPP Release 6

Release 6 adds various features, with HSUPA (high speed uplink packet data) being the key one. HSUPA also goes under the term "enhanced uplink, EUL". The term HSPA (high speed packet access) is normally used to describe a Release 6 network since an HSUPA call requires HSDPA on the downlink.

The downlink of Release 6 remained the same as in HSDPA of Release 5. The uplink data rate of the HSUPA system can go up to 5.76 Mbps with 2 ms TTI used in the network and devices. The practical uplink data rates deployed are 1.4 and 2 Mbps. It is worth noting that there is a dependence between the downlink and uplink data rates. Even if the user is only downloading data at a high speed, the uplink needs to cope with the packet acknowledgments at the same high speed. Therefore any data rate evolution in the downlink needs to have an evolved uplink as well.

HSUPA, like HSDPA, adds functionalities to improve packet data which include:

- **Flexible coding** – HSUPA has the ability to dynamically change the coding and therefore improves the efficiency of the system.
- **Fast power scheduling** – A key fact of HSUPA is that it provides a method to schedule the power to different mobiles. This scheduling can use either a 2 or 10 ms TTI. 2 ms usually reveals a challenge on the uplink interference and coverage when compared to 10 ms TTI operation. Hence, a switch between the two TTI is possible within the same EUL data call.
- **HARQ** – Like HSDPA, HSUPA also utilizes HARQ concepts in lower layers. The main difference is the timing relationship for the retransmission and the synchronized HARQ processes.

1.1.5   3GPP Release 7

The main addition to this release is HSPA+, also known as evolved HSPA. During the commercialization of HSPA, LTE system development has been started, promising a more enhanced bandwidth and system capacity. Evolution of the HSPA system was important to keep up with any competitor technologies and prolong the lifetime of UMTS systems.

HSPA+ provides various enhancements to improve PS data delivery. The features in HSPA+ have been introduced as add-ons. The operators typically evaluate the best options of HSPA+ features for deployment interests, based on the traffic increase requirements, flexibility, and the cost associated for the return of investment. HSPA+ in Release 7 includes:

- **64 QAM** – This is added to the downlink and enables HSPA+ to operate at a theoretical rate of 21.6 Mbps.
- **16 QAM** – This is added to the uplink and enables the uplink to theoretically achieve 11.76 Mbps.

- **MIMO (multiple input multiple output) operation** – This offers various capacity benefits including the ability to reach a theoretical 28.8 Mbps data rate in the downlink.
- **Power and battery enhancements** – Various enhancements such as CPC (continuous packet connectivity) have been included. CPC enables DTX (discontinuous transmission) and DRX (discontinuous reception) functions in connected mode.
- **Less data packet overhead** – The downlink includes an enhancement to the lower layers in the protocol stack. This effectively means that fewer headers are required, and in turn, improves the system efficiency.

1.1.6 3GPP Release 8

On the HSPA+ side, Release 8 has continued to improve the system efficiency and data rates by providing:

- **MIMO with 64 QAM modulation** – It enables the combination of 64 QAM and MIMO, thus reaching a theoretical rate of 42 Mbps, that is, 2×21.6 Mbps.
- **Dual cell operation** – DC-HSDPA (dual cell high speed downlink packet access) is a feature which is further enhanced in Releases 9 and 10. It enables a mobile to effectively utilize two 5 MHz UMTS carriers. Assuming both are using 64 QAM (21.6 Mbps), the theoretical data rate is 42 Mbps. DC-HSDPA has gained the primary interest over other Release 8 features, and most networks are currently either supporting it or in the deployment stage.
- **Further power and battery enhancements** – deploys a feature known as enhanced fast dormancy as well as enhanced RRC state transitions.

The 3GPP Release 8 defines the first standardization of the LTE specifications. The evolved packet system (EPS) is defined, mandating the key features and components of both the radio access network (E-UTRAN, evolved universal terrestrial radio access network) and the CN (evolved packet core, EPC). Orthogonal frequency division multiplexing is defined as the air interface with the ability to support multi-layer data streams using MIMO antenna systems to increase spectral efficiency.

LTE is defined as an all-IP network topology differentiated over the legacy CS domain. However, the Release 8 specification makes use of the CS domain to maintain compatibility with the 2G and 3G systems utilizing the voice calls circuit switch Fallback (CSFB) technique for any of those systems.

LTE in Release 8 has a theoretical data rate of 300 Mbps. The most common deployment is 100 to 150 Mbps with a full usage of the bandwidth, 20 MHz. Several other variants are also deployed in less bandwidth and hence with lower data rates. The bandwidth allocation is tied to the amount of spectrum acquired by the LTE network operators in every country.

The motivations and different options discussed in 3GPP for the EPS network architecture have been detailed in several standardized technical reports in [1–4].

1.1.7 3GPP Release 9 and Beyond

Even though LTE is a Release 8 system, it is further enhanced in Release 9. There are a number of features in Release 9. One of the most important is the support of additional frequency bands and additional enhancements to CSFB voice calls from LTE.

On the HSPA+ side, Release 9 and beyond continued to build on the top of previous HSPA+ enhancements by introducing DC-HSUPA, MIMO + DC-HSDPA, and multi-carrier high speed downlink packet access (MC-HSDPA). The downlink of HSPA+ in this release is expected to reach 84 Mbps, while the uplink can reach up to 42 Mbps.

Release 10 includes the standardization of LTE Advanced, the 3GPP's 4G offering. It includes modification to the LTE system to facilitate 4G services. The requirements of ITU are to develop a system with increased data rates up to 1 Gbps in the downlink and 500 Mbps in the uplink. Other requirements of ITU's 4G are worldwide roaming and compatibility of services. LTE-Advanced is now seeing more interest, especially from the operators who have already deployed LTE in early stages.

As discussed in this 3GPP evolution, the 4G system is designed to refer to LTE-Advanced. However, since UMTS has been widely used as a 3G system, investing in and building up an ecosystem for an LTE network using the same "3G" term would have been misinterpreted. Hence, regulators in most countries have allowed the mobile operators to use the term "4G" when referring to LTE. This book considers the term 4G when referring to an LTE system, especially for the concepts that are still common between LTE and LTE-Advanced.

This chapter describes the overall architecture of an LTE CN, radio access protocols, and air interface procedures. This chapter and the upcoming parts of the book focus on Release 8 and 9 of the 3GPP specifications. The last chapter of the book gives an overview of the features beyond Release 9.

1.2   Radio Interface Techniques in 3GPP Systems

In wireless cellular systems, mobile users share a common medium for transmission. There are various categories of assignment. The main four are FDMA (frequency division multiple access), TDMA, CDMA, and OFDMA (orthogonal frequency division multiple access). Each of the technologies discussed earlier in the chapter utilizes one of these techniques. This is another reason for distinguishing the technologies.

1.2.1   Frequency Division Multiple Access (FDMA)

In order to accommodate various devices on the same wireless network, FDMA divides the available spectrum into sub-bands or channels. Using this technique, a dedicated channel can be allocated to a user, while other users occupy other channels or frequencies.

FDMA channels can suffer from higher interference. They cannot be close together due to the energy from one transmission affecting the adjacent or neighboring channels. To combat this, additional guard bands between channels are required, which also reduces the system's spectral efficiency. The uplink or downlink receiver must use filtering to mitigate interference from other users.

1.2.2   Time Division Multiple Access (TDMA)

In TDMA systems the channel bandwidth is shared in the time domain. It assigns a relatively narrow spectrum allocation to each user, but in this case the bandwidth is shared between a set of users. Channelization of users in the same band is achieved by a separation in both frequency and time. The number of timeslots in a TDMA frame is dependent on the system. For example, GSM utilizes eight timeslots.

TDMA systems are digital and therefore offer security features such as ciphering and integrity. In addition, they can employ enhanced error detection and correction schemes including FEC (forward error correction). This enables the system to be more resilient to noise and interference and therefore they have a greater spectral efficiency than FDMA systems.

1.2.3 Code Division Multiple Access (CDMA)

The concept of CDMA is slightly different to that of FDMA and TDMA. Instead of sharing resources in the time or frequency domain, the devices are able to use the system at the same time and using the same frequency. This is possible because each transmission is separated using a unique channelization code.

UMTS, cdmaOne, and CDMA2000 all use CDMA as their air interface technique. However, the implementation of the codes and the bandwidths used by each technology is different. For example, UMTS utilizes a 5 MHz channel bandwidth, whereas cdmaOne uses only 1.25 MHz.

Codes are used to achieve orthogonality between the users. In the HSDPA system, for example, the channel carrying the data to the user has a total of 16 codes in the code tree. If there are multiple users in the system at the same timeslot of scheduling, the users will share the 16 codes, each with a different part of the code tree. The more codes assigned to the HSDPA user, the higher the data rate becomes. There are limitations on the code tree and hence capacity is tied to the code allocation. Voice users and control channels get the highest priority in code assignment, and then the data users utilize the remaining parts of the tree.

WCDMA systems are also interference limited since all users are assigned within the same frequency in the cell. Hence, power control and time scheduling are important to limit the interference impacting the users' performance.

1.2.4 Orthogonal Frequency Division Multiple Access (OFDMA)

OFDMA uses a large number of closely spaced narrowband carriers. In a conventional FDM system, the frequency spacing between carriers is chosen with a sufficient guard band to ensure that interference is minimized and can be cost effectively filtered.

In OFDMA, the carriers are packed much closer together. This increases spectral efficiency by utilizing a carrier spacing that is the inverse of the symbol or modulation rate. Additionally, simple rectangular pulses are utilized during each modulation symbol. The high data rates are achieved in OFDM by allocating a single data stream in a parallel manner across multiple subcarriers.

The frame structure and scheduling differences between CDMA and OFDMA are discussed in the next chapter.

1.3 Radio Access Mode Operations

3GPP radio access for UMTS and LTE system is designed to operate in two main modes of operation; FDD and TDD. The focus of this book is on FDD mode only.

FDD is the common mode deployed worldwide for UMTS and LTE. Spectrum allocation is also tied to the choice of FDD over TDD. For example, operators with WiMAX deployed prior to LTE have utilized the WiMAX spectrum for investing in LTE TDD rather than FDD. However, with device availabilities as well as simplicity of deployment, FDD is still the main choice of deployment worldwide.

1.3.1 Frequency Division Duplex (FDD)

In FDD, a separate uplink and downlink channel are utilized, enabling a device to transmit and receive data at the same time. The spacing between the uplink and downlink channel is referred to as the duplex spacing.

The uplink channel operates on the lower frequency. This is done because higher frequencies suffer greater attenuation than lower frequencies and, therefore, it enables the mobile to utilize lower transmit levels.

1.3.2 Time Division Duplex (TDD)

TDD mode enables full duplex operation using a single frequency band and time division multiplexing the uplink and downlink signals.

One advantage of TDD is its ability to provide asymmetrical uplink and downlink allocations. Other advantages include dynamic allocation, increased spectral efficiency, and the improved usage of beamforming techniques. This is due to having the same uplink and downlink frequency characteristics.

1.4 Spectrum Allocation in UMTS and LTE

One of the main factors in any cellular system is the deployed frequency spectrum. 2G, 3G, and 4G systems offer multiple band options. This depends on the regulator in each country and the availability of spectrum sharing among multiple network operators in the same country.

The device's support of different frequency bands is driven by the hardware capabilities. Therefore, not all bands are supported by a single device. The demand of multi-mode and multi-band device depends on the market where the device is being commercialized.

Tables 1.1 and 1.2 list the FDD frequency bands defined in 3GPP for both UMTS and LTE.

LTE uses a variable channel bandwidth of 1.4, 3, 5, 10, 15, or 20 MHz. Most common worldwide network deployments are in 5 or 10 MHz, given the bandwidth available in the allocated spectrum for the operator. LTE in 20 MHz is being increasingly deployed, especially in bands like 2.6 GHz as well as 1.8 GHz after frequency re-farming.

LTE-FDD requires two center frequencies, one for the downlink and one for the uplink. These carrier frequencies are each given an EARFCN (E-UTRA absolute radio frequency channel number). In contrast, LTE-TDD has only one EARFCN. The channel raster for LTE is 100 kHz for all bands. The carrier center frequency must be an integer multiple of 100 kHz.

Table 1.1   UMTS FDD frequency bands

Operating band and [band name]		Uplink operating band (MHz)	Downlink operating band (MHz)
I	[UMTS2100]	1920–1980	2110–2170
II	[UMTS1900]	1850–1910	1930–1990
III	[UMTS1800]	1710–1785	1805–1880
IV	[UMTS1700]	1710–1755	2110–2155
V	[UMTS850]	824–849	869–894
VI	[UMTS800]	830–840	875–885
VII	[UMTS2600]	2500–2570	2620–2690
VIII	[UMTS900]	880–915	925–960
IX	[UMTS1700]	1749.9–1784.9	1844.9–1879.9
X	[UMTS1700]	1710–1770	2110–2170
XI	[UMTS1500]	1427.9–1452.9	1475.9–1500.9
XII	[UMTS700]	698–716	728–746
XIII	[UMTS700]	777–787	746–756
XIV	[UMTS700]	788–798	758–768

Table 1.2   LTE FDD frequency bands

Operating band	Uplink operating band (MHz)	Downlink operating band (MHz)
1	1920–1980	2110–2170
2	1850–1910	1930–1990
3	1710–1785	1805–1880
4	1710–1755	2110–2155
5	824–849	869–894
6	830–840	875–885
7	2500–2570	2620–2690
8	880–915	925–960
9	1749.9–1784.9	1844.9–1879.9
10	1710–1770	2110–2170
11	1427.9–1452.9	1475.9–1500.9
12	698–716	728–746
13	777–787	746–756
14	788–798	758–768
17	704–716	734–746

In UMTS, the nominal channel spacing is 5 MHz, but can be adjusted to optimize performance in a particular deployment scenario, such as in UMTS900 to re-farm fewer carriers from GSM900. The channel raster is 200 kHz, which means that the center frequency must be an integer multiple of 200 kHz. The carrier frequency is designated by the UTRA absolute radio frequency channel number (UARFCN).

1.5   LTE Network Architecture

1.5.1   Evolved Packet System (EPS)

3GPP cellular network architecture has been progressively evolving. The target of such evolutions is the eventual all-IP systems; migrating from CS-only to CS and PS, up to PS-only all-IP systems. Figure 1.1 summarizes the network architecture evolutions in 3GPP networks.

In the 3G network and prior to the introduction of the HSPA system, the network architecture is divided into CS and PS domains. Depending on the service offered to the end-user, the domains interact with the corresponding CN entities. The CS elements are mobile services switching center (MSC), visitor location register (VLR), and Gateway MSC. The PS elements are serving GPRS support node (SGSN) and Gateway GPRS support node (GGSN).

Furthermore, the control plane and user plane data are forwarded between the core and access networks. The RAT in the 3G system uses the WCDMA. The access network includes all of the radio equipment necessary for accessing the network, and is referred to as the universal terrestrial radio access network.

UTRAN consists of one or more radio network subsystems (RNSs). Each RNS consists of an RNC and one or more NodeBs. Each NodeB controls one or more cells and provides the WCMDA radio link to the UE.

After the introduction of HSPA and HSPA+ systems in 3GPP, some optional changes have been added to the CN as well as mandatory changes to the access network. On the CN side, an evolved direct tunneling architecture has been introduced, where the user data can flow between GGSN and RNC or directly to the NodeB. On the access network side, some of the RNC functions, such as the network scheduler, have been moved to the NodeB side for faster radio resource management (RRM) operations.

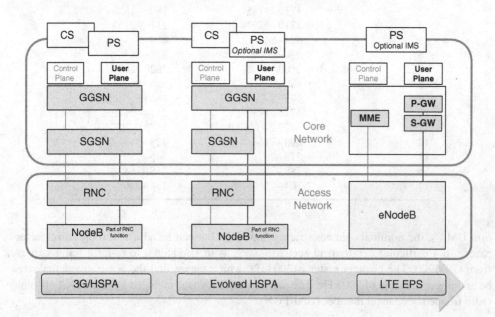

Figure 1.1   Simplified network architecture evolutions.

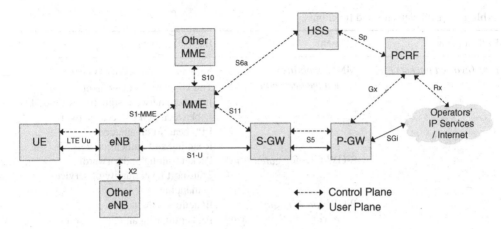

Figure 1.2   Basic EPS entities and interfaces.

Additionally, the IP-multimedia subsystem (IMS) has been defined, earlier before the introduction of LTE, as a PS domain application control plane for the IP multimedia services. It represents only an optional layer/domain that can be used in conjunction with the PS domain/CN.

The LTE network was then introduced as a flat architecture, with user plane direct tunneling between the core and access networks. The EPS system is similar to the flat architecture option in HSPA+. Similar to the 3G system, the LTE system consists of core and access networks, but with different elements and operations.

EPS consists of an E-UTRAN access network and EPC CN. EPS can also interconnect with other RAN; 3GPP (GERAN (GSM/EDGE radio access network), UTRAN) and non-3GPP (CDMA, WiFi, WiMAX).

Though the CS domain is not part of the EPS architecture, 3GPP defines features to allow interworking between EPS and CS entities. This interworking allows traditional services, CS voice speech call, to be set up directly via traditional or evolved CS domain calls, known as CS fallback.

Figure 1.2 shows the basic EPS entities and interfaces. Table 1.3 summarizes the functions of the EPS core and access networks.

1.5.2   Evolved Packet Core (EPC)

EPC includes an MME (mobility management entity), an S-GW (serving sateway), and an P-GW (packet gateway) entities. They are responsible for different functionalities during the call or registration process. EPC and E-UTRAN interconnects with the S1 interface. The S1 interface supports a many-to-many relation between MMEs, S-GWs, and eNBs (eNodeBs) [5].

MME connects to E-UTRAN by means of an S1 interface. This interface is referred to as S1-C or S1-MME [5]. When a UE attaches to an LTE network, UE-specific logical S1-MME connections are established. This bearer, known as an EPS bearer, is used to exchange UE specific signaling messages needed between UE and EPC.

Each UE is then assigned a unique pair of eNB and MME identifications during S1-MME control connection. The identifications are used by MME to send the UE-specific S1 control

Table 1.3 EPS elements and functions

EPS element	Element	Basic functionality
EPC (*evolved packet core*)	MME (*mobility management entity*)	Signaling and security control
		Tracking area management
		Inter core network signaling for mobility between 3GPP access networks
		EPS bearer management
		Roaming and authentication
	S-GW (*serving gateway*)	Packet routing and forwarding
		Transport level quality of service mapping
	P-GW (*packet data network (PDN) gateway*)	IP address allocation
		Packet filtering and policy enforcement
		User plane anchoring for mobility between 3GPP access networks
E-UTRAN (*evolved universal terrestrial radio access network*)	eNodeB (*evolved node B*)	Provides user plane protocol layers: PDCP, RLC, MAC, physical, and control plane (RRC) with the user
		Radio resource management
		E-UTRAN synchronization and interface control
		MME selection

messages and by E-UTRAN to send the messages to MME. The identification is released when the UE transitions to idle state where the dedicated connection with the EPC is also released. This process may take place repetitively when the UE sets up a signaling connection for any type of LTE call.

MME and E-UTRAN handles signaling for control plane procedures established for the UE on the S1-MME interface including:

- Initial context set-up/UE context release,
- E-RAB (EPS-radio access bearer) set-up/release/modify,
- Handover preparation/notification,
- eNB/MME status transfer,
- Paging,
- UE capability information indication.

MMEs can also periodically send the MME loading information to E-UTRAN for mobility management procedures. This is not UE-specific information.

S-GW are connected to E-UTRAN by means of an S1-U interface [5]. After the EPS bearer is established for control plane information, the user data packets start flowing between the EPC and UE through this interface.

Inside the EPC architecture, MME and S-GW interconnects through the S11 interface. The S11 links the MME with the S-GW in order to support control plane signaling [6]. The S5 interface links the S-GW with the PDN-GW (packet data network-gateway) and supports both a control and user planes. This interface is used when these elements reside within the same PLMN (public land mobile network). In the case of an inter-PLMN connection, the interface between these elements becomes S8 [7].

The details of all the interfaces in EPC and E-UTRAN are further discussed in Section 1.6.

1.5.3 Evolved Universal Terrestrial Radio Access Network (E-UTRAN)

E-UTRAN consists of the eNB. The eNB typically consists of three cells [8]. eNB can, optionally, interconnect to each other via the X2 interface. The interface utilizes functions for mobility and load exchange information [9].

eNB connects with the UE on the LTE-Uu interface. This interface, referred to as the air interface, is based on OFDMA.

E-UTRAN provides the UE with control and user planes. Each is responsible for functions related to call establishment or data transfer. The exchange of such information takes place over a protocol stack defined in UE and eNB. Over the interface between the UE and the EPS, the protocol stack is split into the access stratum (AS) and the non-access stratum (NAS).

1.5.4 LTE User Equipment

Like that of UMTS, the mobile device in LTE is termed the user equipment and is comprised of two distinct elements; the USIM (universal subscriber identity module) and the ME (mobile equipment).

The ME supports a number of functional entities and protocols including:

- **RR (radio resource)** – this supports both the control and user planes. It is responsible for all low level protocols including RRC (radio resource control), PDCP (packet data convergence protocol), RLC, MAC (medium access control), and PHY layers. The layers are similar to those in the eNB protocol layer.
- **EMM (EPS mobility management)** – is a control plane entity which manages the mobility states of the UE: LTE idle, LTE active, and LTE detached. Transactions within these states include procedures such as TAU (tracking area update) and handovers.
- **ESM (EPS session management)** – is a control plane activity which manages the activation, modification, and deactivation of EPS bearer contexts. These can either be default or dedicated EPS bearer contexts.

The PHY layer capabilities of the UE may be defined in terms of the frequency bands and data rates supported. Devices may also be capable of supporting adaptive modulation including QPSK, 16QAM, and 64QAM. Modulation capabilities are defined separately in 3GPP for uplink and downlink.

The UE is able to support several scalable channels, including 1.4, 3, 5, 10, 15, and 20 MHz, while operating in FDD and/or TDD. The UE may also support advanced antenna features such as MIMO with a different number of antenna configurations.

The PHY layer and radio capabilities of the UE are advertised to EPS at the initiation of the connection with the eNB in order to adjust the radio resources accordingly. An LTE capable device advertises one of the categories listed in Table 1.4 according to its software and hardware capabilities [10]. Categories 6, 7, and 8 are considered part of LTE-advanced UE's capabilities.

Table 1.4 LTE UE categories

UE category	3GPP release	Downlink		Uplink	
		Maximum data rate (Mbps)	Maximum number of layers	Maximum data rate (Mbps)	Support for 64QAM
Category 1	Release 8/9	10	1	5	No
Category 2	Release 8/9	51	2	25	No
Category 3	Release 8/9	102	2	51	No
Category 4	Release 8/9	150	2	51	No
Category 5	Release 8/9	300	4	75	Yes
Category 6	Release 10	301	2 or 4	51	No
Category 7	Release 10	301	2 or 4	102	No
Category 8	Release 10	3000	8	1500	Yes

1.6 EPS Interfaces

This section summarizes the EPS interfaces and relevant protocols, with reference to the overall architecture in Figure 1.2. The main protocols used inside EPS interfaces are summarized as follows:

- **S1 application protocol (S1-AP)** – Application layer protocol between the eNB and the MME.
- **Stream control transmission protocol (SCTP)** – This protocol guarantees delivery of signaling messages between MME and eNB (S1). SCTP is defined in [11].
- **GPRS tunneling protocol for the user plane (GTP-U)** – This protocol tunnels user data between eNB and the SGW, and between the SGW and the PGW in the backbone network. GTP will encapsulate all end-user IP packets.
- **User datagram protocol (UDP)** – This protocol transfers user data. UDP is defined in [12].
- **UDP/IP** – These are the backbone network protocols used for routing user data and control signaling.
- **GPRS tunneling protocol for the control plane (GTP-C)** – This protocol tunnels signaling messages between SGSN and MME (S3).
- **Diameter** – This protocol supports transfer of subscription and authentication data for authenticating/authorizing user access to the evolved system between MME and HSS (home subscriber service) (S6a). Diameter is defined in [13].

1.6.1 S1-MME Interface

This interface is the reference point for the control plane between eNB and MME [5]. S1-MME uses S1-AP over SCTP as the transport layer protocol for guaranteed delivery of signaling messages between MME and eNodeB. It serves as a path for establishing and maintaining subscriber UE contexts. One or more S1-MME interfaces can be configured per context. Figure 1.3 illustrates the interface nodes.

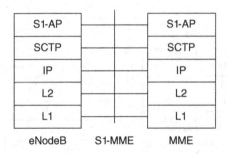

Figure 1.3 Control plane for eNB (S1-MME). (Source: [5] 3GPP TS 2010. Reproduced with permission of ETSI.)

One logical S1-AP connection per UE is established and multiple UEs are supported via a single SCTP association. The following functionalities are conducted at S1-AP:

- Set up, modification and release of E-RABS.
- Establishment of an initial S1 UE context.
- Paging and S1 management functions.
- NAS signaling transport functions between UE and MME.
- Status transfer functionality.
- Trace of active UEs, and location reporting.
- Mobility functions for UE to enable inter- and intra-RAT HO.

1.6.2 LTE-Uu Interface

The radio protocol of E-UTRAN between the UE and the eNodeB is specified in [14]. The user plane and control plane protocol stacks for the LTE-Uu interface are shown in Figures 1.4 and 1.5, respectively. The protocols on E-UTRAN-Uu (RRC, PDCP, RLC, MAC, and the PHY LTE layer) implements the RRM and supports the NAS protocols by transporting the NAS messages across the E-UTRAN-Uu interface.

The protocol stack layer and air interface functions are described in detail in Chapter 2.

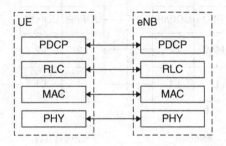

Figure 1.4 User-plane protocol stack. (Source: [14] 3GPP TS 2009. Reproduced with permission of ETSI.)

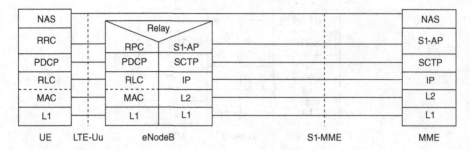

Figure 1.5 Control-plane protocol stack. (Source: [14] 3GPP TS 2009. Reproduced with permission of ETSI.)

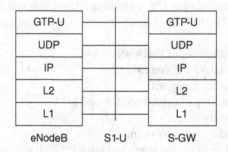

Figure 1.6 User plane of S1-U. (Source: [15] 3GPP. Reproduced with permission of ETSI.)

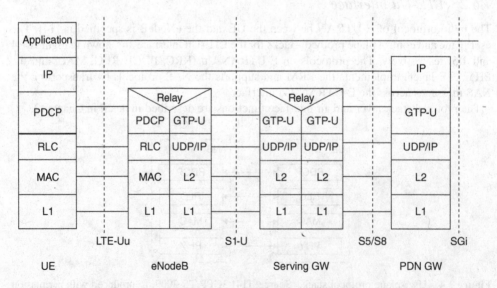

Figure 1.7 User plane protocol stack.

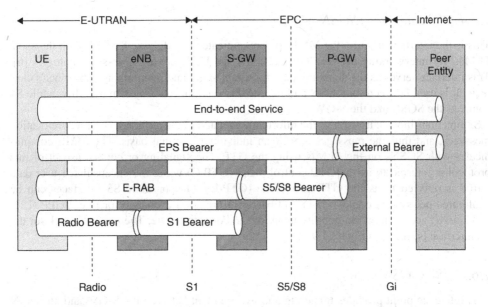

Figure 1.8 EPS bearer service architecture. (Source: [14] 3GPP TS 2009. Reproduced with permission of ETSI.)

1.6.3 S1-U Interface

This interface between E-UTRAN and S-GW is used for user plane tunneling and inter-eNB path switching during handover [15]. The user plane for S1-U is illustrated in Figure 1.6. In addition, the end-to-end protocol stack for the user plane is shown in Figure 1.7. The S1-U carries the user data traffic between the eNB and S-GW. S1-U also implements the DSCP (differentiated services code point). The 6 bit DSCP value assigned to each IP packet identifies a pre-determined level of service and a corresponding priority, which is used to implement the appropriate QoS for the users' data. More details on DSCP are provided in Chapter 7.

The EPS bearer service layered architecture is depicted in Figure 1.8 [14], where:

- A radio bearer transports the packets of an EPS bearer between a UE and an eNB. There is a one-to-one mapping between an EPS bearer and a radio bearer.
- An S1 bearer transports the packets of an EPS bearer between an eNB and the S-GW.
- An S5/S8 bearer transports the packets of an EPS bearer between the S-GW and the P-GW.
- UE stores a mapping between an uplink packet filter and a radio bearer to create the binding between SDFs (service data flows) and a radio bearer in the uplink, described later in this chapter.
- P-GW stores a mapping between a downlink packet filter and an S5/S8 bearer to create the binding between an SDF and an S5/S8 bearer in the downlink.
- An eNB stores a one-to-one mapping between a radio bearer and an S1 to create the binding between a radio bearer and an S1 bearer in both the uplink and downlink.
- An S-GW stores a one-to-one mapping between an S1 bearer and an S5/S8 bearer to create the binding between an S1 bearer and an S5/S8 bearer in both the uplink and the downlink.

1.6.4 S3 Interface (SGSN-MME)

This is the interface used by the MME to communicate with Release 8 SGSNs, on the same PLMN, for interworking between GPRS/UMTS and LTE network access technologies [6]. This interface serves as the signaling path for establishing and maintaining subscriber's contexts. It is used between the SGSN and the MME to support inter-system mobility, while S4 connects the SGSN and the S-GW.

S3 functions include transfer of the information related to the terminal, handover/relocation messages, and thus the messages are for an individual terminal basis. The MME communicates with SGSNs on the PLMN using the GTP. The signaling or control aspect of this protocol is referred to as the GTP control plane (GTP-C) while the encapsulated user data traffic is referred to as the GTP user plane (GTP-U). One or more S3 interfaces can be configured per system context. User and bearer information exchange for inter 3GPP (LTE and 2G/3G) access network mobility in an idle and/or active state. The protocol stack for the S3 interface is shown in Figure 1.9.

1.6.5 S4 (SGSN to SGW)

This reference point provides tunneling and management between the S-GW and an SGSN [6, 15]. It has equivalent functions to the S11 interface and supports related procedures for terminals connecting via EPS. It provides related control and mobility support between the GPRS core and the 3GPP anchor function of S-GW.

This interface supports exclusively GTPv2-C and provides procedures to enable a user plane tunnel between SGSN and S-GW if the 3G network has not enabled a direct tunnel for user plane traffic from RNC to S-GW. The control plane and user plane of the S4 interface are shown in Figure 1.10.

The end-to-end protocol stack for user data of 2G subscribers that camped on the 2G network is illustrated in Figure 1.11. Protocols on the Um and the Gb interfaces are described in [16]. The end-to-end protocol stack for user data of 3G subscribers that camped on the UTRAN network is illustrated in Figure 1.12a. This protocol is used between the UE and the P-GW user plane with 3G access via the S4 interface. SGSN controls the user plane tunnel establishment, providing a direct tunnel between UTRAN and SGW. An alternative approach for UTRAN is via a direct tunnel between UTRAN and SGW via the S12 interface, as illustrated in Figure 1.12b. The protocols on the Uu, the Iu, the Um, and the Gb interfaces are described in [16].

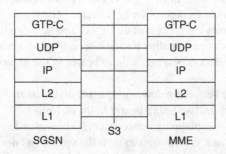

Figure 1.9 Protocol stack for S3 interface between MME and SGSN. (Source: [6] 3GPP TS 2011. Reproduced with permission of ETSI.)

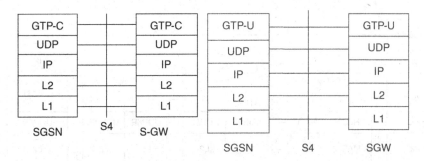

Figure 1.10 Protocol stack of S4 interface (user plane and control plane). (Source: [6] 3GPP TS 2011. Reproduced with permission of ETSI.)

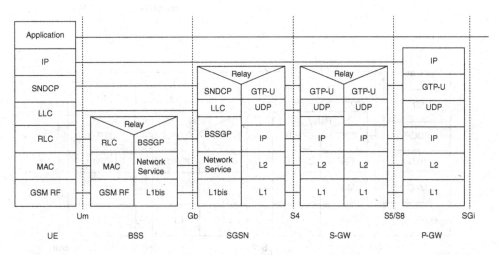

Figure 1.11 UE – user plane for A/Gb mode and for GTP-based S5/S8. (Source: [16] 3GPP TS. Reproduced with permission of ETSI.)

1.6.6 S5/S8 Interface

This reference point provides tunneling (bearer channel) and management (signaling channel) between the S-GW and the P-GW [6, 15]. The S8 interface is used for roaming scenarios. The S5 interface is used for non-roaming scenarios where it provides user plane tunneling and management between S-GW and P-GW. It is used for S-GW relocation during UE mobility and when the S-GW needs to connect to a non-collocated P-GW for the required PDN connectivity. Figure 1.13 illustrates this interface.

There are two protocol options to be used in the S5/S8 interface:

- **S5/S8 over GTP** – Provides the functionality associated with creation, deletion, modification, or change of bearers for an individual user connected to EPS.
- **S5/S8 over PMIPV6** – Provides tunneling management between the SGW and PGW.

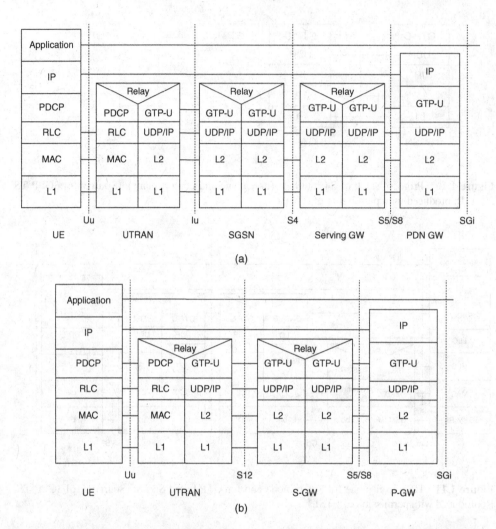

Figure 1.12 (a) UE – user plane with UTRAN for GTP-based S5/S8 via the S4 interface. (b) User plane with UTRAN for GTP-based S5/S8 and direct tunnel on S12. (Source: [16] 3GPP TS. Reproduced with permission of ETSI.)

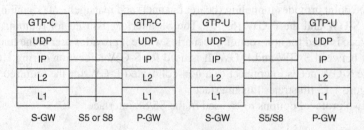

Figure 1.13 Control plane and user planes for S5/S8 interfaces. (Source: [16] 3GPP TS. Reproduced with permission of ETSI.)

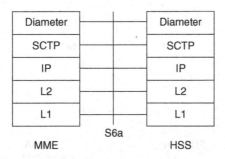

Figure 1.14 Control plane for S6a interface between MME and HSS. (Source: [17] 3GPP TS. Reproduced with permission of ETSI.)

1.6.7 S6a Interface (Diameter)

This is the interface used by the MME to communicate with the HSS, as illustrated in Figure 1.14 [17]. The HSS is responsible for transferring the subscription and authentication data for authorizing the user access and UE context authentication. The MME communicates with the HSSs on the PLMN using the Diameter protocol. One or more S6a interfaces can be configured per system context.

The following list summarizes the functions of S6a:

- Exchange the location information
- Authorize a user to access the EPS,
- Exchange authentication information,
- Download and handle changes in the subscriber data stored in the server,
- Upload the P-GW identity and APN (access point name) being used for a specific PDN connection,
- Download the P-GW identity and APN pairs being stored in HSS for an already ongoing PDN connection.

1.6.8 S6b Interface (Diameter)

This reference point, between a PGW and a 3GPP AAA (access authorization and accounting) server/proxy, is used for mobility-related authentication [18]. It may also be used to request parameters related to mobility and to retrieve static QoS profiles for UEs (for non-3GPP access). Figure 1.15 illustrates the layout of this interface.

The S6b interface is defined between the P-GW and the 3GPP AAA server (for non-roaming case, or roaming with home routed traffic to P-GW in home network) and between the P-GW and the 3GPP AAA proxy (for roaming case with P-GW in the visited network).

The S6b interface is used to inform the 3GPP AAA server/proxy about current P-GW identity and APN being used for a given UE, or that a certain P-GW and APN pair is no longer used. This occurs, for example, when a PDN connection is established or closed. This S6b interface protocol is based on Diameter and is defined as a vendor specific Diameter application, where the vendor is 3GPP.

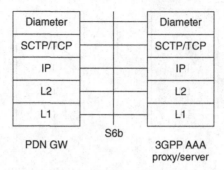

Figure 1.15 Control plane for S6b interface between P-GW and 3GPP AAA. (Source: [18] 3GPP TS. Reproduced with permission of ETSI.)

1.6.9   S6d (Diameter)

It enables transferring the subscription and authentication data for authorizing the user access to the evolved system (AAA interface) between SGSN and HSS [17]. S6d is the interface between S-GW in VPLMN (visited public land mobile network) and 3GPP AAA proxy for mobility related authentication, if needed. This is a variant of S6c for the roaming (inter-PLMN) case. Figure 1.16 illustrates the layout of this interface.

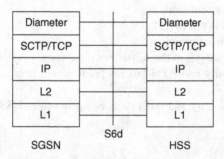

Figure 1.16 Control plane for S6d interface between SGSN and HSS. (Source: [17] 3GPP TS. Reproduced with permission of ETSI.)

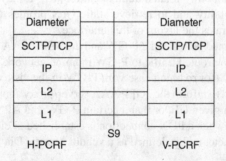

Figure 1.17 S9 interface protocol stack. (Source: [17] 3GPP TS. Reproduced with permission of ETSI.)

1.6.10 S9 Interface (H-PCRF-VPCRF)

The S9 interface is defined between the PCRF (policy and charging rules function) in the home network policy and charging rules function (H-PCRF) and a PCRF in the visited network policy and charging rules function (V-PCRF), as shown in Figure 1.17. S9 is an inter-operator interface and is only used in roaming scenarios. The main purpose of the S9 interface is to transfer policy decisions (i.e., policy charging and control, PCC, or QoS rules) generated in the home network to the visited network and transport the events that may occur in the visited network to the home network. The protocol over the S9 interfaces is based on Diameter. This interface will allow the users when roamed on visited network to be treated with same QoS and same PCC subject to the operators agreement.

1.6.11 S10 Interface (MME-MME)

This is the interface used by the MME to communicate with another MME in the same PLMN or on different PLMNs, see Figure 1.18. This interface is also used for MME relocation and MME-to-MME information transfer or handover. One or more S10 interfaces can be configured per system context. The main function of the GTP-C layer, within this interface, is to transfer the contexts for individual terminals attached to EPC and thus sent on a per UE basis.

1.6.12 S11 Interface (MME–SGW)

This interface provides communication between MME and S-GW for information transfer using GTPv2 protocol, see Figure 1.19. One or more S11 interfaces can be configured per system context. In the case of handover, the S11 interface is used to relocate the S-GW when appropriate, or establish an indirect forwarding tunnel for user plane traffic and to manage use data traffic flow.

1.6.13 S12 Interface

This is the reference point between UTRAN and S-GW for user plane tunneling when a direct tunnel is established. It is based on the Iu-u/Gn-u reference point using the GTP-U protocol, as defined between SGSN and UTRAN or between SGSN and GGSN. The usage of S12 is

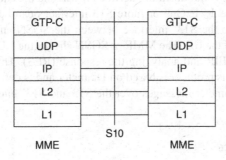

Figure 1.18 Control plane for S10 interface between MMEs. (Source: [17] 3GPP TS. Reproduced with permission of ETSI.)

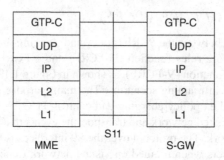

Figure 1.19 Control plane for S11 interface between MME and S-GW. (Source: [17] 3GPP TS. Reproduced with permission of ETSI.)

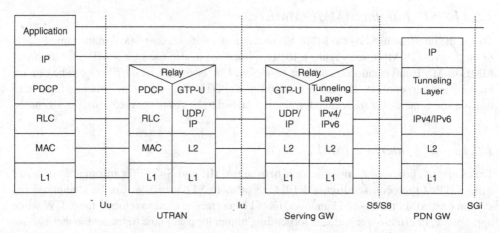

Figure 1.20 UE and PDN-GW user plane with 3G access via direct tunnel on S12 interface. (Source: [17] 3GPP TS. Reproduced with permission of ETSI.)

an operator configuration option. Figure 1.20 demonstrates the UE and P-GW user plane with 3G access via a direct tunnel on the S12 interface.

1.6.14 S13 Interface

This interface provides the communication between MME and the equipment identity register (EIR), as shown in Figure 1.21. One or more S13 interfaces can be configured per system context. This is similar to the S13' interface between the SGSN and the EIR and they are used to check the status of the UE. The MME or SGSN checks the UE identity by sending the equipment identity to an EIR and analyzing the response (RES). The same protocol is used on both S13 and S13'. This protocol is based on Diameter and is defined as a vendor specific Diameter application. Diameter messages over the S13 and S13' interfaces use the SCTP as a transport protocol.

1.6.15 SGs Interface

The SGs interface connects the databases in the VLR and the MME to support CS fallback scenarios [19]. The control interface is used to enable CSFB from E-UTRAN access to

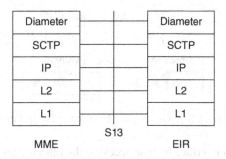

Figure 1.21   Control plane for S13 interface between MME and EIR.

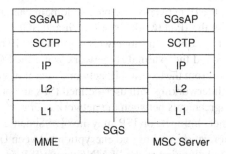

Figure 1.22   SGs interface. (Source: [19] 3GPP TS. Reproduced with permission of ETSI.)

UTRAN/GERAN CS domain access. The SGs-AP protocol is used to connect an MME to an MSC server (MSS), as illustrated in Figure 1.22.

CSFB in the EPS enables the provisioning of CS-domain services (e.g., voice call, SMS, location services (LCS), or supplementary services) by reusing the CS domain when the UE is served by E-UTRAN.

The SGs interface connects the databases in the VLR and the MME to coordinate the location information of UEs that are IMSI (international mobile subscriber identity) attached to both EPS and non-EPS services. The SGs interface is also used to convey some CS related procedures via the MME. The basis for the interworking between a VLR and an MME is the existence of an SGs association between those entities per UE. The SGs association is only applicable to UEs with CS fallback capability activated. The behavior of the VLR and the MME entities related to the SGs interface is defined by the state of the SGs association for a UE. Individual states per SGs association, that is, per UE with CS fallback capability activated, are held at both the VLR and the MME. Chapter 4 provides more details on CSFB and it is performance.

1.6.16   SGi Interface

This is the reference point between the P-GW and the PDN, see Figure 1.23. It can provide access to a variety of network types, including an external public or private PDN and/or an internal IMS service-provisioning network.

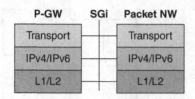

Figure 1.23 Protocol stack for SGI interface between PGW and the packet data network.

The functions of the SGi interface include access to the Internet, Intranet, or an ISP (Internet service provider) and involve functions such as IPv4 address allocation, IPv6 address auto configuration, and may also involve specific functions such as authentication, authorization, and secure tunneling to the intranet/ISP.

When interworking with the IP networks, the packet domain can operate IPv4 and/or IPv6. The interworking point with the IP networks is at the Gi and SGi reference points. Typically in the IP networks, the interworking with subnetworks is done via IP routers. The Gi reference point is between the GGSN and the external IP network while the SGi is between the P-GW and the external IP network. From the external IP network's point of view, the GGSN/P-GW is seen as a normal IP router. Interworking with user-defined ISPs and private/public IP networks is subject to interconnect agreements between the network operators.

The access to the Internet, Intranet, or ISP may involve specific functions, such as user authentication, user's authorization, end-to-end encryption between UE and intranet/ISP, allocation of a dynamic address belonging to the PLMN/intranet/ISP addressing space, and IPv6 address autoconfiguration. For this purpose the packet domain may offer either direct transparent access to the Internet; or a non-transparent access to the intranet/ISP. In this case the packet domain, that is, the GGSN/PGW, takes part in these functions.

1.6.17 Gx Interface

The Gx reference point lies between the PCRF and the PCEF (policy and charging enforcement function) as illustrated in Figure 1.24. This signaling interface supports the transfer of policy control and charging rules information (QoS) between the PCEF in the P-GW and a PCRF server. The Gx application has an own vendor specific Diameter application [20]. With regard to the Diameter protocol defined over the Gx interface, the PCRF acts as a Diameter server, in the sense that it is the network element that handles PCC rule requests for a particular area.

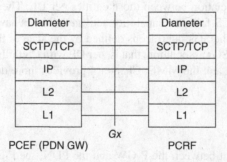

Figure 1.24 Protocol stack for Gx interface between PGW/PCEF and PCRF. (Source: [20] 3GPP TS. Reproduced with permission of ETSI.)

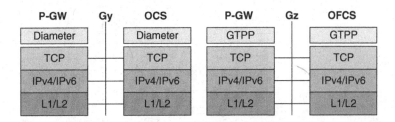

Figure 1.25 Protocol stack for Gy and Gz interfaces.

The PCEF acts as the Diameter client, in the sense that is the network element requesting PCC rules in the transport plane network resources. The main purpose of the Gx interface is to support PCC rule handling and event handling for PCC. PCC rule handling over the Gx interface includes the installation, modification, and removal of PCC rules. All these three operations can be made upon any request coming from the PCEF or due to some internal decision in the PCRF. The event handling procedures allows the PCRF to subscribe to those events. The PCEF then reports the occurrence of an event to the PCRF.

1.6.18 Gy and Gz Interfaces

The Gy reference interface enables online accounting functions on the P-GW in accordance with 3GPP Release 8 specifications. The Gy reference point for online flow-based bearer charging (i.e., OCS, online charging system). On the other hand, the Gz reference point is for offline flow-based bearer charging (i.e., OFCS, offline charging system), see Figure 1.25.

The Gz reference interface enables offline accounting functions on the P-GW. The P-GW collects charging information for each mobile subscriber UE pertaining to the radio network usage. The Gz reference point enables transport of SDF-based offline charging information. The Gz interface is specified in [21].

1.6.19 DNS Interface

MME supports the DNS (domain name system) interface for MME, SGW, PGW, and SGSN selection in the EPC CN. The MME uses the tracking area list as a fully qualified domain name (FQDN) to locate the address relevant to the call. One or more DNS interfaces can be configured per system context (refer to the addresses in Table 1.8).

1.6.20 Gn/Gp Interface

Gn interfaces facilitate user mobility between 2G/3G 3GPP networks. They are used for intra-PLMN handovers [16, 22]. The MME supports pre-Release 8 Gn interfaces to allow interoperation between EPS networks and 2G/3G 3GPP networks. Roaming and inter-access mobility between Gn/Gp 2G and/or 3G SGSNs and an MME/SGW are enabled by:

- Gn functionality, as specified between two Gn/Gp SGSNs, which is provided by the MME and
- Gp functionality, as specified between Gn/Gp SGSN and Gn/Gp GGSN that is provided by the P-GW.

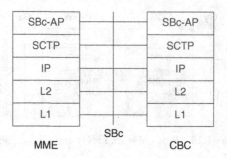

Figure 1.26 Protocol stack for SBc interface between MME and the CBC.

1.6.21   SBc Interface

The SBc application part (SBc-AP) messages are used on the SBc-AP interface between the MME and the cell broadcast center (CBC) [23]. According to Figure 1.26, the SBc-AP interface is a logical interface between the MME and the CBC. All the SBc-AP messages require an SCTP association between the MME and the CBC.

The MME and the CBC support IPv6 [24] and/or IPv4 [25]. The IP layer of SBc-AP only supports point-to-point transmission for delivering SBc-AP messages. SBc-AP consists of elementary procedures (EPs). An EP is a unit of interaction between the MME and the CBC. These EPs are intended to be used to build up complete sequences in a flexible manner. Examples of using several SBc-APs together with each other and EPs from other interfaces can be found [26].

1.6.22   Sv Interface

The Sv is the interface between the MME/SGSN and MSC Server to provide SRVCC (single radio voice call continuity) [27]. The Sv interface, as shown in Figure 1.27, is between the MME or the SGSN and 3GPP MSC server enhanced for SRVCC.[1] The Sv interface is

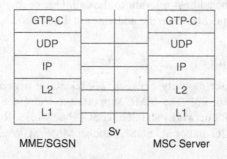

Figure 1.27 Protocol stack for Sv interface between MME/SGSN and the MSS.

---

[1] Refer to Chapter 7 for the detailed description of VoLTE SRVCC.

used to support inter-RAT handover from VoIP/IMS over EPS to a CS domain over 3GPP UTRAN/GERAN access. The Sv messages are based on GTP protocol.

1.7 EPS Protocols and Planes

1.7.1 Access and Non-Access Stratum

Over the interfaces between UE and EPS, protocols are split into AS and NAS. Figure 1.28 describes the LTE entities involved for both NAS and AS procedures. The NAS and AS layers exist equally in the UE and EPS to handle the related control and user plane procedures.

The AS resides between the UE and E-UTRAN and consists of multiple protocol layers: RRC, PDCP, RLC (radio link control), MAC, and the PHY layers. The AS signaling provides a mechanism to deliver NAS signaling messages intended for control plane procedures, as well as the lower layer signaling and parameters required to set up, maintain, and manage the connections with the UE.

The NAS layer between the UE and EPC is responsible for handling control plane messaging related to the CN. NAS includes two main protocols: evolved mobility management (EMM) and evolved session management (ESM) [28]. Tables 1.5 and 1.6 summarize the functions of each of these NAS entities.

1.7.2 Control Plane

The protocol stack of an EPS system is designed to handle both control and user planes, as shown previously in Figure 1.2. The control plane is responsible for signaling message exchange between the UE and the EPC or E-UTRAN.

When the UE is in LTE coverage, there are two control planes set up to carry the signaling messages between the EPS and the UE. The first is provided by RRC and carries signaling between the UE and the eNB. The second carries NAS signaling messages between the UE and the MME.

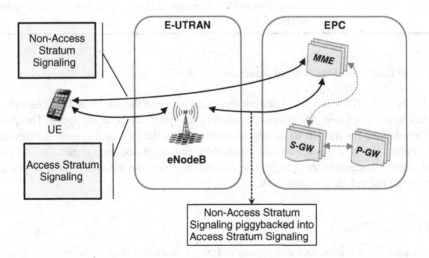

Figure 1.28   LTE NAS and AS.

Table 1.5   Summary of NAS EMM

EMM procedures	Description
Attach	Used by the UE to attach to EPC for packet services in the EPS. It can also be used to attach to non-EPS services, for example, CSFB/SMS
Detach	Used by the UE to detach from EPS services. It can also be used for other procedures such as disconnecting from non-EPS services
Tracking area updating	Initiated by the UE and used for identifying the UE location at eNB level for paging purposes in idle mode
Service request (PS call)	Used by the UE to get connected and establish the radio and S1 bearers when uplink user data or signaling is to be sent
Extended service request (CSFB)	Used by the UE to initiate a circuit switched fallback call or respond to a mobile terminated circuit switched fallback request from the network, that is, non-EPS services
GUTI allocation	Allocate a GUTI (globally unique temporary identifier) and optionally to provide a new TAI (tracking area identity) list to a particular UE
Authentication	Used for AKA (authentication and key agreement) between the user and the network
Identification	Used by the network to request a particular UE to provide specific identification parameters, for example, the IMSI (international mobile subscriber identity) or the IMEI (international mobile equipment identity)
Security mode control	Used to take an EPS security context into use, and initialize NAS signaling security between the UE and the MME with the corresponding NAS keys and security algorithms
EMM status	Sent by the UE or by the network at any time to report certain error conditions
EMM information	Allows the network to provide information to the UE
NAS transport	Carries SMS (short message service) messages in an encapsulated form between the MME and the UE
Paging	Used by the network to request the establishment of a NAS signaling connection to the UE. Is also includes the circuit switched service notification

The main functions of the control plane are

- To facilitate the NAS and AS signaling messages between the concerned interfaces.
- To define the NAS and AS system parameters and protocol layer mapping. The parameters are defined for the UE to be able to connect with the EPS and control all subsequent procedures. The NAS parameters define the EPS bearer-related procedures. The AS parameters define the mechanisms to maintain and manage the connection and the user plane data transfer on the uplink and downlink.

1.7.3   User Plane

The user plane is used for forwarding any uplink or downlink data between the UE and the EPS. In particular, it is used for the delivery of IP packets to and from the S-GW and PDN-GW.

Table 1.6 Summary of NAS ESM

ESM procedures	Description
Default EPS bearer context activation	Used to establish a default EPS bearer context between the UE and the EPC
Dedicated EPS bearer context activation	Establish an EPS bearer context with specific QoS (quality of service) between the UE and the EPC. The dedicated EPS bearer context activation procedure is initiated by the network, but may be requested by the UE by means of the UE requested bearer resource allocation procedure
EPS bearer context modification	Modify an EPS bearer context with a specific QoS
EPS bearer context deactivation	Deactivate an EPS bearer context or disconnect from a PDN by deactivating all EPS bearer contexts
UE requested PDN connectivity	Used by the UE to request the set-up of a default EPS bearer to a PDN
UE requested PDN disconnect	Used by the UE to request disconnection from one PDN. The UE can initiate this procedure to disconnect from any PDN as long as it is connected to at least one other PDN
UE requested bearer resource allocation	Used by the UE to request an allocation of bearer resources for a traffic flow aggregate
UE requested bearer resource modification	Used by the UE to request a modification or release of bearer resources for a traffic flow aggregate or modification of a traffic flow aggregate by replacing a packet filter
ESM information request	Used by the network to retrieve ESM information, that is, protocol configuration options, APN (access point name), or both from the UE during the attach procedure
ESM status	Report at any certain error conditions detected upon receipt of ESM protocol data

The user plane is established when the UE is in connected mode where the data can flow across the protocol layers. The user plane primarily utilizes the AS of the protocol. The NAS layer only provides the information of mapping of upper layer channels needed for the data to flow. Additionally, NAS provides the user plane with the required parameters including QoS. The UE and eNB then utilize these NAS configurations to exchange the user plane data.

1.8 EPS Procedures Overview

1.8.1 EPS Registration and Attach Procedures

When the UE enters the LTE coverage or powers up, it first registers with the EPS network through the "initial EPS attach" procedure [28]. This attach procedure is used to:

- Register the UE for packet services in EPS,
- Establish (at a minimum) a default EPS bearer that a UE could use to send and receive the user application data,
- Allocate IPv4 and/or IPv6 addresses.

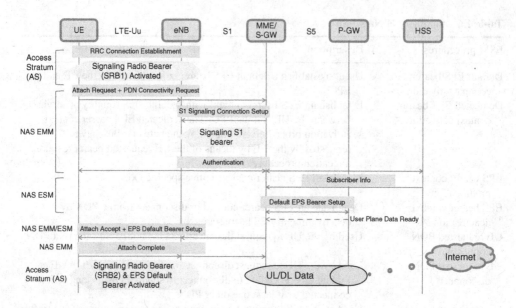

Figure 1.29 EPS attach procedure overview.

The overview of the attach procedure is illustrated in Figure 1.29.

The attach procedure usually starts when the UE initiates the request. After establishing an RRC connection, the UE can send an attach request message to the MME. UE also requests PDN connectivity along with the attach request.

After all necessary signaling connections are established, EPC may trigger security functions. HSS downloads user subscriber information to the MME, which processes the UE request for default EPS bearer set-up. After the default EPS bearer and QoS are negotiated and agreed to among the MME and S-GW/P-GW, the MME forwards the default bearer set-up request to the eNB and the UE.

The eNB and the UE then acknowledge the default bearer set-up, and communicate the attach accept messages to the EPC. The EPS bearer is finally active and data can flow between the UE and the IP network, in both uplink and downlink directions.

At this point, UE typically registers with a default APN, as per the subscription policies. If additional APN is available, the process needs to continue setting up another EPS bearer.

1.8.1.1 Signaling Radio Bearer (SRB)

In order for the control plane information messages in EPS to flow between the UE and the EPC or E-UTRAN, SRBs (signaling radio bearers) are set up at the initial connection request.

Three SRBs are used to transfer RRC and NAS messages to/from the UE:

- **SRB ID 0** – used to establish the RRC connection request when the UE has transitioned into connected mode. SRB0 carries common control information required to establish the RRC connection.

- **SRB ID 1** – used for RRC messages, as well as RRC messages carrying high priority NAS signaling.
- **SRB ID 2** – used for RRC carrying low priority NAS signaling. Prior to its establishment, low priority signaling is sent on SRB1.

Once the SRBs are established, control plane messages and parameters are sent to the UE from the EPC and/or E-UTRAN. The UE will adhere to these parameters to continue the protocol procedures on the AS. The parameters sent to the UE in the SRB messages will control all protocol layers for the data transmission.

1.8.1.2 Default EPS Data Radio Bearer (Default DRB)

One of the significant changes introduced in LTE is that when the mobile device connects to the network it also implicitly gets an IP address. This is called "default EPS bearer activation" [28]. This concept is different from the conventional 3G system of packet data protocol (PDP) context activation.

In 3G systems, the mobile registers to the network first. Then, based on downlink or uplink activities, the IP address allocation procedure starts as part of the "PDP context activation." This procedure is referred to in 3G systems as establishing PS data call. The procedure of PS data call set-up follows the same as that in CS. When the user initiates or receives a call, the CS, or PS call is established and all resources are then allocated at the call set-up stage.

With the default bearer activation in LTE, the packet call is established at the same time as when the UE attaches to the EPS. This is the concept that makes the LTE's connectivity be known as "always-on".

This procedure, opposed to 3G, can provide a significant signaling reduction on the protocol layers and also improves the end-user experience in terms of data re-activation delays after a certain period of inactivity. In 3G, when the user disconnects the data call and then re-initiates a new one, the PDP context activation may start all over again. However, in LTE, if the same procedure is done by the user, the call set-up time for a data call is reduced because the default DRB (Data Radio Bearer) has been already assigned to the user when first attached to the EPS system.

1.8.1.3 Dedicated EPS Data Radio Bearer (Dedicated DRB)

Even though the default DRB is enough for the downlink and uplink data transfer in an EPS network, the default bearer comes without any QoS guarantees. For real-time streaming applications, QoS may be needed, especially on the air interface. Such IP packets associated with these types of applications may need to be assigned with a higher priority than other packets, especially when the bandwidth is limited.

To exploit the services differentiation, LTE has also introduced another EPS bearer known as a "dedicated EPS data bearer" which is initiated for an additional data radio bearer [28].

The dedicated bearer becomes important in order to support different types of applications in EPS network. Dedicated DRB can be set up right after default DRB in the procedures shown in Figure 1.29.

The dedicated DRB does not necessarily require an extra IP address. The protocol stack uses the traffic flow template (TFT) information to decide what to do with each IP packet. Uplink and downlink traffic are mapped onto proper bearers based on TFT filters configured at the UE and P-GW.

This concept makes the dedicated bearer activation similar to the secondary PDP context activation in 3G that can be used by the IMS, for example, to ensure real-time data is delivered promptly.

Due to the mapping between the radio bearer and lower layer logical channels, up to eight DRBs can be set up to carry user plane data connected to multiple PDN. They are divided into only one default EPS bearer and seven dedicated EPS bearers.

1.8.2   EPS Quality of Service (QoS)

In order to support a mixture of non-real-time and real-time applications, such as voice and multimedia, the delay and jitter may become excessive if the flows of traffic are not coordinated. Packet Switches should be able to classify, schedule, and forward traffic based on the destination address, as well as the type of media being transported. This becomes possible with QoS-aware systems.

The QoS for data radio bearers is provided to the eNB by the MME using the standardized QoS attributes. Based on these configured attributes by the EPS, the protocol layers between the UE and eNB can manage the ongoing scheduling of uplink and downlink traffic.

Various parameters are used to control and identify the QoS. The overall QoS parameters are shown in Figure 1.30.

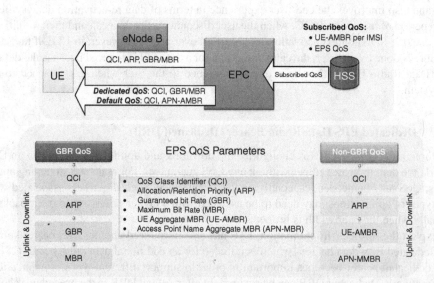

Figure 1.30   EPS QoS definitions and parameters.

1.8.2.1 EPS Bearer QoS

EPS bearer QoS depends on the resource type; either guaranteed bit rate (GBR) or non-guaranteed bit rate (non-GBR). The default DRB is always set up as a non-GBR. A dedicated DRB can be either GBR or non-GBR [29].

As illustrated in Figure 1.30, the GBR-based EPS bearer consists of two distinct parameters; GBR and MBR. The GBR indicates the bit rate that can be expected to be provided by a GBR-based bearer, while the MBR limits the bit rate that can be expected to be provided by this EPS bearer.

The GBR-based QoS parameters provide the eNB with information on the uplink and downlink rates for an E-RAB. E-RAB transports the packets of an EPS bearer between the UE and the EPC based on these QoS parameters indicating the E-RAB's maximum downlink bit rate, maximum uplink bit rate, guaranteed downlink bit rate, and E-RAB's guaranteed uplink bit rate.

Non-guaranteed EPS bearers are subject to control through an AMBR (aggregate maximum bit rate). The AMBR applies to both the subscriber and the APN associated with the subscriber, and is defined as follows:

- **UE-AMBR** – value applies to the total bit rate that can be allocated to a subscriber for all its non-GBR services. The UE-AMBR limits the aggregate bit rate across all non-GBR bearers of a UE (excess traffic may get discarded by a rate-shaping function).
- **APN-AMBR** – value applies to the total bit rate that can be allocated to the subset of a subscriber's services associated with a particular APN. The APN-AMBR limits the aggregate bit rate across all non-GBR bearers and across all PDN connections of the same APN (excess traffic may get discarded).

Similar to GBR-based QoS, the non-GBR parameters have uplink and downlink components.

1.8.2.2 ARP and QCI

The ARP (allocation and retention priority) controls the priority in bearer establishment, modification, or bearer release if resources are limited. In addition, it may be used to indicate which bearers are dropped when there is congestion in the network. This parameter can be used for GBR or non-GBR QoS.

The priority level of an ARP ranges from 0 to 15. The value 15 means "no priority," whereas the value 1 is the highest level of priority, with the value 0 being reserved. In addition, ARP provides preemption capability on other E-RABs. This indicates whether the E-RAB will not preempt other E-RABs or the E-RAB may preempt other E-RABs.

QCI (QoS class indicator) is another common QoS parameter in both GBR and non-GBR EPS bearers. It provides a mapping from an integer value to specific QoS parameters that controls how bearer level packets are forwarded.

QCI controls the packet forwarding, such as scheduling weights, admission thresholds, queue management thresholds, and link layer protocol configuration. QCI values for an E-RAB are typically pre-configured by the operator. QCI are categorized into nine different indicators, as shown in Table 1.7 [29].

Table 1.7   Standardized QCI characteristics

QCI	Resource type	Priority	Packet delay budget (PDB) (ms)	Packet error loss rate (PELR)	Examples of services
1	GBR	2	100	10^{-2}	Conversational voice
2		4	150	10^{-3}	Conversational video (live streaming)
3		3	50	10^{-3}	Real-time gaming
4		5	300	10^{-6}	Non-conversational video (buffered streaming)
5	Non-GBR	1	100	10^{-6}	IMS signaling
6		6	300	10^{-6}	Video (buffered streaming, TCP-based (www, e-mail, ftp, p2p file sharing)
7		7	100	10^{-3}	Voice, video, interactive gaming
8		8	300	10^{-6}	Same as QCI 6 but used for
9		9	300	10^{-6}	further differentiation

Standardized QCI characteristics are not signaled on any interface. They are guidelines for the pre-configuration of node-specific parameters for each QCI. They also ensure that applications or services mapped to a given QCI receive the same minimum level of QoS in multi-vendor network deployments and in the case of roaming. The typical QCI configured by LTE's operators with default EPS bearers carrying best effort traffic is 6 or 9.

An EPS bearer can include multiple SDFs. SDFs mapped to the same EPS bearer receive the same bearer level packet forwarding treatment: scheduling policy, queue management policy, rate-shaping policy, RLC configuration.

Every QCI (GBR and non-GBR) is associated with a priority level. Priority level 1 is the highest priority level. Scheduling between different SDF aggregates should primarily be based on the PDB (packet delay budget). For E-UTRAN, the priority level of a QCI may be used as the basis for assigning the uplink priority per radio bearer.

The purpose of the PELR (packet error loss rate) is to allow appropriate link layer protocol configurations at RLC and HARQ in E-UTRAN. For a certain QCI the value of the PELR is the same in uplink and downlink.

1.8.3   EPS Security Basics

In all 3GPP systems, security is needed to protect the user and control planes data. The security procedures take place at different levels of the connection. In LTE, the EPS security functions are [30]:

- **Authentication and key agreement (AKA)** – to prevent fraud that occurs when a third party obtains a copy of a subscriber's network identification information and uses it to fraudulently access the system.
- **Ciphering** – used to protect all user data and signaling from being overheard by an unauthorized entity.

- **Integrity** – protects signaling information from being corrupted. It is a message authentication function that prevents a signaling message from being intercepted and altered by an unauthorized device.

1.8.3.1 Authentication

The MME initiates the AKA procedure by sending the authentication request message to the UE, as shown in Figure 1.29. The MME sends the random challenge RAND and an authentication token, AUTN, for the network's authentication [30].

Upon receipt of this message, the UE verifies whether AUTN can be accepted. If AUTN is acceptable, the UE's USIM produces a RES and computes CK and IK (ciphering protection key and integrity protection key).

Once the NAS security context is created, the UE (EMM) generates an authentication RES message and includes RES in it. This NAS message is carried by the RRC signaling to the eNB. The eNB forwards the message to the MME.

AKA involves interworking with the subscriber's HSS in order to obtain AAA information to authenticate the subscriber. During AKA, keys are created for AS and NAS integrity protection and ciphering.

1.8.3.2 Integrity and Ciphering

The integrity and ciphering procedures involve both NAS and AS [30]:

- **NAS security context activation** – provides both integrity protection and ciphering for NAS signaling. The procedure takes place between UE and MME.
- **AS security context activation** – provides integrity and ciphering protection for RRC signaling in addition to ciphering for user plane data to be sent over the air interface. The procedure takes place between UE and eNB.

Both authentication and NAS security context activation are not mandatory to occur in every UE attach attempt. However, the AS security context is mandatory to take place for every connection the UE initiates with EPS. In 3GPP, integrity protection is mandated, but the ciphering is only recommended. Figure 1.31 shows the signaling flow of these procedures.

Both UE and EPS negotiate the integrity and ciphering algorithms capabilities indicated as part of "UE network capability" of the EMM attach request message. These algorithms are [30]:

- EEA0 Null ciphering algorithm
- 128-EEA1 SNOW 3G-based ciphering algorithm
- 128-EEA2 AES (advanced encryption standard)-based ciphering algorithm
- EIA0 Null integrity algorithm
- 128-EIA1: SNOW 3G-based integrity algorithm
- 128-EIA2: AES-based integrity algorithm.

MME selects a NAS integrity algorithm and a NAS ciphering algorithm for the UE. The MME is expected to select the NAS algorithms that have the highest priority according to the ordered lists. The selected algorithm is indicated in the NAS security mode command message

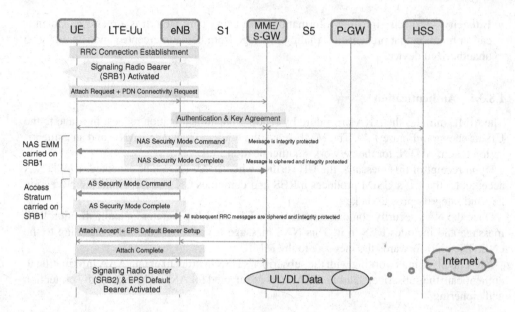

Figure 1.31 NAS and AS security context activation.

to the UE and also includes the UE security capabilities in that message. This message is integrity protected by MME with the selected algorithm.

The UE verifies that the message from the MME contains the correct UE security capabilities. This enables detection of attacks if an attacker has modified the UE security capabilities in the initial NAS message.

The UE then generates NAS security keys based on the algorithms indicated in the NAS security mode command and replies with an integrity protected NAS security mode complete message. NAS security is activated at this point.

After this point, eNB creates the AS security context when it receives the keys from the MME. The eNB generates the integrity and encryption keys and selects the highest priority ciphering and integrity protection algorithms from its configured list that are also present in the UE's EPS security capabilities.

Upon reception of the AS security mode command, the UE generates integrity and encryption keys and sends an AS security mode complete message to the eNB.

1.8.4   EPS Idle and Active States

After the UE attaches to the EPS network, the data activity controls the states in which the UE operates in the EPS network.

There are several states in each of the EPS entities, depending on the connection status. The states are categorized in AS and NAS, as shown in Figure 1.32.

The user plane data can only flow when all the AS and NAS signaling connections and bearers are in active/connected states.

On the air interface, the UE typically transitions into the RRC-idle state after successfully attaching to the LTE system. UE remains in this state as long as there are no radio interface

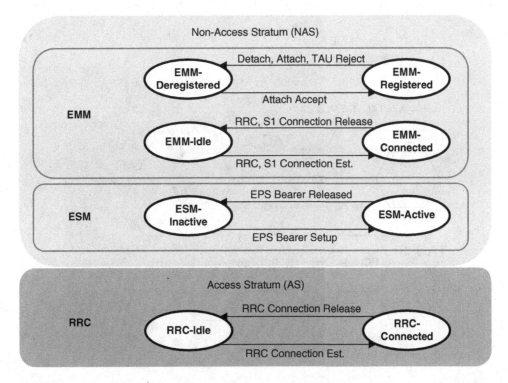

Figure 1.32   EPS idle and active states for NAS and AS.

downlink or uplink packet activities with eNB. When a data activity is initiated by a user or an application installed in the device, the UE immediately transits into RRC-connected state and remains in this state until the packet connectivity timer, known as the "user inactivity" timer, expires. The timer is configured in eNB and used to monitor the data activity for a user within a timed window. When the timer expires, the eNB releases the RRC connection and immediately triggers a UE's state transition to the RRC-idle state.

The same concepts of NAS and AS states are also available in 3G systems. In the UMTS air interface, the RRC states can be in either connected or idle mode. In connected mode, the UE can be served in four different states: Cell_DCH (data channel), Cell_FACH (forward access channel), Cell_PCH (paging channel), or URA_PCH. However, the state transitions in the LTE air interface are simplified to only idle and connected mode, avoiding all the timers and optimizations.

The RRC level state transition from connected to idle mode targets an improved battery lifetime of the device. The battery consumption is expected to be more efficient in the idle state when there is no connectivity or dedicated resource between the device and the eNB.

1.8.5   EPS Network Topology for Mobility Procedures

After the UE camps on an E-UTRAN cell, it uses the NAS procedure to register its presence in a TA of the camped cell. This allows the EPC to page the user in the registered TA(s) while UE is in idle mode.

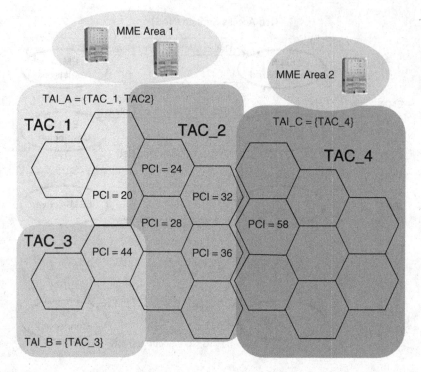

Figure 1.33   EPS network topology for mobility procedure.

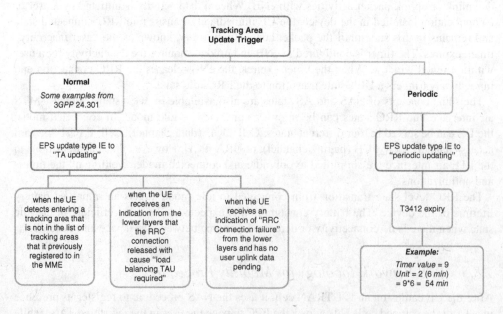

Figure 1.34   Tracking area updating trigger conditions.

Figure 1.33 shows an example of the different locations in which UE can register during the mobility between different eNB cells.

A TA corresponds to the concept of the routing area (RA) used in UMTS. The TA consists of a cluster of eNBs having the same tracking area code (TAC). The TAC provides a way to track UE location in idle mode. TAC information is used by the MME when paging idle UE to notify them of incoming data connections.

The MME sends the tracking area identity (TAI) list to the UE during the TA update procedure. TA updates occur periodically or when a UE enters a cell with a TAC not in the current TAI list. The TAI list makes it possible to avoid frequent TA updates due to ping-pong effects

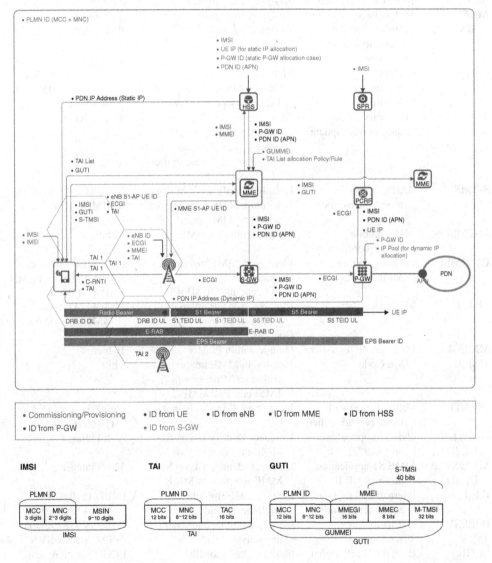

Figure 1.35 EPS identifiers. (Source: [31]. Reproduced with permission of NMC group.)

Table 1.8 EPS identifier definitions

Identifier	Description	Assignment	Definition
IMSI	International mobile subscriber identity	Unique identification of mobile (LTE) subscriber Network (MME) gets the PLMN of the subscriber	IMSI (not more than 15 digits) = PLMN ID + MSIN = MCC + MNC + MSIN
PLMN ID	Public land mobile network identifier	Unique identification of PLMN	PLMN ID (not more than 6 digits) = MCC + MNC
MCC	Mobile country code	Assigned by regulator	3 digits
MNC	Mobile network code	Assigned by regulator	2–3 digits
MSIN	Mobile subscriber identification number	Assigned by operator	9–10 digits
GUTI	Globally unique temporary UE identity	Identify a UE between the UE and the MME on behalf of IMSI for security reasons	GUTI (not more than 80 bits) = GUMMEI + M-TMSI
TIN	Temporary identity used in next update	GUTI is stored in TIN parameter of UE's MM context. TIN indicates which temporary ID to use in the next update	TIN = GUTI
S-TMSI	SAE temporary mobile subscriber identity	Locally identify a UE in short within a MME group (unique within an MME pool)	S-TMSI (40 bits) = MMEC + M-TMSI
M-TMSI	MME mobile subscriber identity	Unique within an MME	32 bits
GUMMEI	Globally unique MME identity	Identify an MME uniquely in global GUTI contains GUMMEI	GUMMEI (not more than 48 bits) = PLMN ID + MMEI
MMEI	MME identifier	Identify an MME uniquely within a PLMN Operator commissions at eNB	MMEI (24 bits) = MMEGI + MMEC
MMEGI	MME group identifier	Unique within PLMN	16 bits
MMEC	MME Code	Identify an MME uniquely within an MME group S-TMSI contains MMEC	8 bits
C-RNTI	Cell-radio network temporary identifier	Identify a UE uniquely in a cell	0 x 0001 ~ 0 x FFF3 (16 bits)
eNB S1AP UE ID	eNB S1 application protocol UE ID	Uniquely identify UE on S1-MME interface in eNB	32-bit integer
MME S1AP UE ID	MME S1 application protocol UE ID	Uniquely identify UE on S1-MME interface in MME	32-bit integer
IMEI	International mobile equipment identity	Identify a ME (mobile equipment) uniquely	IMEI (15 digits) = TAC + SNR + CD
IMEI/SV	IMEI/software version	Identify a mobile equipment uniquely	IMEI/SV (16 digits) = TAC + SNR SVN
ECGI	E-UTRAN cell global identifier	Identify a cell globally EPC can know UE location based on ECGI	ECGI (not more than 52 bits) = PLMN I D + ECI

Table 1.8 (*continued*)

Identifier	Description	Assignment	Definition
ECI	E-UTRAN cell identifier	Identify a cell within a PLMN	ECI (28 bits) = eNB ID + Cell ID
Global eNB ID	Global eNodeB identifier	Identify an eNB globally in the network	Global eNB ID (not more than 44 bits) = PLMN ID + eNB ID
eNB ID	eNodeB identifier	Identify an eNB within a PLMN	20 bits
P-GW ID	PDN GW identifier	Identify a specific PDN-GW HSS assigns P-GW for PDN connection of each UE	IP address (4 bytes) or FQDN (variable length)
TAI	Tracking area identity	Identify tracking area Globally unique	TAI (not more than 32 bits) = PLMN ID + TAC
TAC	Tracking area code	Indicate eNB to which tracking area the eNB belongs (per Cell) Unique within a PLMN	16 bits
TAI List	Tracking area identity list	UE can move into the cells included in TAL list without location update (TA update) Globally unique	Variable length
PDN ID	Packet data network identity	Identify a PDN (IP network), a mobile data user wants to communicate with PDN identity (APN) used to determine P-GW and point of interconnection with a PDN With APN as query parameter to the DNS procedures, the MME will receive a list of candidate P-GWs, and then a P-GW is selected by MME with policy	PDN Identify = APN = APN.NI + APN.OI (variable length)
EPS Bearer ID	Evolved packet system bearer identifier	Identify EPS bearer (default or dedicated) per UE	4 bits
E-RAB ID	E-UTRAN radio access bearer identifier	Identify an E-RAB per an UE	4 bits
DRB ID	Data radio bearer identifier	Identify a DRB per an UE	4 bits
LBI	Linked EPS bearer ID	Identify the default bearer associated with a dedicated EPS bearer	4 bits
TEID	Tunnel end point identifier	Identify the end point of a GTP tunnel when the tunnel is established	32 bits

(Source: [31]. Reproduced with permission of NMC group.)

along TA borders. This is achieved by including the old TAC in the new TAI list received at TA update. When the MME pages a UE, a paging message is sent to all cells in the TAI list.

In the example shown in Figure 1.33, if the UE performs EPS registration from TAI_A, the MMEs send TAC_1, and TAC_2 in the TAI List, implying that the UE can roam around in the eNBs with the TACs belonging to this TAI list without having to re-register with the EPS network. This procedure saves on the signaling load. The UE re-registers with a TAU procedure if the UE enters into the coverage areas of eNB that are part of TAC_3 (in TAI_B) and TAC_4 (in TAI_C).

The TA dimensioning and planning in the network are performed in the optimization stage. The TA planning can prevent the ping-pong effect of TAU to achieve optimization between paging load, registration overhead, UE battery, and improved paging success rate. In the same example, the paging area for UE served in TAI_A will be for all cells belonging to TAC_1 and TAC_2, but the registration area will be limited to TAI_A only.

TA updating can be either periodical or based on the mobility conditions of the device. Figure 1.34 summarizes the triggering conditions of the TAU procedure [28].

The MME area is the part of the network served by an MME. The MME area consists of one or more tracking areas. All cells served by an eNB are included in an MME area. There is no one-to-one relationship between an MME area and an MSC/VLR area. Multiple MMEs may have the same MME area (pool area) and MME areas may overlap each other.

1.8.6 EPS Identifiers

The LTE system is designed to simplify the procedures carried on EPS. This is possible by designing and assigning the required identifiers at different interfaces within the EPS system.

The different identities defined in the EPS system are shown in Figure 1.35 and each is defined in Table 1.8 [31]. Different types of identifiers are needed between the eNB and the UE as part of the RNTI (radio network temporary identifier). These RNTIs are used for different procedures such as paging, random access, and system information on the air interface. They are not shown in the figure, but are discussed in detail in the next chapter.

References

[1] 3GPP (2008) All-IP Network (AIPN) Feasibility Study. TR 22.978.
[2] 3GPP (2008) 3rd Generation Partnership Project; Technical Specification Group Services and System Aspects; 3GPP System Architecture Evolution: Report on Technical Options and Conclusions. TR 23.882 V8.0.0.
[3] 3GPP (2008) Study on Evolved UTRA and UTRAN. TR 25.912.
[4] 3GPP (2008) Requirements for Evolved UTRA (E-UTRA) and Evolved UTRAN (E-UTRAN). TR 25.913.
[5] 3GPP (2010) 3rd Generation Partnership Project; Technical Specification Group Radio Access Network; Evolved Universal Terrestrial Radio Access Network (E-UTRAN); S1 Application Protocol (S1AP). TS 36.413 V8.10.0.
[6] 3GPP (2011) 3rd Generation Partnership Project; Technical Specification Group Core Network and Terminals; 3GPP Evolved Packet System (EPS); Evolved General Packet Radio Service (GPRS) Tunnelling Protocol for Control Plane (GTPv2-C). TS 29.274 V8.11.0.

[7] 3GPP (2010) 3rd Generation Partnership Project; Technical Specification Group Core Network and Terminals; Proxy Mobile IPv6 (PMIPv6) based Mobility and Tunnelling Protocols. TS 29.275 V8.8.0.

[8] 3GPP (2010) 3rd Generation Partnership Project; Technical Specification Group Radio Access Network; Evolved Universal Terrestrial Radio Access Network (E-UTRAN); Architecture Description. TS 36.401 V8.8.0.

[9] 3GPP (2008) 3rd Generation Partnership Project; Technical Specification Group Radio Access Network; Evolved Universal Terrestrial Radio Access Network (E-UTRAN); X2 General Aspects and Principles. TS 36.420 V8.1.0.

[10] 3GPP (2012) 3rd Generation Partnership Project; Technical Specification Group Radio Access Network; Evolved Universal Terrestrial Radio Access (E-UTRA); User Equipment (UE) Radio Access Capabilities. TS 36.306 V10.7.0.

[11] IETF (2007) Stream Control Transmission Protocol. RFC 4960.

[12] IETF (1980) User Datagram Protocol. RFC 768.

[13] IETF (2003) Diameter Base Protocol. RFC 3588.

[14] 3GPP (2009) Evolved Universal Terrestrial Radio Access (E-UTRA) and Evolved Universal Terrestrial Radio Access Network (E-UTRAN); Overall Description. TS 36.300 V8.5.0.

[15] 3GPP (2010) General Packet Radio System (GPRS) Tunnelling Protocol User Plane (GTPv1-U). TS 29.281V9.3.0.

[16] 3GPP (2010) General Packet Radio Service (GPRS); Service Description. TS 23.060 V9.6.0.

[17] 3GPP (2012) Evolved Packet System (EPS); Mobility Management Entity (MME) and Serving GPRS Support Node (SGSN) Related Interfaces Based on Diameter Protocol. TS 29.272 V9.9.0.

[18] 3GPP (2011) Evolved Packet System (EPS); 3GPP EPS AAA Interfaces. TS 29.273 V9.7.0.

[19] 3GPP (2010) Mobility Management Entity (MME) – Visitor Location Register (VLR) SGs Interface Specification. TS 29.118 V9.3.0.

[20] 3GPP (2011) Policy and Charging Control Over Gx Reference Point. TS 29.212 V9.6.0.

[21] 3GPP (2010) Charging Management; Charging Architecture and Principles. TS 32.240 V9.1.0.

[22] 3GPP (2011) Interworking between the Public Land Mobile Network (PLMN) supporting Packet Based Services and Packet Data Networks (PDN). TS 29.061 V10.4.0.

[23] 3GPP (2011) Cell Broadcast Centre interfaces with the Evolved Packet Core. TS 29. 168 V 10.0.0.

[24] Deering, S. and Hinden, R., IETF (1998) Internet Protocol, Version 6 (IPv6) Specification. RFC 2460.

[25] IETF (1981) Internet Protocol (STD 5). RFC 791.

[26] 3GPP (2009) Technical Realization of Cell Broadcast Service (CBS). TS 23.041 V8.2.0.

[27] 3GPP (2010) Sv interface (MME to MSC, and SGSN to MSC) for SRVCC. TS 29.280 V9.2.0.

[28] 3GPP (2011) 3rd Generation Partnership Project; Technical Specification Group Core Network and Terminals; Non-Access-Stratum (NAS) Protocol for Evolved Packet System (EPS). TS 24.301 V8.10.0.

[29] 3GPP (2012) 3rd Generation Partnership Project; Technical Specification Group Services and System Aspects; Policy and Charging Control Architecture. TS 23.203 V10.8.0.

[30] 3GPP (2011) 3rd Generation Partnership Project; Technical Specification Group Services and System Aspects; 3GPP System Architecture Evolution (SAE); Security Architecture. TS 33.401 V8.8.0.

[31] NMC http://www.nmcgroups.com/files/download/NMC.LTE%20Identifiers.v1.0.pdf (accessed 23 September 2013).

2

LTE Air Interface and Procedures

Mohamed A. El-saidny

2.1 LTE Protocol Stack

The LTE (long term evolution) air interface provides connectivity between the user equipment (UE) and the eNB (eNodeB). It is split into a control plane and a user plane, as described in Chapter 1. Among the two control plane signalings, the first is provided by the access stratum (AS) and carries signaling between the UE and the eNB. The second carries non-access stratum (NAS) signaling messages between the UE and the MME (mobility management entity), which is piggybacked into an RRC (radio resource control) message. The user plane delivers the IP (Internet protocol) packets to and from the EPC (evolved packet core), the S-GW (serving gateway), and the PDN-GW (packet data network gateway).

The structure of the lower layer protocols for the control and user planes in AS are the same. Both planes utilize the protocols of PDCP (packet data convergence protocol), RLC (radio link control), and MAC (medium access control), as well as the PHY (physical layer) for the transmission of the signaling and data packets [1].

NAS is the layer above the AS layers. There are also two planes in NAS; the higher layer signaling related to the control plane and the IP data packets of the user plane. NAS signaling exists in two protocol layers, EMM (EPS mobility management) and ESM (EPS session management), as discussed in Chapter 1. The NAS user plane is IP-based. The IP data packets pass directly into the PDCP layer for processing and transmission to or from the user.

Figure 2.1 illustrates the radio interface protocol stack. The protocol stacks reside in both the UE and the E-UTRAN (evolved universal terrestrial radio access network). Control and user plane data flow on the entire stack based on the type of traffic being exchanged from or to the UE. It is illustrated in the figure that the NAS signaling uses the services of RRC, which is then mapped into the PDCP. On the user plane, IP packets are also mapped into the PDCP layer and then delivered down to the lower layers for transmission.

This chapter describes the air interface of LTE, focusing on the AS protocol layers. It then provides an overview of the PHY layer structure and how it utilizes OFDMA (orthogonal frequency division multiple access) for transmission. The chapter concludes with an end-to-end

Design, Deployment and Performance of 4G-LTE Networks: A Practical Approach, First Edition. Ayman Elnashar, Mohamed A. El-saidny and Mahmoud R. Sherif.
© 2014 John Wiley & Sons, Ltd. Published 2014 by John Wiley & Sons, Ltd.

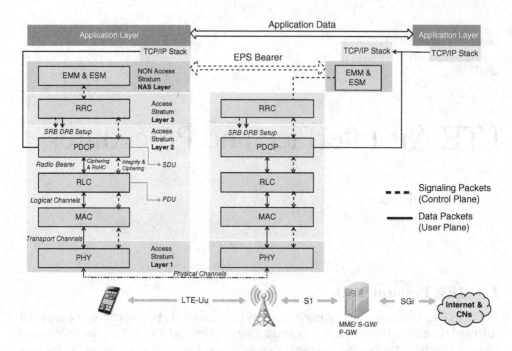

Figure 2.1 LTE protocol stacks.

procedure as when the UE powers-up in an LTE network, interchanging data with the network and mobiles around the eNBs. A comparison with the HSPA(+) PHY layer and procedures is also provided to clarify the concepts of the LTE channels.

2.2 SDU and PDU

Each layer within the protocol stack uses the services of the layer below it and offers services to the layer above it. For example, RRC uses the services of RLC and offers services to the NAS layer. Additionally, each layer in the UE communicates with its peer layer in E-UTRAN, as shown in Figure 2.1.

The user plane IP packets are typically sizable (in bytes, for example, one IP packet is 1500 bytes). The control plane signaling can also contain a larger size message than the air interface can handle in certain radio conditions. Therefore, the packets are not exactly transmitted to the lower layers as received from the upper layers. The packets are usually segmented into smaller units for over-the-air transmission, to maintain the bandwidth of the air interface as well as the radio conditions of the UE.

The unit of data exchanged between entities' peer layers is called a protocol data unit (PDU). For example, the RRC layers of the UE and E-UTRAN communicate with each other via signaling messages that are encapsulated in a PDU. Figure 2.2 demonstrates the definitions of PDU.

To send an uplink PDU from the UE's RRC to the E-UTRAN's RRC layer, it passes down the UE's protocol stack to PDCP, RLC, MAC, and PHY, and then up into the E-UTRAN's

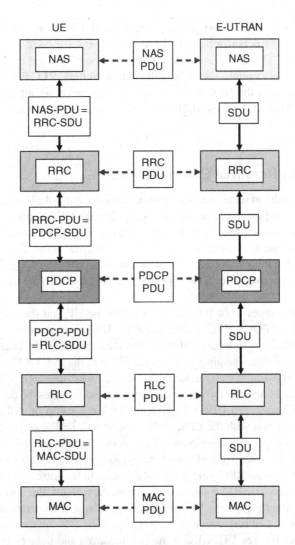

Figure 2.2 Protocol layer SDUs and PDUs.

protocol stack to the RRC layer. On the other hand, the data sent up or down the protocol stack in one of these entities is called a service data unit (SDU), as indicated in Figure 2.2.

For example, the UE's RRC layer sends an RRC signaling message to the PDCP as an SDU. The PDCP transforms this SDU to the RLC layer down the stack. The RLC layer converts them into one or more RLC PDUs after performing functions such as segmentation, concatenation, and RLC headers addition. These RLC PDUs are then constructed as MAC SDUs. The MAC SDU consists of multiple MAC PDUs coming from each part of the stack, for example, when there are user data to be sent in parallel with signaling. The MAC PDUs are then mapped to the transport block (TB) to be sent on the PHY layer and then over the air to the serving cell of the eNB.

In the receiving direction, the eNB decodes the TB which contains the MAC PDUs. After a reordering mechanism is performed, the PDUs are assembled back into RLC SDUs after removing the headers. Then, RLC delivers the SDUs to the PDCP and then to the RRC layer. At the RLC layer in this example, one SDU will contain multiple RLC PDUs, and the delivery to the PDCP and then to the RRC will only occur after receiving all the RLC PDUs that construct the final RRC SDU. This SDU now contains the original RRC signaling message sent by the UE.

2.3   LTE Radio Resource Control (RRC)

The RRC constitutes the main air interface protocol for the control plane signaling messages [1]. In general, signaling messages are needed to regulate the UE behavior in order to comply with the network procedures. Each signaling message the EPS (evolved packet system), sends to the UE, or vice versa, is comprised of a set of system parameters. For example, the eNB needs to communicate the parameters related to mobility procedures when the UE needs to hand over from one cell to another. These parameters will be sent to the UE in a specific RRC message.

In order for the messages to be transferred between the UE and the eNB, the RRC layer uses the services of the PDCP, RLC, MAC, and PHY. During the course of this mapping, the packets are directed on a radio bearer, referred to as the signaling radio bearer, SRB.

The RRC handles all the signaling between the UE and the E-UTRAN. Additionally, the core network NAS signaling is also carried by a dedicated RRC message. When carrying NAS signaling, the RRC does not alter the information but instead provides the delivery mechanism.

As described in Chapter 1, a UE in the LTE network can camp on different states. For a UE with active connection with the EPS, its RRC state will be RR-connected. The types of signaling messages and parameters exchanged in this state handle the UE for mobility, that is, handover, and all the associated radio bearer configurations for the data transmission. For a UE with inactive connection, its RRC state will be RRC-idle. In this case, the UE would not have a dedicated radio bearer (DRB) for data transmission. As a result, the only signaling needed in this state would target paging the UE for incoming calls, or parameters related to mobility in the idle state, that is, cell reselection.

Table 2.1 describes the key LTE RRC signaling messages and their corresponding UMTS ones that closely match the purpose of such messages. The table does not show the NAS-related message names.

The eNB typically uses two methods for confirming that the message has reached the UE. One method is when the UE sends a complete RRC message in response to the RRC message from eNB. The eNB treats this "complete message" as an RRC acknowledgment. For example, when the eNB sends "RRC connection reconfiguration," it waits to receive the UE's "RRC connection reconfiguration complete" in order to complete the RRC procedure. There are RRC timers controlling the timeout duration before the eNB decides to tear down the RRC connection. If the connection is disconnected due to an incomplete RRC procedure, the call is also considered as a dropped call. Hence, dropped calls due to incomplete RRC signaling procedure is an area of optimization in the LTE network, likewise UMTS.

Another RRC confirmation method is a lower layer acknowledgment at RLC and PHY (a joint operation with MAC). This method is required to ensure that any missing segments of

Table 2.1 RRC signaling messages

LTE RRC message name	RRC state/direction	Similar UMTS RRC message name	Main purpose of message
System/master information blocks (SIBs and MIB)	Idle/from eNB	System/master information blocks	Carries parameters for • UE to identify the network (PLMN) and cell (tracking area) • idle mode mobility procedures for cell reselection • RACH procedures • paging procedures
Paging	Idle/from eNB	Paging type 1 *Paging type 2 used for paging in connected mode*	Paging the UE from idle mode for any service
RRC connection request	Connecting/from UE	RRC connection request	UE identity Call establishment cause
RRC connection setup	Connecting/from eNB	RRC connection set-up	Carries parameters for: • SRB1 mapping to all lower layers • RLC parameters for SRB1 • initial physical layer parameters
RRC connection set-up complete	Connected/from UE	NAS message, that is, initial direct transfer	• MME ID • Piggybacked EMM NAS signaling message
UE capability information	Connected/from UE	RRC connection set-up complete	UE capabilities: RAT supported, bands, LTE capabilities
Security mode command	Connected/from eNB	Security mode command	Ciphering and integrity
RRC connection reconfiguration	Connected/from eNB	Measurement control message	Carries connected mode mobility parameters (handover) and neighbor cell information
Measurement report	Connected/from UE	Measurement report	Carries the measurements by the UE for the serving and neighbor cells, depending on the parameters in "RRC connection reconfiguration"
RRC connection reconfiguration	Connected/from eNB	Radio bearer set-up, radio bearer reconfiguration, physical channel reconfiguration	Carries parameters needed for: • establishing DRB and mapping to lower layers • SRB2 mapping for lower layers

(continued overleaf)

Table 2.1   (*continued*)

LTE RRC message name	RRC state/direction	Similar UMTS RRC message name	Main purpose of message
			• any lower layer parameter reconfiguration • physical layer parameters and reconfigurations
RRC connection release	Connected/from eNB	RRC connection release	• Release the RRC connection. It transitions the UE from RRC-connected to RRC-idle • It can also redirect the UE to select another RAT after the release (i.e., LTE to UMTS redirection)

the RRC message on the air interface, in deteriorating RF conditions, is being retransmitted in a timely manner before being delivered to the RRC layer.

Not all RRC messages require an RRC complete message from the UE; it depends on the procedure being carried out. However, all RRC-connected mode messages require lower layer acknowledgments. The RRC complete messages are also delivered from the lower layers, and hence only require lower layer acknowledgments. For any dropped call due to signaling message timeout, both the RRC layer and all lower layers are potential areas of investigation.

2.4   LTE Packet Data Convergence Protocol Layer (PDCP)

The PDCP layer is responsible for the following key functions [2]:

1. It transfers the control and user plane data to and from the upper layers. It receives SDUs from the upper layers and sends PDUs to the lower layers. In the other direction, it receives PDUs from the lower layers and sends SDUs to the upper layers.
2. It is responsible for security functions. It applies ciphering for user and control plane bearers, if configured. It may also perform integrity protection for control plane signaling messages, both RRC and NAS.
3. It performs header compression services to improve the efficiency and performance of over the air transmissions. The header compression is based on robust header compression (ROHC).
4. It is responsible for in-order delivery of packets and duplicate detection services to upper layers between the source and target eNB during the handover procedure in the RRC-connected state.

2.4.1 PDCP Architecture

For a UE in the RRC-connected state, the PDCP acts as the first AS layer to exchange the control and user planes packets. Figure 2.3 illustrates the functions of the PDCP layer.

2.4.2 PDCP Data and Control SDUs

While being constructed from the upper layer SDUs, a PDU holds the data field, the SDU being sent from upper layers, in addition to other needed header information.

The PDCP's PDU header includes a 5-bit sequence number (SN) space for the control plane. For the user plane PDCP PDUs, 12 and 7-bit SN are supported and configured on a per DRB basis. For example, the 12-bit SN format applies only to RLC acknowledged mode (AM) operation of DRB. Additionally, a header field referred to as "D/C" indicates whether the PDU carries user plane data or control information generated at the PDCP layer. The D/C field enables the receiving entity to direct the received PDCP PDU to the intended radio bearer.

The control plane packets are integrity protected, as shown in Figure 2.3. Therefore, a 32-bit MAC-I (message authentication code for integrity) is attached to the PDCP's PDUs.

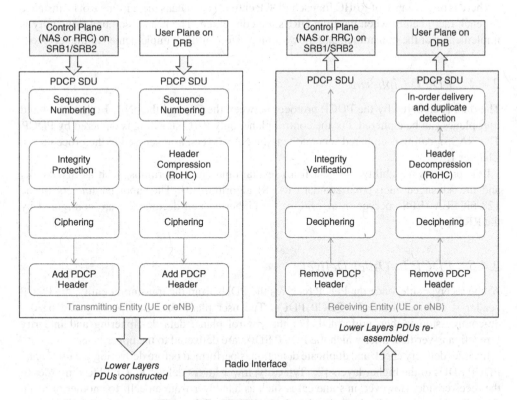

Figure 2.3 PDCP layer, functional view.

The MAC-I is calculated using the transmitted message, an integrity protection key for the user, and other time varying parameters depending on the EPS integrity algorithm (EAI) configured by the RRC layer. Chapter 1 describes the integrity protection architecture, and more information is provided in 3GPP (third generation partnership project) [3]. The PDCP attaches the MAC-I to the end of the control plane PDU. The integrity protection stage does not apply to user plane data being transmitted through the PDCP layer.

2.4.3 PDCP Header Compression

At this point, from the flow shown in Figure 2.3, PDCP PDU has been constructed for the user plane data packet with the proper PDCP headers, that is, SN. PDCP receives IP packets for transmission from the application layer. On top of IP, the transport protocol may be TCP (transmission control protocol), UDP (user datagram protocol), or RTP (real-time transport protocol) generating additional large headers. The header overhead may be reduced by using compression techniques such as RoHC supported in LTE, based on [4]. Header compression also reduces transmission delay and packet loss rate. By this definition, this stage does not apply to the PDUs of the control plane at PDCP.

There is one instance of RoHC for each PDCP entity. That means there is one RoHC instance for each radio bearer when multiple DRBs are configured. The header compression entity is implemented at the transmitter, and the decompression entity is implemented at the receiver.

2.4.4 PDCP Ciphering

The data is ciphered by the PDCP protocol between the UE and the eNB. Both control and user planes can be ciphered. For the control plane, only RRC signaling is ciphered by PDCP. The NAS signaling is ciphered separately at the NAS layers, as discussed in the procedure in Chapter 1.

LTE provides the ability to select from several ciphering algorithms, such as SNOW 3G and the advanced encryption standard (AES) algorithm [3]. The input parameters to the 128-bit EEA (EPS encryption algorithm) for ciphering and deciphering are configured by the RRC layer.

2.4.5 PDCP In-Order Delivery

At the receiver side, once the PDCP receives the PDUs from the transmitter entity, the PDCP header is removed from the PDCP PDUs. The user plane data de-ciphering and header de-compression is then preformed. For the control plane, data de-ciphering and integrity protection is verified, after which the PDCP SDUs are delivered to the upper layers.

In-order delivery check and duplicate detection is performed before delivering the user plane PDCP SDUs to the higher layers [2]. Typically, lower layers deliver PDCP PDUs in-order to the receiver side. However, in some cases such as handover from an eNB to another or when the call is being re-established after a drop in degrading RF conditions, the PDCP PDUs may experience holes in the SN when the lower layers resets.

SN will be tracked in the PDCP layer from the headers attached to the PDCP PDUs being received. In the case of detecting SN holes, PDCP will request a PDCP status report from

Table 2.2   LTE and HSPA PDCP function comparison

Criteria	PDCP in LTE	PDCP in HSPA
PDCP entity	In eNB	In RNC
PDCP used for control plane signaling	Yes	No
PDCP used for user plane data	Yes	Yes
PDCP performs ciphering and integrity protection	Yes	No
RoHC header compression support	Yes	Yes
In sequence delivery and duplicate detection support	Yes	No

the transmitter, indicating a retransmission request of the lost PDCP PDU. If the packets are received out of order within the reordering window, the PDCP performs reordering before delivering the packets to the upper layers. The reordering window size is 2048. If any sequential PDU does not arrive within this window, the PDCP layer sends the remaining PDUs in-order to the upper layer.

Due to this nature of delivery, PDCP provides discard timer functionality and the ability to retransmit missing PDCP PDUs. The transmitter entity will maintain a buffer to store trans-mitted PDUs to support this functionality. The timer is configured by the RRC layer. The SDU is finally discarded when the timer expires.

2.4.6   PDCP in LTE versus HSPA

LTE implements PDCP in both the user plane and the control plane. This is different than UMTS, where the PDCP layer is only designed for the user plane. The main reason for this difference is that the PDCP in LTE takes on the role of security, encryption and integrity. This is one of the main differences of LTE's PDCP layer. For HSPA (high speed packet access), ciphering is performed in the RLC layer and integrity protection is performed in the RRC layer.

Both LTE and HSPA support PDCP header compression, although HSPA supports multi-ple compression techniques such as IP Header Compression as well as RoHC for the user plane data.

The main reason for PDCP being implemented in eNB in LTE is that the header compres-sion parameters are reset during handover. In HSPA, the header compression parameters are transferred across RNCs (radio network controllers), if lossless SRNC (serving radio network controller) relocation is required. This is the reason for PDCP being implemented in the RNC for the HSPA system. The major differences are summarized in Table 2.2.

2.5   LTE Radio Link Control (RLC)

The RLC is part of the protocol layers in both the UE and the eNB. In the downlink, it uses the services of PDCP and offers services to the MAC layer. In the uplink, it uses the services of the MAC layer and offers services to the PDCP layer. In some cases, the RLC uses the services of the RRC directly. This occurs during the establishment of RRC connection prior to SRB1 or SRB2 set-up. In this case, the RRC message will be sent on SRB0, a common control channel

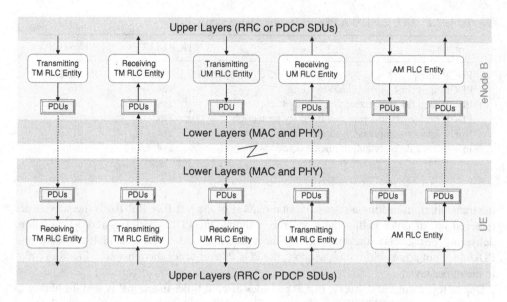

Figure 2.4   Overview model of the RLC sublayer.

(CCCH) discussed later in this chapter, right at the time the UE is connecting; that is, moving from RRC-idle to RRC-connected.

As all other protocol layers, the RLC functions support both control and user plane packets, including [5]:

1. Transfer of upper layer PDUs.
2. Error correction through ARQ (automatic repeat request).
3. Concatenation, segmentation, and re-assembly of SDUs.
4. Resegmentation of RLC PDUs.
5. In-sequence delivery and duplicate detection of RLC PDUs.
6. Protocol error detection and recovery mechanism.
7. RLC SDU discarding mechanism.

2.5.1 RLC Architecture

The RLC layer receives packets from the upper layer radio bearers, signaling, or data, as SDUs. The transmission entity in the RLC layer converts them into RLC PDUs after performing the key functions: segmentation, concatenation, and adding RLC headers, depending on the RLC mode. In the other direction, the receiving entity decodes the RLC PDUs from the MAC layer. After performing reordering, the PDUs are assembled back into RLC SDUs and delivered to the upper layer. Figure 2.4 illustrates the model of the RLC layer [5].

The RLC PDUs are of variable sizes and can be formatted based on the TB available in MAC from the underlying PHY channel. MAC notifies the RLC when a transmission opportunity becomes available, including the total number of RLC PDUs that can be transmitted in the current transmission opportunity.

2.5.2 RLC Modes

RLC is a pivotal layer in the PDU transmission across the protocol stack, and is therefore called the "radio link" control. One of the main functions of the RLC is to provide ARQ operation of the RLC PDUs between the UE and the eNB.

ARQ is a procedure that controls the retransmission of the missing PDUs. The PDU retransmission in the protocol stack is mainly handled by MAC, jointly with the PHY and RLC layers. In MAC, the retransmission is handled by H-ARQ (hybrid automatic repeat request) discussed in Section 2.6. Packet retransmissions at both of these layers protect the control and user plane data for a reliable and quality connection.

There are different layers and services that go through the RLC including SRB and DRB data packets, the retransmission mechanism is not necessarily required for all types of these packets. Therefore, RLC PDUs can operate in three different modes: transparent mode (TM), unacknowledged mode (UM), or Acknowledged Mode (AM).

The mode of operation controls the applicability and functionality of the RLC. TM is only applicable for control plane signaling related RLC packets. AM or UM can be used for control or user plane RLC packets. The chosen mode is controlled by the RRC and conveyed to the UE in the RRC messages at the time of establishing the corresponding radio bearer, SRB or DRB. Since each of these modes has its own functions, pros and cons, the chosen mode is typically up to the eNB implementation.

EPS QoS (quality of service) is one of the important drivers for the choice of DRB being mapped on RLC AM or UM. The eNB can link the choice of the RLC mode to certain QCIs (QoS class indicators) in order to maintain the desired QoS, described in Chapter 1. For example, if DRB in EPS is configured to use QCI 7, some eNB implementation can link this QCI to the usage of RLC UM instead of RLC AM. When QCI 6 or 9 is set by EPS for best effort DRB, the eNB may default the RLC mode to be AM. The relationship between QoS and RLC packet delivery mode is logical because QCI can differentiate the service requirement for error sensitivity and delay tolerance. The RLC mode can therefore maintain these QoS requirements, to some extent.

2.5.2.1 RLC Transparent Mode (TM)

TM can be regarded as null RLC since it is simply a pass through. None of the major RLC functions are applicable to this mode. The RLC layer does not add any header or other overhead. There is also no PDU retransmission occurring for this mode. Hence, ARQ operation does not apply.

The use of TM is limited to the common signaling channels responsible for paging, system information block (SIB) transmission, or initial RRC connection establishment. As seen, these procedures do not necessarily require any re-transmission at the RLC layer in particular, and hence are mapped into RLC TM.

For example, the paging message being sent from the eNB to the UE is mapped into the TM mode. If the UE is at the edge of coverage and radio link conditions do not allow the paging message to be delivered to the UE from the eNB, then retransmission takes place at the message level itself. In this case, either the EPC or the eNB may trigger a paging repetition attempt after the paging timer expires. The re-paging attempt is going to be a new trial to reach the UE and not an RLC retransmission.

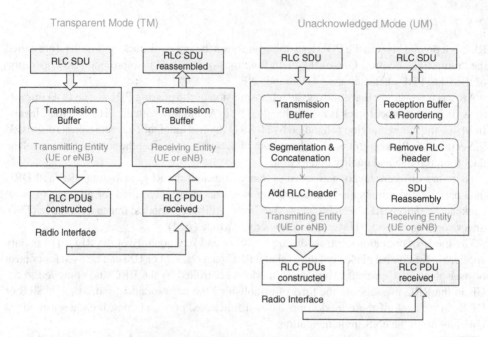

Figure 2.5   RLC TM and UM mode.

The TM mode entity consists simply of a transmission buffer to hold the RLC SDUs until a transmission opportunity becomes available at the lower layers [5]. There is no other processing done by the transmitting RLC entity. The receiving TM RLC entity simply passes the received PDU to higher layers.

The TM layers do not segment or concatenate RLC SDUs. Therefore, each RLC SDU is an RLC PDU. Figure 2.5 illustrates the structure of the TM.

2.5.2.2   RLC Unacknowledged Mode (UM)

Figure 2.5 shows the structure of the RLC UM compared to the RLC TM.

The UM transmitting entity places the received RLC SDUs in the transmission buffer. When a transmission opportunity becomes available, it may perform segmentation or concatenation of RLC SDUs, depending on the SDU size and on the size of the transmission opportunity. After segmentation/concatenation, an RLC header is added. The RLC header includes information such as a SN and length indicators (LIs), described later. The resulting RLC PDU is passed to the MAC layer for transmission.

The UM receiving entity holds the received PDUs in the reception buffer. The PDUs may be out of order due to lower layer retransmissions. As a result, PDUs are reordered based on their SN. After removing the RLC headers, the data fields of the RLC PDUs are assembled back into SDUs, undoing any segmentation and concatenation, and delivered to the upper layers.

2.5.2.3   RLC Acknowledged Mode (AM)

Figure 2.6 shows the structure of the RLC AM. In this mode, a retransmission mechanism is allowed to recover any missing RLC PDU, due to radio conditions, for example. The retransmission mechanism is based on ARQ.

The transmitting AM entity places the received RLC SDUs in the transmission buffer. When a transmission opportunity is available, the SDUs in the transmission buffer are segmented or concatenated. This depends on the size of the underlying transmission opportunity. An RLC header is added to each PDU prior to passing them to the MAC layer for transmission. The RLC PDUs are also placed in the retransmission buffer in case retransmission is necessary.

When the receiver sends an ACK (positive acknowledgment) or NAK (negative acknowledgment) PDU to indicate the status of the PDUs in the reception buffer based on SN, the transmitting entity makes a retransmission decision. If an ACK is received, then that RLC PDU is flushed from the retransmission buffer. If an NAK is received for a part of a PDU or an entire PDU, the transmitting entity schedules a retransmission. If the size of transmission opportunity does not allow the entire RLC PDU to be resent, then resegmentation is possible, whereby a single PDU can be divided into multiple segments. Each segment can then be transmitted as a separate PDU with the RLC header indicating how the segmentation was carried out. Otherwise, the entire RLC PDUs are scheduled for retransmission.

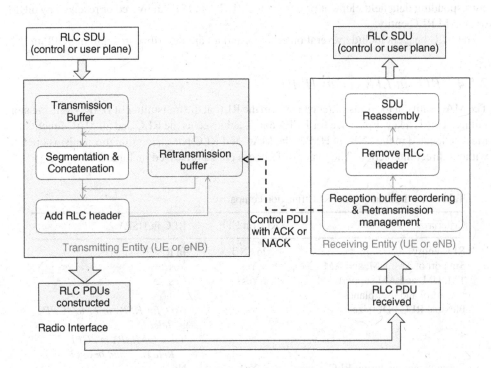

Figure 2.6   RLC Acknowledged Mode (AM).

The receiving RLC entity accumulates the received RLC PDUs in the reception buffer. It performs the reordering before passing a complete SDU to higher layers. status PDUs, defined as control PDUs in 3GPP, are sent by the receiving entity, acknowledging the received PDUs and indicating missing PDUs or parts of the missing PDU segments.

2.5.3   Control and Data PDUs

An RLC PDU can either be an RLC data PDU or an RLC control PDU [5]. A data PDU refers to any control or user plane RLC PDUs carrying the information related to signaling or user data packets. The RLC data PDU is used by TM, UM, and AM RLC entities to transfer upper layer PDUs, whereas the control PDU is the status PDU used only by the RLC AM entity for ACK or NACK retransmission of ARQ procedures.

For AM and UM RLC data PDUs, the RLC headers are added to the PDU as part of the construction of the final PDU to be delivered to lower layers, including the actual data bits. The header typically includes the SN field indicating the SN of the corresponding UM or AM data PDU. For an AM data PDU segment, the SN field indicates the SN of the original AM data PDU from which the AM data PDU segment was constructed. The SN is incremented by one for every UM data or AM data PDU. SN is 10 bits for an AM data PDU and 5 or 10 bits for a UM data PDU.

Another part of the header is the LI field. The LI field indicates the length in bytes of the corresponding data field element present in the RLC data PDU delivered or received by a UM or an AM RLC entity.

The RLC headers include several other fields and all are described in detail in 3GPP in [5].

2.5.4   RLC in LTE versus HSPA

The MAC entity at the transmitter can inform the RLC at the transmitter of HARQ transmission failure. This is a key difference for HSPA and is achieved as the RLC and all MAC functionalities are located in the eNB. In HSPA, the MAC and RLC retransmission mechanisms operate without direct interaction. The major differences are summarized in Table 2.3.

Table 2.3   LTE and HSPA RLC function comparison

Criteria	RLC in LTE	RLC in HSPA
RLC entity	In eNB	In RNC
Support of TM, UM, and AM	Yes	Yes
TM, UM, and AM supports control and user planes	Yes[a]	Yes
Flexible RLC PDU size	Yes	No – *for HSPA prior to 3GPP Release 7* Yes – *for HSPA in 3GPP Release 7 and beyond*
Resegmentation during RLC retransmission	Yes	No
RLC performs ciphering	No	Yes

[a]TM in LTE only supports control plane.

Another important difference is that the TM in UMTS can perform segmentation which is not there in the LTE TM. The TM in UMTS is used to carry both control and user plane whereas in LTE it only carries the control plane. The TM in UMTS is used to carry voice packets in CS (circuit switch) calls, and this is the reason why segmentation is needed in TM.

In HSPA, the RLC PDU sizes are semi-statically configured at the RRC layer. Any change must be initiated through signaling. This is the case for up to 3GPP Release 6. In HSPA+, introduced in Release 7, flexible PDU sizes are supported. LTE supports flexible PDU sizes right from when LTE was introduced in 3GPP. This allows variable size PDUs to be created in order to match the size of the transmission opportunity at the PHY layer and reduce the overhead created by RLC headers.

Ciphering is no longer performed at the RLC layer in LTE. Alternatively, it is done in the PDCP layer, as described in the previous section. The RLC in HSPA performs ciphering for UM and AM modes.

2.6   LTE Medium Access Control (MAC)

MAC is another part of the protocol layers in the UE and the eNB. It provides the interface between the RLC and the PHY layer. MAC performs the following functions [6]:

1. **Channel mapping** – The MAC layer maps logical channels carrying RLC PDUs to transport channels (TrChs). These channels and their mapping are discussed later in this chapter.
2. **Multiplexing** – The information provided to the MAC will come from an RB (radio bearer) or multiple RBs. The data can be multiplexed in the MAC for delivery by the PHY layer.
3. **Scheduling** – The MAC layer performs all scheduling related functions in both the uplink and downlink and thus is responsible for transport format selection associated with all TrChs. Additionally, the MAC is responsible for reporting scheduling related information, such as UE buffer occupancy.
4. **RACH (random access channel) procedures** – MAC is responsible for parts of the Random Access procedures in the uplink during call establishment or handover procedures.
5. **Uplink timing maintenance** – UE needs to maintain timing synchronization with the cell at all times. The MAC layer performs the required procedures for periodic synchronization.

The MAC layer operation is tightly linked to the PHY layer operation. Several of the functions discussed above need close coordination with PHY layer procedures. Therefore, more MAC operation is discussed in Sections 2.7 and 2.9.

2.7   LTE Physical Layer (PHY)

The LTE PHY layer, referred to as L1, provides a new channel structure. The main functions provided by the PHY layer in LTE are described in Table 2.4 [7]. This section begins with a description of the HSPA PHY layer and then introduces an overview of the LTE PHY layer.

2.7.1   HSPA(+) Channel Overview

The PHY layer of the HSPA system is based on WCDMA (wideband code division multiple access) radio access. WCDMA is a code division multiple access system. Spreading is the

Table 2.4 Main PHY layer functions in LTE

Physical layer function	Brief description
Services with higher layers	Error detection on the transport channel and indication to higher layers
	EC encoding/decoding of the transport channel
	Hybrid ARQ soft-combining
	Rate matching of the coded transport channel to physical channels
	Mapping of the coded transport channel onto physical channels
Power control	Power weighting of physical channels
Radio link	Modulation and demodulation of physical channels
	Frequency and time synchronization
	Radio characteristics measurements and indication to higher layers
	RF signal processing
Multiple input, multiple output (MIMO)	MIMO antenna processing
	Transmit diversity (TX diversity)
	Beamforming

process by which information at a lower rate, that is, lower bandwidth, is spread across a wider bandwidth. Uplink and downlink data streams are spread to the chip rate of 3.84 Mcps using orthogonal codes; orthogonal variable spreading factor (OVSF) codes. All OVSF at a given spreading factor (SF) are orthogonal to each other. OVSF codes form a tree such that multiple SFs can be used. The different or variable SFs allow supporting users at different data rates.

In order to separate the signals coming from different cells in the downlink, and the signals coming from different users in the uplink, scrambling codes are used on top of the channelization (OVSF) codes. Gold codes have been chosen as scrambling codes in UMTS. Gold codes simulate a random noise process, known as pseudorandom noise (PN) sequences. Gold codes have good cross-correlation properties, which is good for separating cells and users. The chosen PN codes on the downlink are defined as primary scrambling codes (PSCs), and on the uplink are scrambling codes.

In summary, OVSF codes are used to separate or channelize users on the downlink and separate dedicated channels on the uplink. PSCs separate cells on the downlink for the users to be able to identify a cell from which a radio link is established. Hence, each cell is assigned a different PSC. Meanwhile, the scrambling codes used on the uplink to separate users where each is assigned a unique scrambling code. There is a total of 512 PSCs used for all cells on the downlink, and approximately 17 million scrambling codes for the users on the uplink. Cells are pre-configured with their distinct PSC, while uplink scrambling codes are dynamically assigned by UTRAN's RRC layer for every call a user initiates.

Figure 2.7 shows a possible allocation of the PHY layer downlink channels into the OVSF code tree. In this figure, each channel is assigned a separate OVSF code. For example, the HSDPA (high speed downlink packet access) channel is assigned SF 16. All lower SF below the used codes of SF 16 will be blocked as they would not maintain channel orthogonality. Consequently, SF allocation between the channels is important to ensure all channels and users are allocated a separate code when a call is initiated in the cell.

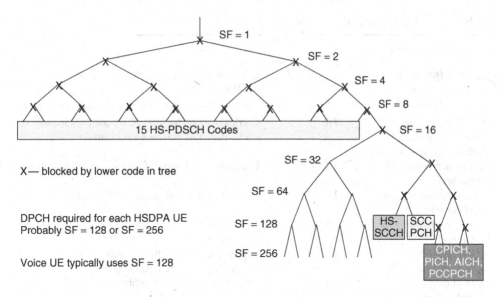

Figure 2.7   OVSF allocation in HSPA/UMTS systems.

2.7.1.1   General UMTS Physical Channels

There are many PHY layer channels in UMTS. Each one has a purpose and usage either in connected or idle modes. Table 2.5 summarizes the UMTS channels. The HSPA channels are discussed in the next section.

HSDPA is mainly introduced to replace the dedicated PHY channels, DPDCH (dedicated physical data channel), with shared PHY channels on the user plane. The motivation behind this channel allocation is to save the OVSF codes and power between multiple users in the same cell. The increase in the user data rate requires using the upper side of the OVSF code which will block the lower side. Dynamic code allocation in the HSDPA system in particular allows the increase of the data rate whilst the dedicated channels used for voice users are minimally impacted.

2.7.1.2   HSDPA Channels

In 3GPP Release 99, the PS (packet switched) service data rate can range from 64 to 384 kbps. When a PS data call is initiated, UTRAN assigns a UE downlink and uplink DPDCH channels with a SF that is suitable for the user's data rate. For example, the downlink with rate of 64 kbps can use SF 64 while 384 kbps utilize SF 8. For multiple users of 64 kbps, they all may get assigned the branches of the code tree on SF 64. The dedicated channel with all its code remains allocated for the user even when there is no data activity. This dedicated code allocation can waste the bandwidth without providing higher data rates than 384 kbps. However, dedicated channels are still suitable for CS voice calls because the voice packets require a dedicated connection between the UE and UTRAN. CS voice calls can utilize SF 128 for the CS data rate of 12.2 kbps.

Table 2.5 Summary of UMTS physical layer channels

Physical layer channel	Direction	Main functions
PCCPCH (*primary common control physical channel*)	DL	Carries RRC broadcast messages such as SIBs or MIB
		Carries SFN used for timing (system frame number)
SCCPCH (*secondary common control physical channel*)	DL	Carries paging channel (PCH) and forward access channel (FACH) transport channels
SCH (*synchronization channel*)	DL	Used to identify the PSC, frame, and slot timing
CPICH (*common pilot indicator channel*)	DL	Used for cell signal quality estimation
PICH (*paging indicator channel*)	DL	An indicator for notifying the user of any incoming paging from the network
AICH (*acquisition indicator channel*)	DL	Carries acquisition channel indicators used during RACH procedure acknowledgments
PRACH (*physical random access channel*)	UL	Carries the RACH preambles
DPCCH (*dedicated physical control channel*)	DL and UL	Used for DPCH synchronization, transmit power control (TPC) commands
		Carries transport format combination indicator (TFCI) to identify the packet block size
DPDCH (*dedicated physical data channel*)	DL and UL	Carries actual control and user planes packets (voice, data, or signaling)

HSDPA has been introduced where dedicated channels are no longer needed on the user plane. The user is allocated any code branch from SF 16 with up to a total of 15 codes. The 16th branch of SF 16 will be free to open up the branches for the next SF used for other UMTS channels, such as CPICH (common pilot indicator channel), or PCCPCH (primary common control physical channel), as shown in Figure 2.7.

In the case of a single HSDPA user in near cell conditions, the network scheduler assigns it the entire 15 codes of SF 16. When multiple HSDPA users are active in the same cell, the 15 codes of SF 16 are split between them. At any scheduling instance a user has low or no downlink data activities, the codes are adjusted or released to serve the other users. This is the concept of a shared channel introduced in release 5 and utilized in all subsequent 3GPP releases for HSDPA and its evolved versions.

The HSDPA PHY layer works as illustrated in Figure 2.8 with the following channels:

- **High speed shared control channel (HS-SCCH)** – A downlink PHY channel that carries downlink control information related to HSDPA transmission. The UE monitors this channel continuously to determine when to read its data from the HSDPA, and the modulation scheme used on the assigned PHY channel. This channel also carries HARQ information and the number of codes assigned to the user for HS-PDSCH.

- **High speed physical downlink shared channel** (**HS-PDSCH**) – A downlink PHY channel shared by several UEs. It supports QPSK (quadrature phase shift keying) and 16-QAM (quadrature amplitude modulation) and 64-QAM (in 3GPP release 7 and beyond). It is a multi-code transmission with up to 15 codes. It is allocated to a user at 2 ms time intervals.
- **High speed dedicated physical control channel** (**HS-DPCCH**) – An uplink PHY channel that carries a feedback from the UE to assist the NodeB's scheduling. The feedback includes a channel quality indicator (CQI) and a ACK/NAK of a previous HSDPA transmission as part of the HARQ transmission or retransmission.

Consecutive HS-PDSCH assignments to a single UE in time and code domains allow the theoretical maximum HSDPA data rate to be achieved. The procedure of the HSDPA PHY layer in Figure 2.8 is:

1. The UE measures the downlink channel quality and sends a CQI report on the HS-DPCCH.
2. If the NodeB decides to schedule data to the UE, it will send information on the HS-SCCH to assign the PHY channel and give the UE information about how the data is being encoded.
3. The UE then starts decoding the data on HS-PDSCH with all related control information in HS-SCCH.
4. After the UE decodes the data, it sends an ACK or NAK on the HS-DPCCH. The UE sends the ACK or NAK, depending on the decoding result of the HS-PDSCH. In the case of failed HS-PDSCH decoding, the UE sends a NAK. The NodeB may schedule the data retransmission during a later time slot. A CQI report is also included in this transmission for the scheduling of all subsequent HS-PDSCH.

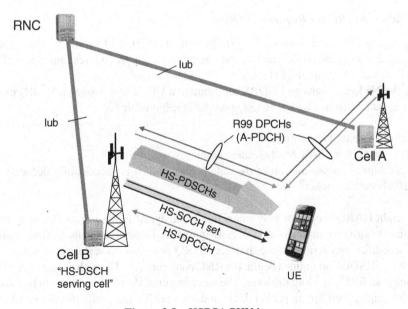

Figure 2.8   HSDPA PHY layer.

Channel Quality Indicator (CQI)

The CQI is a metric that reflects the quality of the downlink channel as measured by the UE. Depending on the UE's implementation and its receiver architecture, it may perform better or worse than another UE under the same channel conditions. Advanced receivers implemented in devices nowadays allow better CQI estimations and hence enhance the user's throughput and cell capacity or coverage.

The NodeB uses the UE's CQI reports in its scheduling algorithm. The details of this scheduling are implementation dependent. The CQI value reported is an index to a table with a range of 0–30, where each row of the table maps to a combination of transport block size (TBS), number of HS-PDSCH codes, modulation scheme (QPSK, 16-QAM, or 64-QAM) and reference power adjustment.

The CQI reported corresponds to the highest data rate that the UE can decode with an error rate less than 10%, assuming the channel conditions and transmit power stay at the same level as in the reference period. With this rule of thumb, the scheduler can adjust the TBS based on the CQI reported to meet an average of 10% block error rate (BLER).

The constant changes in radio environments, caused by multipath effects and UE mobility, lead to fluctuating channel quality. Additionally, UE's receivers may perform differently in similar RF conditions. Under these circumstances, choosing a TBS based only on the reported CQI makes it difficult to always achieve the optimum downlink throughput.

A common scheduling algorithm, referred to as CQI adjustment, allows the UE that is over- or under -estimating the CQI to get a TBS that meets the 10% average BLER in varying radio conditions. The NodeB's scheduler monitors the channel quality fluctuations for HSDPA users in a cell in real time and dynamically determines an appropriate TBS to achieve higher downlink throughput for HSDPA users and higher cell throughput, while the BLER target is controlled within the 10% BLER. The same concept is also utilized in the LTE system.

Hybrid Automatic Repeat Request (HARQ)

To support consecutive assignments, HSDPA defines an HARQ protocol. This protocol is implemented in both the NodeB and the UE, and consists of procedures implemented in both the MAC-hs sub-layer and the PHY layer.

When the NodeB assigns an HSDPA subframe to a UE, it also assigns a HARQ process to handle the data transfer. The UE HARQ process is responsible for

- Decoding the initial transmission
- Sending an ACK or NAK for the transmission
- Soft-combining retransmission of the data packet until it is successfully decoded or until NodeB aborts the packet.

Up to eight HARQ processes may run simultaneously. At least six simultaneous processes are required to sustain consecutive HSDPA assignments. Depending on its implementation, the NodeB scheduler may require more than six HARQ processes to sustain consecutive assignments. When HSDPA operations begin, the RNC configures the UE with the number of HARQ processes in an RRC signaling message. The mechanism of HARQ transmission is also utilized in HSUPA (high speed uplink packet data), and later in LTE but with different requirements.

HSDPA Mobility

Unlike Release 99 operation, HSDPA does not support soft handover. There is only one HSDPA service cell at a time for each UE. Once the serving cell quality degrades, the UE and NodeB perform a serving cell change procedure to another cell, depending on the UE's reported CPICH measurements of each cell.

During an HSDPA call, the dedicated Release 99 channels, DPDCH and DPCCH, are still allocated to the UE for several purposes. One reason is there may be another concurrent CS call in parallel with the HSDPA call. This is a common case in smartphones where the user is in voice call while a data transfer is active. Another reason is that the control plane signaling packets are transmitted between UE and UTRAN on the Release 99 DPDCH and DPCCH channels, referred to as the associated-dedicated physical channel (A-DPCH).

An option for minimizing the usage of dedicated channels in an HSDPA call is to map the signaling into the HSDPA channel, a feature known as SRB over HSDPA. It has been introduced in Release 6 and further enhanced in Release 7. The feature substitutes the A-DPCH with an enhanced fractional-dedicated physical channel channel (EF-DPCH) shared among up to 10 users.

If signaling is mapped to the Release 99 PHY channel, the UE would support soft handover between multiple cells only for the DPCH channels while only one of these cells is serving the HSDPA, as shown in Figure 2.8.

2.7.1.3 HSUPA Channels

HSUPA has been introduced in 3GPP Release 6 to improve the data rate to a maximum of 5.76 Mbps. Figure 2.9 illustrates the HSUPA PHY layer operation.

- **Enhanced dedicated physical control channel (E-DPCCH)** – An uplink PHY channel for control information associated with E-DPDCH. It carries information about the transport format and the HARQ retransmission. It also includes one bit to support scheduling decisions at the NodeB, happy bit.
- **Enhanced dedicated physical data channel (E-DPDCH)** – An uplink PHY channel that carries uplink data for the HSUPA channel. Up to four channels can be used to carry the uplink data in a multi-code transmission scheme.
- **E-DCH absolute grant channel (E-AGCH)** – A downlink PHY channel that carries scheduler grant information from the serving cell. The absolute grant directly indicates to the UE the traffic-to-pilot (T/P) ratio required to be used for scheduled transmissions.
- **E-DCH relative grant channel (E-RGCH)** – A downlink PHY channel that carries scheduler grant information from cells belonging to the serving NodeB as well as to non-serving cells in the E-DCH Active Set. The relative grant instructs the UE to increase, decrease, or maintain the current T/P ratio from the level of the last received grant (could be from the last absolute grant received).
- **E-DCH hybrid ARQ indicator channel (E-HICH)** – A downlink PHY channel that carries feedback (ACK/NAK) from the NodeB on the previous data transmission, to support HARQ retransmission. Since soft handover is supported for HSUPA, each cell belonging to the E-DCH active set transmits the E-HICH information.

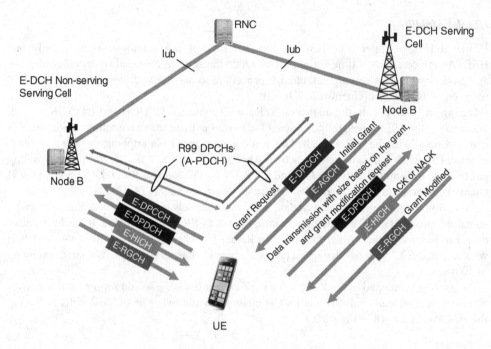

Figure 2.9   HSUPA PHY layer.

The procedures of the HSUPA PHY layer in Figure 2.9 are:

1. The UE asks the NodeB for a grant to transmit data on uplink.
2. If the NodeB allows the UE to send data, it indicates the grant in terms of the T/P ratio. The grant is valid until a new grant is provided.
3. After receiving the grant, the UE can transmit data starting at any TTI (time transmission interval) and may or may not include further requests. Data are transmitted according to the selected transport format based on the grant T/P value. The transport format is then signaled to the NodeB in E-DPCCH.
4. After the NodeB decodes the data, it sends an ACK or NAK back to the UE. If the NodeB sends a NAK, the UE sends the data again with a retransmission in the same HARQ in the next round-trip opportunity.

Serving Grant (SG)

The grant is determined based on the uplink interference situation (rise-over-thermal noise, RoT) at the NodeB receiver, taking into account the UE's transmission requests and level of satisfaction.

The Node B indicates the T/P ratio to the UE by means of the E-AGCH grant value. The grant is valid until a new grant is given through E-AGCH, or until it is modified through an E-RGCH command.

Grant is simply a power allocation for the UE to send its data on E-DPDCH. This power is an offset from the DPCCH power on the uplink. Once the UE receives the grant value, it interpolates or extrapolates the power value of the grant into a maximum number of bits that can be sent on E-DPDCH. The relationship between the grant and uplink power is needed because the UMTS system is uplink-interference limited. The higher the data rate, the higher the uplink interference becomes.

Therefore, the SG (serving grant) assignment depends on the UE's reports of its power and buffer calculations as well as the total interference level measured by the NodeB on the uplink for all users. The granularity of SG assignments makes the HSUPA a method of enhanced uplink (EUL) power control.

UE Transmission Request for SG

The UE requests a grant from the NodeB by means of the scheduling information (SI), which is determined according to the UE's power (power headroom) and data buffer availability. The power headroom reporting (PHR) is added in HSUPA mechanisms to address the cases when the UE is located in cell-edge conditions. Hence, the NodeB scheduler is made aware of the remaining power available for EUL channels in order to schedule the UE accordingly and control the uplink interference. Power and buffer feedback are sent to the NodeB from the UE in the E-DPDCH channel.

An additional scheduling feedback, referred to as happy bit, is sent by the UE in the E-DPCCH. The E-DPCCH is a channel always transmitting, thus, the grant request can be sent in this bit at all times. The UE's happy bit is set to 1 or "happy" when, for the assigned grant, the uplink data buffer is estimated to be fully emptied within a pre-configured timer by the RRC, defined as "happy bit delay condition". If, with the assigned grant, the UE cannot empty the buffer within this time, the happy bit is then set to 0, indicating "unhappy".

The scheduler may take both happy bit and SI into account when scheduling a grant on E-AGCH or E-RGCH. The down- or up-sizing of the grant depends on both of these feedback mechanisms as well as the RoT level measured by the serving cell, or by the neighboring cells within the UE's active set.

HSUPA Mobility

Unlike HSDPA, HSUPA supports soft handover between multiple cells. This is to achieve macro diversity for a more efficient uplink data rate. If one cell in the UE's active set receives the uplink data sent on the E-DPDCH, this is enough to consider that data is received. In this case, a retransmission is not required, even if the other cell does not decode the uplink data. The uplink data received by any of the cells is then forwarded to the RNC for upper layer processing.

An additional reason for having the soft handover in HSUPA is governing the uplink inter-ference across cells. If, for example, soft handover had not been supported in HSUPA, the UE sending data to the HSUPA serving cell could cause uplink interference to a neighboring cell not in control of the grant assignment. Thus, the interference on the neighbors would have increased. Since there is no direct interface between NodeB in the UMTS, the soft handover for interference control is required in the HSUPA.

There are three categories of cell in soft handover during an HSUPA call:

- **Serving E-DCH cell** – The cell from which the UE receives E-AGCH. The UE can receive E-RGCH and E-HICH from this cell as well.

- **Serving (E-DCH) RLs** – Set of cells that contain at least the serving cell and from which the UE can receive and combine the serving E-RGCH. The UE can receive E-HICH from these cells. The cells can also increase, decrease, or hold the grant. There is no E-AGCH possible from this set.
- **Non-serving RL** – Cell(s) that belong to the E-DCH active set but does not belong to the serving RLs and from which the UE can receive E-RGCH. The UE can receive E-HICH from this cell. This set can only decrease or hold the grant. The main functions of these cells are to control the interference from the UE and decode the data for macro diversity gains.

There are rules applying to multiple E-RGCH and E-HICH coming from the different cells. Once a UE receives any E-HICH ACK from any of the cells, the UE treats this as an acknowledgment of valid uplink E-DPDCH reception, and hence no retransmission happens. For E-RGCH, any SG Down command overrides any Up command to control the interference. Up or Down commands from E-RGCH will increase or decrease the SG received initially in E-AGCH by a certain index in the SG table. The SG updating process continues on every TTI.

Similar to HSDPA, control plane signaling on the uplink can be mapped into HSUPA channels in what is referred to as SRB over HSUPA. The signaling can also be mapped into Release 99 uplink channels requiring the presence of dedicated channels, DPDCH and DPCCH. The fewer uplink channels, the better the control over the interference.

Table 2.6 Summary of LTE physical layer channels

Physical layer channel	Direction	Main functions	Similar channel in UMTS
PBCH (*physical broadcast channel*)	DL	Carries RRC broadcast messages such as SIBs or MIB	PCCPCH
		Carries SFN used for timing (system frame number)	
SCH (*synchronization channel*)	DL	Used to identify the Cell ID, frame and slot timing	SCH
DL-RS (*downlink reference signal*)	DL	Used for cell signal quality estimation	CPICH
DM-RS (*demodulation references signal*)	UL	Channel estimation for uplink coherent demodulation/ detection of the uplink control and data channels	DPCCH
SRS (*sounding reference signal*)	DL	Used to provide uplink channel quality estimation feedback to uplink scheduler for channel dependent scheduling at the eNB	None
PRACH (*physical random access channel*)	UL	Carries the RACH preambles	PRACH

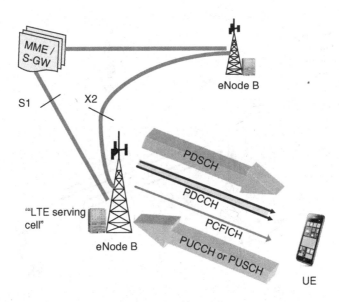

Figure 2.10   LTE downlink PHY layer.

2.7.2   General LTE Physical Channels

Similar to UMTS, there are many PHY layer channels in the LTE system. Each one has a purpose and usage either in connected or idle modes. Table 2.6 summarizes the LTE PHY channels. The channels related to LTE downlink or uplink transmissions are described in detail in the next sections.

2.7.3   LTE Downlink Physical Channels

The concept of downlink LTE PHY channels is similar to that in HSDPA. Figure 2.10 illustrates the overall structure of LTE PHY channels.

The LTE downlink PHY layer channels are [1]:

- **Physical control format indicator channel (PCFICH)** – Informs the UE about the number of OFDM symbols used for the PDCCHs. UMTS does not have an equivalent channel.
- **Physical downlink control channel (PDCCH)** – Informs the UE about:
  - The resource allocations of LTE downlink data in addition to HARQ information. It works similarly to HS-SCCH in HSDPA.
  - The uplink scheduling grant works similarly to E-AGCH and E-RGCH in HSUPA.
- **Physical downlink shared channel (PDSCH)** – Carries information related to:
  - The downlink data of the control and user planes. This works similarly to HS-PDSCH in HSDPA and also DPDCH when carrying signaling messages.
  - Paging, which is similar to the SCCPCH (secondary common control physical channel) in UMTS.
  - SIBs which are similar to PCCPCH in UMTS. The LTE PBCH (physical broadcast channel) carries some of the SIBs, and the PDCCH carries other ones.

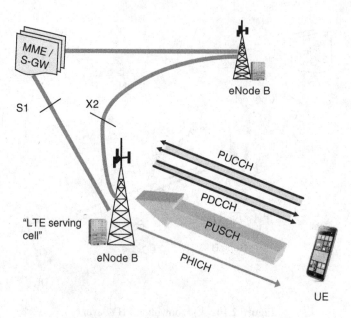

Figure 2.11   LTE uplink PHY layer.

The procedures of the LTE downlink PHY layer in Figure 2.10 are:

1. UE reports the CQI and the RI (rank indicator) or the PMI (precoding matrix indicator) in the PUCCH (physical uplink control channel), or the PUSCH (physical uplink shared channel) if multiplexed with uplink data.
2. The scheduler dynamically allocates downlink resources to the UE. The UE reads the PCFICH every subframe to determine the number of OFDM symbols occupied by the PDCCH. The UE then reads the PDCCH to determine the assigned resources (modulation and resource blocks).
3. The eNB sends user data in the PDSCH.
4. The UE attempts to decode the PDSCH and sends ACK or NACK using the PUCCH, or PUSCH if multiplexed with uplink data.

Further details about the procedures of each of these steps are described in the next sections.

2.7.4   LTE Uplink Physical Channels

The concept of downlink LTE PHY channels is similar to that in HSUPA. Figure 2.11 illustrates the overall structure of the LTE PHY channels.

The LTE uplink PHY layer channels are [1]:

- **PUCCH** – Informs the eNB about several functions including:
  - Control information for uplink data transmission which works similarly to E-DPCCH in HSUPA.

- HARQ ACK/NAKs in response to downlink transmission. It also carries the CQI, RI, and PMI. This works similarly to HS-DPCCH in HSDPA.
- Carries scheduling request (SR) for uplink grant assignments which work similarly to E-DPCCH in HSUPA.
- **PUSCH** – carries the following:
  - Dedicated control and traffic data for both control and user planes. This works similarly to E-DPDCH in HSUPA for user plane data and DPDCH for signaling packets.
  - CQI, RI, ACK/NACK, buffer, and power status which works similarly to HS-DPCCH and E-DPDCH in HSPA.

The procedures of the LTE uplink PHY layer in Figure 2.11 are:

1. When UE does not have uplink grant, it sends an SR on PUCCH. If the UE does not have any PUCCH resources, it goes through an RACH procedure to request uplink resources.
2. The eNB scheduler allocates resources to the UE in terms of uplink grant on PDCCH. Alternatively, the random access response (RAR) message carries the UL-SCH (uplink shared channel) grant.
3. Based on the assigned resources (grant = physical resource blocks (PRBs), modulation and coding scheme (MCS)), the UE sends user data on the PUSCH.
4. If the eNB decodes uplink data successfully, it sends ACK on the PHICH (physical hybrid automatic repeat request indicator channel), otherwise NACK is sent for a UE's retransmission procedure.

2.8   Channel Mapping of Protocol Layers

Previous sections have so far covered the details of the protocol layers for LTE and shown the differences compared to UMTS and HSPA. The control and user plane data flowing across these protocol layers are usually mapped into different channels in each layer before being transmitted over the air. This section summarizes the channel mapping from the radio bearers all the way to the PHY layer channels.

2.8.1   E-UTRAN Channel Mapping

The concept of "channels" is not new. Both GSM (global system for mobile communications) and UMTS define various channel categories, however, LTE terminology is closer to UMTS. Figure 2.12 illustrates the overall structure of the LTE channel mapping across all layers.

There are three main categories of channels [1] in addition to the radio bearer configured by the RRC and NAS, described in detail in Chapter 1.

2.8.1.1   Logical Channels

The interface between the MAC and the RLC provides the logical channels. Logical channels are classified as either control logical channels, which carry control data such as RRC signaling, or traffic logical channels which carry user plane data.

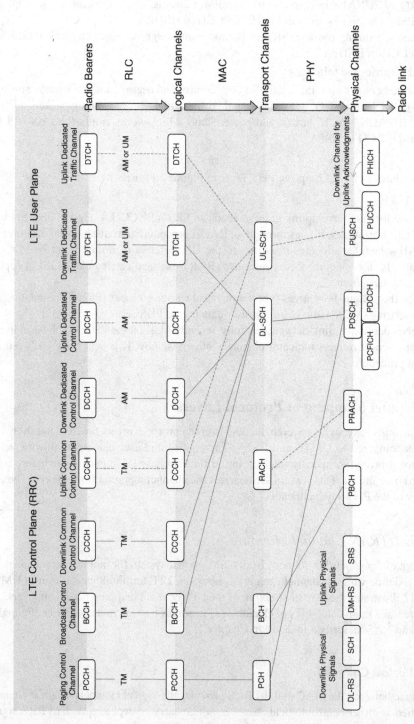

Figure 2.12 E-UTRAN uplink and downlink channel mapping.

- **Control logical channels** – The various forms of these channels include
  - **BCCH (broadcast control channel)** – A downlink channel used to send SI (system information) messages from the eNB. These are defined by RRC.
  - **PCCH (paging control channel)** – A downlink channel used by the eNB to send paging information.
  - **CCCH** – Used to establish a RRC connection, SRB. The SRB is also used for re-establishment procedures after any call drop. SRB 0 maps to the CCCH. These are for both uplink and downlink.
  - **DCCH (dedicated control channel)** – Provides a bidirectional channel for signaling. There are two DCCH activated. One is used for SRB1 carrying RRC messages, as well as high priority NAS signaling. The other is used for SRB2 carrying low priority NAS signaling piggybacked into RRC messages. Prior to its establishment low priority signaling is sent on SRB1.
- **Traffic logical channels** – The DTCH (dedicated traffic channel) is used to carry DRB IP packets. The DTCH is configured for uplink and downlink.

2.8.1.2 Transport Channels

In UMTS, TrChs were split between common and dedicated channels. However, LTE has moved away from dedicated channels in favor of the common/shared channels. As the data being constructed at TrCh, the MAC builds the TB based on the size of the packets coming from the RLC. Multiple RLC PDUs can go into one TB and hence the MAC decides on the TBS to be sent into the PHY layer.

The main LTE TrChs include:

- **BCH (broadcast channel)** – A fixed format channel which occurs once per frame and carries the MIB (master information block). Note that the majority of system information messages are carried on the DL-SCH (downlink-shared channel).
- **PCH (Paging channel)** – Used to carry the PCCH, that is, paging messages.
- **DL-SCH** – This is the main downlink channel for data and signaling. It supports dynamic scheduling, as well as dynamic link adaptation. In addition, it supports HARQ operation to improve performance. As previously mentioned, it also facilitates the system information messages.
- **RACH** – Carries limited information and is used in conjunction with PHY channels and preambles to provide contention resolution procedures.
- **UL-SCH** – Similar to the DL-SCH, this channel supports dynamic scheduling (eNB controlled) and dynamic link adaptation by varying the modulation and coding. In addition, it supports HARQ operation to improve the link performance.

2.8.1.3 Physical Channels

The PHY Layer facilitates transportation of the MAC TrCh, as well as providing scheduling, formatting, and control indicators. The TB coming from the MAC layer is mapped into the corresponding PHY channel to be sent over the air. The TBS is tied to the channel quality feedback from the UE, such as the CQI and RI.

CRC (cyclic redundancy check) bits are added to the TB. The purpose of the CRC is to detect errors which may have occurred when the data is being sent. The UE uses the CRC bits to detect errors on PDSCH, for HARQ retransmissions. The PHY layer performs other functions on the TB, such as channel coding and rate matching, to ensure reliable transmission of the TB over the air [8].

All uplink and downlink PHY channels are described in the previous section.

2.8.2   UTRAN Channel Mapping

Figure 2.13 illustrates the overall structure of UMTS channel mapping across all layers.

The UMTS channel mapping is extensive with many interconnections between channels compared to LTE. However, 3GPP releases for HSPA+ have provided a simplified mapping to match the concepts with LTE. Most dedicated TrCh have been eliminated in HSPA+. In this case, the logical channels will be mapped into HS-DSCH and E-DCH and then to the PHY channels. This change also eliminates a few channels from downlink and uplink which can save the scheduling HSPA power and codes for a better performance.

The HSPA+ features supporting this kind of simplified mapping need a support from network and devices. Hence, there is currently minimal deployment of these features around different markets. With more commercial devices supporting such features, the operators are expected to start rolling out the features.

2.9   LTE Air Interface

The LTE air interface has procedures similar to that in HSPA. The main difference is that LTE uses OFDMA instead of WCDMA. This requires changes in the PHY layer as well as enhancements in some of the MAC and RLC functionality.

2.9.1   LTE Frame Structure

The type 1 radio frame structure in the PHY layer is used for FDD (frequency division duplex) and is 10 ms in duration. It consists of 20 slots, each lasting 0.5 ms. Two adjacent slots form one subframe of length 1 ms. For FDD operation, 10 subframes are available for downlink transmission and 10 subframes are available for uplink transmission, each separated in the frequency domain.

Figures 2.14 and 2.15 illustrate the LTE-FDD and UMTS FDD frame structures, respectively. They highlight the slots and subframe concepts and the mapping between time and frequency for LTE (or code for UMTS).

The type 2 radio frame structure is used for LTE-TDD (time division duplex), not covered in this book.

2.9.2   LTE Frequency and Time Domains Structure

Figure 2.14 depicts the LTE radio frame mapping into time and frequency domains before being transmitted over the air. The E-UTRA (evolved universal terrestrial radio access) downlink is based on OFDMA. It enables multiple devices to receive information at the same time

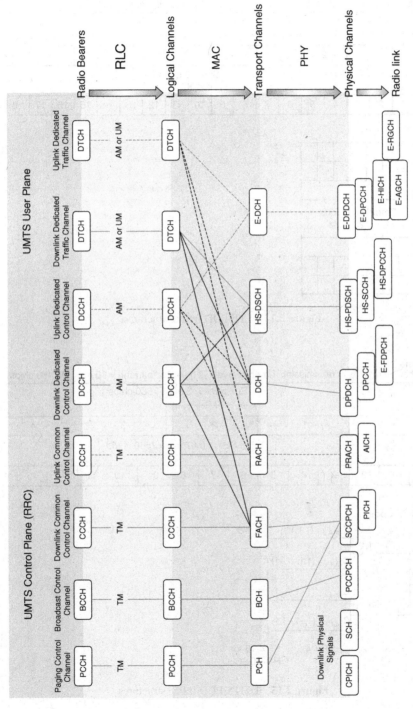

Figure 2.13 UTRAN uplink and downlink channel mapping.

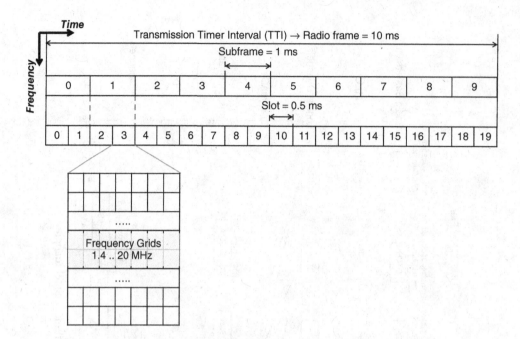

Figure 2.14   LTE FDD frame structures.

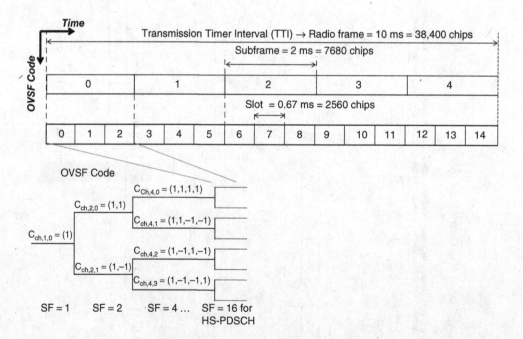

Figure 2.15   UMTS FDD frame structures.

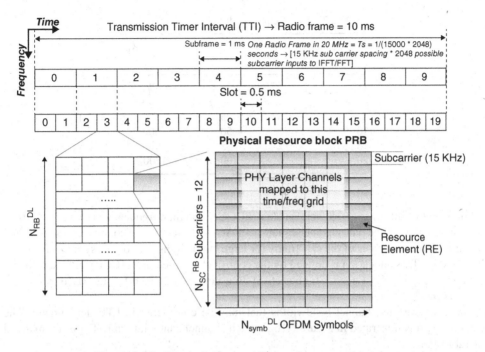

Figure 2.16 Physical resource block and resource element.

but on different parts of the radio channel. In most OFDMA systems this is referred to as a "subchannel," that is, a collection of subcarriers. However, in E-UTRA, the term subchannel is replaced with the term physical resource block. A PRB is used in LTE to describe the PHY resource in the time/frequency grid. Figure 2.16 illustrates the LTE time/frequency grid definitions.

A PRB consists of 12 consecutive subcarriers and lasts for one slot, 0.5 ms. Each subcarrier is spaced by 15 kHz. The N_{RB}^{DL} parameter is used to define the number of RB (resource blocks) used in the downlink. This is dependent on the channel bandwidth. In contrast, N_{RB}^{UL} is used to identify the number of resource blocks in the uplink. Each resource block consists of N_{SC}^{RB} subcarriers, which for standard operation is set to 12 or a total of 180 kHz lasting in a 0.5 ms slot.

The PRB is used to identify an allocation. It typically includes six or seven symbols, depending on whether an extended or normal cyclic prefix is configured.

The term RE (resource element) is used to describe one subcarrier lasting one symbol. This can then be assigned to carry modulated information, reference information, or nothing. This definition is also shown in Figure 2.16.

The RB resources assigned to a specific UE by the eNB scheduler can be contiguous where subcarriers from one RB to another are sequential, referred to as "localized virtual resource block." Alternatively, the RB resources scheduled to a UE can be distributed where some resources are continuous and some are assigned at a pre-defined distance away and referred to as "distributed virtual resource blocks".

Table 2.7 LTE frequency domain configuration

Channel bandwidth (MHz)	N_{RB}	$N_{SC} = 12 N_{RB}$
1.4	6	72
3	15	180
5	25	300
10	50	600
15	75	900
20	100	1200

For a control information channel, such as PDCCH, the time/frequency grid shares parts of the bandwidth with the data channels, such as PDSCH. Therefore, the control region is limited to the first three symbols of the sub-frame for large bandwidths and to four symbols for small bandwidths. This allocation is defined by the control channel element (CCE) and the resource element group (REG). Each CCE consists of 36 usable REs derived from 9 REGs × 4 usable REs per REG.

There are various channel bandwidths that may be considered in LTE deployment. The frequency grids illustrated in Figure 2.16 for each channel bandwidth in LTE are summarized in Table 2.7.

2.9.3 OFDM Downlink Transmission Example

The example in Figure 2.17 shows the time and frequency domain mapping of the PHY channels on the downlink. It illustrates that PDSCH symbols are mapped in a way that avoids the control region and symbols reserved for reference signals.

PDSCH resources in this subframe are shared in the frequency domain, that is, the PRB, between multiple users in the same cell. The first OFDM symbol is used for a cell-specific reference signal, with four transmit antennas in this example. PDCCH is transmitted in the first one to three OFDM symbols carrying resource allocation and MCS to the corresponding allocated UEs. Other resources used in this example are for PHICH carrying HARQ acknowledgments.

The set of REs in an REG depends on the number of cell-specific reference signals configured. In this example, four transmit antennas are used for the transmission of cell-specific reference signals, DL-RS (downlink reference signal). These reference signals, or pilot, per antenna port are transmitted using REs in the time/frequency grid. When multiple reference signals (antenna ports) are used, predefined REs for carrying the signals for each antenna port are allocated. REs used for DL-RS are not considered for any further assignment of REGs carrying the control information.

DL-RS for antenna ports 0 and 1 are inserted in two (the first and the last third) OFDMs symbol of each slot whilst antenna ports 2 and 3 are inserted in only one (the second) OFDM symbol of each slot. Each antenna port uses six subcarrier spacing on the entire subframe.

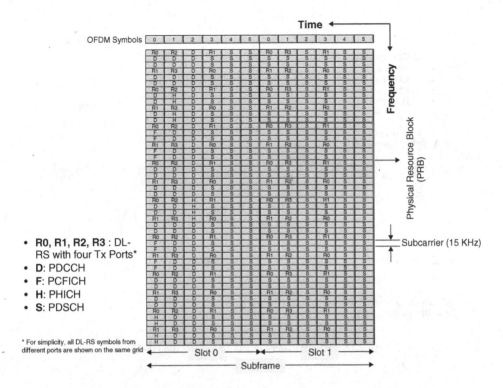

Figure 2.17   Downlink transmission example for OFDM mapping.

2.9.4   Downlink Scheduling

2.9.4.1   HSDPA Scheduling

In HSDPA, HS-SCCH is used in scheduling users on the downlink. When NodeB has data available for a user, it first sends HS-SCCH with the control information for the UE to be able to decode HS-PDSCH carrying the IP packets.

Figure 2.18 explains the HSDPA scheduling mechanism for multiple devices using HS-SCCH. Users #1 and #2 in the figure are scheduled in the same 2 ms TTI instance, while user #3 is not, because there are only two HS-SCCH configured in this example. Both users #1 and #2 are scheduled to share the HS-PDSCH codes for data transmission intended for each. User #3 may get scheduled in the following 2 ms TTI, if one of the transmitted HS-SCCH is scrambled with the H-RNTI (HS radio network temporary identifier) belonging to user #3.

HS-SCCH is a shared channel and there can be up to four HS-SCCH configured by the RRC layer in the RNC, for each cell. Each UE in an HSDPA call reads all four and determines the one assigned to it by using the H-RNTI configured in the RRC. H-RNTI is hence a unique address assigned to a user in each call (during RRC set-up or reconfiguration) that is used to descramble the HS-SCCH. By the time a user decodes HS-SCCH successfully with its

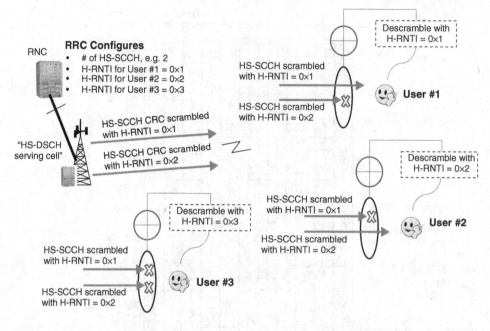

Figure 2.18 HSDPA users scheduling using the HS-SCCH.

H-RNTI, it then reads the HS-PDSCH to retrieve the data packets. This mechanism allows multiple users to share the HS-PDSCH codes in the same 2 ms TTI time scheduling instance.

HS-SCCH is spread with SF 128. The number of HS-SCCH is network-configurable, depending on the HSDPA scheduling implementation and the categories of HSDPA devices. With one HS-SCCH configured in the network, users being scheduled with small payloads, that is, far or low activity users, will occupy the whole TTI with a low number of codes and TBs. As a result, other available codes will not be used for nearby users (with full buffer) who are not assigned in the given TTI. With more than two HS-SCCH, near and far users with different payload sizes can share the codes/time domain, resulting in efficient cell throughput.

2.9.4.2   LTE Scheduling

Similar to HSDPA, the PDCCH control channel in LTE is used to schedule the users on the downlink. However, there is a major difference in LTE scheduling. In HSDPA and prior to 3GPP release 7, HSDPA PHY channels are used to carry only user plane data, and hence HS-SCCH is only scrambled with H-RNTI which simplifies the formats intended for HS-DSCH TrCh. In addition, HS-SCCH is used to schedule users on the downlink only.

On the other hand, and as Figure 2.12 has illustrated, the LTE PHY channels typically carry different types of TrCh in the downlink. For example, DL-SCH can carry information related to control/user plane signaling/data packets on uplink or downlink, paging, broadcast messages, and RACH procedure related messages. Furthermore, PDCCH is used to schedule users on the uplink and informs the UE of the uplink power control commands.

LTE's PDCCH and PDSCH require an indication of the type of data being transmitted in order for the UE to understand the purpose of the procedures being initiated. Therefore, there are two ways to identify the type of data being sent [9]:

- **Downlink control information (DCI)** – The control information sent in PDCCH to indicate what type of uplink or downlink is being sent to the UE.
- **E-UTRAN UE identities for PHY layer** – They indicate the corresponding RNTI used on the PDCCH for either user data or control signaling, paging, broadcast messages, or RACH procedures.

The LTE system uses a set of DCI messages to convey control and scheduling information to devices. The set of DCI of each PDCCH is defined in LTE Release 8 and illustrated in Table 2.8 [9].

The size of the DCI format depends on its function, as well as the system bandwidth. There are various rules associated with the formatting of the DCI messages.

The eNB first encodes the PDCCH message by scrambling the CRC bits with the configured RNTI(s). For each DCI, and based on the purpose of PDSCH transmission, different RNTI configurations are used by the eNB. On the UE side, multiple RNTIs can be associated with one PDCCH decoding and, hence, multiple CRC checks are performed to retrieve the PDSCH data associated with the purpose of transmission.

The list of the commonly used RNTIs for LTE transmissions is summarized in Table 2.9 [1, 9].

From Tables 2.8 and 2.9, the PDCCH search space for the UE becomes huge. The UE needs to trial a different decoding hypothesis with each RNTI and each DCI when decoding PDCCH.

Table 2.8 DCI formats

DCI format	Description
0	PUSCH resource assignment/aperiodic CQI/RI request
1	PDSCH resources with no spatial multiplexing. Scheduling of one PDSCH codeword
1A	Compact scheduling of one PDSCH codeword and random access procedure initiated by a PDCCH order
1B	PDSCH with one codeword transmission and closed-loop SU-MIMO
1C	Special purpose (paging, random-access, system information). Very compact scheduling of one PDSCH codeword
1D	Compact scheduling of one PDSCH codeword in MU-MIMO
2	PDSCH with two codeword transmission and closed-loop spatial multiplexing in SU-MIMO
2A	PDSCH with two codeword transmission and open-loop spatial multiplexing in SU-MIMO
3	Transmission of TPC (transmit power control) commands for PUCCH and PUSCH with 2-bit power adjustments
3A	Transmission of TPC (transmit power control) commands for PUCCH and PUSCH with 1-bit power adjustments

Table 2.9 RNTI configuration

DCI format	Definition	Usage
SI-RNTI	System information-radio network temporary identifier	Used when system information blocks (SIBs) are carried on DL-SCH
P-RNTI	Paging-radio network temporary identifier	Used when paging message is carried on DL-SCH
RA-RNTI	Random access-radio network temporary identifier	Used when random access response is carried on DL-SCH
C-RNTI	Cell-radio network temporary identifier	Uniquely used for identifying RRC connection and user plane scheduling on DL-SCH
TC-RNTI	Temporary cell-radio network temporary identifier	Used for the random access procedure
TPC-PUCCH-RNTI	Transmit power control for PUCCH-radio network temporary identifier	Used for the uplink power control of PUCCH
TPC-PUSCH-RNTI	Transmit power control for PUSCH-radio network temporary identifier	Used for the uplink power control of PUSCH

To limit the PDCCH decoding to a maximum of 44 blind decodes in each subframe, PDCCH UE-specific search space and common search space are defined in 3GPP [9]. The search spaces (also known as control regions) thus limit the number of decodes the UE performs for each PDCCH format combination.

The common search space carries information such as paging, PRACH (physical random access channel) response, system information, and UL TPC (transmit power control) commands for the number of UEs. All UEs monitor the common search space if they are configured by RRC. The UE-specific search space carries information such as control plane signaling or user plane data. The UE-specific search space indicates the starting offset for blind decodes. Different UEs may have different offsets. It is similar to allocating multiple HS-SCCHs (maximum of 4) to several UEs in HSDPA.

For the common and UE-specific search space to be allocated, 3GPP in [9] defines the PDCCH aggregation level. As described earlier in this section, the CCE is the smallest resource unit for transmission on PDCCHs. Each CCE contains nine REGs with each REG containing four REs. Based on coding rates, the eNB can aggregate one, two, four, or eight CCEs to construct a PDCCH, which corresponds to aggregation level 1, 2, 4, or 8, as defined in 3GPP [9]. Therefore, the aggregation level indicates the number of CCEs occupied by a PDCCH.

Aggregation levels 1 and 2 (each with six PDCCH candidates), 4 or 8 (each with two PDCCH candidates) are allocated for the UE-specific search space and those are the CCEs the UE will monitor during decoding UE-specific PDCCH. Aggregation levels 4 (with four PDCCH candidates) or 8 (with two PDCCH candidates) are allocated for the common search space and those are the CCEs the UE will monitor during decoding common PDCCH. Note that the PDCCHs with aggregation level 8 have the lowest coding rate and best demodulation performance.

Table 2.10 PDCCH RNTI to DCI configurations for downlink and uplink

Downlink or uplink	Common		UE-specific		
Downlink	*SI-RNTI, P-RNTI, RA-RNTI*	*C-RNTI, TC-RNTI*	*TC-RNTI*	*C-RNTI*	*MIMO transmission mode[a]*
	DCI 1A, 1C	DCI 1A	DCI 1A, 1	DCI 1A, 1	1
				DCI 1A, 1	2
				DCI 1A, 2A	3
				DCI 1A, 2	4
				DCI 1A, 1D	5
				DCI 1A, 1B	6
				DCI 1A, 1	7
Uplink	*C-RNTI, TC-RNTI*	*TPC-PUCCH-RNTI, TPC-PUSCH-RNTI*	*C-RNTI*		
	DCI 0	DCI 3, 3A	DCI 0		

[a]MIMO transmission modes are discussed in Chapter 4.

Table 2.10 summarizes the search space for the uplink and downlink related data sent on PDCCH. To avoid conflicts between addresses and to simplify the search space further, the valid combinations of RNTIs that can be configured within the same TTI are defined in 3GPP [10]. For example, and from [10], the UE monitors the SI-RNTI (system information-radio network temporary identifier) and the P-RNTI (paging-radio network temporary Identifier) when decoding PDCCH without trialing any other RNTIs. An additional rule is that TC-RNTI (temporary cell-radio network temporary identifier) and the C-RNTI (cell-radio network temporary identifier) cannot be both configured at the same time.

From Table 2.10, in each search space there are two payload sizes to be monitored by the UE,[1] according to the possible DCI combination configured on each search space, as follows:

- **Common search**: DCI 0/1A/3/3A, and 1C.
- **UE-specific search**: DCI 0/1A and one of the following:
  - C-RNTI DCI associated with the configured transmission mode. It can be one of any of the DCIs 1, 2A, 2, 1D, or 1B. This case is used when C-RNTI is configured but TC-RNTI is not configured; or
  - TC-RNTI DCI 1 when TC-RNTI is configured but not C-RNTI.

Putting together the explanations of DCI and RNTI thus far, we can now derive how the number of PDCCH decoding hypotheses explained earlier in the section can be as high as 44. For the user-specific search, a maximum of two payload sizes can be available for decoding with the combination of aggregation levels decoding candidates, creating a search space up to (2 payloads $\times$ (6 + 6 + 2 + 2 PDCCH candidates) = 32). On the common search side, any

---

[1] DCI 0/1A/3/3A share the same payload size, whilst other DCIs have different payload sizes [9].

Table 2.11   RNTI calculation

RNTI	Configuration
SI-RNTI	Pre-configured in 3GPP [6] to 0xFFFF
P-RNTI	Pre-configured in 3GPP [6] to 0xFFFE
RA-RNTI	MAC layer computation based on PRACH configuration
C-RNTI	MAC layer converts C-RNTI from TC-RNTI once RACH procedure completes, or by RRC during re-establishment
TC-RNTI	MAC random access response allocates TC-RNTI during RACH procedure
TPC-PUCCH-RNTI	Configured by RRC [11]
TPC-PUSCH-RNTI	Configured by RRC [11]

RNTI combinations can only use up to two payload sizes, and with the combination of aggregation levels decoding candidates the common search space can be up to $(2 \text{ payloads} \times (4 + 2 \text{ PDCCH candidates}) = 12)$. As a result, there are a total of up to 44 PDCCH decoding hypotheses exercised for each subframe for the entire control regions.

Unlike HSDPA, where H-RNTI is configured by RRC, LTE's RNTIs are mostly derived from the MAC layer, based on computed or pre-configured values [6, 9]. Table 2.11 summarizes the RNTI calculations [6].

For paging procedures, for example, the PCCH is a logical channel that MAC maps to the PCH TrCh according to Figure 2.12. The PCH is transmitted using the PDSCH in a manner similar to the DL-SCH. For the PCH, the unique P-RNTI with a pre-configured value of 0xFFFE (see Table 2.11) is scrambled in the PDCCH allocation message. Multiple UEs may be paged simultaneously using a single paging allocation with the UE identity attached within the paging message itself, for the UE to determine whether the page is directed to it or not. Hence, the usage of P-RNTI in LTE substitutes the need for a PICH (paging indicator channel) PHY channel as in the UMTS system.

2.9.4.3   Data Control Channel Dimensioning

As discussed in this section so far, both HS-SCCH and PDCCH are essential in HSDPA and LTE scheduling, respectively. The number of HS-SCCH in a cell is controlled by the total allocated power to the HSDPA channels as well as the number of OVSF codes for the HSDPA. Generally, when more than one HS-SCCH (up to four) are configured in a cell, HS-PDSCH scheduling, in terms of power and codes, is impacted. Looking back at Figure 2.7, if more than one HS-SCCH is configured in a cell, the maximum number of SF-16 HS-PDSCH codes will be limited to 14 or less. This effectively means that the maximum downlink throughput per user is limited, in a case of best RF conditions.

Additionally, HS-SCCH consumes downlink power from the total allocated to the HSDPA in a cell. In typical deployments, HS-SCCH can be power controlled (with respect to downlink DPCH channel) or assigned with a fixed power. This means that the more HS-SCCH channels configured in a cell, the more power they get assigned; impacting the scheduling of HS-PDSCH per user. Hence, the impact on the user throughput in a cell is important when deciding the maximum number of HS-SCCH.

In the same way, PDCCH in LTE impacts the PDSCH and hence the overall throughput. However, the concept of PDCCH dimensioning is different than HSDPA as there is no limit on the number of PDCCH. The PDCCH is typically a function of the overall LTE bandwidth and the number of antennas in a cell. The PDCCH is located at the beginning of each subframe, and is the primary occupant of total control space. Looking back at Figure 2.17, the entire control space is used by the PDCCH, after excluding REs used for DL-RS and REG used for the PCFICH and PHICH. Hence, more PDCCH CCE allocated in a cell can significantly impact the available capacity for the PDSCH.

As the PDCCH is used for most of the LTE operations (i.e., paging, SIB, scheduling, etc.), its dimensioning therefore depends on many factors. Some factors are uplink/downlink traffic demand, paging load, RACH procedure load, signaling load, and SIB scheduling. The amount of CCE occupied by the PDCCH also depends on the user's RF conditions. As such, if the user is in good RF conditions, then a different number of CCEs could be assigned on the PDCCH, hence impacting the PDSCH resource availability in the frequency-domain.

Another factor impacting the PDCCH dimensioning is the number of symbols allocated to the channel in the time-domain. The PCFICH channel is used to inform the UE about the number of OFDM symbols used for the PDCCH in a subframe. This is mainly done by assigning a certain CFI (control format indicator) in the PCFICH. The CFI indicates to the UE one of three different values. For a cell bandwidth with number of downlink RBs $\leq$ 10, CFI 1, 2, and 3 correspond to 2, 3, and 4 PDCCH symbols, respectively. For a cell bandwidth with number of downlink RBs > 10, CFI 1, 2, and 3 correspond to 1, 2, and 3 PDCCH symbols, respectively.

The number of scheduled OFDM symbols occupied by the PDCCH depends on the user's RF conditions as well as on the number of users in cell-edge conditions. In near cell conditions, it is usually expected that users are assigned with CFI = 1 more often than CFI = 2 or 3. In far cell conditions, it is expected that the eNB scheduler will assign CFI = 2 or 3 more often than CFI = 1, also depending on the load of the cell. At cell-edge, higher CFI can improve the effective coding rate and hence increase the spectral efficiency. However, in near cell conditions using a lower CFI improves the total number of REs assigned to PDSCH, and hence improves the overall user's throughput.[2]

2.9.4.4 LTE Feedback Reporting for Downlink Scheduling

The UE can provide several options of feedback information about the radio channel conditions. The feedback reporting options depend on the MIMO and eNB configurations.

1. **CQI**
   Unlike HSDPA, LTE defines multiple types of CQI. Wideband CQI relates to the entire system bandwidth. Sub-band CQI relates to a value per sub-band. This is defined and configured by the RRC layer and relates to the number of resource blocks in the PHY layer. The CQI per codeword is reported for MIMO spatial multiplexing, depending on the MIMO transmission mode, as discussed in further chapters. CQI can be sent on either the PUCCH or PUSCH in periodic or aperiodic configuration. The CQI can take the value of 0–15, depending on the UE's measured RF conditions.

---

[2] Refer to Chapter 8 for code rate explanations of categories 3 and 4 devices.

2. **RI**

The RI indicates the number of useful transmission layers when MIMO spatial multiplexing is used. It can take the value of 1 or 2, depending on the measured RF conditions. In the case of transmit diversity, the rank becomes equal to 1 (RI = 1). RI is discussed in detail in the MIMO section in Chapter 4.

3. **PMI**

PMI enables the UE to select an optimal precoding matrix. Like sub-band CQI, the eNB defines which resource blocks are related to a PMI report. The PMI reports are used in various modes, including closed loop spatial multiplexing, multi-user MIMO, and closed-loop rank 1 precoding. PMI is discussed in detail in the MIMO section in Chapter 4.

2.9.5   Uplink Scheduling

This section describes procedures related to uplink scheduling and reporting.

2.9.5.1   Uplink Control Information (UCI)

Similar to downlink PDCCH, the uplink PUCCH supports multiple formats. Each format, defined as UCI (uplink control information), is used for a certain type of operation. PUCCH formats are summarized in Table 2.12 [9].

2.9.5.2   Scheduling Request (SR)

Similar to HSUPA, the uplink scheduling is tied first to the SR from the UE on the PUCCH and then to the grant allocation on the DCI 0 on the PDCCH.

   When the UE has data to transmit on the uplink, it sends an SR on the PUCCH. The SR is a bit that is used by the UE when it does not have UL-SCH resources. When the eNB receives the SR on the PUCCH with the UCI 1, the eNB sends the PDCCH grant with the DCI 0. When the grant is allocated, the UE starts transmission of the uplink data on the PUSCH, using the

Table 2.12   UCI formats

UCI format	Description
1	Scheduling request
1a	ACK/NACK
	ACK/NACK + SR
	With 1 bit per subframe
1b	ACK/NACK
	ACK/NACK + SR
	With 2 bits per subframe
2	CQI/PMI or RI
	(CQI/PMI or RI) + ACK/NACK
2a	(CQI/PMI or RI) + ACK/NACK
2b	(CQI/PMI or RI) + ACK/NACK

C-RNTI. The eNB further modifies the grant, depending on the buffer and power information feedback reported by the UE.

The uplink grant contains information for the UE related to the number of PRB assigned for the uplink PUSCH transmission, the TBS, HARQ information, and the modulation type.

The UE follows a configured timer for a maximum number of SR re-transmission on the PUCCH. This means that when the UE does not have an uplink grant and sends the SR on the PUCCH, it can trigger another SR until the maximum number of SR retransmissions configured by the RRC is reached. If the maximum number of SR retransmissions is reached, the UE initiates a random access procedure. If the RACH procedure is not successful, the call eventually drops.

The reason for triggering the RACH procedure is that the RACH response from the network carries an initial grant assignment to the UE.

2.9.5.3 Buffer Status Report (BSR) and Power Headroom Report (PHR)

When the UE has a scheduling grant, the grant needs to be modified in every TTI. The grant modification depends on the UE's data buffer and power headroom. Both the BSR (buffer status report) and the PHR are only reported on the PUSCH.

- The BSR is reported by the UE on the PUSCH and provides the serving eNB with information about the amount of data in the uplink buffer. Then, the eNB adjusts the grant in terms of the PRB, TBS, and modulation scheme based on the reported buffer. The BSR is an index working similarly to the SI sent by the UE in the HSUPA, as described in previous sections.
- The PHR is reported by the UE on the PUSCH and provides the serving eNB with information about the difference between the nominal UE's maximum transmit power, that is, 23 dBm, and the estimated power for UL-SCH transmission. Then, the eNB adjusts the grant in terms of the PRB, TBS, and modulation scheme based on the reported power and buffer. The method is very helpful if the UE is at the edge of coverage where such information allows the eNB to assign the resources suitable for the RF conditions. The PHR is an index working similarly to the SI sent by the UE in the HSUPA, as described in previous sections.

Both the BSR and PHR are data managed by the UE's MAC and PHY layers.

2.9.6 LTE Hybrid Automatic Repeat Request (HARQ)

The HARQ within the MAC layer is designed to transmit and retransmit TBs. HARQ provides a PHY layer retransmission function that significantly improves performance and adds robustness.

The retransmission protocol selected in LTE is SAW (stop and wait). In SAW, the transmitter persists in the transmission of the current TB until it has been successfully received, before initiating the transmission of the next one.

The HARQ mechanism occurs the same way in the LTE downlink and uplink. For LTE-FDD, there are eight HARQ processes in the downlink. The uplink has eight HARQ processes for a non-subframe bundling operation, that is, normal HARQ operation, and four HARQ processes in the uplink for a subframe bundling operation. The TTI bundling operation is described in later chapters.

Combining data in each HARQ process buffer enables the receiver side to recover the data decoded in error. One method for HARQ combining is "chase combining". Chase combining ensures that each retransmission is simply a replica of the data first transmitted. The decoder at the receiver combines these multiple copies of the same information. This type of combining provides time diversity and soft combining gain at a low complexity cost and imposes the least demanding UE memory requirements of all HARQ methods.

Another HARQ combining method is "incremental redundancy (IR)." The IR method ensures that retransmissions include additional redundant information that is incrementally transmitted if the decoding fails on the first attempt. This causes the effective coding rate to increase based on the number of retransmissions sent.

Various HARQ scheduling parameters are required, such as the NDI (new data indicator), TBS, and the HARQ process ID. For UL-SCH transmission, the HARQ information also includes the RV (redundancy version) used for HARQ combining. In the case of MIMO with two TBs sent, the HARQ information comprises a set of NDI and TBS for each TB.

2.10   Data Flow Illustration Across the Protocol Layers

Now that the LTE protocol layers and the air interface have been described in detail, this section provides examples of how the data are flowing across the layers.

2.10.1   HSDPA Data Flow

Figure 2.19 illustrates an example of the data flowing on the downlink HSDPA from the application layer to the PHY layer.

In this simplified figure, the data are coming from application layer maps into RLC packets. After headers are added and segmentation of the IP packets is performed, the RNC sends the

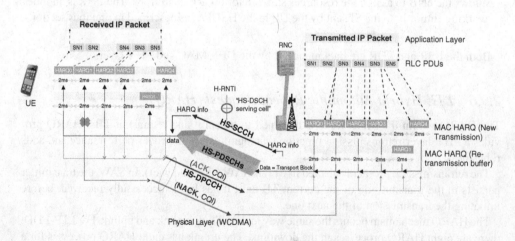

Figure 2.19   HSDPA data flow example.

data into the MAC layer. The MAC maps one or more RLC PDUs into a HARQ process. Each MAC PDU in a HARQ process is sent over the HSDPA channel HS-PDSCH. HS-SCCH informs the UE of the control information of this transmission with user-specific H-RNTI.

Once the UE decodes the HS-PDSCH transmission correctly, based on the CRC attached in the TB, it sends HS-DPCCH ACK alongside the CQI to indicate the channel conditions. NodeB uses this feedback to decide on the size of the packets to send. NodeB retransmits any TB on an HARQ process that failed HS-PDSCH CRC, when it receives NACK from the UE. Any data not recovered in the HARQ process, can then be recovered at the RLC layer by detecting any holes in the PDUs' SN.

2.10.2   LTE Data Flow

Figure 2.20 illustrates an example of the data flowing on the downlink LTE from the application layer to the PHY layer.

In this simplified figure, the data come from application layer maps into the PDCP SDUs for ciphering and then into RLC packets. The RLC adds headers and segments the PDCP SDUs into RLC PDUs. The MAC maps one or more RLC PDUs into a HARQ process, depending on the CQI received from the UE. Each MAC SDU in a HARQ process is sent over the PDSCH channel. The PDCCH informs the UE of the control information (DCI) of this scheduling instance with a user-specific C-RNTI.

Once the UE decodes the PDSCH transmission correctly, it sends the PUCCH or PUSCH ACK alongside the CQI and other MIMO-related channel information. The eNB uses this feedback to decide on the size of the packets. When the eNB receives NACK from the UE on a particular HARQ, a retransmission is performed on the same HARQ process to recover the data first at lower layers. If the lower layer is unable to decode the data on the HARQ level, the RLC can then request a retransmission of the RLC PDUs as a new transmission on the lower layer. The UE sends the successfully decoded packets to the upper layer for processing.

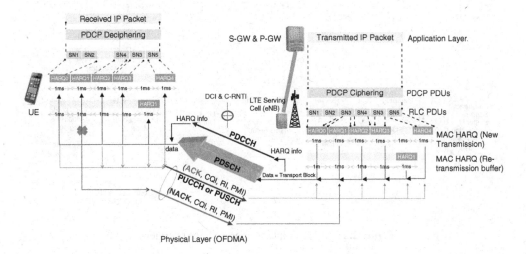

Figure 2.20   LTE downlink data flow example.

2.11   LTE Air Interface Procedures

This section is an introduction to the LTE performance optimization and planning strategies discussed in the remaining chapters of the book.

Each part of the air interface procedure requires optimization and implementation tradeoff to achieve the operator's targeted performance. The optimization aspects of the procedures described in this section are explained in Chapters 3 and 4. Chapter 5 exploits the air interface fundamental concepts of this section into deployment strategies. Chapter 6 discusses the coverage and capacity planning aspects of LTE air interface. Finally, Chapter 7 continues with the evolution into voice over long term evolution (VoLTE).

2.11.1   Overview

The overall LTE air interface procedures are illustrated in Figure 2.21 and discussed in the following sections.

2.11.2   Frequency Scan and Cell Identification

When the LTE device powers on, it needs to perform an LTE attach procedure to connect to the EPC. The EPS attach procedure takes place after the UE accesses a suitable cell from the surrounding LTE eNBs deployed in a network.

In order for the UE to identify the cell and synchronize with the radio frame timing, the eNB sends synchronization signals (SCH) over the center 72 sub-carriers.

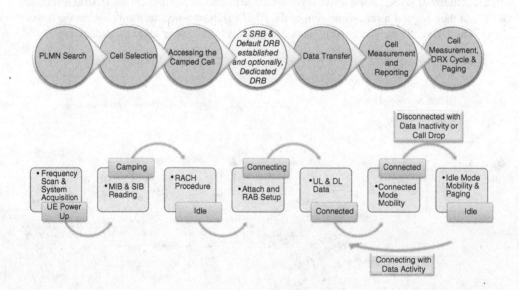

Figure 2.21   LTE air interface procedures overview.

The SCH is comprised of the PSS (primary synchronization signal) and the SSS (secondary synchronization signal). Together they enable the UE to identify the physical cell identity (PCI) and then synchronize any further transmissions. There are 504 unique PCIs, divided into 168 cell identity groups, each containing three cell identities (sectors).

Once a PCI is identified and both slot and frame synchronization is done through the PSS and SSS, the UE acquires the strongest cell measured during this cell search stage, known as the acquisition Stage.

During the eNB planning and deployment, the PCI planning of cells in adjacent clusters is an important topic to avoid any mismatch given the limited number of PCI. A mismatch in the PCI within two nearby cells can typically lead to system acquisition failures, low throughput or eventually call drops. Chapter 6 discusses the PCI planning aspects in detail.

2.11.3   Reception of Master and System Information Blocks (MIB and SIB)

The downlink in LTE is based on scalable OFDMA with channels ranging from 1.4 to 20 MHz. Initially the UE is unaware of the downlink configuration of the cell. To retrieve the cell configuration, the UE needs to monitor the PBCH PHY channel right after cell acquisition. The PBCH carries the MIB.

The MIB repeats every 40 ms and uses a 40 ms TTI. It carries system configuration parameters:

- The downlink bandwidth – 6, 15, 25, 50, 75, or 100 resource blocks
- The PHICH configuration parameter and cyclic prefix information
- The SFN (system frame number) is used by the UE to know the subframe number for synchronization of all PHY channels transmitted.

Additional SIB messages which carry the system parameters the UE uses in idle mode need to be decoded. SIBs are carried in system information (SI) messages which are then transmitted on the DL-SCH based on various system parameters. Other than SIB1, the UE uses the SI-RNTI to decode the SIBs carried on the PDSCH. SIB1 on the other hand, is broadcast with 80 ms periodicity. Scheduling of all other SIBs is specified in SIB1, except for SIB2 which is always contained in the first SI.

Table 2.13 summarizes the common information sent in each SIB. Other SIBs not shown in the table are SIB 8 optionally configured for IRAT (inter-radio access technology) reselection from LTE to 3GPP2 technologies; SIB 9 that is also optional for the HeNB (home eNodeB); and finally, SIB 10 and 11 for ETWS (earthquake and tsunami warning system) notifications.

After the UE decodes the required SIBs successfully, it camps on the selected cell. However, to finally verify a measured cell's levels, the UE performs a procedure known as "cell selection." The criteria for this process are based on the UE's downlink measurements and threshold values configured in the SIBs. Cell selection is needed to ensure that the UE that passes the acquisition stage (UE-dependent implementation), can only camp on the cell within coverage thresholds only conveyed in the SIBs. A cell measured with levels lower than the thresholds is not suitable for selection and the UE may then try another LTE acquisition or another PLMN (public land mobile network) search on other technologies.

Table 2.13 System information blocks

LTE SIB	Common information broadcasted	Equivalent SIB in UMTS
SIB 1	PLMN list, tracking area code, intra frequency reselection, closed subscriber group, frequency band indicator, SI periodicity, and mapping information	MIB and SIB 1/2
SIB 2	RACH information, reference signals information, paging channel information, uplink PHY channel information, access timers, and constants	SIB 1/5/7
SIB 3	Cell reselection information	SIB 3/4
SIB 4	Neighboring cell related info only for intra-frequency cell reselection. It includes cells with specific reselection parameters and blacklisted cells	SIB 11/12
SIB 5	Contains information relevant only for inter-frequency cell reselection	SIB 11/12
SIB 6	Contains information relevant only for inter-RAT cell reselection to UTRAN	SIB 19 for UTRAN to E-UTRAN cell reselection
SIB 7	Contains information relevant only for inter-RAT cell reselection to GERAN	SIB 3/11/12

GERAN, GSM/EDGE radio access network.

2.11.4 Random Access Procedures (RACH)

At the end of a successful cell selection, the UE performs a random access (RACH) procedure on the PRACH/RACH. This is used to access the cell and establish the RRC connection where the SRBs are assigned to the UE.

RACH procedure concepts are similar to UMTS. An open loop power control process is used as the UE sends preambles to the eNB with a certain power. When the eNB acknowledges the preamble, the RACH procedure is successful and the UE thus accesses the cell to establish the RRC connection.

In LTE, the RACH procedure can also be performed in different parts of the call, during handover, or after unsuccessful uplink grant request by the UE. This is different than UMTS. Therefore, LTE defines two different RACH procedures: contention-based RACH, or contention-free RACH.

Figures 2.22 and 2.23 illustrate the RACH procedure for each type.

In contention-based RACH, it is possible that more than one UE could select the same preamble signature, and the eNB would assign the same PRB to both UEs for UL transmission of a message. A contention resolution procedure is hence needed to resolve possible collisions. A result of this contention resolution is that only one UE is allowed to continue with the RACH process. It is typically used at call set-up when the UE is preparing to access the system. This type of RACH allows more preamble indexes which eases the PRACH congestion.

In the contention-free RACH procedure, the eNB explicitly signals which random access preambles index to use. This avoids more than one UE using the same signature sequence and eliminates resource contention while sending data on the uplink. Hence, this RACH procedure

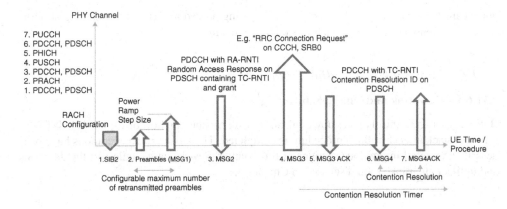

Figure 2.22 Contention-based random access.

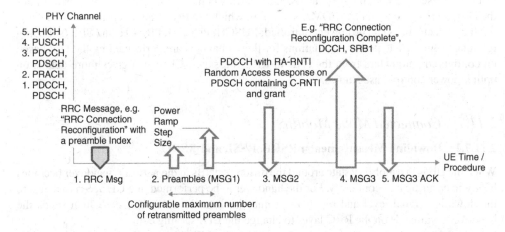

Figure 2.23 Contention-free random access.

is typically used in handover. This is because it is faster to complete when the preamble is reserved for handover. The choice of each procedure is network vendor dependent. Chapter 3 explains the performance comparison and optimization process for the RACH types.

2.11.5   Attach and Registration

Once the UE completes the RACH procedure, it begins to establish the RRC connection, SRBs, followed by DRBs, with the EPS. This stage is part of the UE's attach and registration with the EPS. Upon a successful attach procedure in the NAS and RRC layer, the UE gains access to the user plane where the IP packets can flow on uplink or downlink.

Devices with voice-centric or data-centric capabilities typically attach with EPS differently. The data-centric device attaches to EPS services only whilst the voice-centric UE is required to attach to the EPS as well as the non-EPS (i.e., the legacy 3GPP core) for circuit-switch fallback voice calls (CSFB). Chapter 1 describes the signaling procedure, EPS attach and registration. Chapter 4 explains the CSFB attach procedures in detail.

2.11.6   Downlink and Uplink Data Transfer

2.11.6.1   Downlink and Uplink Scheduling

Both the user and control planes downlink and uplink packets can flow on the entire E-UTRAN protocol stack after the attach procedure is completed. The air interface operations have been described in detail in previous sections of this chapter. Several case studies for the downlink and uplink data transfer are discussed in Chapter 3.

2.11.6.2   Uplink Power Control

Similar to UMTS, LTE requires a power control procedure. It manages the eNB interference levels in the serving and, possibly, across neighboring cells. Uplink power control determines the average power over an SC-FDMA symbol in which the PHY channel is transmitted.

All uplink channels can be power controlled: PUSCH, PUCCH, PRACH, and SRS (sounding reference signal). The power calculations for these channels are performed by the UE, based on configured parameters from the eNB in RRC messages. Chapter 3 gives more details on uplink power control case studies.

2.11.7   Connected Mode Mobility

2.11.7.1   Downlink Measurements: RSRP, RSSI, and RSRQ

When the UE is in a mobility state around the different eNBs in the network, handover becomes handy in ensuring call continuity. For the handover to be performed, the UE needs to measure the downlink signal level and report the values back to the eNB. The eNB then sends the handover commands on the RRC layer to change the LTE serving cell.

Table 2.14 summarizes the measurement types the UE performs for connected and idle mode mobility [12]. The mobility procedures are normally controlled by downlink level measurements.

The measurements listed in the table are derived from the DL-RS. The reference signals are spread over the entire bandwidth with REs' assignments depending on the number of antennas configured, see Figure 2.17 for an example of DL-RS mapping in the time/frequency grid.

RSRP

From the example in Figure 2.17, RSRP (reference signal received power) is calculated as a linear average over the power from the first antenna port, indicated as R0. Within the measurement cycle (or bandwidth), the RSRP is estimated over all REs where R0 is transmitted. In this example, the RSRP is estimated over all R0 in each slot where each RB has two REs with R0 per OFDMA symbol. If the UE can reliably detect that R1 is available, it may use R1 in addition to R0 to determine the RSRP in each transmission symbol over the measurement cycle [12]. In this case, each antenna RSRP will accumulate the energy over the two antenna ports separately in the corresponding OFDM symbol (i.e., four REs in one RB over one OFDM

Table 2.14   UE measurements

LTE measurement	Definition	Similar WCDMA measurement
RSRP (*reference signal received power*)	Linear average over the power contributions of the REs that carry cell-specific RS within the considered measurement frequency bandwidth	RSCP (received signal code power)
RSSI (*received signal strength indicator*)	The linear average of the total received power (W) observed only in OFDM symbols containing reference symbols for antenna port 0	RSSI (*received signal strength indicator*)
RSRQ (*reference signal received quality*)	Defined as the ratio N × RSRP/(E-UTRA carrier RSSI), where N is the number of RBs of the E-UTRA carrier RSSI measurement bandwidth	Ec/No (*energy per chip divided by the power density in the band*)

symbol for R0 and R1). RSRP is measure in dBm with upper and lower ranges of −44 and −140 dBm, respectively.

An example of deriving the maximum and minimum possible RSRP value is shown in [13]. Furthermore, assume the shortest measurement bandwidth (MBW) of 6 RB (i.e., 72 REs) transmitting with 43 dBm (i.e., total downlink Tx power per cell). This means that the RSRP is 1/72 of the total power. Assuming all REs are going through a similar path loss of −100 dB, then the RSRP can be derived as follows:

$$RSRP = 43 - 100 - 10 \log (72) = -75.6 \text{ dBm}$$

RSSI

The RSSI (received signal strength indicator) is measured only in OFDM symbols containing DL-RS for antenna port 0, in the MBW [12]. From the example in Figure 2.17, RSSI can be derived from the power[3] distributed over all REs in the symbols where R0 is transmitted (e.g., symbol 0 in one RB). The RSSI reference sensitivity power level[4] depends on the channel bandwidth and the frequency band. In the 20 MHz channel in the LTE-FDD band 3, the reference sensitivity is set to −91 dBm [14].

For example, assuming in band 3 with 20 MHz (i.e., = 1200 REs) with a receiver sensitivity of −91 dBm, then for a case where traffic load is present and all REs have the same power level, the reported RSRP can be derived as

$$RSRP = -91 - 10 \log (1200) = -121.8 \text{ dBm}$$

---

[3] RSSI includes power from all sources: co-channel serving and non-serving cells, adjacent channel interference, thermal noise [12].
[4] The reference sensitivity power level REFSENS is the minimum mean power applied to both the UE antenna ports at which the throughput shall meet or exceed the requirements for the specified reference measurement channel [14].

RSRQ

The RSRQ (reference signal received quality) is the ratio between the RSRP and the RSSI, depending on the MBW, that is, resource blocks. Consider an ideal interference[5] and noise-free cell where reference signals and subcarriers carrying data are of equal power over one RB (i.e., 12 REs). Then, over the 100 RBs in the 20 MHz system in the example in Figure 2.17, for one OFDM symbol with R0, then RSRQ is estimated as

$$RSRQ = 10 \log \left(\frac{100 * 1RE}{100 * 12RE} \right) = -10.79 \, dB$$

If traffic data is not present in the RB, then over R0 for one OFDM symbol

$$RSRQ = 10 \log \left(\frac{100 * 1RE}{100 * 2RE} \right) = -3.01 \, dB$$

The RSRQ has upper and lower ranges of -3 and $-19.5 \, dB$, respectively.

In UMTS, E_c/N_0 holds a close definition to RSRQ where both are load dependent measurements. Consider a configuration where the CPICH power is set to $-10 \, dB$ with three equal power cells overlapping while there is only one cell in the UE's active set. N_0 will be equivalent to $(\hat{I}_{or} + I_{oc})$, where $\hat{I}_{or}$ is the own-cell interference and I_{oc} is the other-cell interference. Accordingly, $\left(\frac{E_c}{N_0} = \frac{E_c}{\hat{I}_{or} + I_{oc}} \right)$. Dividing by $\hat{I}_{or}$, then $\left(\frac{E_c}{N_0} = \frac{E_c/\hat{I}_{or}}{1 + (I_{oc}/\hat{I}_{or})} \right)$.

Considering the terms $\left(\frac{E_c}{\hat{I}_{or}} = \text{the assumed serving cell CPICH power of } -10 \, dB \right)$, and the term $\left(\frac{I_{oc}}{\hat{I}_{or}} = \frac{1}{G} \right)$, where G, the geometry is the ratio of the power densities received from the target cell to the interference from all other cells then, with the assumption that the three cells are all of equal power, that is, $G = \frac{1}{3}$

$$\frac{E_c}{N_0} = 10 \log \left(\frac{E_c}{\hat{I}_{or}} \right) - 10 \log \left(1 + \frac{1}{G} \right) = -10 - 10 \log (1 + 3) = -16 \, dB.$$

This is the estimated E_c/N_0 in the cell-edge scenario, as the assumption is that the UE in the border of a three cells area.

The serving and neighbor cell measurements in the LTEs connected or idle mode may vary depending on the actual cell bandwidth, as previously explained. In some network deployments, neighboring cells may have different transmission bandwidths. In this case, the RRC informs the UE of the allowed MBW for the UE to use while estimating the cells' reference signals [11, 15]. The MBW can be configured in each eNB, depending on the deployments of the serving and neighbor cells.

For example, in a network with heterogeneous bandwidths where some cells are deployed with 20 MHz and others with 15 MHz, the MBW can be set to the least carrier bandwidth (i.e., MBW of 75 RBs for the 15 MHz, see Table 2.7). Other eNB implementation may always set the MBW to the lowest level of 6 (i.e., measurements over 6 RBs only) to limit all cells to the same MBW. The MBW of 6 is used in this case to limit it to the same bandwidth occupied

---

[5] Inter-cell interference, which in practice would decrease the results. Inter-cell interference would appear as a wideband RSSI increment impacting the denominator in the RSRQ calculations.

Table 2.15 Mobility measurement reporting in connected mode

Handover event	Definition
A1	Serving becomes better than threshold. UE measurements can be based on RSRP and/or RSRQ
A2	Serving becomes worse than threshold. UE measurements can be based on RSRP and/or RSRQ
A3	Neighbor becomes offset better than serving. UE measurements can be based on RSRP and/or RSRQ
A4	Neighbor becomes better than threshold. UE measurements can be based on RSRP and/or RSRQ
A5	Serving becomes worse than threshold1 and neighbor becomes better than threshold2. UE measurements can be based on RSRP and/or RSRQ
B1	Inter RAT neighbor becomes better than threshold. UE measurements can be based other RAT measurements values (i.e., WCDMA CPICH RSCP and/or Ec/No)
B2	Serving becomes worse than threshold1 and inter RAT neighbor becomes better than threshold2. UE measurements can be based on LTE and other RAT measurements

by PSS and SSS channels, that is, the center 72 sub-carriers (SCH bandwidth is 1.25 MHz), referred to as narrowband measurements.

The UE performs other measurements which may not be specific to handover. These are CQI, RI, and SINR (signal to interference noise ratio). These measurements are discussed more in Chapters 3 and 4.

2.11.7.2 Handover

The eNB executes the handover based on the UE measurements. The eNB's RRC layer requests the UE to measure either intra-frequency cells, inter-Frequency cells, or IRAT cells belonging to other 3GPP or non-3GPP systems.

The reporting by the UE is controlled by the eNB through periodic or event-based measurements [11]. The list of events configured by the eNB's RRC layer is summarized in Table 2.15. The performance and optimization of those parameters is provided in Chapter 3.

In UMTS, the same concept of event-based measurement applies, but because UMTS supports handover between multiple cells, the events have different definition scope, such as event 1A used to add a cell into an active set, or 1B used to remove a cell from an active set.

2.11.8 Idle Mode Mobility and Paging

When the eNB releases the RRC connection for the UE, it transitions from connected mode into idle mode. The idle mode is used when there is no uplink or downlink activity for the UE. This state provides the UE with a battery saving option where it is only required to perform measurements or monitor the paging message.

2.11.8.1 Downlink Measurements

The UE measures the RSSI, RSRP, and RSRQ in idle mode, similarly to those in connected mode, as discussed in the previous section. In connected mode, the events used for handover can be defined independently for the RSRP, RSRQ, or both. In idle mode, however, 3GPP Release 8 defines cell reselection criteria only for RSRP measurements. Release 9 has introduced the RSRQ criteria as part of the cell reselection procedure [16]. In Release 9, the RSRQ criteria can also be used for cell selection and absolute priority-based reselection.

2.11.8.2 Idle Mode Cell Reselection

As all other 3GPP technologies, LTE supports intra-frequency, inter-frequency, and IRAT cell reselection in idle mode. During the cell reselection procedure, the UE reads the MIB and SIBs on the target cell. This is because the cells can configure different system parameters broadcast.

The details of the reselection operation and procedures are provided in Chapter 3. CSFB-related cell reselection techniques and optimization are further discussed in Chapter 4.

2.11.8.3 Tracking Area Updating

As discussed in Chapter 1, the UE in idle mode may camp on cells belonging to different tracking areas (TAs). In this case, the UE is required to perform a tracking area updating (TAU) process. The UE retrieves the tracking area code (TAC) from SIB1 and compares it to the saved one from the last visited TA. If they do not match, the UE performs the TAU procedure. This process keeps the EPC updated with the location of the device for paging purposes.

In addition to the TA explanation in Chapter 1, Chapter 6 discusses the TA planning and dimensioning with several illustrative case studies.

2.11.8.4 Paging Procedure

While the UE is in idle mode, it monitors the paging message in a pre-defined cycle, referred to as the paging cycle or discontinuous reception (DRX). The paging cycle is configured in SIB2 or other RRC messages. It is mainly used in the UE for battery saving as it shuts down its transceiver for the duration of the DRX cycle. The DRX mechanism is discussed in Chapter 4.

The paging message is first sent from the EPC to the E-UTRAN. There can be multiple repaging attempts by each of these entities if there is no response from the UE. The eNB configures the UE with various parameters which enable it to identify a time when it should listen to the page, referred to as PO (paging occasion). The PO where the UE monitors the page on is a function of parameters such as the range of the DRX cycle, and IMSI (international mobile subscriber identity). The page indication is sent on the PDCCH using the P-RNTI. Then, the actual paging message is transmitted on the PDSCH.

References

[1] 3GPP (2010) 3rd Generation Partnership Project; Technical Specification Group Radio Access Network; Evolved Universal Terrestrial Radio Access (E-UTRA) and Evolved Universal Terrestrial Radio Access Network (E-UTRAN); Overall Description. TS 36.300 V8.12.0.

[2] 3GPP (2009) 3rd Generation Partnership Project; Technical Specification Group Radio Access Network; Evolved Universal Terrestrial Radio Access (E-UTRA); Packet Data Convergence Protocol (PDCP) Specification. TS 36.323 V8.6.0.

[3] 3GPP (2011) 3rd Generation Partnership Project; Technical Specification Group Services and System Aspects; 3GPP System Architecture Evolution (SAE); Security Architecture. TS 33.401 V8.8.0.

[4] IETF (2007) The RObust Header Compression (ROHC) Framework. RFC 4995.

[5] 3GPP (2009) 3rd Generation Partnership Project; Technical Specification Group Radio Access Network; Evolved Universal Terrestrial Radio Access (E-UTRA) Radio Link Control (RLC) Protocol Specification. TS 36.322 V8.7.0.

[6] 3GPP (2009) 3rd Generation Partnership Project; Technical Specification Group Radio Access Network; Evolved Universal Terrestrial Radio Access (E-UTRA) Medium Access Control (MAC) Protocol Specification. TS 36.321 V8.8.0.

[7] 3GPP (2009) 3rd Generation Partnership Project; Technical Specification Group Radio Access Network; Evolved Universal Terrestrial Radio Access (E-UTRA); LTE Physical Layer – General Description. TS 36.201 V8.3.0.

[8] 3GPP (2009) 3rd Generation Partnership Project; Technical Specification Group Radio Access Network; Evolved Universal Terrestrial Radio Access (E-UTRA); Multiplexing and Channel Coding. TS 36.212 V8.8.0.

[9] 3GPP (2009) 3rd Generation Partnership Project; Technical Specification Group Radio Access Network; Evolved Universal Terrestrial Radio Access (E-UTRA); Physical Layer Procedures. TS 36.213 V8.8.0.

[10] 3GPP (2011) 3rd Generation Partnership Project; Technical Specification Group Radio Access Network; Evolved Universal Terrestrial Radio Access (E-UTRA); Services Provided by the Physical Layer. TS 36.302 V8.2.1.

[11] 3GPP (2010) 3rd Generation Partnership Project; Technical Specification Group Radio Access Network; Evolved Universal Terrestrial Radio Access (E-UTRA); Radio Resource Control (RRC). TS 36.331 V8.9.0.

[12] 3GPP (2009) 3rd Generation Partnership Project; Technical Specification Group Radio Access Network; Evolved Universal Terrestrial Radio Access (E-UTRA); Physical Layer – Measurements. TS 36.214 V8.7.0.

[13] 3GPP (2008) Reporting Range of RSRP. R4-080625.

[14] 3GPP (2011) 3rd Generation Partnership Project; Technical Specification Group Radio Access Network; Evolved Universal Terrestrial Radio Access (E-UTRA); User Equipment (UE) Radio Transmission and Reception. TS 36.101 V8.15.0.

[15] 3GPP (2009) Bandwidth of Mobility Related Measurements in E-UTRAN. R4-070192.

[16] 3GPP (2012) 3rd Generation Partnership Project; Technical Specification Group Radio Access Network; Evolved Universal Terrestrial Radio Access (E-UTRA); User Equipment (UE) Procedures in Idle Mode (Release 9). TS 36.304 V9.11.0.

3

Analysis and Optimization of LTE System Performance

Mohamed A. El-saidny

Optimization is a crucial step in the deployment and ongoing commercial operation of any cellular system. The overall goal of the process is to ensure that the coverage, capacity, and performance requirements defined during the initial design phase are met in the deployed network. This is achieved by making either physical or parametric modifications.

The key challenge of optimization is defining the loading and traffic distribution in the network. For this reason, effective processes must be in place to optimize the system performance iteratively at every stage of network deployment, from network planning to ongoing commercial operation. Effective optimization maximizes positive user experience and minimizes investment in infrastructure. Hence, adequate optimization processes should enable solutions, whether temporary or permanent, to be developed within the framework and restrictions of network operation.

At all stages of deployment, the optimization process typically considers a set of network's key performance indicators (KPIs) and a set of KPIs reflecting on the subscriber's perceived performance. These can relate to the RF environment in terms of coverage and interference, or to performance targets associated with specific services. KPI targets and focus will change based on the optimization process phase of the network.

To ensure a successful network optimization and performance assessment, a clear overall process is necessary that addresses every stage of a typical network deployment. Testing must be appropriate and feasible. Assessment of the required resources is also necessary.

Network acceptance between a network operator and an infrastructure vendor is typically defined by the adherence of the network to a set of KPI targets. In addition to defining the process of optimization and which criteria will be utilized to assess performance, the specific tools that will be employed must also be defined. Initially, drive testing is the predominant method of data collection. As the optimization cycle advances, there is an increasing reliance on network counters and statistics.

Design, Deployment and Performance of 4G-LTE Networks: A Practical Approach, First Edition. Ayman Elnashar, Mohamed A. El-saidny and Mahmoud R. Sherif.
© 2014 John Wiley & Sons, Ltd. Published 2014 by John Wiley & Sons, Ltd.

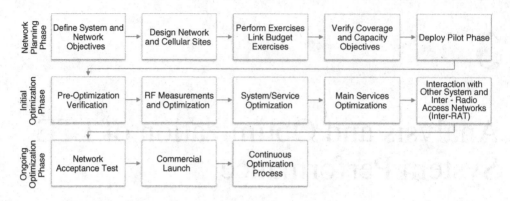

Figure 3.1   Deployment and optimization processes.

Figure 3.1 shows the overall deployment and optimization process from the network planning phase until the system is commercial, after which, continuous optimization is employed to adapt the network to changes and growth of traffic patterns and user experience.

During the network planning phase, the majority of effort is focused on determining appropriate site locations and modifying the physical configuration to meet the network objectives defined in the initial design planning and link budget. Initial optimization begins after the network has been deployed in a pilot location defined to carry out the testing and verification steps. Physical and parametric changes are implemented to ensure that a set of RF and service KPIs are met. The coverage and capacity aspects can be planned between different types of radio access technologies during the initial optimization stage and then fine tuned during the last steps of the optimization process, and once the system is deployed on a commercial phase.

This chapter focuses on the continuous optimization process after LTE (long term evolution) has been deployed in a network. The optimization process, including performance measurements and troubleshooting mechanisms, in addition to common issues and case studies is demonstrated. Moreover, once LTE is commercially deployed, re-evaluation of the performance optimization to that of legacy radio access deployed in the network becomes important. The troubleshooting mechanisms and case studies in this chapter are understood to be impacting the 4G network performance, even if they are related to the 3G networks. This is because there is an inter-connectivity and seamless handovers between the systems across the networks based on the coverage areas and the design aspects of the network. Other chapters give detailed aspects of the network planning and initial optimization phases in the context of network planning, network design, and link budget, with specific focus on LTE.

3.1   Deployment Optimization Processes

After the initial optimization phase, an acceptance test procedure is typically carried out in a controlled environment to ensure that all contractual obligations of the infrastructure vendors have been met. This stage is considered as an opportunity for extensive testing before commercial launch. Following commercial launch, the network operator has limited control over the test conditions and all ongoing optimization will be carried out with live traffic on the network.

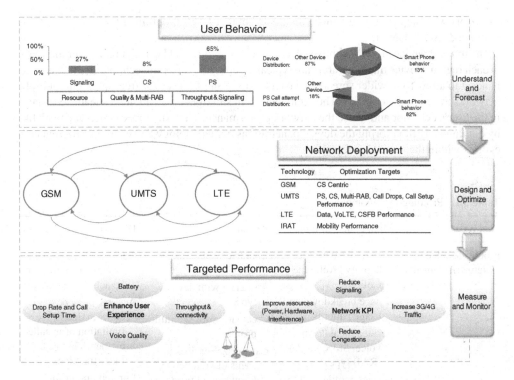

Figure 3.2 Continuous optimization phase.

During the live network testing, the necessary optimization actions will be highly dependent on user traffic. There could be significant differences between forecasted and actual traffic patterns. In addition, new licensed features and services may be needed to cope with the actual traffic from the original estimations. Hence, continuous optimization attempts to maximize the utilization of deployed equipment while adequately planning for necessary capital investments.

Figure 3.2 illustrates the different areas the optimization engineers could assess during decisions on modifying the original design, or adding new features to enhance the performance. All of these factors used during the ongoing optimization process are described in detail in the next section.

3.1.1 Profiling the Device and User Behavior in the Network

The traffic patterns and user distribution may differ from those used during the network planning phase and from assumptions made during initial optimization. Continuous optimization is an ongoing process that adapts the network to the changes in subscriber usage, behavior, distribution, and growth. It attempts to keep the KPIs within acceptable targets.

As the smartphone market has significantly expanded in recent years and is expected to grow more in the years to come, the user's profiling tools become important in network operations. The tools are used to understand the behavior of the subscriber utilizing the different network radio accesses, and the applications running on the devices without a direct interface from the user. With the vast investments in the 4G ecosystem, user, and application behaviors will keep changing which can create another level of optimization process to cope with the device behavior.

The trend of device growth within each cellular technology will keep changing over the years in each region. Device manufacturers are typically deploying different technologies in the same handset to allow for a wider spread of the technology and for users to freely roam in multiple networks in different countries. The rapid change in smartphone technologies leads to change in the user's behavior and hence the network performance. These factors can become readable by continuously monitoring the device penetrations per technology and even for different categories within the same technology.

Table 3.1 Profiling tools comparison

Tool type	Main collected inputs	Advantages	Disadvantages
Probes installed at network elements	User activity profile: activity level, type of calls Device breakdown and usage patterns Service breakdown: voice, SMS, PS data (RAB type) Service profile: call volume, duration, data volume, and throughput	Characterize user activity level Associate with device types Create user/device traffic profile Categorize services in terms of volume, frequency of usage Profile data user behavior in terms of volume, throughput, and signaling	Generate huge amount of data that needs to be actively processed Accuracy can be a concern due to limitation of installed elements and cost of more functions Limited output for application level profiling No information about specific device functions like battery usage
Profiling application installed in devices		Characterize user activity level Associate with device types Create user/device traffic profiles Categorize services (volume and frequency) Profile data user behavior (volume, throughput, and signaling) Profile information about battery, CPU usage Detailed information about application usage and impacts	Challenges on development on different mobile operating systems For collecting representative data, a good amount of users needed to download applications Interaction from user can be required Impact on battery and data usage of devices Security concerns in case of location information needed

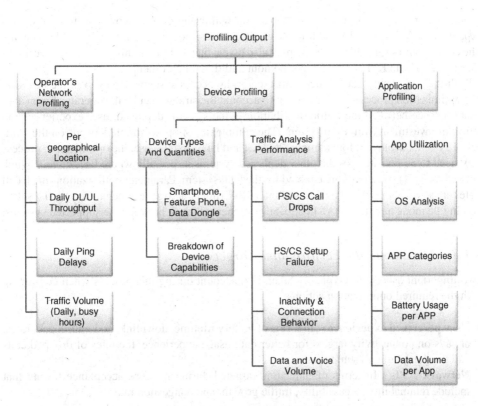

Figure 3.3   Output metrics and statistics of profiling tools.

As the smartphone profiling and behavior studies become important to feed the designer with requirements to improve the network, different tools can be used to assess the device and application level behaviors. Table 3.1 illustrates two of the methods currently used by the network operators to retrieve device and application behaviors.[1]

These tools can be used by the network operators to estimate the users, applications, and services behaviors based on the location and usage of each element. They also allow operators to collect a representative amount of readings from the users for virtually every mobile type, service, and application in the network.

The outcomes of the profiling tools are categorized in Figure 3.3. Those types of statistics can then be used to employ both physical and parametric changes to the network, to ensure that a set of targeted KPIs are met, as shown previously in the network deployment block in Figure 3.2. Optimization processes are expected to be employed before making capital expenditure decisions, such as adding sites or carriers.

3.1.2 Network Deployment Optimization Processes

Network optimization is the process of improving the service performance based on actual subscriber distribution. In addition to maintaining the service KPIs, continuous optimization

---

[1] Tools are provided by third-party companies or infra-vendors.

is also required to adapt to network expansion within the existing coverage footprint. This expansion includes steps like additional sites to increase coverage depth or available capacity. The continuous optimization process may also reveal that additional infrastructure is necessary and that a specific KPI cannot be met without additional investment.

With the complexity added due to different radio access systems deployed in the network, this optimization process needs to take into account the strategy set by the operator for an adequate balance between the traffic distribution in the systems deployed, user experience, and return of investment from each system. The complexity of optimization depends on the number of systems deployed, for example, 2G, 3G, and LTE. In this case, the optimization process may need to be revisited for the other deployed systems, especially where LTE circuit switch fallback (CSFB) voice calls are enabled for the LTE system. Parameter optimizations in all call states are also to be revisited for cases of inter-RAT (inter-radio access technology) cell reselection in idle mode and inter-RAT handover in connected mode between the supported systems.

3.1.3   Measuring the Performance Targets

Two important aspects are typically taken into account during this process when coming up with the required optimization actions:

- **User perceived experience** – In terms of battery lifetime, downlink, and uplink data rates, always-on connectivity targets for better data usage experience, the rates of dropped calls and the delays of call setup time.
- **Network KPIs** – In terms of indicators targeted during network acceptance testing that include retainability, accessibility, traffic growth, and congestion rates.

All optimization activities, regardless of phase, are driven by a set of target KPIs. The targets and the specific KPIs are defined according to the criteria initiated during the network design. During each phase of network optimization, different KPIs are utilized for RF or service performance. For the 4G system, the related KPIs, for both user and network, can be categorized as shown in Table 3.2.

Table 3.2   Key performance indicator (KPI) categories

KPI type	Target	When to use?
RF KPI	RF optimization to measure versus planned	Network planning, network rollout, and initial optimization phase
Service KPI	Evaluate the quality of service expected to be seen by the users for different services	Optimization and commercial introduction phase, and for debugging specific problems
Operation KPI	Continuously collected and trended to set the network performance and behavior for further optimization processes	At all network optimization stages

Table 3.3 RF KPI for LTE and HSPA+

Test scenario	LTE RSRP, UMTS RSCP	LTE RSRQ, UMTS E_c/N_o	LTE SINR	(LTE CQI), (UMTS CQI)
Near cell in good RF	RSRP/RSCP $> -50\,$dBm	RSRQ $> -8\,$dB $E_c/N_o > -10\,$dB	$> 20\,$dB	(12–15) (26–30)
Mid cell in medium RF	$-80\,$dBm $<$ RSRP/ RSCP $< -70\,$dBm	$-12\,$dB $<$ RSRQ/ $E_cN_o < -10\,$dB	$10\,$dB $<$ SINR $< 15\,$dB	(7–11) (20–25)
Far cell in poor RF	$-100\,$dBm $<$ RSRP/ RSCP $< -90\,$dBm	$-15\,$dB $<$ RSRQ/ $E_cN_o < -12\,$dB	SINR $< 5\,$dB	(<6) (<20)

3.1.3.1 RF Measurement and Optimization

RF optimization targets and KPIs are collected through drive tests, preferably using scanner tools. Unlike service KPIs, they are not tracked continuously over the entire network lifecycle. When service KPIs are first collected, RF KPIs are not the best indicator of network performance. For specific user complaints after the commercial stage, network optimizers typically monitor the RF measurements in terms of debugging whether further RF related fixes can be deployed in an observed situation.

Table 3.3 indicates possible targets in different RF conditions. Though the focus of this chapter is on the LTE system, a reference to HSPA/HSPA+ (high speed packet access) is used for comparison and benchmarking purposes. The RF comparisons between LTE and UMTS (universal mobile telecommunications system) systems, such as RSCP (received signal code power) and CQI (channel quality indicator) are described in later sections in this chapter and in Chapter 6. RF measurements are usually factors of the bandwidth for which each system is deployed, as well as the cells' transmit and pilot signal power. The definition of each RF measurement is described in detail in Chapter 2.

During RF scanning measurements, the distribution of each RF KPI is the common representative way of detecting RF related problems and also in understanding the nature of the test location. Figure 3.4 shows the probability density function (PDF) and the cumulative density function (CDF) of the measured RSRP (reference signal received power) in a sample test route in mobility conditions. With a map representation of the RSRP as well as the distribution statistics, the RF planner can proceed with identifying areas of possible RF optimization, especially in locations where RSRP showed a low range of distribution.

Areas of low RSRP typically point to symptoms of pilot pollutions, overshooting, lack of dominant cells, differences in cell coverage area, "out of service" locations, neighbor list issues, or coverage holes. Solutions for each RF problem symptom vary depending on the site configuration and the morphology of the test location (i.e., dense urban, urban, suburban, or rural areas).

The RSRQ (reference signal received quality) measurement provides additional information when RSRP alone is not sufficient to make reliable RF optimization decisions. As described in Chapter 2, RSRQ is the ratio between the downlink reference signal RSRP and the RSSI (received signal strength indicator), depending on the measurement bandwidth. As RSRQ combines signal strength as well as interference level, this KPI provides an additional indicator for identifying mobility issues related to quality measures. Figure 3.5 shows the PDF and CDF of the measured RSRQ in the same test route.

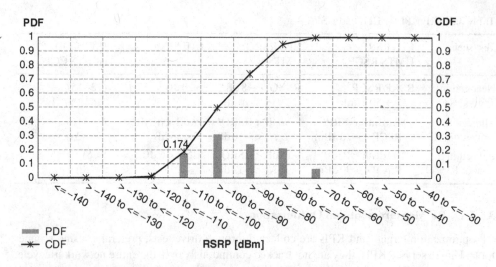

Figure 3.4 PDF and CDF of the measured LTE RSRP in the mobility route.

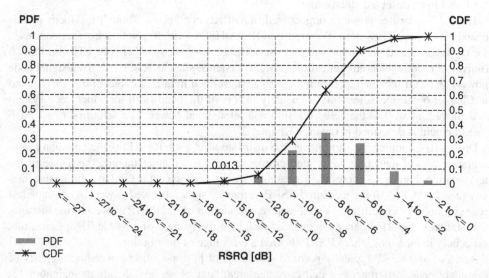

Figure 3.5 PDF and CDF of the measured LTE RSRQ in the mobility route.

The figure shows very minimal samples with RSRQ below −15 dB. This is an indication of good coverage, low load, and low interference from the serving cells in the mobility route. The behavior in RSRQ is hence different than that in RSRP shown before, and in the same mobility route. Therefore, correlating RSRQ statistics with RSRP becomes important in identifying areas where low load and good RF conditions exist, based on RSRQ measurement, but with lower RSRP. As a result, this may indicate possibilities of underlying RF issues (e.g., pilot pollutions) rather than coverage holes.

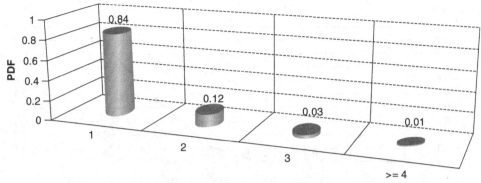

Figure 3.6 PDF of serving cell dominance in the mobility route.

Correlating Figures 3.4 and 3.5, the areas of RSRP ≤ -100 dBm show a PDF of 17% while the area of RSRQ ≤ -15 has a PDF of only 1%. These PDF samples are chosen to reflect poor RF conditions according to Table 3.3. Therefore, the quality measure in this mobility route, reflected by RSRQ, shows very strong coverage areas while the received power, reflected by RSRP, shows 17% of edge of coverage conditions.

An important fact about this test is that it is conducted in connected mode. One basic method to identify RF problems is to run idle mode RF statistics and compare. Or alternatively, use the connected mode data in plotting the distributions of the number of neighbor cells within a certain threshold from the serving cell. Figure 3.6, in the same mobility route, shows the statistics of the number of neighbor cells with RSRP within "+4 dB" from the serving cell. The "4 dB" used in this KPI reflects the handover threshold configured by the network parameters in connected mode (i.e., event A3 threshold). The figure confirms that 16% of the samples are showing two or more cells with RSRP of "4 dB" higher than the serving cell.

This is a strong indication of lack of dominance or pilot pollution in some areas. One way to fix such issues is by adjusting the cells' antenna electrical tilting, inter-site distance, or the handover parameters. Modifying the handover parameters can be done after proper RF solutions. By adjusting handover parameters to reduce the threshold (e.g., to 3 dB instead of 4 dB), the user can thus perform the handover faster and hence reduce the number of low RSRP samples. This RF optimization in return improves the end-user experience in mobility conditions.

Another good method to benchmark the LTE RF conditions is by comparing them to those in UMTS; especially since UMTS coverage is presumably mature. Though the two systems have different configurations in terms of cells' and pilot power, as well as being in different frequency bands, the comparison is beneficial because the basic RF optimization techniques are very similar. In the same mobility route where the RSRP has been collected, the UMTS RSCP is also recorded and shown in Figure 3.7. In this test, all sites are collocated LTE and UMTS cells (i.e., no standalone LTE sites). The area is fully covered with both LTE and UMTS, as designed during the network planning phase.

According to the RF targets in Table 3.3, the edge of coverage in UMTS is reflected by an RSCP ≤ -100 dBm. In Figure 3.7, the percentage of samples with RSCP ≤ -100 dBm is 0%, much lower than the LTE RSRP in the same range shown in Figure 3.4. It is noteworthy that

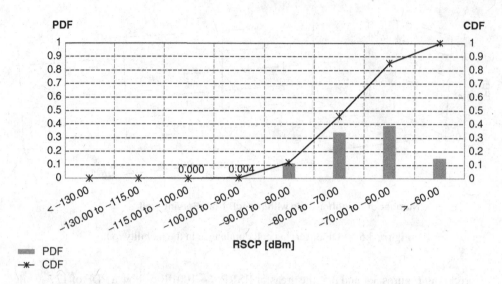

Figure 3.7   PDF and CDF of the measured UMTS RSCP in mobility route.

LTE is deployed with a lower frequency band (LTE in 1800 MHz, and UMTS in 2100 MHz) in this tested cluster, hence it is expected that the LTE coverage distribution will be even stronger than for UMTS. This is another strong indication of the previously discussed underlying RF conditions in LTE coverage. These methods all lead to optimizing LTE RF coverage to target the KPIs and improve the end-user performance.

RSRQ depends on the data traffic and load in the serving cell. In practice, the quality of the channel and the ability of the receiver to decode the data depend not only on the data traffic from its own cell, but also the data traffic and interference from neighboring cells. Hence, RSRQ measurements are a good representation of the serving cell signal quality, but not necessary the channel quality. As illustrated in Chapter 2 when discussing the RSRQ definition, it has been shown that RSRQ can fluctuate in the same conditions, depending on the data traffic increase or decrease in the serving cell, and without necessarily any change in the capability of the receiver to decode the data (i.e., the same channel quality). Thus, an additional RF measure in LTE is the signal to interference noise ratio (SINR) that is used to provide a robust reference of the channel quality. Figure 3.8 shows the distribution of SINR in the same mobility route where the RSRP and RSRQ have been shown previously.

Correlating the RSRP, RSRQ, and SINR in Figures 3.4, 3.5, and 3.8 shows that the serving cells are experiencing very low load represented by the RSRQ but still with low areas of the RSRP and SINR. The distribution in Figure 3.8 shows a 16% SINR CDF with samples of 4 dB or less, representing the edge of coverage defined in Table 3.3. This correlates well with the poor RSRP areas in Figure 3.4 (17% of samples indicating cell edge conditions) even though the RSRQ is showing ranges outside any edge of coverage representation. Thus, resolving lack of dominance in this mobility route, previously discussed, is a possible way to fix the low RSRP areas, which will in turn improve the channel quality represented by the SINR. Improving the SINR greatly improves the user's capability to report better downlink channel conditions to the eNB (eNodeB). Hence, the eNB can schedule the UE (user equipment) more efficiently, enhancing the end-user experience.

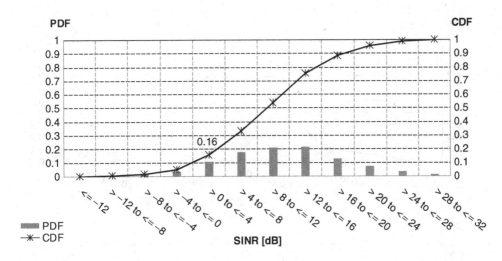

Figure 3.8 PDF and CDF of the measured LTE SINR in the mobility route.

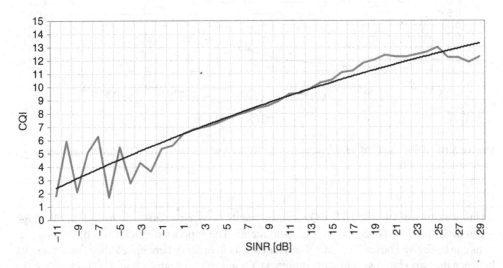

Figure 3.9 Measured SINR versus the reported CQI in the mobility route.

In addition, the CQI is used to estimate the downlink channel quality used for dynamic scheduling. The CQI is used by the network scheduler to derive the best modulation and coding scheme (MCS) achieving a block error rate (BLER) of < 10% [1]. The CQI effectively represents the MCS that the UE can utilize in the measured RF conditions. From the same drive test, it can be concluded that there is a linear relation that determines the reported CQI from the measured SINR, in low load conditions, as illustrated in Figure 3.9. Hence, enhancements to RF conditions will reflect directly on both SINR and CQI reporting.

The CQI is a channel quality indicator that is also used in HSPA+. In the same mobility route, Figure 3.10 shows the reported CQI for each system. The measured CQI in LTE has an average index of 9 while in HSPA+, the average CQI index obtained in the same route is 20.

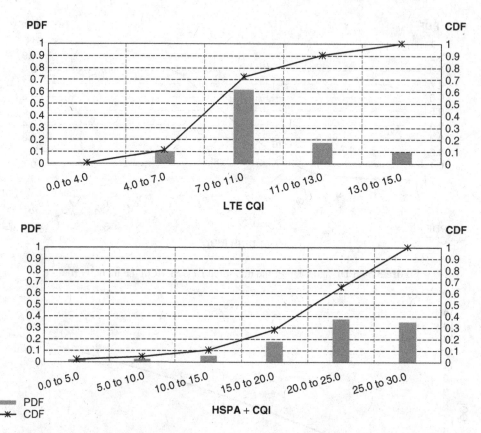

Figure 3.10 PDF and CDF of the measured LTE CQI and HSPA+ (both DC-HSDPA carriers) CQI in the mobility route.

One way to validate the LTE CQI measurements is by correlating them to HSDPA (high speed downlink packet access) CQI measurements. The supported efficiencies of the two systems can be derived based on the CQI measurements in order to benchmark the channel quality between the two systems. While the reported CQI indices for both LTE and HSPA+ differ, the relation between the CQI index and the efficiency can be used as an indicator to compare RF conditions. The efficiency is defined in the context of modulation and coding rate. More specifically, the reported CQI by the UE corresponds to the supportable modulation and coding rate (i.e., spectral efficiency in bits per second per hertz) that it can receive with a transport block error probability of less than 10%. For HSPA+, it is derived based on the coding rate for each CQI, while for LTE the mapping table is available in [1].

In both systems, higher efficiency leads to higher order modulation and a higher coding rate. As provided in the 3GPP standard [1], the LTE efficiency of each CQI index is demonstrated in Figure 3.11. Similarly for HSPA+ and as per the 3GPP standard [2, 3], the efficiency derived from the CQI index is shown in Figure 3.11.

Figure 3.11 [4] indicates that the practical estimated average CQI indices for both LTE and HSPA+ systems (measured as 9 and 20, respectively) yield the same average efficiency, about

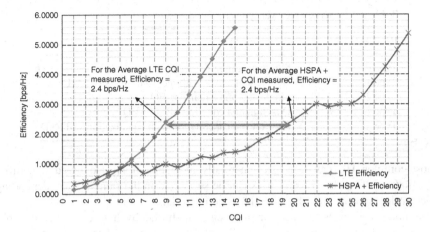

Figure 3.11   Efficiencies of LTE and HSPA+ systems as a function of CQI.

2.4 bps/Hz. Therefore, this method is an assuring point that the RF conditions and power distributions for data channels for both systems are very close, which yields a rational field benchmarking and KPI validations.

The drive routes used for RF measurements are typically drafted to reflect the expected traffic distribution and service area. The routes should stay within the intended coverage area, ultimately considering a building penetration loss that corresponds to the network planning values.

Lastly, the benchmarking data must also consider the device capabilities. Choosing the device for field test matching the commercial ones, ensures the validity of RF KPI targets. 3GPP classifies the device capability in different categories and technologies. Figure 3.12 lists

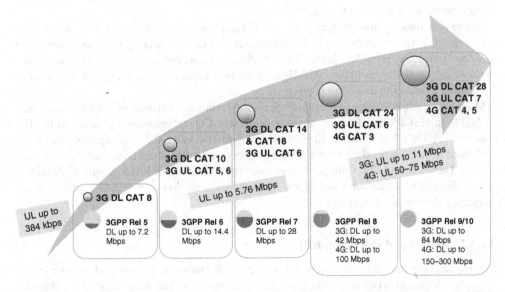

Figure 3.12   UE categories and maximum supported speed in commercial devices, for each 3GPP release.

Table 3.4 Defined receiver types for 3G devices

Receiver type	Basic description
Type 0	Single antenna rake only (rake)
Type 1	Dual antenna rake with received diversity (RxD)
Type 2	Single antenna equalizer (EQ)
Type 3	Dual antenna equalizer (RxD + EQ)
Type 3i	Interference cancellation with equalization and RxD

the most common UE categories commercially available. It is also noted that different combinations of 3G and LTE categories can be supported in the same device, depending on the price or the market in which the device is commercial.

Furthermore, 3G devices in particular are categorized further by their receiver type. Table 3.4 lists the receiver types defined in 3GPP [5]. 3GPP does not mandate any specific receiver implementation, and it is mainly left to chipset vendors' proprietary designs. However, the specifications only give performance guidelines for each type.

Types 1, 2, 3, and 3i are often referred to as advanced receivers. The rake receiver is the basic architecture for multipath combining. The equalizer (EQ) attempts to flatten the received spectrum by enhanced multi-antenna diversity. The EQ complements the rake receiver performance in high geometry and multi-path interference, dominating scenarios where the rake is sub-optimal. On the other hand, RxD (received diversity) provides a second antenna branch allowing the collection of more signal energy while exploiting spatial diversity. RxD complements the EQ, especially in scenarios with low geometry and/or low delay spread with limited multi-paths. Receivers with interference cancellation capabilities tend to improve the downlink signal and channel quality levels by cancelling the interference from neighboring cells. The dynamic switching between each type within a call is possible, depending on the device implementations in different RF conditions.

The reason for emphasizing receiver and UE categories is because each type can provide different levels of improvements to RF conditions and channel quality reporting. Advanced receivers ultimately provide a higher CQI, leading to improved spectrum efficiency. On the network side, with improved downlink signal reception, capacity, and coverage, both increase, in addition to reducing downlink power.

Moreover, each HSDPA category provides a different interpretation of the CQI according to 3GPP. In HSDPA and HSPA+, there are different CQI tables used according to the UE category [2]. On the other hand, LTE's CQI interpretation is related to being reported on the wideband or per sub-band for each PDSCH (physical downlink shared channel) transmission mode.[2] The wideband CQI value is obtained over the entire bandwidth for frequency-selective scheduling (FSS), while sub-band CQI reflects the transmission over selected sub-bands for frequency-diverse scheduling [1].

3.1.3.2 Service and Operation KPI

Service and operation KPIs are collected over the entire network through performance counters, considering all types of traffic. The initial deployment of the network needs to consider

---

[2] Refer to Chapter 4 for details on PDSCH transmission modes.

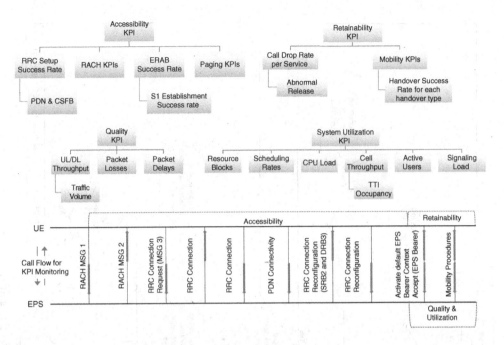

Figure 3.13 Typical LTE operation KPIs.

the cost aspect, thus a limited number of sites is deployed. As the network matures and starts to generate revenue, additional sites can be added to the network. From a network planning point of view, this translates to increased cell edge confidence and increased area reliability. As a result, the network performance is expected to differ from what was initially defined.

The other aspect in a mature network is that the user distribution is known, unlike during the initial planning. This allows increasing the site density, or directing the optimization effort where needed the most, thus where it improves the performance the most.

In LTE systems, the essential operational KPIs are highlighted in Figure 3.13. These KPIs differ from one vendor to another, but the main categories remain the same. Network optimizers typically use these KPIs in order to set the performance targets and identify problematic situations for further case-by-case troubleshooting. The KPIs are extracted from each part of the LTE call flow. Thus, understanding the call flow helps greatly in defining the KPIs when identifying the failure causes.

Accessibility KPIs usually identify the issues at the EPS (evolved packet system) attach procedures, call setup, or during tracking area updating as part of LTE mobility. Whereas, the retainability KPIs are captured during calls in connected mode and mobility conditions between different LTE frequencies or RATs. The system utilization KPIs give a statistical view of the cell capacity and traffic usage. They are continuously monitored during any optimization process or new feature introduction. Low levels of system utilization KPIs typically draw attention toward network dimensioning, planning, or even deployment strategy modifications. Vendor licenses and features for additional hardware or software resolution can be an important outcome from these KPIs.

The user and eNB performance is usually measured based on the service and quality KPIs related to uplink or downlink throughput. Low levels of any of the KPIs trigger detailed

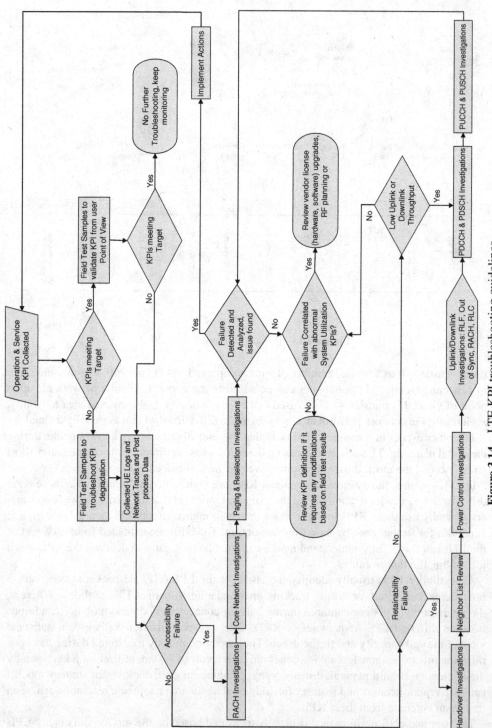

Figure 3.14 LTE KPI troubleshooting guidelines.

troubleshooting mechanisms by means of field testing. In addition to having sufficient samples to ensure statistical KPI validity, it is also important to collect samples over the entire area under consideration during the field testing.

The definitions of the various LTE KPIs are vendor-specific and subject to changes, depending on the optimization process. However, the overall objective is collecting satisfactory indications of possible issues to debug. Then, based on the abnormality of each KPI retrieved, the optimizer can follow basic guidelines for troubleshooting. These are discussed next.

3.1.4   LTE Troubleshooting Guidelines

LTE optimization guidelines are essential in established troubleshooting methodologies. Troubleshooting guidelines are typically defined based on the iterative experience of the common issues observed from field testing and KPI monitoring. Figure 3.14 illustrates a sample guideline of a troubleshooting sequence useful in debugging each of the areas showing low or unexpected KPI levels for one or a group of LTE cells. From this guideline, the troubleshooting effort can be further steered toward the case that matches the area of the failures.

The main areas while investigating accessibility issues usually revert to PRACH (physical random access channel) troubleshooting. Figure 3.15 shows the general guidelines that can be followed in troubleshooting accessibility problems. As explained in Chapter 2, the key objective of the PRACH is to allow the UE to perform cell access in different call stages.[3] The UE selects a preamble signature sequence, out of the possible 64, for preamble transmission, referred to in LTE as MSG1. The eNB detects this preamble and responds with the RA-RNTI (random access-radio network temporary identifier) on the PDCCH (physical downlink control channel). Corresponding to this PDCCH, the random access response (RAR or MSG2) is then transmitted on the PDSCH. The RAR explicitly identifies which time–frequency resource was utilized by the UE to transmit the preamble. In this way the UE knows that its preamble sequence has been received successfully by the eNB. The RAR also includes the uplink grant, time alignment information, and C-RNTI (cell-radio network temporary identifier). The UE at this point can send data or signaling messages on the uplink. The UE sends this data with the C-RNTI it received in the RAR message. During the call setup process, contention resolution may be needed to avoid collisions between users at cell access.

RACH (random access channel) problems can occur at any of these explained steps. Securing a good RACH performance is essential in improving the overall performance. RACH problems in general can vary from RF issues to suboptimal parameters. In addition, they can occur due to high cell congestion. Hence, correlating RACH problems with system utilization KPIs is another required step while investigating the accessibility problems.

Another type of accessibility failures is related to core network rejections or failures of specific procedures. Table 3.5 summarizes the most common core network rejection or failures, and the associated UE behavior according to 3GPP specifications [6]. Depending on the situation of occurrence, core network behavior can be unexpected in some failed scenarios and hence each case can be debugged in a case-by-case way. Core network related issues usually impact a large number of users and hence it is expected these different scenarios will be comprehensively verified ahead of the commercial stage.

---

[3] Refer to Figures 2.22 and 2.23 in Chapter 2 for more detail.

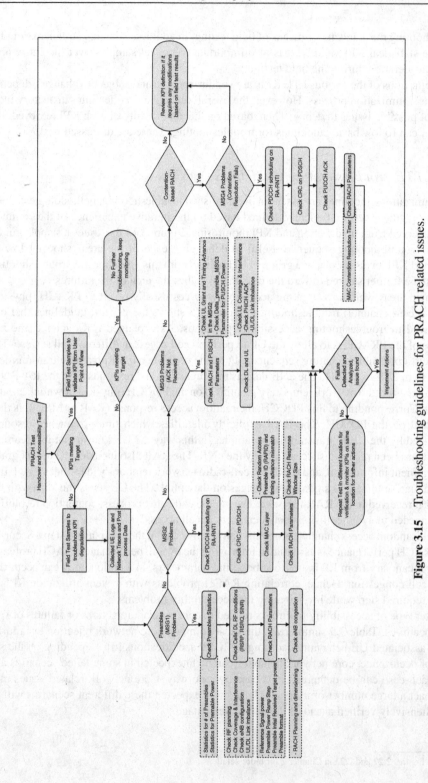

Figure 3.15 Troubleshooting guidelines for PRACH related issues.

Table 3.5 Summary of core network response and UE behavior to attach procedure at system accessibility stage

Core network failures during access procedure	UE behavior
Attach reject with cause code #3, #6 ("Illegal UE/Illegal ME")	USIM is declared invalid for CS and PS services. No full service acquisition allowed on any 3GPP RAT
Attach reject with cause code #7 ("EPS services not allowed")	USIM is invalid for PS service on all RATs. UE will disable LTE RAT and search for service on other RAT
Attach reject with cause code #9, #10 ("UE identity cannot be derived by the network," "Implicitly detached")	Immediate re-attach on same LTE network
Attach reject with cause code #11 ("PLMN not allowed")	PLMN is blocked for all RATs. UE will do PLMN selection, searching for service on other networks
Attach reject with cause code #12 ("TA not allowed")	Current TAI is forbidden. PLMN is not forbidden. There is no attempt to acquire service by UE. If UE reselects to non-forbidden TA or LA, registration will be attempted
Attach reject with cause code #13 ("Roaming not allowed in this TA")	Current TAI is added to forbidden TA for roaming list. UE will no longer camp on this TAI. UE will initiate PLMN selection starting from HPLMN
Attach reject with cause code #14 ("EPS services not allowed in this PLMN")	PLMN is blocked for all RATs. UE will do PLMN selection, searching for service on other networks
Attach reject with cause code #15 ("No suitable cells in tracking area")	Current TAI is added to forbidden list. UE will no longer camp on this TAI. UE will initiate PLMN selection starting from last camped PLMN
Attach reject with cause codes other than the above	For other cause codes, UE will stay camped on the LTE cell, and will retransmit attach requests five times. Each subsequent attach request is separated by 10 s. After five times, UE will start 12 min timer and search for service on other networks
No response from network after sending attach request	Retry sending attach request up to five times, each time waiting for response for 15 s, then wait another 10 s before retry. If all are unsuccessful (after 115 s), block the PLMN for 12 min and search for service on other networks
RL failure during attach procedure before/after security established, or connection release during attach procedure	Attach attempt counter is incremented

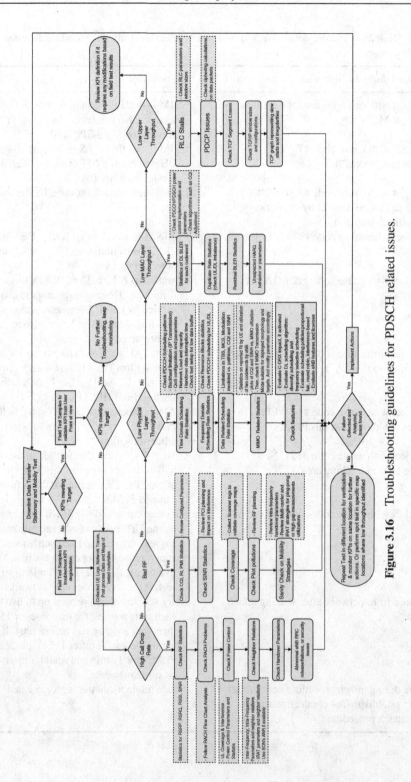

Figure 3.16 Troubleshooting guidelines for PDSCH related issues.

On the other hand, in order to troubleshoot retainability, mobility, and low throughput issues, the main guidelines concentrate on the PHY layer channels; PDSCH, PDCCH, PUCCH (physical uplink control channel), and PUSCH (physical uplink shared channel). Figures 3.16 and 3.17 illustrate the sample process and guidelines for such debugging. Case studies for sample failures related to these guidelines are described in the next sections.[4]

3.2   LTE Performance Analysis Based on Field Measurements

This section provides a practical performance analysis of LTE based on field test results from a commercially deployed 3GPP Release 8 network. The analysis demonstrates the downlink and uplink throughputs in mobility conditions in addition to handover performance. Likewise, a comparison is presented between LTE and Release 8 HSPA+ with Dual-Cell high speed downlink packet access (DC-HSDPA) feature.

The performance testing in this section was conducted in collocated (i.e., same physical sites) LTE/HSPA+ commercial networks. Full commercial networks are used for the analysis consisting of LTE in 1800 MHz with a 20 MHz channel and UMTS in 2100 MHz with DC-HSDPA using 2×5 MHz adjacent carriers. A commercial dual mode data terminal connected to a laptop with logging and tracing tools is used. The LTE CAT 3 terminal is capable of downlink 100 Mbps and uplink 50 Mbps, with downlink 2×2 MIMO (multiple input multiple output) and 64QAM and uplink 16QAM. The same dual mode terminal is used with HSPA+ Category 24 capable of downlink 42 Mbps and uplink 5.76 Mbps (i.e., HSUPA (high speed uplink packet data) category 6) with downlink 64QAM on both carriers without MIMO.

The drive test mobility route covers ~35 km with an urban morphology. The drive test was conducted at speed of 80 km/h and once at a time as the LTE and HSPA+ networks are both sharing the same backhauling. Also, the testing was conducted at low traffic time and the two networks are optimized to have a rational benchmarking. The output power of HSPA+ cells is 40 W (i.e., each carrier is 40 W) while the LTE cells of 60 W per antenna branch (2×60 W with MIMO 2×2). The configured downlink primary common pilot physical channel (P-CPICH) power of the HSPA+ cells is 33 dBm while the configured reference signal power of the LTE cells is 18 dBm.

The main focus of the LTE performance analysis in this section is on the air interface protocol of the user and control planes. As a result, from the different mechanisms performed at the different layers, the total uplink, and downlink throughputs will be maintained separately at each layer. For TCP/IP (transmission control protocol/internet protocol) type of applications, the expected theoretical uplink and downlink throughputs of LTE and HSPA+ are provided in Figure 3.18, with the configurations shown in Tables 3.6 and 3.7. The throughputs are obtained via simulation using the generalized protocol parameters specified in [1, 7, 8]. It is worth mentioning that the end-user throughput is reflected by the application layer data rate (i.e., TCP/IP, for file transfer protocol (FTP) type applications), and is impacted directly by the physical layer mechanisms and performance.

During the assessment of downlink and uplink throughput performance, several quality, and system KPIs become important in order to benchmark the performance or troubleshoot certain issues raised during field testing. Table 3.8 lists several of these quality KPIs. Different tools can be used to collect these KPIs either from the UE or network tracing point of view.

---

[4] Refer to Chapter 2 for the description of PDCCH, PDSCH, PUCCH, and PUSCH and the protocol layers techniques shown in these troubleshooting guidelines.

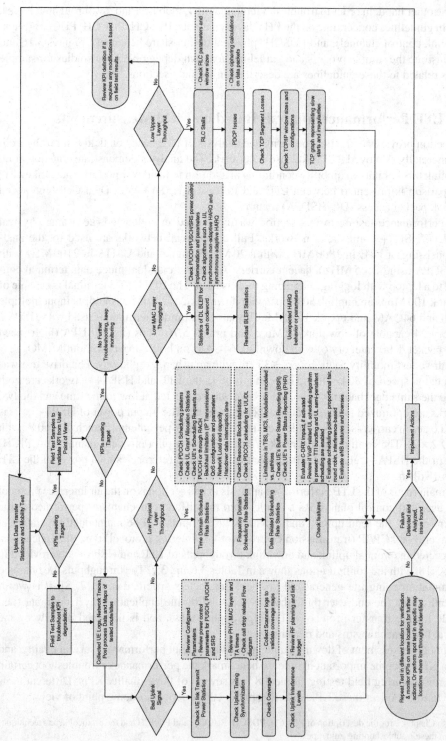

Figure 3.17 Troubleshooting guidelines for PUCCH and PUSCH related issues.

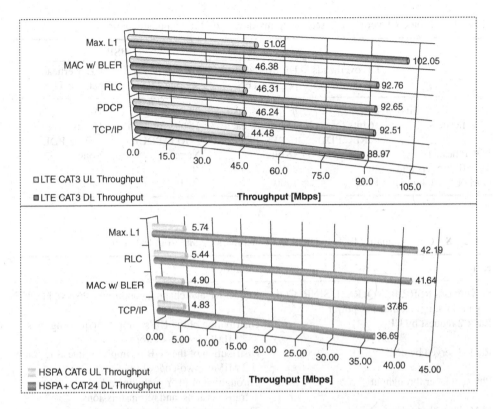

Figure 3.18   Theoretical LTE/HSPA(+) uplink and downlink throughputs.

Table 3.6   General parameter assumptions for theoretical throughput calculations

Inputs	LTE		HSPA(+)	
	Downlink	Uplink	Downlink	Uplink
System bandwidth (MHz)	20	20	2 × 5	5
UE Category	3	3	24	6
Number of transport blocks	2 (MIMO 2 × 2)	1 (MIMO not configured but 2Rx are used on UL	2 (MIMO not configured but 1 TB per each carrier)	1 (MIMO not configured but 2Rx are used on UL)
Maximum transport block size (TBS) (bits)	51 024	51 024	42 192 per carrier	11 484
Physical layer block error rate (BLER) target (%)	10	10	10	10
User physical layer scheduling rate (%)	100	100	100	100

Table 3.7 Protocol layer inputs used for uplink/downlink throughput simulations

Inputs	LTE		HSPA(+)	
	DL overheads per one TBS	UL overheads per one TBS	DL overheads per one TBS	UL overheads per one TBS
MAC headers (bits)	8	8	24	18
RLC headers	16 bits for each RLC PDU	16 bits for each RLC PDU	16 bits for each RLC PDU	16 bits for each RLC PDU
PDCP headers	16 bits	16 bits	None	None
TCP/IP headers and MTU size (bytes)	40 1500	40 1500	40 1500	40 1500

Table 3.8 General quality PKIs for field test benchmarking and troubleshooting

KPI	Definition
Serving cell RSRP, RSRQ, RSSI, SINR, CQI, path loss, uplink transmit power	Reflect RF and channel conditions observed by UE
Rank 2 request by UE	Distribution of the UE's samples requesting rank 2 MIMO (two-codewords)
Rank 2 served by eNB	Distribution of the eNB's samples requesting Rank 2 MIMO (two-codewords)
Physical layer throughput	Throughput at PHY including all new transmissions, and retransmissions
MAC layer throughput	Throughput at MAC including only new transmissions, and excluding MAC headers
Upper layer throughput	Throughput at RLC, PDCP, and application layers. Examples are shown in Figure 3.18
Scheduling rate in time domain	PDCCH scheduling rate in time domain (samples of TTIs for how frequent the UE receiving PDCCH with C-RNTI for data scheduling) for either uplink or downlink
Normalized physical throughput	PHY layer throughput/scheduling rate
BLER on all transmissions	Distributions of BLER on all transmissions over all packets with C-RNTI (PDSCH on downlink and PUSCH on uplink)
Number of RBs scheduled	Distribution of the samples of resource blocks assigned by eNB scheduler for uplink or downlink
MCS	Modulation scheme assigned by eNB for uplink or downlink

3.2.1   Performance Evaluation of Downlink Throughput

This section provides guidelines on some useful methodologies for evaluating and benchmarking the LTE downlink throughput performance. The mobility route and KPIs used in the throughput analysis have been described in previous sections. This evaluation aims at demonstrating several ways of assessing the LTE performance in steps in order to reach a good understanding of possible areas of optimization. It is always useful to measure the HSPA+ performance in parallel in order to benchmark the network performance in general and compare specific functionalities.

The physical layer throughput distributions measured in LTE and DC-HSDPA networks are shown in Figure 3.19. The average LTE downlink physical layer throughput is estimated to be ~33 Mbps with instantaneous peaks of ~96 Mbps, while DC-HSDPA has an average of ~9 Mbps with an instantaneous peak of 25 Mbps.

To compensate the effect of the loading in the DC-HSDPA network,[5] a normalized downlink throughput is used to estimate the expected throughput of the unloaded network in the mobility route. The average network scheduling rate is 73% (estimated from the number of successful HS-SCCH (high speed shared control channels) decoded by the UE[6]), thus the normalized average DC-HSDPA throughput (i.e., physical layer throughput divided by scheduling rate) is estimated to be ~12.3 Mbps. This estimation represents a single user scheduled in the network under the given RF conditions discussed in Section 3.1.3.1 collected in the same route. As a result, the average spectral efficiency of the LTE is 1.65 bps Hz^{-1} while it is 1.23 bps Hz^{-1} for HSPA+ with 34% improvement during the mobility conditions in this field test. A similar throughput measurement methodology can be followed for any mobility route.

In another aspect, the LTE channel quality is typically characterized in terms of achieved SINR rather than signal level (i.e., reference signal power). Figure 3.20 shows the normalized

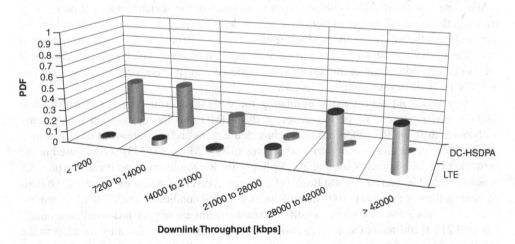

Figure 3.19 Measured LTE and DC-HSDPA (Carrier1+Carrier2) downlink physical layer throughputs.

---

[5] Normalization for DC-HSDPA is used since the UMTS network is typically more loaded than LTE, especially in the initial deployment stage. The same normalization concept can apply to LTE scheduling rate.
[6] Refer to Figure 2.18 in Chapter 2 for details on the HSDPA scheduling rate.

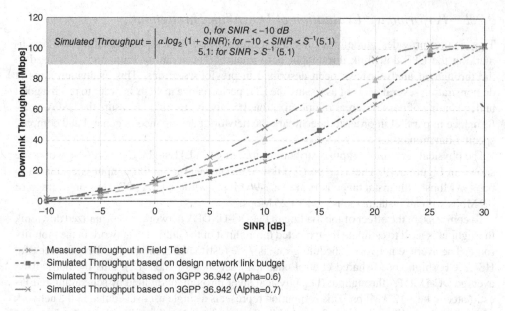

$$\text{Simulated Throughput} = \begin{cases} 0, & \text{for } SNIR < -10 \ dB \\ \alpha.\log_2(1 + SINR); & \text{for } -10 < SINR < S^{-1}(5.1) \\ 5.1: & \text{for } SINR > S^{-1}(5.1) \end{cases} \quad (5.1)$$

--*-- Measured Throughput in Field Test

--■-- Simulated Throughput based on design network link budget

—▲— Simulated Throughput based on 3GPP 36.942 (Alpha=0.6)

—✕· Simulated Throughput based on 3GPP 36.942 (Alpha=0.7)

Figure 3.20   Downlink throughputs versus SINR for field measurement results and simulations.

measured downlink throughput as a function of SINR compared to simulated values. The simulated curves correspond to the expected throughput based on link budget calculations in the network,[7] in addition to TS 36.942 of 3GPP standards [9] that provide a model for LTE link level performance.

Also note that the modeled channel types considered in the standard may not necessarily represent the expected user conditions in an actual deployment, but the curves are used for the proximity and the validity of the field test results. Figure 3.20 demonstrates that the measured throughput is well presented by the simulated throughput based on link budget estimations. It shows that the link budget in this situation is a little pessimistic compared to the simulations in 3GPP (with different attenuation levels).

Another important aspect while evaluating the downlink throughput is the MIMO and 64QAM performance. The performances of the MIMO and the modulation schemes are analyzed in different RF conditions to benchmark the gains and utilizations of these features.

In the study from this mobility drive test, Transmission Mode 3 (TM3) is configured in the network.[8] The UE requests Rank 1 or 2 according to the RF conditions. Specifically, in poor RF conditions, the UE tends to request Rank 1 (i.e., transmit diversity where the same data streams are sent on the two antenna ports). Meanwhile, in good RF conditions, the UE tends to request Rank 2 (i.e., space diversity where two different data streams are sent on the two antenna ports). Figure 3.21 [4] illustrates the average two-codewords usage as scheduled by the eNB in the mobility route as a function of the CQI. It can be concluded that the two-codewords usage is very strong in the entire mobility route, starting to fall to 50% at CQI = 8 and diminishing at CQI < 3. The usage of the two-codewords in the route is estimated to be ~62%, on average

---

[7] Refer to Chapter 6 for details on link budget calculations.

[8] Refer to Chapter 4 for details on comparisons between different MIMO transmission modes.

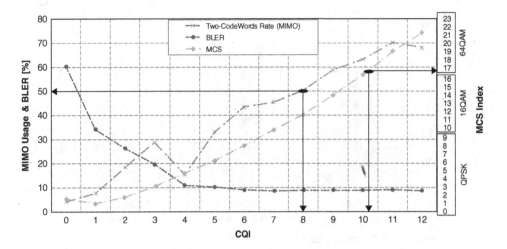

Figure 3.21   LTE MIMO, 64QAM usage and BLER as a function of CQI.

Table 3.9   CQI, modulation, and TBS index
table for PDSCH from, 3GPP Release 8

CQI index	MCS range	Modulation
1	0–9	QPSK
2		QPSK
3		QPSK
4		QPSK
5		QPSK
6		QPSK
7	10–16	16QAM
8		16QAM
9		16QAM
10	17–28	64QAM
11		64QAM
12		64QAM
13		64QAM
14		64QAM
15		64QAM

representing the mix of one and two codewords for MIMO (i.e., transmit diversity or spatial multiplexing in varying RF).

As for the higher order modulation, based on the reported CQI, the eNB scheduler decides on the MCS to be assigned to the user. MCS indices of 0–31 are allowed according to the 3GPP standard [1]. The range of MCS index and allowed modulation mapped for each CQI level are summarized in Table 3.9, as already given by 3GPP [1].

As shown in Table 3.9, the range of MCS 0–9 allows QPSK (quadrature phase shift keying) modulation, MCS 10–16 allows 16QAM modulation, and MCS 17–28 allows 64QAM

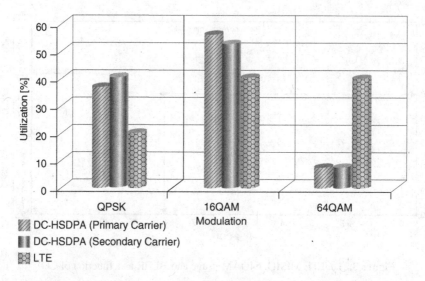

Figure 3.22   LTE and HSPA+ modulation usage in mobility route.

modulation usage, while the range of MCS 29–31 is reserved for special operation during retransmissions. The MCS index is then used by the UE to derive the transport block size (TBS) in bits for each stream.

Figure 3.22 illustrates that 64QAM usage (i.e., MCS $\geq$ 17) is strong in the entire route with an average measured usage of 40%, minimized at CQI < 10 as also correlated with Figure 3.21. The 3GPP standard (Table 3.9) expects 64QAM usage for CQI $\geq$ 10 (i.e., MCS index $\geq$ 17) to maintain the 10% BLER [1]. Hence, the test results in Figures 3.21 and 3.22 demonstrate a good match between CQI and 64QAM usage in the mobility route, in line with the 3GPP requirements. Meanwhile, the BLER is maintained at 10% for the entire range of CQI (except at the edge of the coverage when CQI $\leq$ 4). This illustrates a good implementation of the network scheduler in general.

On the other hand, 64QAM usage for HSPA+ is 7.4%, per carrier, for the entire route, as depicted in Figure 3.22. 3GPP, as listed in Table 3.10, expects strong usage of 64QAM at CQI $\geq$ 25 [2], for a category 24 UE. HSDPA transport block scheduling is typically based on the number of HS-PDSCH (high speed physical downlink shared channel) codes and power. Hence, a high modulation can still be used with low TBS, and vice versa.

Table 3.10   CQI, modulation mapping table for HS-PDSCH from Release 8 3GPP 25.214

CQI index	Modulation
1..15	QPSK
16..25	16QAM
26..30	64QAM

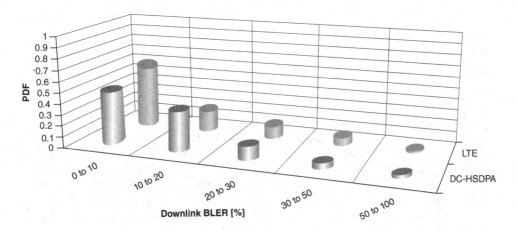

Figure 3.23   Downlink BLER measurements for DC-HSDPA and LTE from mobility route.

Referring to Figure 3.11 illustrating the CQI versus efficiency, the relation between 64QAM and CQI efficiencies can be set. With LTE CQI 10, the starting point of 64QAM usage is expected at an efficiency 2.7 bps/Hz, but with DC-HSDPA CQI 26, the starting point of 64QAM usage is expected at an efficiency 3.3 bps/Hz. Thus, LTE is expected to see higher 64QAM usage than DC-HSDPA in general, and this is confirmed in this route with the average CQI measured in each system. This is an important factor in RF planning as the network optimizer should target higher CQI by means of RF optimization to achieve better 64QAM utilization for both systems.

The measured CQI for both technologies yields different coding rates and hence different data rates. However, Figure 3.23 demonstrates the DC-HSDPA sub-block error rate (sBLER) and LTE downlink BLER are maintained at 10%, regardless of how much LTE throughput is higher than DC-HSDPA. This reflects that both systems have the same positive effect on the user's observed throughput when the packets are transported from the physical layer to the MAC (medium access control) layer with controlled BLER levels. This is also the case here because the vendors tested in both systems deploy CQI adjustment that helps to control the target BLER with any level of CQI.

3.2.2   *Performance Evaluation of Uplink Throughput*

This section provides guidelines on some useful methodologies for evaluating and benchmarking the LTE uplink throughput performance. The mobility route and KPIs used in the analysis of this section have been described in previous sections. Benchmarking with HSUPA is not necessarily required because of the substantial throughput difference between uplink in LTE and HSUPA (with the commonly commercial HSUPA category 6).

Similar to the methodology we used in the downlink, Figure 3.24 illustrates the distribution of the LTE uplink throughput with an average of ~14 Mbps and an instantaneous peak of 50 Mbps. Therefore, the average mobility spectral efficiency of the uplink is 0.7 bps/Hz. A similar throughput measurement methodology can be followed for any mobility route.

Similar to the previously discussed targets in the downlink, Figure 3.25 [4] illustrates that uplink BLER is also maintained at levels of 10% most of the time. Meanwhile, the figure

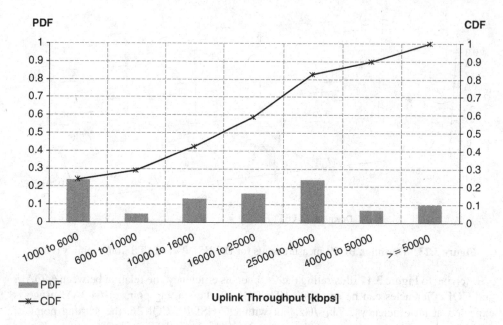

Figure 3.24   LTE uplink throughput in the mobility route.

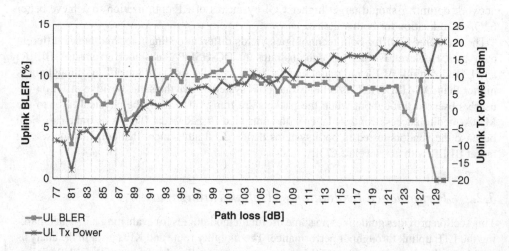

Figure 3.25   Uplink Tx power and BLER in the mobility route.

shows that the uplink power is essentially a function of the path loss and power control parameters. From this field data, the UE hits high uplink power (>15 dBm) at a path loss of 114 dB (i.e., RSRP −96 dBm as reference signal power set at 18 dBm), which is close to the edge of coverage scenario.

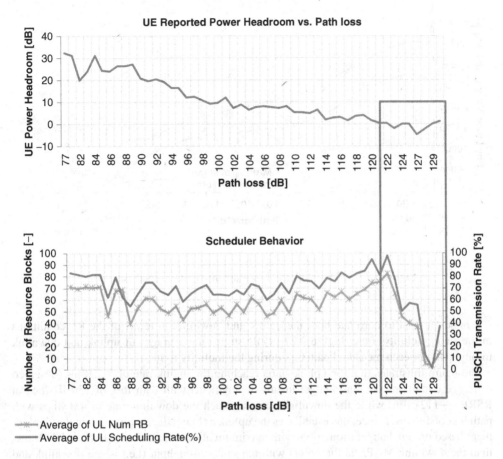

Figure 3.26 Uplink scheduling behavior relative to UE's power headroom and measured path loss.

Unlike the downlink throughput, the uplink is usually limited by a number of factors, such as the assigned eNB grant, buffer occupancy, and the UE's power headroom. As discussed in detail in Chapter 2, and as part of the uplink scheduling procedures, the UE reports its buffer and power status to the eNB to facilitate uplink scheduling. It is important to assess and understand the uplink scheduling behavior in terms of the UE's power headroom. Figure 3.26 thus shows the scheduled resources by the eNB defined in terms of the number of uplink resource blocks (RBs) and time-domain scheduling rates (i.e., assigned grants in the PDCCH). It is then plotted versus the UE's reported power headroom on the levels of path loss, depicting the uplink, and downlink RF conditions.

The figure shows that the network scheduler definitely takes the UE's power headroom into account when assigning the uplink resources in terms of number of RBs and uplink grant, when the buffer was full (i.e., continuous FTP upload ongoing). It is shown that the scheduling rate and RBs by eNB start reducing when the UE reports low power headroom values (i.e., the UE has reached maximum Tx power), as expected. Uplink power starts to increase, reaching its maximum at a path loss of > 122 dB (~RSRP of −104 dBm).

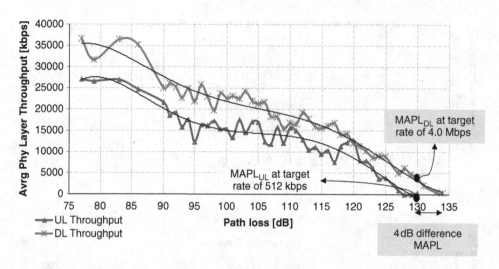

Figure 3.27 LTE uplink and downlink throughput versus path loss.

It is next useful to correlate both the uplink and downlink in terms of the RF conditions in the same mobility route. Figure 3.27 [4] illustrates the averages of uplink and downlink throughputs for each measured path loss during the mobility route.

The figure illustrates that the uplink starts reaching the uplink rate of 512 Kbps (designed cell-edge conditions per link budget in this deployment) with path loss of 130 dB (i.e., at RSRP = −112 dBm) while the downlink starts to reach the downlink rate of 4.0 Mbps with path loss of 130 dB. Hence, this establishes the uplink as being the limiting link in this deployment based on link budget estimation, with maximum allowed path loss (MAPL) of 4 dB less than the downlink MAPL at the points with the same throughput (i.e., where downlink and uplink are providing 512 kbps). In practice, determining the limiting link (downlink or uplink) based on MAPLs helps to estimate the cell radius per morphology and, therefore, the number of sites required to achieve the network coverage requirements.

3.3 LTE Case Studies and Troubleshooting

This section covers several case studies from several issues observed in a deployed LTE network. It is generally understood that the nature of issues impacting the user throughput and system performance vary in details from one network to another. However, the main objectives of this section are to provide practical troubleshooting mechanisms and a know-how methodology to identify and fix the problematic areas. The case-study examples in this section are expected to motivate the reader to follow similar level of investigations while troubleshooting similar types of LTE issues.

The previous section has discussed how to benchmark LTE performance and the key metrics to use during monitoring the optimization process. This section covers case studies of LTE downlink and uplink throughput, and LTE intra-frequency handover. The next section covers case studies related to the inter-RAT and inter-frequency. All case studies covered in this section have the same network configuration explained previously.

3.3.1 Network Scheduler Implementations

So far, methodologies for evaluating the LTE's uplink and downlink throughput performance have been discussed. As shown in these discussions as well as the overview procedures in Chapter 2, it is important to understand the overall mechanisms of network schedulers.[9]

Network schedulers for uplink and downlink are the main differentiator for the performance of one infra-vendor over another. Hence, each vendor tends to utilize different mechanisms when assigning the data for a user or group of users. There are many policies to ensure fair scheduling among users, taking into account the available resources. The most common are round robin (RR) and proportional fair (PF). RR typically ensures both time/frequency domain scheduling and allocation in sequential patterns. The RR typically aims at ensuring user fairness with limited inputs to the algorithm in terms of considering the CQI reported by different UEs. Hence, the main disadvantage is that RR may lead to lower system capacity with more unutilized resources, while fairness is guaranteed at best levels. On the other hand, PF typically aims at making a trade-off between system capacity and user fairness. This is achieved by taking the overall data available for the UE in the buffer proportionally to the actual channel quality (i.e., CQI). There are other ways of scheduling depending on the available cell power and QoS-aware (quality of service) schedulers.

However, in most cases, the same inputs to the scheduler for each user remain the same regardless of vendor implementations and scheduler policies. Figure 3.28 illustrates the basic and most important inputs to an LTE scheduler while determining the data to be allocated to each user.

The case studies discussed next require a basic understanding of the scheduler policies according to the overview in Figure 3.28.

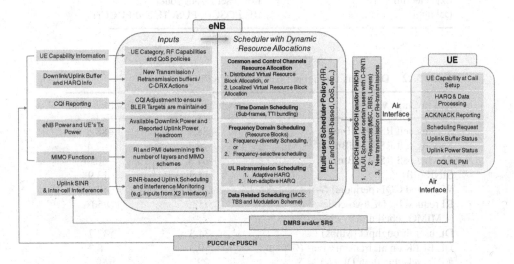

Figure 3.28 Basic uplink and downlink scheduling implementations.

---

[9] Refer to Section 2.9 and Figure 2.20 in Chapter 2 for an overview of LTE uplink and downlink scheduling.

3.3.2 LTE Downlink Throughput Case Study and Troubleshooting

Following the troubleshooting flow diagrams in Figures 3.16 and 3.17, the issue described in Table 3.11 is accordingly debugged and analyzed. The troubleshooting of downlink and uplink related channels is necessary once such low throughput is observed. As a first step, the optimizer can look at the general statistics and quality KPIs in the route, taking some benchmarking data into account, preferably with similar RF conditions. Table 3.12 shows the overall statistics of areas impacting the throughput, where the troubled area and the benchmarked statistics are both compared.

After checking all statistics for RF conditions, MIMO and 64QAM utilizations, the throughput difference between the troubled and benchmarked areas is not justified by the statistics alone. Hence, a deeper troubleshooting is needed. One of the methods that can be used in debugging such a case is to check the UE's post-processed traces, and the scheduler implementation in each instance the downlink packets are received from the eNB.

Troubleshooting in this situation can be detailed and, therefore, a good knowledge of the end-to-end system is required. The debugging can start from the physical layer and up. Once the physical layer is proven to show the expected performance, the area of backhauling and upper layer investigations can be carried out as next steps.

Table 3.11 Low LTE downlink throughput summary

Drive test conditions	Mobility test
Measurements	Downlink data transmissions
Issue symptom	High downlink duplicate retransmissions
Expected impact	Low user throughput
Debugging guidelines	RF, PDSCH, PUSCH, and PUCCH

Table 3.12 Downlink throughput performance KPIs

KPI (average values)	Benchmarked route	Troubled route
Serving cell RSRP (dBm)	−88.9	−88.6
Serving cell RSRQ (dBm)	−6.7	−7.4
Serving cell RSSI (dBm)	−61.8	−60.9
SINR (dB)	12.3	13.9
Wideband CQI (periodic) (−)	9.9	10.4
RI request by UE/two-codewords MIMO scheduled by eNBs (%)	1.5/52%	1.9/50%
DL user throughput (Mbps)	34.3	32.7
DL BLER on all transmissions (%)	9.3	8.3
# RBs scheduled on DL (−)	72	69
DL MCS (−)	13.6	13.1

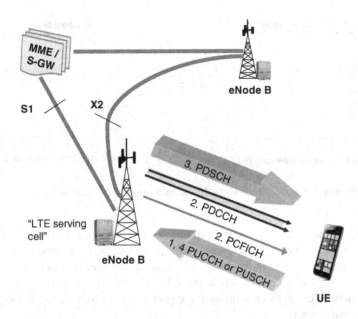

Figure 3.29   Duplicate downlink retransmissions concept.

One of the related physical layer concepts to validate is the behavior of HARQ (hybrid automatic repeat request) data retransmission between the UE and the eNB for the received downlink packets.[10] Figure 3.29 shows the physical layer transmission and retransmission mechanism, and it is summarized in the following steps:

1. The UE reports the CQI in the PUCCH (or PUSCH if multiplexed with uplink data).
2. The scheduler dynamically allocates downlink resources to the UE, and then the UE reads the PDCCH to determine assigned resources.
3. The eNB sends user data in the PDSCH.
4. The UE attempts to decode the PDSCH and sends an acknowledgment to the eNB to indicate whether the data are correctly received or a retransmission is needed. ACK indicates positive acknowledgment, and NACK indicates negative acknowledgment. The UE uses the PUCCH (or PUSCH if multiplexed with uplink data) in sending these acknowledgments.

If at step 4, the UE sends ACK for the correctly decoded PDSCH, and in the next HARQ round trip time, the UE receives the same packet again; the packet is thus discarded by the UE for being a duplicate retransmission. The end result impacts the total user throughput. This retransmission mechanism is essential in validating the physical layer and HARQ procedures.

The duplicate retransmission can then be statistically represented by a distribution over the entire route. Figure 3.30 illustrates the distribution of the duplicate downlink retransmission in the entire troubled drive route and compares it to the benchmarked.

---

[10] A good reference is Figure 2.20 in Chapter 2 listing the overall HARQ procedures.

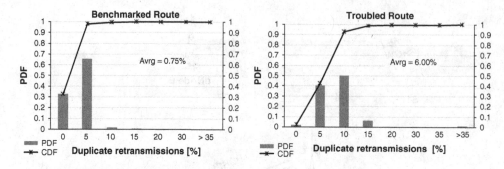

Figure 3.30 Distributions of duplicate downlink retransmissions rates.

As observed in the figure, the average duplicate retransmission in the troubled route is 6% while the benchmarked route showed stable duplicate retransmissions. The 6% higher duplicate retransmission directly means a 6% loss in end-user throughput because these retransmissions are not necessary when the UE has already acknowledged the packets. However, the scheduler keeps on sending them over again, occupying the HARQs intended for the new transmissions.

The reasons for duplicate retransmissions can vary: from scheduler implementation issues, uplink power issues, to underlying RF issues causing imbalance between the uplink and downlink. The uplink power issue means that the UE sends the acknowledgments, but the eNB is unable to decode them. The imbalance between uplink and downlink means the downlink RF conditions from the serving LTE cell are better than those of any neighboring cells and hence no handover is triggered, while the uplink is experiencing higher interference causing the eNB not to decode the acknowledgments. The uplink power investigations are described in the next case studies. However, the imbalance between uplink and downlink can be checked in terms of handover parameters, undesired interference created in the cell configuration (feeder losses), or site/equipment configurations.

In this particular case, the issue is determined to relate to the network scheduler. For two code-word MIMO transmissions, the eNB scheduler always retransmits both TBs (transport blocks), even if only one of them is NAKed by the UE, creating the duplicate retransmission instances. After the eNB scheduler is fixed, the duplicate retransmissions revert back to normal values and user throughput is improved to match the benchmarked levels. It is also noteworthy that the benchmarked route is tested with a different infra-vendor, and hence the issue has not been observed. Such a scheduler issue, if it exists, is expected to impact any test route conducted for the same infra-vendor.

The same issue can also occur in HSDPA networks. In UMTS, one of the main RF optimization methods is adjusting the pilot power (i.e., CPICH (common pilot physical channel) power) to limit the cell overlapping and number of cells in the active set. However, different CPICH power across neighboring cells can cause inconsistent HSDPA serving cell change where the downlink is strong, while the uplink is weak. The issue manifests itself when the NodeB does not decode the uplink ACKs, and thus duplicate retransmission occurs for data that have been successfully decoded by the UE. Hence, one RF optimization method can very well cause additional downlink throughput losses for the users.

3.3.3 LTE Uplink Throughput Case Studies and Troubleshooting

3.3.3.1 Limited Uplink Throughput

Following the troubleshooting flow diagrams in Figure 3.17, the issue described in Table 3.13 is accordingly debugged and analyzed. The troubleshooting of uplink related channels is necessary once such low throughput is observed. As a first step, the optimizer can look at the general statistics and quality KPIs in the route, taking some benchmarking data into account. The benchmarked route in this study is conducted on a different infra-vendor but in similar RF conditions. Table 3.14 shows the overall statistics of areas impacting the throughput, where the troubled area and the benchmarked statistics are both compared.

As listed in Table 3.14, the two routes show strong RF conditions in terms of RSRP, RSRQ, CQI, and UE uplink transmit power. The uplink throughput in the troubled area however is 10 Mbps less than the benchmarked area. Therefore, the troubleshooting intends to find the reasons for the low throughput.

As long as the RF conditions are very close, it is expected that the network scheduler will utilize the same uplink MCS and RBs for fair scheduling. As shown in the comparison table of statistics, the troubled area is showing limited MCS of 22 with a maximum of 90 RBs, while the benchmarked area reaches an MCS of 24 with a maximum of 96 RBs. One can take these values and investigate the details of the throughput distributions in order to detect any limitation on the network scheduler, as shown in Figure 3.31.

Table 3.13 Low LTE uplink throughput

Drive test conditions	Mobility test
Measurements	Uplink data transmissions
Issue symptom	Limited maximum uplink throughput
Expected impact	Low user throughput compared to benchmarked values
Debugging guidelines	RF, PUSCH, and PUCCH

Table 3.14 Uplink throughput performance KPIs

KPI (average values)	Benchmarked route	Troubled route
Serving cell RSRP (dBm)	−76.7	−58.5
Serving cell RSRQ (dBm)	−4.2	−5.4
SINR (dB)	25.3	29.5
Wideband CQI (−)	12.5	13.6
Rank 2 request by UE	2.0	2.0
UL Tx power (dBm)	1.07	−3.8
UL user throughput (Mbps)	46.4	35.9
UL BLER on all transmissions (%)	1.3	8.5
Num of PUSCH RBs scheduled on DL (−)	92.7 (Max = 96)	87.3 (Max = 90)
UL MCS (−)	23.9 (Max = 24)	21.9 (Max = 22)

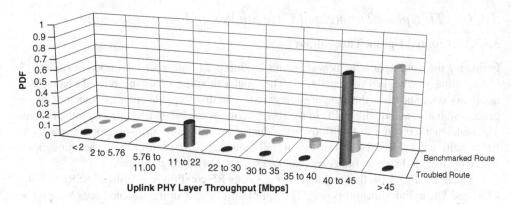

Figure 3.31   Uplink throughput distribution comparison.

From Figure 3.31 it can be observed that there is a limitation on the maximum uplink throughput at the physical layer. The benchmarked values are shown to be reaching >45 Mbps, where the troubled area is mainly limited to <45 Mbps. This can be an indication of a hard limit on the uplink throughput. One of these capabilities impacting the throughput is the scheduled uplink MCS, which reflects the maximum TBS and modulation scheme to use. 3GPP Release 8 standard allows up to MCS 24 to be used for UE category 3 where 16QAM is the highest modulation order allowed. The difference in this case needs further investigation as explained next.

Figure 3.32 simulates the expected uplink throughput in the 20 MHz channel given all factors in the LTE configuration discussed in Tables 3.6 and 3.7. The maximum theoretical uplink PHY layer throughput with MCS limited to 22 and 90 RBs in the troubled drive route is 42.37 Mbps, compared to 51.02 Mbps if the MCS is 24 with 96 RBs (as in the benchmarked area). Therefore, it is clear that the main contributor to the limited throughput is the maximum uplink MCS and RBs used by the network scheduler. Hence, understanding the relation between limiting the MCS and the RBs in the troubled area is important.

In 20 MHz, the maximum allowed number of recourse blocks (RBs) is 100. The PUCCH is never transmitted simultaneously with the PUSCH from the same UE. However, the PUCCH requires reserved resources, depending on the parameters given by higher layers. If six of these RBs are used for the PUCCH channel, then this leaves 94 RBs for the PUSCH. Moreover, with prime factor reduction as defined in 3GPP [10], then the UE can be assigned 90 RBs from the 94. With 90 RBs, the theoretical achieved peak rate with an uplink MCS of 22 is 42.37 Mbps, as simulated in Figure 3.32. The PUCCH is hence a suitable contributor to limit the number of RBs. In the benchmarked route, the PUCCH resource bandwidth only reserves 4 RBs and hence the PUSCH can allocate a maximum of 96 RBs. The trade-off of using higher or lower PUCCH resources depends on the number of UEs accessing the cells and typically dimensioned during the network planning phase. By reducing the PUCCH overhead, more resources can be used for PUSCH transmission, which is typically done in a low load situation. By increasing the PUCCH resources, more UEs can be admitted to the cell, but with less PUSCH resources, and thus this is typically done in a high load situation.

In addition, another major factor that causes lowering of the uplink peak rate, in this troubled route for the same UE category, is the presence of the SRS (uplink sounding reference

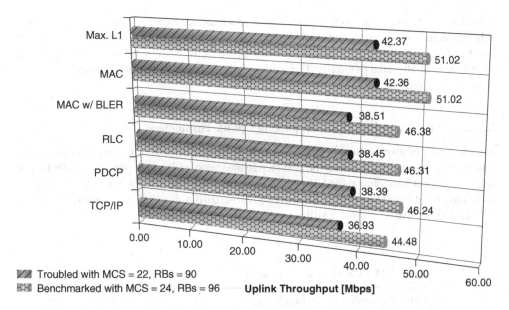

Figure 3.32 Simulated theoretical uplink throughput with uplink MCS 22 and MCS 24.

signal) channel. It consumes additional uplink resources (depending on the SRS configuration) from the PUSCH. The uplink SRS is a standard channel that eNB uses to perform the FSS. It thus allows the eNB to estimate the uplink SINR and exploit the RF channel response variation across different frequency regions of the wide-band carrier.

LTE's RF channel response in different frequency portions of the carrier bandwidth can be noticeably different. SRS transmissions facilitate the eNB to analyze the real-time uplink channel frequency and allow uplink data transmissions to be allocated in the frequency portion that has the best channel response. As a result, the RF spectrum efficiency is improved, especially under multi-user traffic conditions. Using SRS for uplink scheduling can potentially yield performance gains in a multi-user network environment, which is what the network scheduler in this troubled area has been implemented for. The eNB typically determines the SRS bandwidth based on the PUCCH and total system bandwidths. Additionally, the eNB can adjust the time-domain resources for SRSs based on the cell loading conditions.

However, the benefit of SRS comes with some impacts. The reservation of the SRS resource leads to losses of uplink PUSCH symbols in terms of coding rate. To cope with this, the eNB can then adjust the MCS based on the cell-specific SRS subframe configuration. As a result, MCS is capped at 22 in the troubled route. The reduction of PUSCH resources leads to 16% average uplink throughput loss for one user. Therefore, with the PUCCH and SRS bandwidth resource reservations, the maximum assigned uplink RBs and MCS can be limited.

It can be concluded in this case that with 16QAM, up to MCS 22, and 90 RBs may be supported with a specific SRS and PUCCH configuration. On the other hand, MCS 24 with up to 96 RBs may be supported without SRS configuration and lower PUCCH bandwidth reservation. The throughput loss is likely notable in a demonstration scenario, although the impact in the real deployment is much less due to the small probability of peak data rate when many users are served in the cell. For implementations without SRS, the demodulation reference signal

(DM-RS) can be used for uplink channel quality estimation. However, the uplink DM-RS is only transmitted on subcarriers assigned to the UE and, therefore, does not provide sufficient wideband channel quality information for frequency selective resource allocation, particularly for RBs not yet allocated to the UE. The implementation trade-offs can be assessed during the deployment stage.

3.3.3.2 Excessive Uplink Power Impacting Uplink Throughput

The main area of investigation in this case study is the high uplink transmit power measured at the UE, as listed in Table 3.15. The first step in the investigation is to plot the distribution of uplink Tx power measured at the UE, shown in Figure 3.33. The distribution in general shows more samples at high Tx power, with an average of ~10 dBm, for a UE with a power amplifier (PA) capable of a maximum of 23 dBm.

Table 3.15 Uplink power case study

Drive test conditions	Mobility test
Measurements	Uplink data transmissions
Issue symptom	High uplink Tx power in general in this route
Expected impact	High uplink interference
Debugging guidelines	PUCCH, PUSCH, and SRS

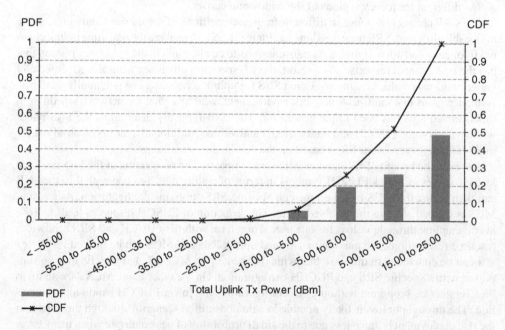

Figure 3.33 Total uplink transmit power.

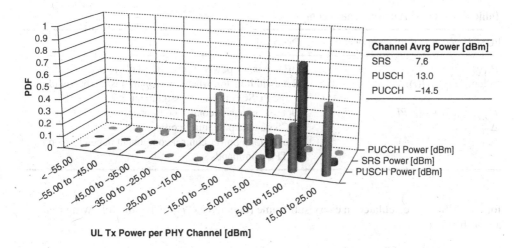

Channel Avrg Power [dBm]	
SRS	7.6
PUSCH	13.0
PUCCH	−14.5

Figure 3.34   Uplink Tx power distribution per channel.

Plotting the power distribution from each uplink channel is illustrated in Figure 3.34. The figure shows that the PUSCH power is dominating the total UE's Tx power. SRS is not continuously transmitted, but maintaining a good uplink Tx power is expected due to the significance of this channel. The third component is the PUCCH channel that carries control information with less bandwidth and, therefore, the uplink power contribution of this channel to the total power is less than the other two channels.

In CDMA-based (code division multiple access) systems all users in the uplink share the same frequency resources. As a result, the Node-B needs to manage the power of individual UEs so that each UE can maintain a specific level of link quality. Thus, the use of power control is needed to manage the rise of signal level over the thermal noise level (i.e., rise-over-thermal, RoT) for stable operation, when the increase of other users' power causes additive uplink interference.

In OFDMA (orthogonal frequency division multiple access) systems, individual UEs use different time/frequency resources for transmission. As a result, intra-cell interference and managing RoT are less critical factors, but the inter-cell interference becomes an important one. However, a mechanism is still needed to maintain a specific link level quality. This is achieved through uplink power control. The LTE power control is applied on the uplink channels using two mechanisms:

1. Open-loop power control (OLPC) that provides a basic operating point for the UE's Tx power and contributes to the measured downlink path loss and system parameters,
2. Closed-loop power control (CLPC) that is used by the eNB to further adjust the uplink power based on uplink quality using transmit power control (TPC) commands. The TPC commands sent by the eNB dynamically adjust the power by a certain factor (in dB) depending on the required BLER target and link quality.

One of the troubleshooting mechanisms is to verify the network's power control parameter settings that could impact the UE's Tx power, specifically, the PUSCH channel. Uplink power

Table 3.16 PUSCH power control component

PUSCH power control component	Description
P_{MAX}	UE's max Tx power (typically, 23 dBm)
$10\log_{10}(M_{\text{PUSCH}}(i))$	Bandwidth factor, where $(M_{PUSCH}(i))$ is the bandwidth of the PUSCH resource assignment
$P_{\text{O\_PUSCH}}(j) + \alpha(j).PL$	Open loop power control component
$\Delta_{\text{TF}}(i)$	MCS dependent control component
$f(i)$	Closed loop power control component depending on the TPC commands

for the PUSCH is calculated in every sub-frame (denoted as "i"), using the following relation, and Table 3.16:

$$P_{\text{PUSCH}}(i) = \min\{P_{\text{MAX}}, 10\log_{10}(M_{\text{PUSCH}}(i)) + P_{\text{O\_PUSCH}}(j) + \alpha(j).PL + \Delta_{\text{TF}}(i) + f(i)\}$$

OLPC provides a basic operating point for the UE Tx power. Further adjustments are made dynamically on top of this operating point. For PUSCH, $P_{\text{o\_PUSCH}}$ is a UE-specific parameter providing a base level for PUSCH data transmissions to allow for different BLER operating points.

"α" is the fraction, between 0 and 1, controlling the sensitivity to path loss. This can be used to control the UE transmit power. For example, the UEs at the cell-edge may transmit at very high power, reducing the average cell capacity of the network by causing very high inter-cell interference. α allows a trade-off between cell-edge data rates and cell capacity. This fraction is cell-specific and is conveyed to the UE via RRC (radio resource control) messages. PL represents the measured path loss by the UE according to:

$$PL = \text{reference signal power} - \text{higher layer filtered RSRP}$$

where the reference signal power is provided by a higher layer, and higher layer filtered RSRP is the measured value by the UE depending on the DL-RS (downlink reference signal).

For the PUSCH, Δ_{TF} is an MCS-dependent component based on the selected TBS and the number of resource elements used for transmission. The Δ_{TF} component is calculated based on the eNB's broadcasted UE-specific parameter referred to as the deltaMCS. Finally, the value $f(i)$ represents the TPC command the UE receives on the PDCCH to increase, decrease, or hold the PUSCH power.

In this drive route, PUSCH power control parameters are configured based on cell-specific parameters (in SIB 2 (system information block)) and UE-specific (in the RRC connection setup). The parameters configured in the system are highlighted in Table 3.17.

Let us estimate the UE's uplink power based on the parameters in Table 3.17, and by applying the PUSCH power control equation shown before. Taking an example of RSRP at -74 dBm, the number of PUSCH resource blocks assigned to the UE in the given TTI (time transmission interval) is 90, the last $f(i)$ is -2 dB (the initial power of the PUSCH) and the TPC command sent by the eNB as 0 dB, the uplink Tx power is then calculated by the UE as follows:

$$P_{\text{PUSCH}}(i) = \min\{23, 10\log_{10}(90) + (-67 + 0) + (0.7 \times 92) + 0 + (-2)\} = 15 \text{ dBm}$$

Table 3.17 PUSCH power control parameters in the drive route

Parameter used in PUSCH	Cell-specific (SIB-2)	UE-specific (RRC connection setup)
$P_{o\_PUSCH} = p0$-Nominal PUSCH $+ p0$-UE-PUSCH	p_0-NominalPUSCH = $-67\,$dBm	p_0-UE-PUSCH $= 0\,$dB
α	0.7	–
Δ_{TF}	–	deltaMCS-Enabled $=$ en0 $\rightarrow$ Ki $= 0 \rightarrow \Delta_{TF} = 0$
Close loop power control mode *accumulation Enabled*	–	True (use accumulate TPC of $-1, 0, 1,$ or $3\,$dB)
Path loss	Calculated by UE depending on RSRP and reference signal power (set to $18\,$dBm in SIB2)	
M_{PUSCH}	Assigned number of uplink RBs in PDCCH	
$f(i)$	Depends on TPC command bits in PDCCH where it is the sum of the last PUSCH power function (it could be the initial value) in the previous sub-frame and the TPC command sent to the UE	

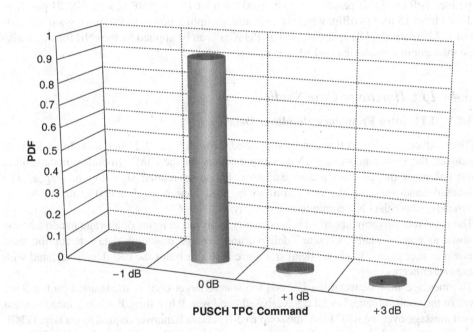

Figure 3.35 PUSCH TPC commands in the drive route.

Hence, at very high RSRP ($-74\,$dBm), the UE Tx power reaches $15\,$dBm. This shows an expectedly high uplink power in good RF conditions. For the entire route, the eNB does not attempt to reduce the UE's UL Tx power either, since the TPC command distribution observed in the entire route, in Figure 3.35, is showing 90% usage of $0\,$dB TPC commands.

By looking thoroughly at the impacts of the OLPC and CLPC on the uplink Tx power, it can be noted that the main contributor to the adjustments of the PUSCH power is the OLPC component. In this case, the uplink power is mainly driven by the UE's measured downlink path loss and the configured parameters by the eNB. This result is expected in a lightly loaded system where a typical uplink scheduling mechanism tends to increase the data rate and hence the UE Tx power, to maximize the SINR, and the uplink throughput in all RF conditions. This is concluded based on the total power distribution per each uplink channel in the test route, as well as the contribution coming from the CLPC TPC commands. All this illustrates that the eNB scheduler rarely requires the immediate adjustment of power by sending a majority of 0 dB TPC commands to the UE. However, with an increasing load, some modification to these parameters is required to control every UE's initial and adjusted uplink power.

One of the important factors is the high setting of cell-specific "p0-NominalPUSCH" as -67 dBm in this drive route. Higher settings will improve the PUSCH reception at the eNB, but will also drive higher UE transmit power, leading to an increasing interference to neighboring cells, and leading to lowering the overall cell throughput. Higher settings will be needed if α is set to a very low value so that the base PUSCH power can be increased when there is lack of path loss compensation. The setting of this parameter typically takes into account the eNB sensitivity (i.e., noise floor) and the PUSCH targeted RoT. For the same RF conditions explained before, if this parameter is changed from -67 to -96 dBm, the PUSCH power is reduced from 15 to -14 dBm, which is a reasonable uplink power in the given good RF conditions. Alternatively, UE-specific "p0-UE-PUSCH" can be applied by the eNB for users with high throughput demand in good RF conditions.

3.3.4   LTE Handover Case Studies

3.3.4.1   LTE Intra-Frequency Handover Delays

LTE introduces handovers that can be supported through S1 or X2 interfaces. In an S1-type handover, the source and target eNBs communicate via the MME (mobility management entity) through the S1 interface to exchange handover related signaling messages. The X2-based handovers are the most commonly used handovers due to the need for less overheads, where the eNB communicate directly.

The handover performance of LTE is normally analyzed in terms of average handover time (known as control plane "C-plane" delays), and average downlink data interruption time (known as user plane "U-plane" delays). Figure 3.36 illustrates the call flow associated with these two concepts.

The message flow between the UE and the source/target eNB is transmitted on the RRC layer of the control plane. For LTE's X2-based handover, when the UE sends a measurement report message over the RRC layer, the source eNB sends a handover request to the target eNB, including a list of the bearers to be transferred and whether the downlink data forwarding is proposed. Then, the source eNB sends the RRC connection reconfiguration message to the UE over the RRC layer. At the same time, the downlink packets received at the source eNB from the S-GW (serving gateway) are forwarded to the target eNB. Then, the UE synchronizes with the target eNB using a random access procedure, after which, the UE sends the RRC connection reconfiguration complete message over the newly established RRC with the target eNB. The UE then starts collecting the SIBs from the target eNB carrying the required information about the cell parameters. The target eNB sends the UE context release message to the source eNB

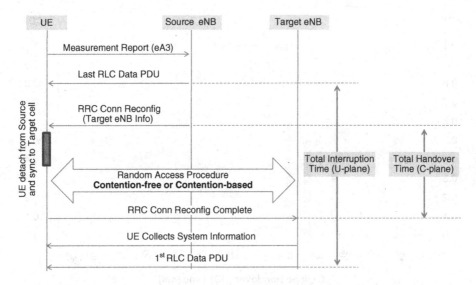

Figure 3.36   LTE handover measurement flow, C-Plane and C-plane delays.

confirming the successful handover and enabling source eNB resources to be released. Finally, the data start flowing directly from the new serving eNB into the UE.

Figure 3.37 demonstrates the average LTE X2-based handover delays in the C-plane, collected from the same mobility route discussed before. The delays measured include the total processing time and the air interface packet latencies needed to process the handover request sent by the eNB. The field measurements are collected from two implementations. Each one uses a different RACH procedure during handovers (one with contention-based and the other with contention-free RACH). This case study compares the performance of each handover procedure.

As shown in Table 3.18, during the handover, the contention-based RACH procedure requires a higher RACH time because the UE needs to wait for a contention resolution. However, in the contention-free procedure, this contention resolution is not needed to complete the process once the UE receives the RAR message from the eNB. As a result, the contention-based RACH adds an extra 7 ms of delays to the C-plane handover. The choice of the RACH procedure versus the C-plane delays is a trade-off to consider in a deployed LTE network. The trade-off has been discussed in Chapter 2.

On the other hand, Figure 3.38 illustrates the LTE downlink data interruption time observed during handovers in the route. The interruption time is mainly calculated based on the RLC (radio link control) packets received from the source and target eNBs during the handover (i.e., the last RLC packet on the source eNB to the first RLC packet on the target).

As shown in Table 3.19, it is expected that the U-plane delays in the contention-free procedure will provide a better interruption time than the contention-based one since the C-plane delays are shorter (see Table 3.18). This requires further troubleshooting to improve the U-plane performance in this scenario. In general, the expected time delay difference between the C-plane and U-plane is within 10 ms, in typical RF conditions. Table 3.20 lists some of the impacts of the handover delays.

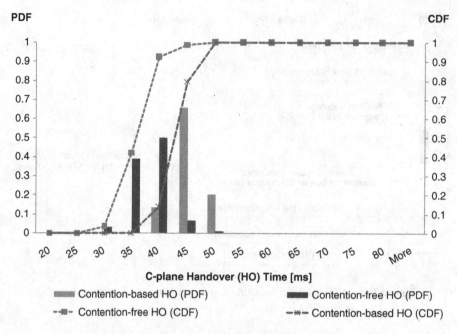

Figure 3.37   **LTE** C-plane handover delays distribution.

Table 3.18   Summary of C-plane handover delays

RACH procedure during handovers	Mean C-plane delay (ms)	Min, Max C-plane delay (ms)	Average RACH time (ms)
Contention-based	43.4	38.0, 49.0	25.2 (Max = 30, Min = 20)
Contention-free	36.4	30.0, 47.0	17.9 (Max = 29, Min = 12)

Table 3.19   Summary of U-plane interruption delays

RACH procedure during handovers	Mean U-plane delay (ms)	Min, Max U-plane delay (ms)
Contention-based	50.0	39.0, 80.0
Contention-free	62.3	40.0, 139.0

In this case, the higher than expected U-plane delays are related to the parameters configured at the eNB. The parameters are observed to allow handovers to other cells with weak RF conditions. This has introduced higher BLER on the data packets arriving on the target cell. With high BLER, the U-plane delays increase, impacting the user-perceived throughput during handover. However, this has not necessarily impacted the C-plane delays.

One step of optimization in this scenario is to review the handover parameters. The targeted parameters are set based on the trade-offs between a stable handover success rate, and lower

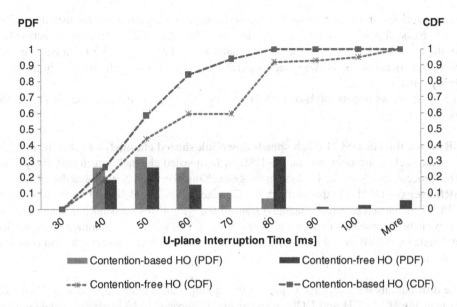

Figure 3.38 LTE U-plane interruption delays distribution during handovers.

Table 3.20 Summary of causes for higher than expected C-plane and U-plane delays

Factors increasing HO delays	PDSCH BLER in source cell	RACH	PDSCH/PUSCH BLER in target cell
C-plane delay	–	RACH preamble retransmissions	–
U-plane delay	Downlink RRC handover message retransmissions or high BLER on source cell data packets	RACH preamble retransmissions	Uplink RRC handover complete message retransmissions, high BLER on target cell data packets

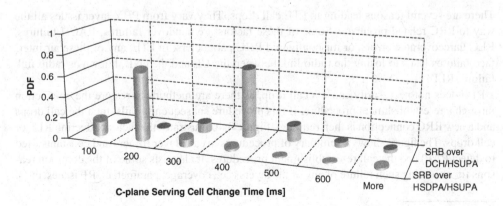

Figure 3.39 HSPA(+) C-plane serving cell change time.

U-plane latencies. In the commercial deployment stage, LTE parameters are usually audited on a site-basis, depending on the RF conditions and the targeted KPIs set in every network.

On the other hand, to benchmark the LTE handover delays, Figure 3.39 demonstrates the C-plane time delay of the HSPA+ network (referred to as serving cell change), in the same mobility route.

The figure shows two methods of SRB (signaling radio bearer) mapping on the HSPA network[11]:

- **SRB over the HS-DSCH (high speed- downlink shared channel)** – In this method, the C-plane packets are delivered on the HSDPA high speed channel transmitted every 2 ms. This process speeds up the C-plane message exchange between NodeB and the UE.
- **SRB over the DCH** – In this method, the C-plane packets are delivered on the Release 99 DPCHs (dedicated physical channels) transmitted with 40 ms transmission intervals. The delays in this method are expected to be the highest. This is the most common configuration at this stage, as SRB over the HS-DSCH is being slowly deployed in networks and is subject to device support and capabilities.

The distributions show average C-plane serving cell change delays of 158 and 389 ms for SRB over the HS-DSCH and DCH, respectively. Compared to LTE handover delays from Table 3.18, it is clear that LTE handover delays outperform those in HSPA+. The X2-based handover in LTE provides significant improvements, while the HSPA+ control plane still needs to go through the RNC (radio network controller). This improves the LTE data interruption time in mobility and hence the end-user experience, compared to HSPA+. This also makes the LTE system suitable for VoIP (voice over internet protocol) deployments (i.e., VoLTE, voice over long term evolution) where the data interruption time is essential in mobility conditions with different inter-arrival times of the voice packets.

It is important to note that, the HSPA+ system in this trial has not implemented the enhanced serving cell change (E-SCC), defined in 3GPP Release 8. With E-SCC, the interruption time in the C-plane and U-plane are expected to improve significantly for the HSPA+ network and end user, also for a suitable VoIP deployment in the UMTS roadmaps [11].

3.3.4.2   LTE Handover Failure Examples and Troubleshooting

There are several reasons leading to LTE call drops. They vary from PHY layer issues all the way to RRC related problems. Some of these factors are handover failures, RACH failures, RLC unrecoverable errors, or misconfigured RRC parameters. In LTE, any of these air interface failures leads to losing the radio link between the UE and the eNB, known as radio link failure (RLF).

RLF does not necessarily cause a call drop, as there are methods to restore the connection through a re-establishment procedure. If this procedure subsequently fails, then the call drops and a new RRC connection is then required. Figure 3.40 illustrates the factors affecting RLF or call drops. The figure shows a summary of procedures in each layer (timers and constants used to detect RLF) and the call re-establishment procedures. It also lists some of the common reasons for observing such failure of any of the layers (i.e., coverage, parameters, RF issues, etc.).

---

[11] Refer to Chapter 4 for a comprehensive discussion of these methods as part of CSFB optimizations.

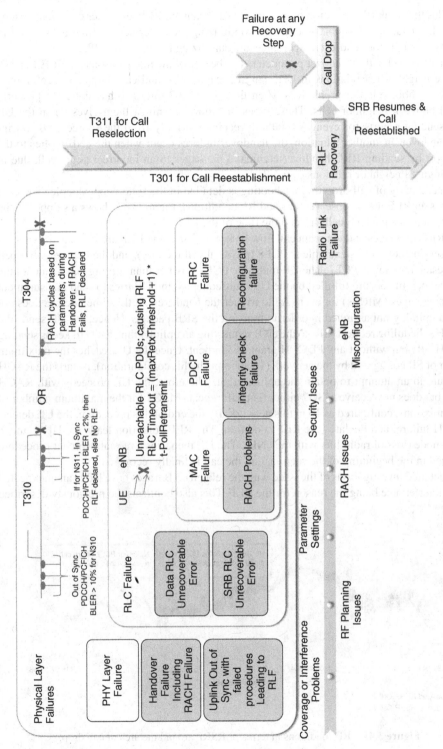

Figure 3.40 Common reasons and procedures for radio link failures (RLF) or call drops in LTE.

Besides the weak RF condition or coverage issues causing RLF, the other common reasons are handover failures. Troubleshooting and optimizing handover success rate is essential in ensuring a satisfactory end-user experience and stabilizing the network KPIs.

Similar to UMTS, the handover parameters can be one of the main reasons for RLF. In addition, the neighbor list relations in LTE can also cause RLF and call drops. One commonly missed neighbor relation problem is when the logical X2 interface has not been properly defined between neighbor cells. These types of failures manifest themselves when the UE keeps sending reports for event A3 without receiving a reply from the source eNB to trigger a handover. In another situation, the handover failures occur when the eNB replies to the UE report by sending RRC handover command messages to an incorrect target cell, due to misconfigured neighbor relations over X2.

One case study of call drop troubleshooting is detailed in the following example. The case describes an RLF that is observed in good RF conditions. Figure 3.41 shows a snippet of the RF conditions of the LTE cells around the time of the RLF.

The RRC messages and the main call flow before and after the RLF are listed in Table 3.21.

The serving cell in this example is PCI 8 (physical cell identity), and the best neighbor cell in the tested location is PCI 7. The UE finds that PCI 7 is better than serving PCI 8, and all the handover parameters are fulfilled by the UE. Subsequently, the UE tries to send measurement reports messages (MRMs) for event A3 to trigger the handover. At this time, the UE does not have an uplink grant assigned in order to transmit the SRB packets. Hence, the UE sends the PUCCH scheduling request (SR) to the eNB requesting an uplink grant. The UE keeps sending PUCCH SRs but without any PDCCH grant assignment. Once the UE reaches the maximum number of SR configured by the network (16 attempts in this configuration), it starts the RACH procedure in an attempt to obtain the uplink grant. Therefore, the UE proceeds with RACH MSG1 but does not receive MSG2 either. The UE repeats MSG1 for the maximum number of retransmissions, configured as 10 in this network. By the end of this procedure, the UE detects a RACH failure, and declares the RLF condition. The RLF condition leads the UE to move to idle mode (loss of radio link with the eNB). The UE then tries a reestablishment procedure described in the beginning of the section, and the call is finally recovered.

The detailed investigations of the issue with the related eNB traces have led to an unexpected uplink interference being the reason for the RLF. This also confirms the previously described

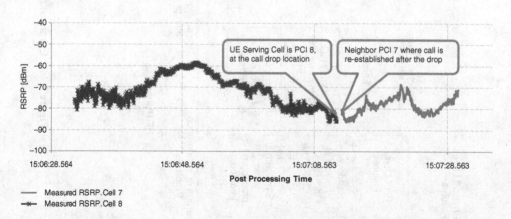

Figure 3.41 RF conditions in terms of RSRP around the area of call drop.

Table 3.21 Call drop failure flow diagram

TimeStamp	RRC message	(UL:<) (DL:>)	Message info
15:06:32.466	RRCconnectionreconfiguration	>	event A3 parameters configured by eNB on PCI 8
15:06:32.469	RRCconnectionreconfiguration complete	<	UE and eNB will use the parameters configured
15:06:33.191	systeminformationblockType1	>	–
15:07:11.756	measurementreport	<	UE sends event A3 to report PCI 7 is better than serving cell PCI 8, according to the parameters configured
–	Grant requests by UE to eNB	<	Grant request procedure triggered by UE but eNB does not assign UE UL grant to send MRM
–	Grant timer expires	<	
–	LTE RACH ACCESS START	<	UE needs to start RACH process
15 : 07 : 11.947	RACH ACCESS FAILS	–	RACH fails after reaching maximum number of attempts without a reply from eNB
15:07:11.947	LTE RRC RADIO LINK FAILURE	–	UE triggers RL Failure
Call Drop		< >	Call drop indicated
15:07:12.189	systeminformationblockType1	>	UE moves to idle mode and reads SIB-1 on the best cell (PCI 7)
15:07:12.205	RRCconnectionreestablishment request	<	Call re-establishment process succeeds and call recovered
15:07:12.232	RRCconnectionreestablishment	>	

call flow where the eNB is unable to reply to any uplink grant request from the UE. The downlink conditions in this example case study are shown to be strong, but the uplink in this case has been impacted by extra losses due to equipment issues.

Lastly, any RLF leads to significant end-user data interruption, and degradation in the network retainability KPIs. While the reasons for call drops can vary in each deployment, this section has shown methodologies that can be beneficial in troubleshooting similar problems.

3.4 LTE Inter-RAT Cell Reselection

The EPS system is designed to be able to interwork with other RATs (i.e., cellular systems). The other 3GPP RATs are GERAN (GSM/EDGE radio access network) or UTRAN (universal terrestrial radio access network). Other non-3GPP systems are C2K, eHRPD (high rate packet data), WiFi, and WiMax. This chapter covers the mobility procedures related to inter-RAT between LTE (referred to in this section as E-UTRAN (evolved universal terrestrial radio access network)) and UMTS (referred to as UTRAN). In addition, it discusses inter-frequency mobility within LTE cells in different bands/frequencies. Each of the subsections covers 3GPP

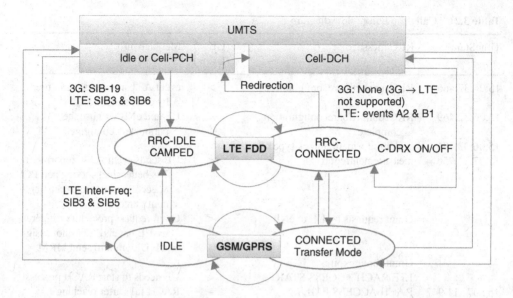

Figure 3.42   E-UTRA states and inter-RAT mobility procedures.

requirements for the procedures in each mobility state, as well as an overview of parameters and troubleshooting case studies.

In Chapters 1 and 2 the different RRC states in E-UTRA were discussed. In addition, Figure 3.42 illustrates the mobility procedures between E-UTRAN, UTRAN, and GERAN from different states in each system. Transitions between connected and idle states in any system are normally triggered based on the data activity, either uplink or downlink. On the other hand, a transition from connected to idle RRC state can be the result of any call drop or loss of coverage (i.e., out of service).

There are many criteria (e.g., coverage, load balancing, terminal capabilities, access restrictions, and subscription attributes) considered at the network design stage, when making a mobility decision between cells in the same RAT, different RATs or frequencies. The mobility decision can be based on UE-assisted measurements or blindly based on pre-defined criteria by the network. In blind mobility, handover in particular, the network takes an anonymous decision to the UE for a handover between systems or frequencies, and sometimes even between cells within the same frequency. Though likely less reliable, these types of mobility handover decisions are typically made without information about the target cell (referred to as blind handover or re-direction), depending greatly on the implementation of the network features and the voice/data traffic load situation in the operator's network.

The inter-RAT and/or inter-frequency optimization efforts can vary in complexity, depending on the network architecture in terms of which RATs are deployed and how many different carriers or bands the network operator owns. The optimization process can be extremely complex when multiple systems are deployed in the same network (e.g., multiple UTRAN frequencies/bands, GERAN, multiple LTE FDD (frequency division duplex) frequencies/bands, and/or LTE TDD (time division duplex)). The legacy UTRAN to GERAN inter-RAT must be revisited to adopt the new frequency(s) deployed for the LTE system. Many of the parameters related to each transition shown in Figure 3.42 require an optimization process by taking into account the trade-off between KPI targets and end-user experience (battery in particular).

3.4.1 Introduction to Cell Reselection

In general, for both LTE and UMTS, the UE adhere to the cell reselection rules based on the parameters broadcast in the SIBs from the LTE eNB or UMTS NodeB.[12] The main reselection concepts have, in most of the cases, remained the same in LTE as in UMTS, more about reselection procedures is given in [12] and [13]. Let us cover the new concepts uniquely introduced in LTE.

The 3GPP release 8 has introduced a priority layer concept. This means that any IRAT (inter-radio access technology) or inter-frequency cell reselection between cells or frequencies (or even bands) is controlled by the assigned priority. The layer priority is not applied to cells from the same frequency as that of the serving cell.

As the complexity of the deployed system topologies increases, such priority-based reselection becomes important. With the diversity of cells deployed in the network (femto, micro, or macro) within the same or different RATs, the priority reselection can assist the operator in enforcing the targeted camping strategy. In this situation, the cell reselection can be layered up by assigning the cells into high, low, or equal priorities. Priorities are typically provided to the UE via system information or RRC release messages.

Table 3.22 summarizes the concept of layer priority and the measurement requirements by the UE, for inter-frequency, and inter-RAT. In LTE deployment, it is expected that the priority of LTE would be higher than that of other RATs, especially as LTE provides a higher data rate than other RATs (e.g., UMTS). In other scenarios, the LTE priority might be lower than non-LTE femto cells (i.e., home NodeB) in the deployed areas. Hence, the priority setting is an optimization choice that depends on the designed camping strategy, LTE deployment coverage, targeted performance, and end-user perceived experience.

Multi-mode capability is one of the key elements in the devices supporting multiple RATs. The modem capability is essential to provide seamless mobility across various RATs. The fact that the different deployed RATs may overlap or set as hotspots makes it even more important to provide uninterrupted services. As the user moves across inter-RAT 2G/3G (UMTS/GERAN/HRPD) and 4G (LTE) coverage areas, the optimization process thus needs to ensure service continuity while maintaining a good trade-off with the device's battery consumption. In thehe multi-mode device, though appealing to users, the battery consumption can be a major concern, depending on the scale of measurements with the configured network parameters.

3.4.2 LTE to WCDMA Inter-RAT Cell Reselection

In the LTE idle state, SIBs are broadcast using the PBCH (physical broadcast channel) and the PDSCH to all UEs. SIBs 2 and 3 contain parameters related to access and cell reselection. Additional cell reselection information for intra-and inter-frequency cells is defined in SIBs 4 and 5, respectively. And the IRAT reselection information is contained in SIBs 6, 7, and 8

---

[12] Refer to Table 2.13 in Chapter 2 for description of LTE's SIBs and the equivalent SIBs in UMTS.

Table 3.22 Layer priority of candidate cells (RATs) relative to serving cell priority

Priority configured by network	No priority	Lower — Serving cell signal power > threshold	Lower — Serving cell signal power < threshold	Equal — Serving cell signal power > threshold	Equal — Serving cell signal power < threshold	Higher — Thresholds to regulate the measurements
LTE inter-frequency	No reselection allowed	Measurements not mandatory	Measurement mandatory	Measurements not mandatory	Measurements not mandatory	Measurement mandatory
Inter-RAT from LTE to WCDMA	No reselection allowed	Measurements not mandatory	Measurement mandatory	Not allowed in standard		Measurement mandatory but regulated for battery saving
Inter-RAT from WCDMA to LTE	No reselection allowed	Measurement mandatory	Measurement mandatory			

for UTRA, GERAN, and CDMA2000, respectively. SIB9 contains information enabling the support of Home eNB.

Unlike UMTS, the LTE neighbor cell list is not provided to the UE in any of the SIBs. However, SIB4, for example, can include a list of neighbor cells only for indicating a per-cell additional reselection "offset" or to mark certain cells as blacklisted.

SIB Type 3 contains information that enables the UE to carry out cell reselection. The main SIB Type 3 parameters are:

1. Q_{hyst} parameter that is applied to a cell's measured RSRP to provide the ranking of the cells.
2. $S_{nonintrasearch}$ and $Thresh_{serving,low}$, which are utilized with the cell priority to control when measurements of other frequencies and/or RATs are performed for reselection. The priority of a cell can be specified between 0 and 7, with 0 indicating the lowest priority.
3. $S_{intrasearch}$ controls the threshold below which a UE carries out intra-frequency measurements for reselection. Reselection timers are specified for normal operation and when mobility-specific reselections are implemented (t-ReselectionEUTRA-SF).

According to 3GPP [12], high and medium mobility are optional features in LTE, whereby reselection parameters can be modified from those associated with the normal state. If configured in SIB3, the UE will count the number of reselections that occur within predefined timers that hence detect the scaling needed for reselection. Upon entering either of the mobility states, the Q_{hyst} and $T_{reselection}$ of the serving cell are adjusted by values defined in the SIBs for each speed state.

For non-intra-frequency searches, the UE behaves differently based on the priority of the neighboring system. The UE performs evaluation of non-intra-frequency systems only if their priority is included in the system information. When the other system has a higher priority, the UE always evaluates the defined neighbor cells. Reselection is performed if the measured quality of the neighbor cell is above a signaled threshold ($Thresh_{x,high}$) regardless of the quality of the serving cell.

For equal priority (inter-frequency only) systems, an identical mechanism to that of intra-frequency reselection is followed whereby the neighbor cell must be ($Q_{hyst} + Q_{offset}$) better than the serving cell for reselection to take place. Note that inter-RAT systems cannot be defined to have the same priority as an E-UTRA system.

For lower priority neighboring systems, reselection can be triggered only if the serving cell falls below a fixed threshold ($Thresh_{serving,low}$) and the neighbor cell is detected above a fixed threshold ($Thresh_{x,low}$). All of the conditions described must be satisfied for a period of time defined by the appropriate reselection timer.

3.4.2.1 E-UTRAN to UTRAN Cell Reselection Rules

Figure 3.43 shows the overall E-UTRAN to UTRAN cell reselection procedure as specified in [12].

SIB Type 6 defines reselection parameters for UTRAN. Reselection from LTE is always based on the comparative priority of the other RAT. LTE and any other RAT cannot be defined to have the same priority, and other RATs will not be considered for reselection unless a priority (cell reselection priority) is specified in SIB 6.

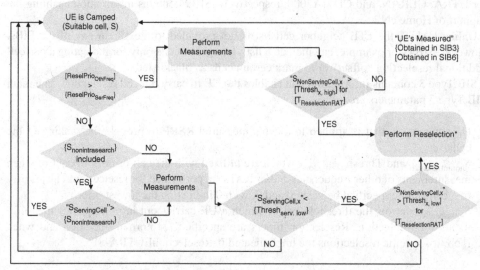

Figure 3.43   Inter-RAT cell reselection procedure.

Based on the priority level defined in SIB 6, the UE will use the RAT-specific value of either $Thresh_{x,high}$ or $Thresh_{x,low}$ (and $Thresh_{serving,low}$) to decide if reselection takes place. Additionally, each RAT will have an associated value for $T_{reselection}$ as well as the optional mobility state related parameters (discussed previously) and maximum allowed transmit power.

Moreover, the maximum UE transmit power for UTRAN is specified, along with the suitability related values of $Q_{rxlevmin}$ and $Q_{qualmin}$ for RSCP and E_c/N_o, respectively.

3.4.2.2 UTRAN Cell Measurement Rules While Camped on E-UTRAN

As shown in Figure 3.43, once the serving cell measurement goes below the $S_{nonintrasearch}$ parameter defined in SIB 3, the UE is required to measure the UTRAN cells, if SIB 6 is defined. The measurement periods are regulated in the standard in ways that avoid unnecessary measurements and maintain the UE battery lifetime.

In addition to the $S_{nonintrasearch}$ parameter, 3GPP defines the rules for the UE to schedule the UTRAN measurements. In general, the UTRAN measurements while the UE is camped on LTE are regulated depending on the LTE DRX (discontinuous reception) cycle length (or paging cycle defined in SIB 2). The 3GPP standard provides the following requirements for UTRAN cell measurements while camped on LTE idle mode, as summarized in Table 3.23 [14].

The UTRAN neighbor cells need to be measured according to the following rules from [14]:

- Inter-RAT cell reselection evaluation will be performed only on those cells for which priority has been assigned
  - In idle mode in the 3GPP Release 8, eNB is only required to provide a list of frequencies on which UTRAN measurements need to be conducted in SIB 6. This means that detailed neighbor list information (PSC, primary scrambling code) is not available to the UE.

- A new UMTS cell needs to be detected by $N_{UTRA}*T_{\text{DetectUTRA,FDD}}$ seconds
- An already detected lower priority UMTS cell needs to be measured at least once every $N_{\text{UTRA}}*T_{\text{MeasureUTRA,FDD}}$ seconds
- An already detected higher priority UMTS cell needs to be measured at least once every $T_{\text{MeasureUTRA,FDD}}$ seconds
- The measurements needs to be filtered using at least two measurements
- Measurements should be at least spaced by $(N_{\text{UTRA}}*T_{\text{MeasureUTRA,FDD}})/2$
- The filtering should be such that the time constant of the filter is $< N_{\text{UTRA}}*T_{\text{EvaluateUTRA,FDD}}$
- N_{UTRA} is the number of carriers measured for UTRAN cells.

Furthermore, the UTRAN cell search by the UE is categorized into two phases. As there is no UTRAN neighbor cell list conveyed to the UE, then the UE is required to search for all 512 PSCs. Hence, one way to limit the amount of full-space search is by defining two search categories as follows:

- **Cell detection phase** – Schedules UTRAN cells for the detection phase trying to find the best cell among the 512 PSCs. The cell is considered detectable when any UTRAN cell CPICH $E_c/I_o \geq -20\,\text{dB}$ and SCH $E_c/I_o > 17\,\text{dB}$. According to Table 3.23 and the filtering rules mentioned before, the UE is required to perform a full search on the scrambling codes:
  - Once every $N_{\text{UTRA}}*48$ DRX cycle (for 0.32 s DRX)
  - Once every 24 DRX cycle (for 0.64 s DRX)
  - Once every 12 DRX cycle (for 1.28 s DRX)
  - Once every 12 DRX cycle (for 2.56 s DRX)
- **Cell measurement phase** – When any cell is considered detectable in the previous phase, the UE is required to keep measuring that particular cell for a possible inter-RAT reselection according to the rules in Figure 3.43. In this phase, the UE schedules searches for each UMTS frequency as follows:
  - Once every 8 DRX cycle (for 0.32 s DRX)
  - Once every 4 DRX cycle (for 0.64 s DRX)
  - Once every 2 DRX cycle (for 1.28 s DRX)
  - Once every 2 DRX cycle (for 2.56 s DRX).

An example of these phases is illustrated in Figure 3.44. This shows that for an LTE DRX cycle of 1.28 s, the UE performs the cell detection phase of the UTRAN cells (for one carrier defined in SIB 6, as in this example) every 12 DRX cycles, and the cell measurement phase on the detected cells is carried out every two DRX cycles.

Table 3.23   Measurement requirements for UTRAN in idle mode

DRX cycle length (s)	$T_{\text{DetectUTRA,FDD}}$ (s)	$T_{\text{MeasureUTRA,FDD}}$ (s)	$T_{\text{EvaluateUTRA,FDD}}$ (s)
0.320	30	5.12	15.36
0.640	30	5.12	15.36
1.28	30	6.4	19.20
2.560	60	7.68	23.04

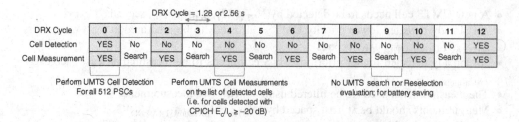

Figure 3.44 Example of UTRAN cell search mechanisms in LTE.

3.4.3 WCDMA to LTE Inter-RAT Cell Reselection

Similar to E-UTRAN to UTRAN cell reselection discussed in the previous section, the 3GPP standard also allows UTRAN to E-UTRAN cell reselection from UMTS idle, cell_PCH, or URA_PCH states. Other RAT to LTE measurements are essential in maintaining mobility back to LTE coverage. During the initial deployment of LTE, LTE cell coverage is expected to be limited to hotspots overlaid on ubiquitous 3G/2G. Under such a scenario, the mobility procedures must ensure that UE leaves the LTE when coverage degrades, and comes back when the LTE coverage conditions improve. IRAT parameters are hence the key player in ensuring the best RAT is selected, in ways not to compromise the end-user perceived performance.

3.4.3.1 UTRAN to E-UTRAN Cell Reselection Rules

The 3GPP standard in [13] provides the cell reselection rules from UTRAN to E-UTRAN. For this type of inter-RAT to take place, the SIBs broadcast in UTRAN need to define the LTE information in order for the UE to monitor the LTE cells and reselect when criteria are fulfilled.

3GPP in [15] defines a new SIB type 19 (SIB-19) for broadcasting the LTE related information. Figure 3.45 explains the fields broadcast in SIB 19 from the serving UTRAN cell while in idle, cell_PCH, or URA_PCH states.

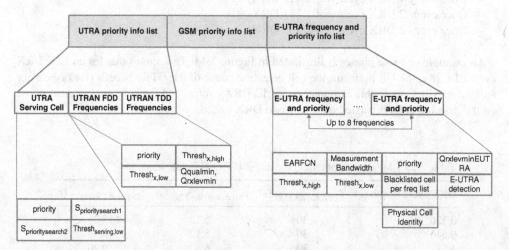

Figure 3.45 System information block type 19, 3GPP.

If SIB-19 is broadcast with a list of E-UTRAN frequencies and priorities, the UE will measure and reselect to the LTE E-UTRAN cells according to the rules shown in Figure 3.46.

As explained in the figure, the cell reselection rules depend on the priority settings of the serving UTRAN frequency and the E-UTRAN priority defined in SIB 19, and referred to as absolute priority cell reselection in [13]. The UE follows the absolute priority reselection according to the criteria defined in [13]:

- **Criterion 1**: the $S_{rxlevnonServingCell,x}$ of a cell on an evaluated higher absolute priority layer is greater than $Thresh_{x,high}$ during a time interval $T_{reselection}$
- **Criterion 3**: $S_{rxlevServingCell} < Thresh_{serving,low}$ or $S_{qualServingCell} < 0$ and the $Srxlev_{nonServingCell,x}$ of a cell on an evaluated lower absolute priority layer is greater than $Thresh_{x,low}$ during a time interval $T_{reselection}$.

Cell reselection to a cell on a higher absolute priority layer than the camped frequency should be performed if criterion 1 is fulfilled. Cell reselection to another UTRAN inter-frequency cell on an equal absolute priority layer to the camped frequency should be performed if criterion 2 is fulfilled (not shown above since not related to inter-RAT).

Reselection to a cell on a lower absolute priority layer than the camped frequency should be performed if criterion 3 is fulfilled. If more than one cell meets the criteria, the UE should reselect the cell with the highest $S_{rxlevnonServingCell,x}$. The UE should not perform cell reselection until more than one second has elapsed since the UE camped on the current serving cell. For UE in RRC connected mode states cell_PCH or URA_PCH the interval $T_{reselections,PCH}$ applies, if provided in SIB4 [15].

The reselection rules discussed here are specified in 3GPP Release 8. 3GPP Release 9 provides slightly different criteria for introducing cell reselection based on RSRQ measurements. However, the same concepts discussed here also apply, but with different parameters.

3.4.3.2   E-UTRAN Cell Measurement Rules while Camped on UTRAN

The 3GPP Release 8 in [16] provides different E-UTRAN measurement rules while UE is camped on UTRAN cells in idle mode, cell_PCH, or URA_PCH. The UE must be able to identify new E-UTRA cells and perform RSRP measurements of identified E-UTRA cells, if carrier frequency information is provided by the serving cell, even if no explicit neighbor list with physical layer cell identities is provided.

There are mainly two different rates of E-UTRAN cell search defined in [16] to prevent high consumption of UE battery life. If $S_{rxlevServingCell} > S_{prioritysearch1}$ and $S_{qualServingCell} > S_{prioritysearch2}$ then the UE should search for E-UTRA layers of higher priority at least every $T_{higherprioritysearch}$ where $T_{higherprioritysearch}$ is "60*N_layer" seconds, where N_layer is the number of high priority frequencies across all RATs. Hence, the purpose of this procedure is to savebattery consumption by reducing the number of UE measurements.

If $S_{rxlevServingCell} \leq S_{prioritysearch1}$ or $S_{qualServingCell} \leq S_{prioritysearch2}$ then the UE must search for and measure E-UTRA frequency layers of higher or lower priority in preparation for possible reselection. In this scenario, the minimum rate at which the UE is required to search/measure higher priority layers is the same as that defined for lower priority layers. This procedure is summarized in Figure 3.47.

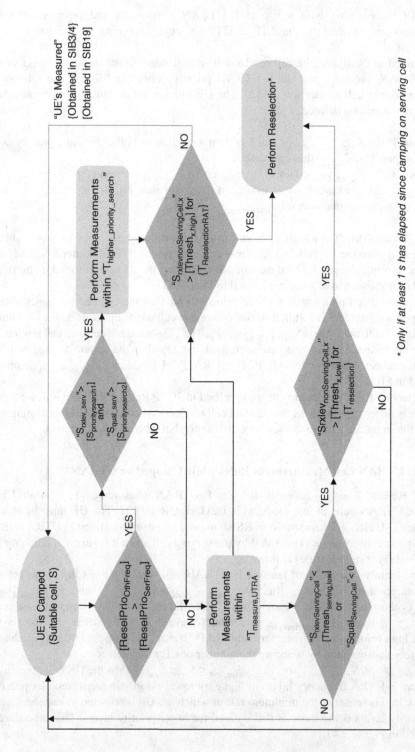

Figure 3.46   Inter-RAT cell reselection procedure.

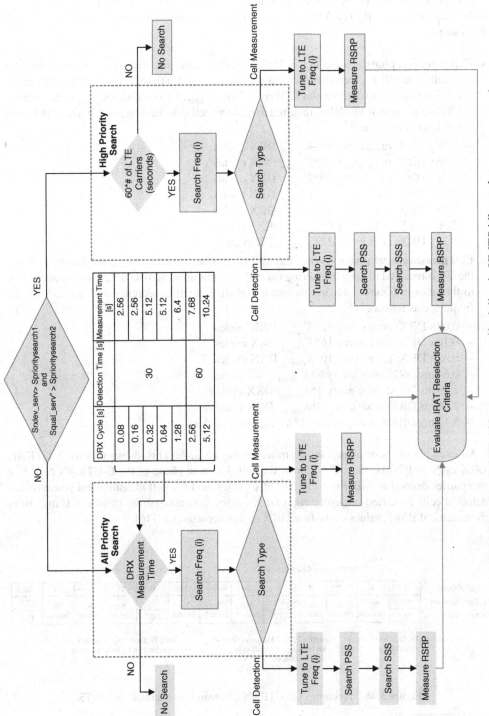

Figure 3.47 E-UTRAN measurement scheduling in UMTS idle mode.

Similar to the search phases described in the previous section for UTRAN, 3GPP also allows different searches for E-UTRAN cells, in addition, to reduce the amount of full-space (504 PCI) searches:

- **Cell detection phase** – Schedules E-UTRAN cells for the detection phase trying to find the best cell. The cell is considered detectable when any E-UTRAN cell RSRP ≥ -123 dBm. The UE is required to detect a new LTE cell within $K_{carrier} * 30$ seconds ($K_{carrier}$ is the number of LTE frequencies). The UE is required to perform cell detection on E-UTRAN cells within the following timelines[13]
  - 0.08 s DRX: once every $96*K_{carrier}$ DRX cycles
  - 0.16 s DRX: once every $48*K_{carrier}$ DRX cycles
  - 0.32 s DRX: once every $24*K_{carrier}$ DRX cycles
  - 0.64 s DRX: once every $24*K_{carrier}$ DRX cycles
  - 1.28 s DRX: once every $12*K_{carrier}$ DRX cycles
  - 2.56 s DRX: once every $6*K_{carrier}$ DRX cycles
  - 5.12 s DRX: once every $3*K_{carrier}$ DRX cycles
- **Cell measurement phase** – When any cell is considered detectable in the previous phase, the UE is required to keep measuring the cell for a possible inter-RAT reselection according to the rules in Figure 3.46. In this phase, the UE schedules searches for each E-UTRAN frequency as follows [16]:
  - 0.08 s DRX: once every $32*K_{carrier}$ DRX cycles
  - 0.16 s DRX: once every $16*K_{carrier}$ DRX cycles
  - 0.32 s DRX: once every $16*K_{carrier}$ DRX cycles
  - 0.64 ms DRX: once every $8*K_{carrier}$ DRX cycles
  - 1.28 ms DRX: once every $4*K_{carrier}$ DRX cycles
  - 2.56 ms DRX: once every $2*K_{carrier}$ DRX cycles
  - 5.12 ms DRX => once every $1*K_{carrier}$ DRX cycles.

An example of these phases is illustrated in Figure 3.48. This shows that for a UTRAN DRX cycle of 0.640 s, the UE performs the cell detection phase of the E-UTRAN cells (for one carrier defined in SIB 19) every 24 DRX cycles, and the cell measurement phase on the detected cells is carried out every eight DRX cycles. It is noteworthy that the UE may filter the measured RSRP values every four DRX cycles according to [16].

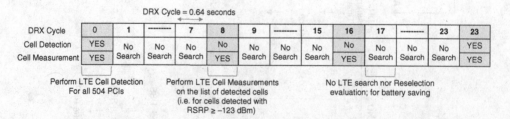

Figure 3.48   Example of E-UTRAN cell search mechanisms in UMTS.

---

[13] The DRX cycle in UTRAN can be different than that in the E-UTRAN case.

3.5   Inter-RAT Cell Reselection Optimization Considerations

3.5.1   SIB-19 Planning Strategy for UTRAN to E-UTRAN Cell Reselection

SIB-19 is one of the SIBs for which the optimization process is profound, in order to control the inter-RAT cell reselection from UTRAN to other priority layers such as LTE. LTE frequencies can be deployed in hotspots or overlaid with other UTRAN frequencies, therefore, confining the UE measurements through SIB-19 planning becomes one of the network planning aspects for a better user experience.

The example in Figure 3.49 illustrates an LTE system deployed in a hot-spot, with ubiquitous UMTS coverage. In this example, the LTE cells are assumed to be configured with a higher absolute priority cell reselection than UMTS frequencies. When the mobile is at location-1 while in LTE idle, UE is camped on the LTE cell from site-1. By the time the mobile moves toward location-2, the idle mode inter-RAT cell reselection from LTE to UMTS is triggered, according to the reselection rules discussed in the previous section.

At the coverage area of location-2, the UE camps to the cell from the UMTS site-2. SIB-19 on the UMTS cell in site-2 is defined and, therefore, when the UE moves from location-2 back to location-1, the inter-RAT cell reselection to LTE can be triggered. Both UMTS site-2 and LTE site-1 have a neighboring relation and hence SIB-19 is expected to be broadcast to allow the UE to evaluate the inter-RAT cell reselection, depending on the mobility movement. This is a case where LTE site-1 and UMTS site-2 overlap in coverage areas.

Alternatively, when the user that is camped on the UMTS cell in the coverage area of site-2 moves toward location-3, it is expected that the UE will perform UMTS intra-frequency cell reselection. At location-3, site-3 is considered in a second tier of LTE coverage from site-1. Hence, LTE coverage is not expected to be strong in this location. A troubled scenario in this location may occur if SIB-19 is configured on site-3. In this case, the device is required to perform a high priority measurement on the LTE cell, as instructed in SIB-19 of site-3. This leads to a negative impact on the users' battery lifetime, as the UE will always search for an LTE cell, but without being able to perform the inter-RAT reselection when the LTE service coverage is very weak at location-3. The issue manifests itself when the LTE layer is set at higher absolute priority than the UMTS cell.

The continuous LTE measurements in that location prove to be costly on device battery, without any possible reselection. Meanwhile, a troubled scenario is magnified if the LTE coverage from site-1 reaches the coverage area of UMTS in site-3. In the latter scenario, the device can

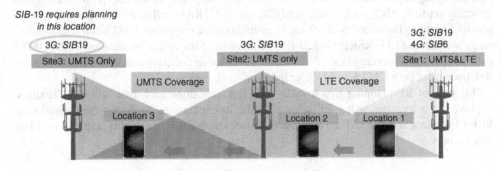

Figure 3.49   WCDMA and LTE coverage example.

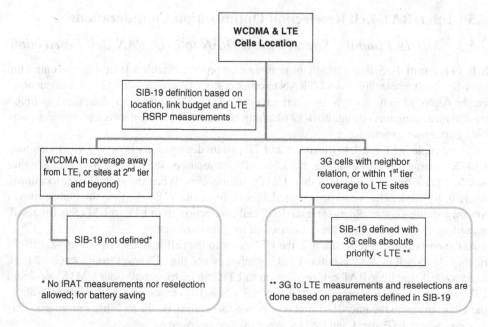

Figure 3.50   WCDMA SIB-19 planning strategy.

reselect from UMTS site-3 to a very weak LTE cell in site-1, causing a significant interruption to data services, and possible voice call failures, once camped on a weak LTE coverage.

In essence, the planning of the SIB-19 is important in such a deployment scenario wherein LTE is a hotspot, while its layer priority is higher. The higher layer priority is expected as a planning method for the users with an LTE capable multi-mode device to remain on the LTE service as long as possible, and come back to LTE as soon as possible. To confine the LTE hotspot coverage in this scenario, the SIB-19 can be defined within the first tier from LTE coverage. In this way, the device camps on the high speed LTE service without compromising the battery lifetime.

Figure 3.50 summarizes the SIB-19 planning in a deployment, to overcome such a problem. The planning of SIB-19 is recommended to be done for each UMTS site. This is possible by performing RSRP measurements within the LTE network. A satisfactory RSRP level is typically set to levels ≥ -120 dBm, matching the E-UTRAN cell detection discussed in the previous section. Therefore, SIB-19 can be enabled on the respective UMTS cell within a coverage area having LTE RSRP ≥ -120 dBm. Otherwise, SIB-19 can be disabled in the UMTS cells having RSRP coverage of < -120 dBm. This planning should also take into account the link budget for both UMTS and LTE in indoor coverage areas.

This type of RF planning tied to SIB-19 is typically done each time a new LTE site is deployed in a location, using RF scanner tools. In conclusion, confining the definition of SIB-19 within a suitable LTE coverage area ensures a good service quality and end-user battery savings.

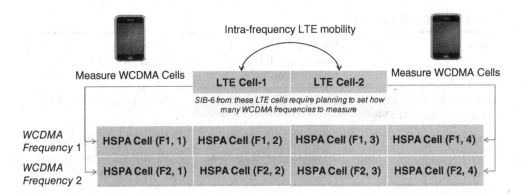

Figure 3.51 WCDMA multi-carrier deployment interaction with LTE coverage.

3.5.2 SIB-6 Planning Strategy for E-UTRAN to UTRAN Cell Reselection

Another viable scenario impacting the device performance during inter-RAT measurements in idle mode is SIB-6 planning. As explained in previous sections, 3GPP allows eNBs to configure multiple UTRAN FDD frequencies (up to 16 of them). Based on this information, the UE is required to monitor all defined UTRAN frequencies for possible inter-RAT reselection.

Assuming a deployment scenario in Figure 3.51, where two UMTS carriers are deployed in one-to-one blanket overlay (i.e., two carriers are present in all UMTS sites depicted in this example as frequency F1, and frequency F2). When UE camps on LTE cells, such as LTE cell-1 or cell-2, then SIB-6 from any of these cells are broadcast to the UE to perform inter-RAT reselection, once moving away from the LTE coverage area. The typical question raised in this situation is "how many WCDMA (wideband code division multiple access) carriers need to be configured in LTE's SIB-6 for inter-RAT measurements?".

Release 8 inter-RAT reselection criteria depend on the measurements of the LTE serving cell RSRP as well as the candidate WCDMA cell's RSCP, and according to this example of deployment, the two WCDMA carriers are overlaid, hence cells in both carriers collocated in the same site are expected to radiate the same RF, in return, a similar measured RSCP. Thus, defining only one of the WCDMA carriers for inter-RAT reselection evaluation from LTE can be sufficient to avoid extensive device battery drainage.

From the example under discussion, a troubled scenario can occur if SIB-6 configures both WCDMA carriers. In this case, the UE is mandated to search (i.e., detect and/or measure) two WCDMA carriers on the configured search occasion. This undesirably leads to negative impact on the battery without a substantial gain in the reselection performance.

The authors have collected field measurement statistics in a deployed LTE network with two UMTS carriers. In one test run, all second-carrier UMTS frequency has been excluded from SIB-6 definitions. In another run, both UMTS frequency carriers have been included in SIB-6. The results of the trial are shown in Table 3.24.

As shown in the results of this trial, as RSCP is the main reselection criteria in 3GPP Release 8, the UE reselects to the first measured UMTS cell, while unnecessarily measuring

Table 3.24 SIB-6 planning trial results

Results	SIB-6 defines two UMTS carriers for UE IRAT measurements	SIB-6 defines only first UMTS carrier for UE IRAT measurements
Average device wake-up time to measure 3G cells	134 ms for each DRX cycle	58 ms for each DRX cycle
Average RSRP for each carrier	3G carrier_1 = − 82.1 dBm 3G carrier_2 = −82.9 dBm	3G carrier_1 = −82.0 dBm 3G carrier_2 = "not defined for measurements"

the other carrier with a similar RSCP (both carriers are 1–1 overlay). The UE therefore needs an average of 76 ms in extra wake-up time in idle mode to measure both carriers, when compared to the other trial configured with only one UMTS carrier. This extra wake-up time translates to extra battery consumption and without any substantial gain in the behavior of the inter-RAT reselection.

LTE Release 9 defines both RSCP and/or E_c/N_o while measuring the UTRAN carriers for inter-RAT reselection. In a Release 9 network, it may be beneficial to enable both UMTS carriers in SIB-6 if the load is significantly imbalanced between the two. However, due to the known fluctuations of E_c/N_o in the presence of highly loading HSDPA activities, then depending on E_c/N_o only for reselection may lead to terminating the LTE coverage sooner than intended. Hence, SIB-6 planning must take into account the trade-off between the number of UTRAN frequency carriers to configure, using reselection criteria based on RSCP or E_c/N_o or both, and the UE battery consumption. The clear recommendation is that any reduction in the UTRAN measurements can give great improvement in the device battery lifetime.

3.5.3   Inter-RAT Case Studies from Field Test

This section covers examples of inter-RAT cell reselection issues in a commercial deployment, as well as a troubleshooting process used to identify and address the troubled case. It is worth noting that the design of the inter-RAT parameters should address both cases of UTRAN to E-UTRAN and vice versa concurrently, to leverage the performance to the expected targets.

3.5.3.1   Case Study 1: Inter-RAT Cell Reselection Ping-Pong

Tables 3.25–3.27 list the inter-RAT cell reselection parameters being evaluated in this case study. They are all the relevant parameters needed for inter-RAT reselection between E-UTRAN and UTRAN, as discussed in previous sections.

Based on the parameter values Tables 3.25–3.27, the UE measures and evaluates inter-RAT cell reselection criteria from E-UTRAN to UTRAN or UTRAN to E-UTRAN according to the sequence of events in Figures 3.52 and 3.53, respectively.

The device measurements in this case study are taken from a multi-mode device supporting LTE and UMTS. In indoor stationary RF conditions, wherein the device does not perform any mobility movement, a high amount of inter-RAT reselection ping-pongs between the two RATs is observed.

Table 3.25 UTRAN to E-UTRAN cell reselection parameters

Parameter	UMTS SIB	Value	Converted value
DRX cycle in idle	1	6	640 ms
$Q_{rxlevmin}$	3/4	−58	−115 dBm
$Q_{qualmin}$	3/4	−18	−18 dB
$Thresh_{serving, low}$ (RSCP)	19	1	2 dB
$Thresh_{x, high}$ (RSRP)	19	9	18 dB
EUTRA $Q_{rxlevmin}$ (RSRP)	19	−70	−140 dBm
$S_{prioritysearch1}$	19	2	4 dB
$S_{prioritysearch2}$	19	2	2 dB
$T_{reselection}$	3/4	1	1 s

Table 3.26 E-UTRAN to UTRAN cell reselection parameters

Parameter	LTE SIB	Value	Converted value
$Q_{rxlevmin}$	6	−58	−115 dBm
$Q_{qualmin}$	6	−22	−22 dB
$P_{MaxUTRA}$	6	33	33 dBm
$S_{non\text{-}intrasearch}$	3	8	16 dB
$Thresh_{serving, low}$	3	7	14 dB
$Thresh_{x, low}$	6	6	12 dB
$Thresh_{x, high}$	6	0	0 dB
$Treselection_{UTRAN}$	6	1	1 s
$Q_{rxlevmin}$	3	−60	−120 dBm
P_{Max}	3	Absent	23 dBm

Table 3.27 E-UTRAN to/from UTRAN priority parameters

Parameter	SIB	Value
E-UTRAN → UTRAN		
WCDMA (per carrier)	LTE 6	4
LTE	LTE 6	5
UTRAN → E-UTRAN		
WCDMA (all carriers)	WCDMA 19	4
LTE	WCDMA 19	5

Figure 3.54 shows the amount of reselections that took place over the duration of the test.[14] With the sequence of events in Figures 3.52 and 3.53, it is observed that in only 7 min of test, UE triggered a total of 15 ping-pong IRAT reselections, at a rate of one IRAT reselection every 2 min. This is clearly impacting the device battery negatively.

---

[14] In the remaining parts of this chapter, L2W refers to "LTE to WCDMA" reselection or handover. And W2L refers to "WCDMA to LTE" reselection or handover.

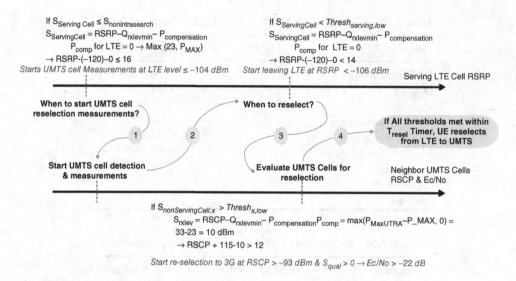

Figure 3.52   E-UTRAN to UTRAN cell reselection behavior.

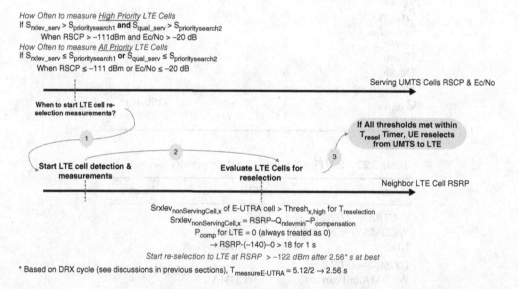

Figure 3.53   UTRAN to E-UTRAN cell ee-selection behavior.

Additionally, every L2W or W2L reselection requires a registration process on the new RAT, which leads to an increase of the signaling traffic on the respective RAT and the user reachability. During each registration process in any of the RAT, the UE needs to register its presence in the core network, and hence the user is unavailable to make or receive data or voice calls.

This ping-pong situation is mainly the cause of suboptimal inter-RAT parameter settings configured in this trial. Upon leaving LTE at a relatively acceptable coverage to go to UMTS,

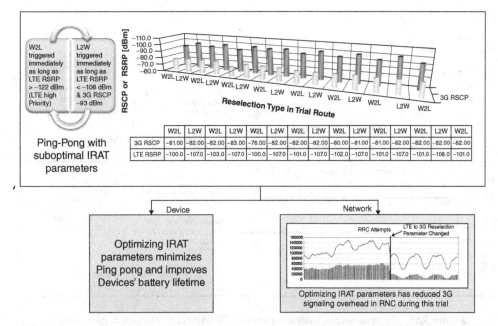

Ping-Pong with suboptimal IRAT parameters

	W2L	L2W	W2L	L2W	W2L	L2W	W2L	L2W	W2L	L2W	W2L	L2W	W2L	L2W	W2L
3G RSCP	–81.00	–82.00	–82.00	–83.00	–76.00	–82.00	–82.00	–82.00	–80.00	–81.00	–81.00	–82.00	–82.00	–82.00	–82.00
LTE RSRP	–100.0	–107.0	–103.0	–107.0	–100.0	–107.0	–101.0	–107.0	–102.0	–107.0	–101.0	–107.0	–101.0	–108.0	–101.0

Device

Optimizing IRAT parameters minimizes Ping pong and improves Devices' battery lifetime

Network

Optimizing IRAT parameters has reduced 3G signaling overhead in RNC during this trial

Figure 3.54 Ping-pong reselection behavior between UTRAN and E-UTRAN in a stationary test location. Positive impact on the UMTS RRC attempts after modifying the network parameters.

the device tends to measure and reselect back to LTE quickly, due to higher LTE absolute priority settings, hence the ping-pong occurs. The case reveals that in an indoor situation where LTE's measured RSRP is relatively less than the WCDMA measured RSCP.

Addressing the inter-RAT parameters in this scenario is significant to reduce the amount of ping-pongs impacting user battery and reachability, in addition to increasing the signaling traffic. Modifying the inter-RAT parameters (see next section for parameter setting trade-off), has proved to give a significant reduction in UMTS signaling attempts, by 50%, when the ping-pongs are reduced, as shown in the lower part of Figure 3.54.

3.5.3.2 Case Study 2: Continuous UMTS Cell Measurements while Device Camped on LTE

Using the same parameters in Table 3.26 (detailed further in Figure 3.52), another troubled scenario is analyzed in this case study. This scenario takes place mainly in indoor locations where both LTE and UMTS cell measurements are weak.

The upper part of Figure 3.55 shows the measured LTE's RSRP and UMTS's RSCP, along with the reselection criteria. In this situation, the UE is camped on the LTE cell first. The figure illustrates that when both LTE and 3G are weak, UE performs continuous 3G measurements, but without any possible IRAT reselections to execute, according to L2W reselection parameters used in this trial.

This occurs because the criterion for starting UMTS measures is set at RSRP ≤ -104 dBm, while the reselection criteria for executing the reselection is set at RSCP > -93 dBm. With the measured signals shown in the figure, it effectively means that the UE cannot reselect to the

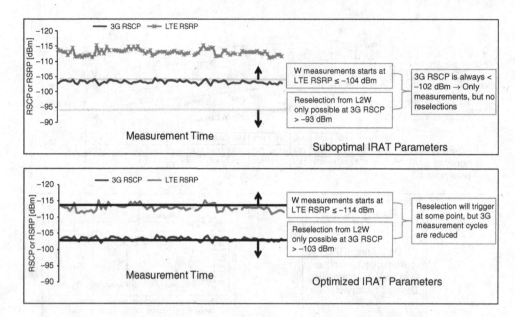

Figure 3.55   Continuous 3G measurements in LTE without reselection in case study 2.

LTE cell as long as its RSCP is ≤ – 93 dBm (average RSCP here is –104 dBm). At the same time, the UE is always measuring the UMTS cell because the criteria for "when to start UMTS measurements" is set to RSRP ≤ – 104 dBm (average RSRP here is –114 dBm).

The cost of measuring UMTS cells while in LTE without any possible reselection clearly impacts the battery lifetime. Hence parameter reviews are necessary in this scenario.

On the other hand, the lower part of Figure 3.55 shows the adjusted parameters recommended to be set in this scenario in order to improve the UE's battery life. This part of the figure shows that by modifying "when to start UMTS measurements" to RSRP ≤ – 114 dBm (average RSRP here is –114 dBm); there is a limited number of UMTS searches, as listed in Table 3.28 (reduced from 79 searches to 28 only in the same location). And by modifying the executing criterion for the reselection to RSCP > – 103 dBm, the IRAT reselection is triggered when the LTE cell becomes unsuitable. Refer to the next section for the list of the parameters that have been modified to fix this issue.

3.5.3.3   Case Study 3: Service Outage in LTE Idle Mode

An additional legitimate consideration, while designing the E-UTRAN to UTRAN inter-RAT cell reselection parameter, is the impact of the LTE intra-frequency parameter on the overall

Table 3.28   The impact of IRAT parameter change on the discussed scenario in this trial

	Before change	After change
Number of WCDMA cell measurements in 2 min	79	28
Number of WCDMA measurements is reduced by > 100%, providing a good saving on battery		

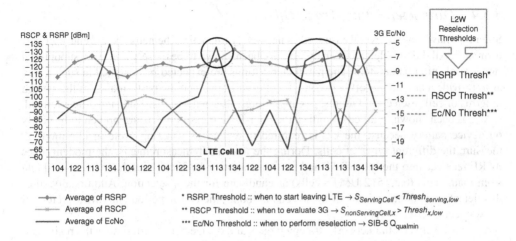

Figure 3.56 Service outage in idle mode example for case study 3.

cell reselection behavior. As long as the LTE layer priority is higher than the UMTS layer priority, LTE's intra-frequency reselection is typically triggered first in the same measurement cycle when other LTE intra-frequency and UMTS cells are ranked higher than the LTE serving cell.

Let us consider the example in Figure 3.56 where the UE is camped on an LTE cell (shown as PCI in the figure). Thus, within a measurement cycle (i.e., LTE DRX cycle) the UE can measure both the intra-frequency and inter-RAT cells, as long as the serving LTE cell is below the measurement thresholds. In this case study, a trouble scenario is analyzed further.

In the figure, the LTE serving cell is very weak, and UMTS RSCP meets the E-UTRAN to UTRAN reselection threshold but, as shown in the circled areas, an inter-RAT reselection is not triggered, instead an intra-frequency reselection (e.g., from PCI 134 to 113) always occurs between cells with very low RSRP. This scenario occurs when the intra-frequency cell reselection criteria allows a low hysteresis (Q_{hyst}). In this example, Q_{hyst} is set to 1 dB. Even if the inter-RAT reselection parameters are optimized but the Q_{hyst} is set to such a low value, the troubled situation can still occur.

When LTE's Q_{hyst} for intra-frequency is set to 1 dB, at the time when the UE starts $T_{reselection}$ for the better ranked UMTS cell (shown in the area in the figure), the $T_{reselection}$ for the LTE intra-frequency has already started. The UE hence performs LTE intra-frequency cell reselection between very weak LTE cells (with many ping-pongs) preventing E-UTRAN to UTRAN inter-RAT reselection from taking place. Eventually, the UE only reselects to the UMTS cell when the LTE serving at RSRP reaches −130 dBm, a level close to out of service. During this time, a user attempting to initiate data service or voice call may very well experience service interruption.

Therefore, when the LTE intra-frequency Q_{hyst} parameter is too small, the UE may perform frequent cell reselection to cells only marginally better than the current serving cell, which may lead to excessive battery consumption, as well as remaining on a weak LTE coverage for a longer time instead of reselection to a better RAT. Considering a higher hysteresis value for the ranking of the serving cell during cell reselection may prevent this type of ping-pong and hence improve the end-user perceived experience.

3.5.4 Parameter Setting Trade-Off

So far, we have shown several case studies negatively impacting the network and end-user performance. All the troubled areas analyzed have been shown to be closely related to suboptimal IRAT reselection parameters. The parameter settings are expected be designed in such a way as to provide a good trade-off between UE battery lifetime and the performance of the system in terms of cell reselection failures, paging failures, user reachability, and signaling overload.

As has been shown in all case studies, the inter-RAT measurements are proven to be costly on device battery because the UE is required to extend its wake-up cycle in idle mode to measure the different types of cells. During the inter-RAT measurements, the modem tunes its RF receiver into the other RAT and performs another RAT cell search, trying to find one from many cells (e.g., 512 UMTS cells) as candidate for the reselection. Additionally, after this detection is done and one of those cells is found, the cell is periodically measured until the reselection is executed.

A balance between the network strategy targets and device battery lifetime is the main contributor to the design of the inter-RAT parameters. The general rule of thumb to follow during the parameter optimization is to allow a good range of hysteresis between the different RATs. These hysteresis values should be defined based on link budget calculations.

Table 3.29 shows the relations between LTE RSRP and UMTS RSCP at different morphologies, assuming a 5 MHz channel (for both UMTS and LTE) in order to use a practical comparison within the same bandwidth. The LTE received signal power is −18 dBm (1800 MHz band) and the UMTS CPICH power is 33 dBm (2100 MHz band). The received signal level values are based on link budget calculations shown in Chapter 6.

The received signal levels can also be validated from a field test. The mobile can be placed at the edge of coverage in each RAT and, using a scanner or UE measurements, the values for such comparisons can be derived. Then, the downlink throughput is measured to assess which of the RATs provides a better end-user experience. These factors are then used during the final decision on the inter-RAT parameter settings that will provide the best data speed, less reachability issues, and best battery consumption levels. Table 3.30 gives an example of an output from such measurements. These measurements are performed in the exact same locations of the troubled areas as previously discussed in the three case studies.

The results in Table 3.30 indicate that the downlink speed at the LTE coverage edge can outperform that of UMTS. Hence, the LTE layer should still be set to higher absolute priority than the UMTS layer(s). The parameter design should ensure that an LTE device will stay on LTE as long as possible and come back to LTE when coverage is reasonable.

Table 3.29 LTE RSRP vs. UMTS RSCP in different morphologies (5 MHz channel)

Morphology	RSRP (dBm)	CPICH RSCP (dBm)
Rural	−122.14	−112.8
Suburban	−115.14	−105.8
Urban	−107.1	−98.3
Dense urban	−107.1	−98.3

Table 3.30 LTE and UMTS downlink speeds in different RF conditions

	Weak LTE coverage, weak UMTS coverage	Weak LTE coverage, good UMTS coverage
UMTS RSCP (dBm)	Average = −103.2, Max = −100, Min = −106	Average = −82.9, Max = −78, Min = −87
LTE RSRP (dBm)	Average = −113, Max = −111, Min = −115	Average = −104, Max = −100, Min = −111
UMTS DL speed (Mbps)	0.662	9.556
LTE DL speed (Mbps)	11.731	19.984

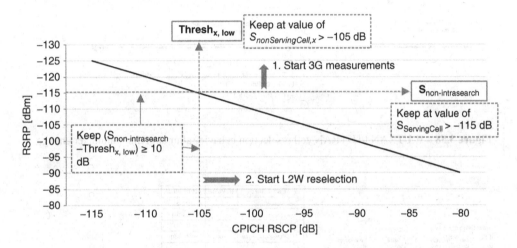

Figure 3.57 When to start 3G measurements and when to reselect to 3G.

Based on all of these discussed methodologies, the main inter-RAT parameter design can be set as shown in Figure 3.57.

In urban hotspot LTE deployment, as shown in Table 3.29, the 10 dB difference between RSRP and RSCP can be used to reflect on the IRAT parameter settings. By keeping the RSRP level of the UMTS cell, and the LTE $RSRP \geq 10$ dB, the UE can then avoid any type of ping-pong or unnecessary measurements without reselection. Also keep in mind that when coming back from UMTS onto LTE, the parameters (such as $Thresh_{x, high}$) must be kept in line to avoid going back to UMTS unnecessarily. Therefore, maintaining a sufficient gap between $Thresh_{serving, low}$ and $Thresh_{x, high}$ within 4 dB or higher, can provide a robust L2W and W2L reselection triggering for reducing ping-pongs.

From the case studies 1 and 2 in previous sections, the troubled scenarios can be fixed by using the parameter settings shown in Figures 3.58 and 3.59, for E-UTRAN to UTRAN and UTRAN to E-UTRAN, respectively. This takes into account the rule of thumb as used in Figure 3.57.

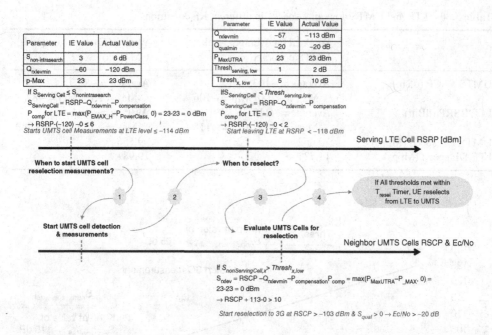

Figure 3.58 E-UTRAN to UTRAN cell reselection behavior after applying parameter change.

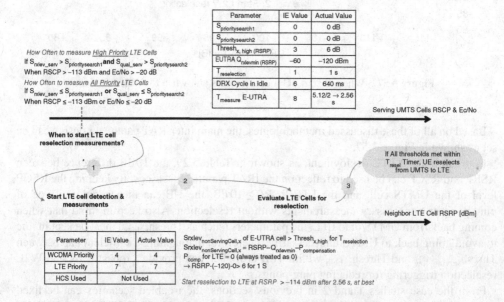

Figure 3.59 UTRAN to E-UTRAN cell reselection behavior after applying parameter change.

Applying the parameter changes shown in these figures has been proved to eliminate the inter-RAT ping-pong situation in case study 1, as well as to reduce the number of UMTS measurements in LTE by > 100%, addressed by case study 2.

While this is an example of optimization enhancements, the same ideas can be applied to any network optimization process for inter-RAT. The conclusion of such rounds of optimization is to maintain a good user experience within LTE coverage without compromising the battery lifetime of a device in the network. As part of such optimization cycles, field measurements should be repeated after each step of optimization, checking areas such as:

- Parameter setting to improve the rates of intra-frequency, and inter-RAT cell reselection failures, and ping-pongs.
- Parameter setting to improve the success rate of system access (e.g., success rate for the transitions from idle mode to connected mode).
- Parameter setting to improve the device battery and user data speeds.
- Parameter setting to improve the paging success rate and user reachability, in addition to signaling overhead in each RAT.

3.6   LTE to LTE Inter-Frequency Cell Reselection

3.6.1   LTE Inter-Frequency Cell Reselection Rules

Inter-frequency reselection is based on absolute priorities. The UE tries to camp on the highest priority frequency available. These priorities are provided in LTE SIB-5 and are valid for all UEs in the serving cell. In addition, specific priorities per UE can be signaled in the RRC connection release message, known as dedicated priority.

Only the frequencies listed in SIB-5 are considered for inter-frequency reselection. This list can contain a maximum of eight inter-frequencies the UE may be allowed to monitor within E-UTRAN. The parameters provided in the SIB-3 are also considered for ranking evaluations.

For inter-frequency neighboring cells, it is possible to indicate the cell specific offset to be considered during reselection. These parameters are common to all cells on a different frequency. Blacklists can be provided to prevent the UE from reselecting to specific intra-and inter-frequency. Cell reselection also can be speed-dependent, as explained in previous sections.

3.6.2   LTE Inter-Frequency Optimization Considerations

3.6.2.1   SIB-5 Planning for LTE Inter-Frequency Cell Reselection

Similar to the planning criteria discussed for SIB-19 and SIB-6, a planning for inter-frequency SIB-5 definitions is necessary and for the same reasons discussed before. The same symptoms of issues (e.g., continual LTE inter-frequency measurements without possible reselection) can typically occur if SIB-5 is defined in areas where LTE inter-frequency cells do not exist.

Figure 3.60 illustrates some of the optimization methodologies to follow for the planning of SIB-5 in a cell-by-cell evaluation. The nature of the other LTE carrier in terms of the location of deployment (indoor or outdoor) is a key factor for such planning effort.

The other aspect of SIB-5 planning is the design of the absolute priority setting on the different LTE inter-frequency layers, and their associations with the intra-frequency and inter-RAT.

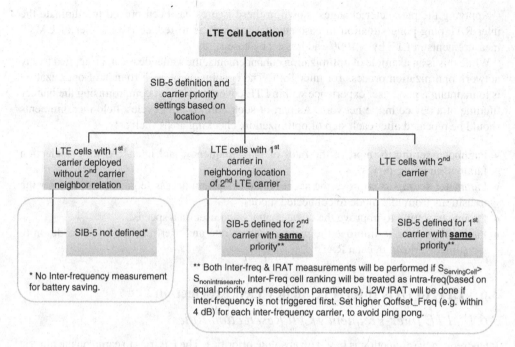

Figure 3.60   SIB-5 planning for LTE inter-frequency cell reselection.

For inter-frequency cell reselection to a frequency that has equal priority to the LTE serving cell frequency, the UE evaluates the relative ranking of the serving and neighbor cells following the intra-frequency rules specified in [12]. LTE inter-frequency can also set higher or lower priorities. Such setting is not generally recommended because of the impact on the device battery and the complexity of deployment. However, high/low priority setting becomes important for home eNBs (femto cells) deployed in a different frequency or band, for which, the cells can be deployed with higher priority, for example.

3.6.2.2 LTE Inter-Frequency Parameter Discussion

For the most part, intra-frequency cell reselection is similar to that of intra-frequency for equal priorities. For a high or low inter-frequency layer, the inter-RAT cell reselection criteria applies depending on the configured $Thresh_{serving,high}$, or $Thresh_{serving,low}$.

Figure 3.61 illustrates an example of LTE inter-frequency reselection parameters for equal priority setting. The UE initially camps on the LTE F1 cell (first frequency) with physical cell ID 1. The only other neighbor defined is the F2 (second frequency) cell with physical cell ID 12. The priorities of the two frequencies F1 and F2 are defined as the same. At point 1, $S_{Servingcell} < S_{nonintrasearch}$ and thus the UE begins the search for an inter-frequency neighbor. The F2 cell is ranked the highest at point 2 because its received level is ($Q_{hyst} + Q_{offset}$) better than the serving F1 cell. The UE reselects to the F2 cell when the timer expires (set as 2 s, at point 3 in this example).

The fact that $S_{nonintrasearch}$ is common to LTE inter-frequency measurements and inter-RAT requires additional planning. One of the ways to handle this situation is by treating sites within

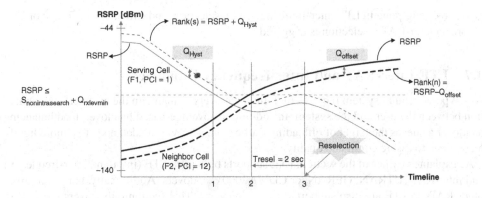

Figure 3.61   Inter-frequency reselection example (equal priority neighbor).

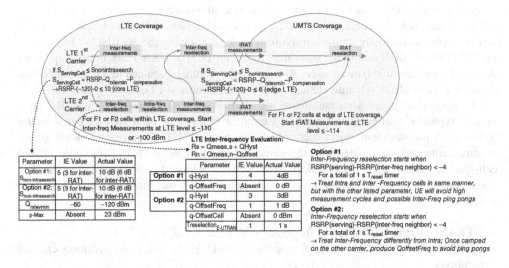

Figure 3.62   LTE Inter-frequency reselection parameters, option #1, with equal priority.

core LTE coverage, inter-frequency in particular, differently than edge LTE sites. Settings of "$S_{nonintrasearch}$" and "LTE SIB-3 $Q_{rxlevmin}$" on a per cell basis avoids situations where the UE is performing concurrent inter-RAT and inter-frequency measurements. This per-cell setting also helps in performing inter-frequency LTE reselection before inter-RAT which allows the UE to remain on LTE coverage longer.

In addition, "$Q_{offsetFreq}$" can be introduced on the inter-frequency LTE layer to treat intra- and inter-frequency cell reselection differently for equal priority inter-frequency reselection. This offset is especially important if no changes to "$S_{nonintrasearch}$" and "LTE SIB-3 $Q_{rxlevmin}$" are made on the serving LTE cell.

Figure 3.62 summarizes the flow when these parameter setting options are exercised alongside inter-RAT.

In these options (option #1 and #2 in the figure), the optimization process is expected to consider different settings from LTE F1 or F2, depending on the locations of the cells. The target is to avoid ping-pong in LTE inter-frequency cell reselection, and extend the coverage of LTE before an inter-RAT reselection is triggered.

3.7   LTE Inter-RAT and Inter-frequency Handover

Any type of cellular system requires mobility handovers to maintain the air interface connection between the user and the system. In addition to coverage-based handover, load balancing handover achieves the target of offloading traffic from a heavily loaded node to a more lightly loaded one to assure continued quality.

As explained earlier in the section, LTE supports both intra-LTE (intra-and inter-frequency) and inter-RAT (UTRAN, GERAN, or CDMA2000) handovers. Additionally, handovers from other RATs to LTE are also supported by the standard. This section only covers the LTE to WCDMA inter-RAT and LTE inter-frequency handovers. UTRAN to E-UTRAN handover in connected mode is not widely deployed because the LTE network is not voice centric (i.e., no circuit switch as in legacy 3G services); hence, handover during voice call from UTRAN cell-DCH state is not possible. Additionally, for PS (packet switched) data calls, it is expected that smartphones normally transmit infrequent packets, allowing the 3G RNC to reconfigure the UE from cell-DCH connected state idle mode (cell-PCH or URA-PCH, depending on the deployed state transitions) more often. As a result, the feature has gained less importance than the inter-RAT reselection feature, at least in the current status of LTE deployments.

E-UTRAN determines what measurements a UE is allowed to perform. The RRC layer in LTE typically configures different measurement objects for each type of event and/or handover. A measurement object corresponds to a single E-UTRA carrier frequency (intra- or inter-frequency). Associated with this carrier frequency, E-UTRAN can configure a list of cell-specific offsets and a list of "blacklisted" cells. Blacklisted cells are not considered in event evaluation or measurement reporting. For inter-RAT, a measurement object is defined as follows:

- **UTRAN**: a single UTRA carrier frequency with a list of PSCs
- **GERAN**: a set of GERAN carrier frequencies (ARFCNs, absolute radio frequency channel numbers)
- **CDMA2000**: a set of cells (defined by PN (pseudorandom noise) offsets and search window size) on a single (HRPD or 1xRTT) carrier frequency.

Inter-frequency and inter-RAT measurement objects can include measurement gaps, if required by the UE (as indicated in UE radio capability), to allow the data communications to be suspended while the UE measures the RF signal on other frequency/RAT cells. Also, for inter-RAT neighbors, E-UTRAN can configure a list of cell-specific offsets and a list of "blacklisted" cells. The details of these procedures are in [17].

E-UTRAN can configure the UE to perform radio quality measurements utilizing different measurement quantities defined for each measurement object. The definition also configures applicable filtering of the measurements that will control when a measurement report is sent to the eNB. The RF metrics (depending on the technology) are also specified in each measurement object. These can be RSRP, RSRQ, RSCP, and/or E_c/N_o.[15]

---

[15] Refer to Chapter 2, Section 2.11.7 for a comprehensive discussion on these measurements.

Measurement reporting can be configured to be either event-triggered or periodic. For event-triggered reporting, five LTE (A1 through A5) and two inter-RAT (B1 and B2) events are defined, as discussed in Chapter 2. Measurement reports are generated when the criteria associated with each event are met. Depending on the terminal's capabilities, some events (e.g., inter-frequency) may require the E-UTRAN to define measurement gaps. For periodic reporting, the defined metric(s) will be reported at an interval defined by E-UTRAN.

The UE sends its capability as part of feature group indicators (FGIs) as defined in [17]. In 3GPP release 8, the UE should include this field in the UE-EUTRA-capability field sent in the RRC message. For a specific indicator, if all functionalities for a feature group listed in Table 3.31 have been implemented and tested, the UE should set the indicator as one (1), otherwise (i.e., if any one of the functionalities in a feature group listed in Table 3.31, has not been implemented or tested), the UE should set the indicator as zero (0). Note that the FGI are continuously changing from one 3GPP release to another, depending on the features being added for the UE in each release.

As an example of the settings related to handover mobility, if the UE advertizes the FGI as [featuregroupindicators "01111111 00001111 11111100 10100000"], then this indicates the UE supports the following related handover features, for which the eNB will react accordingly:

- **Bit 8**: EUTRA RRC_CONNECTED to UTRA CELL_DCH PS handover.
- **Bit 13**: Inter-frequency handover.
- **Bit 14**: Measurement reporting event: Event A4 and Event A5.
- **Bit 15**: Measurement reporting event: Event B1.
- **Bit 22**: UTRAN measurements, reporting, and measurement reporting event B2 in E-UTRA connected mode.
- **Bit 25**: Inter-frequency measurements and reporting in E-UTRA connected mode.
- **Bit 27**: EUTRA RRC_CONNECTED to UTRA CELL_DCH CS handover.

Inter-RAT handover can be blind or non-blind. In the blind scenario, the handover takes place without knowledge of the radio conditions of the other technology. The handover can be triggered by the radio conditions of the serving cell or by other considerations, such as cell or infrastructure loading. The disadvantage of this approach is the unknown coverage quality of the target cell, while the advantage is that the handover can be executed quickly without additional measurements. This is particularly helpful in CSFB voice calls from LTE, as will be shown in the next chapter.

For non-blind IRAT handover, the eNB configures the UE with the one of the following events.

- **Event B1**: Inter RAT neighbor becomes better than threshold.
- **Event B2**: Serving becomes worse than threshold1 and inter-RAT neighbor becomes better than threshold2.

For IRAT handovers, Event B1 is the most commonly used in different deployments, especially when combined with event A2. Figure 3.63 shows an example of how the reporting and execution of the inter-RAT handover is done, based on events A2 and B1.

For the UE to report the UTRAN cells based on event B1, eNB needs to configure the UE with a measurement gap. For a non-blind handover, the UE is configured to make measurements on

Table 3.31   Definitions of feature group indicators, 3GPP Release 8

Index of indicator (bit number)	Definition (description of the supported functionality, if indicator set to one)	Notes	UE support from example shown
1 (leftmost bit)	Intra-subframe frequency hopping for PUSCH scheduled by UL grant DCI format 3a (TPC commands for PUCCH and PUSCH with single bit power adjustments) Multi-user MIMO for PDSCH Aperiodic CQI/PMI/RI reporting on PUSCH: Mode 2–0 – UE selected subband CQI without PMI Aperiodic CQI/PMI/RI reporting on PUSCH: Mode 2–2 – UE selected subband CQI with multiple PMI		Not supported
2	Simultaneous CQI and ACK/NACK on PUCCH, i.e., PUCCH format 2a and 2b Absolute TPC command for PUSCH Resource allocation type 1 for PDSCH Periodic CQI/PMI/RI reporting on PUCCH: Mode 2–0 – UE selected subband CQI without PMI Periodic CQI/PMI/RI reporting on PUCCH: Mode 2–1 – UE selected subband CQI with single PMI		Supported
3	Semi-persistent scheduling	Can only be set to 1 if the UE has set bit number 7 to 1	Supported
	TTI bundling 5 bit RLC UM SN 7 bit PDCP SN		
4	Short DRX cycle	Can only be set to 1 if the UE has set bit number 5 to 1	Supported
5	Long DRX cycle DRX command MAC control element		Supported

Table 3.31   (*continued*)

Index of indicator (bit number)	Definition (description of the supported functionality, if indicator set to one)	Notes	UE support from example shown
6	Prioritized bit rate		Supported
7	RLC UM	Can only be set to 0 if the UE does not support voice	Supported
8	EUTRA RRC_CONNECTED to UTRA CELL_DCH PS handover	Can only be set to 1 if the UE has set bit number 22 to 1	Supported
9	EUTRA RRC_CONNECTED to GERAN GSM_dedicated handover	Related to SR-VCC Can only be set to 1 if the UE has set bit number 23 to 1	Not Supported
10	EUTRA RRC_CONNECTED to GERAN (packet_) idle by cell change order EUTRA RRC_CONNECTED to GERAN (Packet_) idle by cell change order with NACC (network assisted cell change)		Not Supported
11	EUTRA RRC_CONNECTED to CDMA2000 1xRTT CS active handover	Can only be set to 1 if the UE has sets bit number 24 to 1	Not Supported
12	EUTRA RRC_CONNECTED to CDMA2000 HRPD active handover	Can only be set to 1 if the UE has set bit number 26 to 1	Not Supported
13	Inter-frequency handover	Can only be set to 1 if the UE has set bit number 25 to 1	Supported
14	Measurement reporting event: event A4 – neighbor > threshold Measurement reporting event: event A5 – serving < threshold1 and neighbor > threshold2		Supported
15	Measurement reporting event: event B1 – neighbor > threshold	Can only be set to 1 if the UE has set at least one of the bit number 22, 23, 24, or 26 to 1	Supported

(*continued overleaf*)

Table 3.31   (*continued*)

Index of indicator (bit number)	Definition (description of the supported functionality, if indicator set to one)	Notes	UE support from example shown
16	Periodical measurement reporting for non-ANR related measurements		Supported
17	Periodical measurement reporting for SON/ANR	Can only be set to 1 if the UE has set bit number 5 to 1	Supported
	ANR related intra-frequency measurement reporting events		
18	ANR related inter-frequency measurement reporting events	Can only be set to 1 if the UE has set bit number 5 to 1	Supported
19	ANR related inter-RAT measurement reporting events	Can only be set to 1 if the UE has set bit number 5 to 1	Supported
20	If bit number 7 is set to "0"	Regardless of what bit number 7 and bit number 20 is set to, UE should support at least SRB1 and SRB2 for DCCH + 4x AM DRB	Supported
	SRB1 and SRB2 for DCCH + 8x AM DRB	Regardless of what bit number 20 is set to, if bit number 7 is set to "1," UE should support at least SRB1 and SRB2 for DCCH + 4x AM DRB + 1x UM DRB	
	If bit number 7 is set to "1" SRB1 and SRB2 for DCCH + 8x AM DRB SRB1 and SRB2 for DCCH + 5x AM DRB + 3x UM DRB		

Table 3.31 (*continued*)

Index of indicator (bit number)	Definition (description of the supported functionality, if indicator set to one)	Notes	UE support from example shown
21	Predefined intra- and inter-subframe frequency hopping for PUSCH with $N_{sb} > 1$ Predefined inter-subframe frequency hopping for PUSCH with $N_{sb} > 1$		Supported
22	UTRAN measurements, reporting, and measurement reporting event B2 in E-UTRA connected mode		Supported
23	GERAN measurements, reporting, and measurement reporting event B2 in E-UTRA connected mode		Not Supported
24	1xRTT measurements, reporting, and measurement reporting event B2 in E-UTRA connected mode		Not Supported
25	Inter-frequency measurements and reporting in E-UTRA connected mode		Supported
26	HRPD measurements, reporting, and measurement reporting event B2 in E-UTRA connected mode		Not Supported
27	EUTRA RRC_CONNECTED to UTRA CELL_DCH CS handover	Related to SR-VCC Can only be set to 1 if the UE has set bit number 8 to 1	Supported

the target technology since there is only one transceiver at the UE side. If measurement gaps are required, a likely approach is to configure inter-RAT measurements when the coverage of the serving LTE cell reaches some predefined minimal value. In this way the overhead associated with measurement gaps is minimized. In contrast to compressed mode for UMTS, a single LTE measurement gap configuration is defined that will be used for all IRAT measurements.

Only one configuration is possible at any one time. The gap duration is always 6 ms with a repetition period of either 40 or 80 ms (corresponding to 15 or 7.5% of the available subframes, respectively, and referred to as gp0 or gp1). The specific starting SFN (system frame number) and subframe for each gap is defined for the UE via the RRC, enabling diversity of gap location throughout the available time slots. The UE reports when gaps are necessary for either inter-frequency and/or inter-RAT during its capability reporting based on FGI, described earlier.

Although the gap has a duration of 6 ms, during which no downlink transmissions are scheduled for the UE and the UE does not transmit, an additional restriction is that the UE cannot

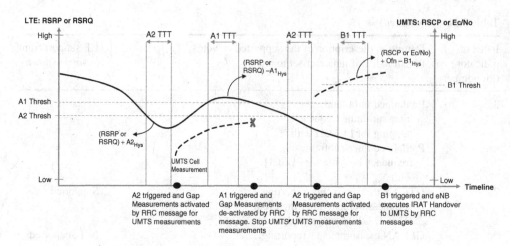

Figure 3.63 E-UTRAN to UTRAN inter-RAT handover based on events A2 and B1.

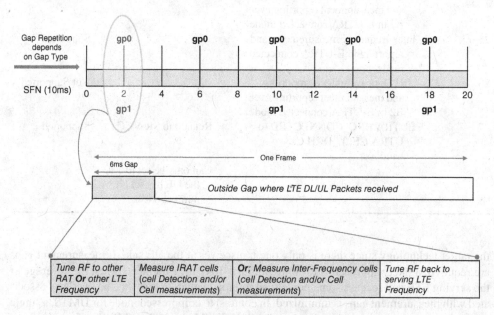

Figure 3.64 An example of measurement gaps timing.

transmit in the first subframe following a measurement gap. This corresponds to a 7 ms silence period for the UE in the uplink. Figure 3.64 shows an example of how the gaps are configured (for either gp0 or gp1) and what measurements are performed in each gap.

Once the UE is camped on UTRAN during the IRAT handover procedure, an RRC connection is set up. A routing area update request is sent to the target SGSN (serving GPRS support node). The SGSN is able to derive the MME address from the existing routing area identity included by the UE in the update request message. The SGSN then queries the source

MME, sending a context request message. UE's context information is returned to the SGSN, including security and EPS bearer information. This procedure ensures data continuity during the IRAT handover.

There are different ways that the inter-RAT handovers can be completed at both ends in E-UTRAN and UTRAN from an air interface perspective. Once UE performs the measurements on the UTRAN cells and the eNB decides to perform the handover, the step of leaving LTE and going to UMTS can be done as:

- **Seamless Handover**: if FGI capability advertised to the eNB shows that the UE supports PS seamless handover, then this mechanism can be triggered on the air interface.
  - In this handover procedure, the target cell is prepared in advance and the device can enter that cell directly in connected mode. IRAT measurements of signal strength may be required while on LTE in this procedure, prior to making the handover.
  - The source eNB sends the mobilityfromEUTRAcommand message to the UE, instructing the UE to hand over to the target access network. This message contains a transparent container (targetRAT-message container) including radio aspect parameters that the target RNC has set up in the handover preparation phase.
  - The advantage of this type of handover is its reliability, since measurements on the target UTRAN cell are mostly required. Additionally, the data interruption time is less in this case since the UE does not need to transition through idle mode to complete the handover. The drawback is that the support of devices is required, as well as the optimization process on the UTRAN neighbor list and parameters to ensure a successful completion of measurements and then handover.
- **Redirection**: The PS redirection procedure is a mix of "RRC connection release → cell reselection → RRC connection establishment".
  - Normally PS redirection is used by networks which are not able to perform PS handover. Indeed, PS redirection is based on following steps:
    - Network connection release (including information for frequency to reconnect, target cell ID to reconnect, target RAT information) while the UE is in idle mode. This will cause temporary service interruption.
    - UE by itself applies cell selection to connect new RAT/new cell based on information (target frequency, target cell ID) included in the connection release message.
    - In the target cell, UE triggers RAU (routing area updating) (e.g., in the case of the UMTS network) and then re-establishes the data session within new RAT.
  - The advantage of this type of handover is that the UE does not require measurements on the target UTRAN cell, making the handover faster. This is more suitable for CSFB type handovers triggered when a user initiates or receives a voice call while in LTE. The drawback is the high service interruption during the redirection process, which includes the UE going to idle mode, acquiring a cell on the target UTRAN carrier, reading the SIBs and then establishing the connection.

3.7.1   Inter-RAT and Inter-Frequency Handover Rules

The 3GPP standard in [14] shows the measurement rules for the UE during inter-RAT or inter-frequency handover. In the cell reselection procedure, the UE acts on the parameters broadcast in the SIBs, while in connected mode handover, the UE acts on the parameters sent in dedicated RRC connected state messages (i.e., RRC connection reconfiguration). Another fact to be considered during the optimization is that the handover measurements require the eNB to define the candidate neighbor cells, in terms of cell IDs (or PSCs), while the reselection can be done without a neighbor list, as only the RAT frequency is defined in the system information, in Release 8.

For each configured carrier, the UE needs to two do phases of UTRAN cell monitoring:

- **Phase #1**: Cell detection
  - Finding a cell with $E_c/N_o \geq -20$ dB
  - To be done on the neighbor cells defined to be monitored
  - Typically, up to four cells are detected in one gap measurement.
- **Phase #2**: Cell measurement
  - Measuring the detected cells found in Phase #1
  - To be done on each detected cell in each frequency
  - The outcome from this phase is to evaluate whether the measured cells meet the handover criteria
  - Typically, the top six cells found in Phase #1 are continuously measured.

Therefore, the optimization process on the measurements in LTE connected mode are of similar importance to those for reselections: to achieve a good balance between user throughput performance, device battery lifetime, and the overall network call drop rate reduction. The next subsection covers the aspects of inter-RAT and inter-frequency handover optimizations.

3.7.2 Inter-RAT and Inter-Frequency Handover Optimization Considerations

One of the different aspects to consider during optimization, unlike cell reselection, is the UTRAN broadcast neighbor relation definitions for inter-RAT handovers. As has already been mentioned, the connected mode handovers require a definition of particular inter-RAT neighbors in the RRC message. If the eNB instructs the UE to measure a UTRAN cell during gap measurement occasions, but the cell is not physically in the same location of nearby LTE coverage, the measurement gaps would be open for a long time, impacting the end-user throughput. This is because, during the gaps, the UE is not allowed to transmit or receive data on the uplink nor downlink.

Another area to consider during the handover optimization is the number of measured UTRAN frequencies for possible handover which can be a factor in how fast to trigger the handover. More UTRAN frequencies configured by the eNB leads to more measurement gaps, thus impacting the PS throughput. The design for how many UTRAN carriers to configure is highly dependent on the topology of the networks. Additionally, reducing the number of UTRAN neighbors per UTRAN frequency provides faster inter-RAT handover. From field measurements, it is proven that the more UTRAN neighbors configured, the more cell detection/measurements are triggered. Thus, keeping the number of UTRAN neighbors configured for the UE to less than 10 is recommended to detect and measure the UTRAN cells/frequencies faster, as per the field measurements trials carried out on this topic.

References

[1] 3GPP (2009) 3rd Generation Partnership Project; Technical Specification Group Radio Access Network; Evolved Universal Terrestrial Radio Access (E-UTRA); Physical Layer Procedures. TS 36.213 V8.8.0.

[2] 3GPP (2011) 3rd Generation Partnership Project; Technical Specification Group Radio Access Network; Physical layer procedures (FDD). TS 25.214 V8.12.0.

[3] 3GPP (2009) 3rd Generation Partnership Project; Technical Specification Group Radio Access Network; Multiplexing and Channel Coding (FDD). TS 25.212 V8.11.0.

[4] Elnashar, A. and El-Saidny, M. A. (2013) Looking at LTE in Practice: A Performance Analysis of the LTE System based on Field Test Results, *IEEE Vehicular Technology Magazine*, **8**(3), 81–92, Sept. 2013.

[5] 3GPP (2012) 3rd Generation Partnership Project; Technical Specification Group Radio Access Network; User Equipment (UE) Radio Transmission and Reception (FDD). TS 25.101 V11.3.0.

[6] 3GPP (2009) 3rd Generation Partnership Project; Technical Specification Group Core Network and Terminals; Non-Access-Stratum (NAS) Protocol for Evolved Packet System (EPS). TS 24.301 V8.10.0.

[7] 3GPP (2011) 3rd Generation Partnership Project; Technical Specification Group Radio Access Network; Evolved Universal Terrestrial Radio Access (E-UTRA); User Equipment (UE) Radio Transmission and Reception. TS 36.101 V8.15.0.

[8] 3GPP (2011) 3rd Generation Partnership Project; Technical Specification Group Radio Access Network; Evolved Universal Terrestrial Radio Access (E-UTRA); User Equipment (UE) Radio Access Capabilities. TS 36.306 V8.8.0.

[9] 3GPP (2010) 3rd Generation Partnership Project; Technical Specification Group Radio Access Network; Evolved Universal Terrestrial Radio Access (E-UTRA); Radio Frequency (RF) System Scenarios.

[10] 3GPP (2009) 3rd Generation Partnership Project; Technical Specification Group Radio Access Network; Evolved Universal Terrestrial Radio Access (E-UTRA); Physical Channels and Modulation. TS 36.211 V8.9.0.

[11] Qualcomm White Paper (2009) HSPA+ R8 Enhanced Serving Cell Change Performance Evaluation, March 2009.

[12] 3GPP (2009) 3rd Generation Partnership Project; Technical Specification Group Radio Access Network; Evolved Universal Terrestrial Radio Access (E-UTRA); User Equipment (UE) Procedures in Idle Mode. TS 36.304 V8.8.0.

[13] 3GPP (2010) 3rd Generation Partnership Project; Technical Specification Group Radio Access Network; User Equipment (UE) Procedures in Idle Mode and Procedures for Cell Reselection in Connected Mode. TS 25.304 V8.9.0.

[14] 3GPP (2012) 3rd Generation Partnership Project; Technical Specification Group Radio Access Network; Evolved Universal Terrestrial Radio Access (E-UTRA); Requirements for Support of Radio Resource Management.

[15] 3GPP (2010) 3rd Generation Partnership Project; Technical Specification Group Radio Access Network; Radio Resource Control (RRC); Protocol Specification. TS 25.331 V8.10.0.

[16] 3GPP (2010) 3rd Generation Partnership Project; Technical Specification Group Radio Access Network; Requirements for Support of Radio Resource Management (FDD). TS 25.133 V8.11.0.

[17] 3GPP (2010) 3rd Generation Partnership Project; Technical Specification Group Radio Access Network; Evolved Universal Terrestrial Radio Access (E-UTRA); Radio Resource Control (RRC). TS 36.331 V8.9.0.

4

Performance Analysis and Optimization of LTE Key Features: C-DRX, CSFB, and MIMO

Mohamed A. El-saidny and Ayman Elnashar

With the growing demand for data services, it is increasingly challenging to meet the required air interface data capacity and cell edge efficiency. On the other side of the cellular ecosystem, the transition to smartphones presents a significant revenue opportunity for operators, as there is substantially higher average revenue per user (ARPU) from smartphone sales and services. While the rollout of 4G LTE (long term evolution) radio networks is proceeding rapidly, smartphone penetration is also increasing exponentially. Hence, network operators need to ensure that the 4G subscribers' experience is as good or even better than with 3G systems, given the inherent high speed broadband deployed with 4G systems. Nonetheless, smartphone users or application behaviors add more demand on the network operators to apply methods and features that stabilize the system's capacity and consequently improve the end-user experience.

Several LTE features were initially introduced in 3GPP (third generation partnership project) Release 8 and further enhanced in later releases addressing the challenges discussed. Utilizing the spectrum and providing a better user reception at the edge of LTE coverage is essential and for this, different types of MIMO (multiple-input, multiple-output) derivatives are introduced: open-loop, closed-loop, or higher rank 4 × 4 MIMO. Additionally, the battery standby time is seen as an imminent topic for optimization. One solution is to utilize the LTE connected mode discontinuous reception (C-DRX) feature that improves the battery lifetime for smartphone users who tend to generate frequent small-size packets.

Although the LTE networks in all their current and future developments are data-centric all-IP (Internet protocol), the interaction with the circuit switch (CS) network for voice services, such as 3G or 2G systems (UMTS (universal mobile telecommunications system) or GSM (global system for mobile communications)); is one of the topics LTE forums have been continuously debating. The voice call solution, circuit switch fallback (CSFB), was introduced

Design, Deployment and Performance of 4G-LTE Networks: A Practical Approach, First Edition. Ayman Elnashar, Mohamed A. El-saidny and Mahmoud R. Sherif.
© 2014 John Wiley & Sons, Ltd. Published 2014 by John Wiley & Sons, Ltd.

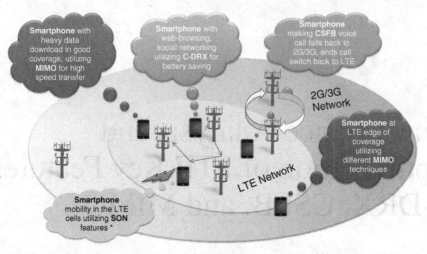

* SON features are explained in chapter 8

Figure 4.1   Representation of advanced features of the 4G system.

in Release 8 to enable the smartphone user to make legacy voice calls on the LTE network by falling back to the underlying 3G or 2G systems. As the voice calls are the main measure of the network performance, this area is a candidate for ongoing optimization from 3GPP Release 8 and beyond. Figure 4.1 summarizes these features and illustrates several situations the user could experience along the LTE network.

It is seemingly imperative to meet the increasing capacity challenges and user's experience improvements before transitioning into LTE-Advance in Release 10, and beyond. As a result, the current deployment of the LTE system still involves several levels of optimizations and enhancements to offer to the users in the network. This chapter focuses on the LTE features of 3GPP's Release 8 and 9 standards. The performance aspects of the following topics are explained[1]:

1. LTE C-DRX for smartphone power and battery saving,
2. CSFB solution for LTE voice calls, and
3. MIMO techniques.

4.1   LTE Connected Mode Discontinuous Reception (C-DRX)

The new generation of mobile communication systems aims to give users a revolutionary mobile experience, providing higher data rates and lower latencies that can transform the overall industry into a new wireless ecosystem of smartphone devices and applications. However, smartphone applications put a disproportionate burden on networks because of their many connections and low data volume transfers per connection. In addition, the energy demands of battery-powered devices to serve the transactions for each application may cause the battery to drain out, even without a direct interaction by the end user. The development of new

---

[1] Refer to Chapters 1 and 2 for fundamentals of LTE's air interface and protocols.

architectures and procedures to build power-aware systems has become a feature in the design of LTE. Alongside the introduction of the LTE smartphones and tablets and other handheld high-speed LTE services, battery consumption becomes a major bottleneck to the end user.

LTE utilizes the concept of discontinuous reception (DRX) to efficiently save mobile battery power during the user's inactivity periods. To exploit maximum battery savings, the LTE system allows DRX in RRC (radio resource control) idle[2] and/or connected[3] modes on the radio interface between the eNB (eNodeB) and the UE (user equipment) [1]. The DRX mechanisms are mainly used at the radio interface to save the mobile and eNB power pertinent to modem activities when the end user or applications are indeed inactive.

The use of C-DRX can inevitably lead to extra delays in transmitting/receiving data, which may be unacceptable for certain applications. Thus, it is essential to achieve a balance between packet delays and battery savings. This section evaluates the performance and optimization aspects of C-DRX and their impact on the battery standby time and packet delays.

4.1.1   Concepts of DRX for Battery Saving

In the LTE system, the UE can be served in RRC idle-camped state after successfully attaching to the LTE system. The UE remains in this state as long as there are no radio interface downlink (DL) or uplink (UL) packet activities. The UE immediately transits into RRC connected mode for any DL or UL activities. It remains in this state until the packet connectivity timer, known as the user-inactivity timer, expires. When the timer expires, the eNB releases the RRC connection and immediately triggers a transition to the idle-camped state again. Figure 4.2 describes the RRC state transitions. The figure also shows RRC idle-not-camped to which the

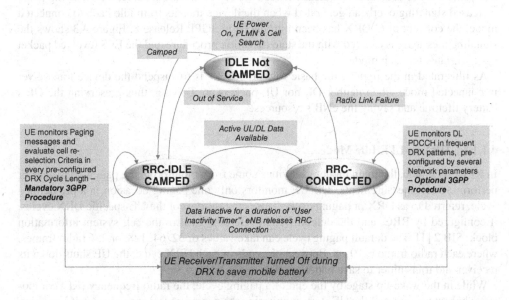

Figure 4.2   DRX and state transition mechanisms.

---

[2] RRC-idle mode DRX is referred to as paging cycle in this chapter.
[3] RRC-connected mode DRX is referred to as C-DRX in this chapter.

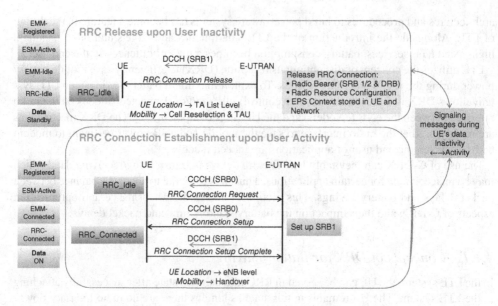

Figure 4.3   Transition between RRC idle and connected.

UE transitions when it goes out of service or drops the call. The UE is initially served in this state when searching for LTE service at power-on.

The state transition targets an improved battery lifetime and network resources when there is no connectivity or dedicated resources between the device and the eNB. However, owing to the increased signaling overhead generated when the device transits from idle back to connected mode, the concept of C-DRX has been introduced in 3GPP Release 8. Figure 4.3 shows the signaling messages associated with the state transition procedure and the EPS (evolved packet system) states for each mode.

As illustrated in the figures, the basic role of C-DRX is to suspend the device transceiver in connected mode when neither DL nor UL packets are flowing, thus preserving the UE's battery lifetime and saving the eNB's resources.

4.1.1.1 DRX in LTE Idle Mode

In RRC-idle mode, the main modem activities come from monitoring the paging messages or performing cell measurements. The UE monitors only one paging occasion in a pre-defined cycle, referred to as DRX or paging cycle. This is the shortest of the UE-specific DRX cycles, if configured by RRC, and the default DRX cycle broadcast in the cell system information block; SIB 2 [1]. The default paging cycle can take values of 32, 64, 128, or 256 radio frames, where each radio frame is 10 ms. Between the ends of each DRX cycle, the UE shuts down its receiver and transmitter to save battery.

While in the wake-up stage by the end of a paging cycle, the radio frequency (RF) components become active, and the UE starts monitoring the paging occasion to identify if it is being paged. In addition, the UE evaluates the surrounding neighbor cells for a possible change of serving cell, when the current cell's level degrades, as part of the cell reselection procedure [2]. The duration of each wake-up cycle depends on the types of measurements needed, and this is where the battery power is essentially consumed in idle mode. During the wake-up cycles and based on the network multi-layer topology, the UE may be instructed to perform intra-frequency, inter-frequency, or inter-system (i.e., IRAT, inter-radio access technology) measurements. The UE may also decide not to perform any neighbor measurements if the serving cell is measured better than pre-defined thresholds, to preserve the battery [2].

4.1.1.2   C-DRX in Connected Mode

In RRC-connected mode where all EPS states are active, the data packets flow in DL, UL, or bidirectionally between the eNB and UE. Thus, the concept of DRX becomes different than idle mode.

In connected mode, the UE reads the shared physical downlink control channel (PDCCH) in every sub-frame (1 ms) to find out whether resources are allocated in DL and/or UL. When there is no data activity flowing, this procedure consumes much unnecessary power from the battery. To overcome the unnecessary battery drainage, C-DRX defines the periods of inactivity while monitoring the physical layer channels [3].

C-DRX is an optional feature in LTE. When the eNB indicates that C-DRX is enabled, the UE monitors the PDCCH in pre-defined sub-frames, referred to as, active duty. The UE subsequently turns off its receiver for a predefined scheduling period, referred to as sleep duty. In return, C-DRX provides substantial savings to the device battery and increases the network efficiencies.

The ideal balance between power saving and data latencies is substantial, thus, requiring system performance studies. There are several factors taken into account to find this sort of balance:

- The operator's assumed and forecasted traffic generated by the smartphones is key information to reach high performance with a parameter set integral to the data dynamics.
- The network topology and architecture in terms of the radio frequencies deployed, and different technologies the UE is instructed to measure periodically. In this aspect, the more measurements the UE is required to perform in mobility conditions, the greater the impact on the battery.

4.1.2   Optimizing C-DRX Performance

4.1.2.1   Key C-DRX Parameters

The C-DRX functions are controlled by the RRC parameters timeline illustrated in Figure 4.4 [4]. C-DRX functions according to a timer-based mechanism, as well as to several conditions of the MAC (medium access control) layer buffer status, or RACH (random access channel) process being in progress.

The main parameters which are defined by 3GPP in [1, 3] and controlling C-DRX operation are summarized as follows:

- **DRX cycles** – The UE maintains two DRX cycles, short and long DRX cycles, which have different durations. The short DRX cycle is optional and, if configured, the UE starts with a short DRX cycle (2–640 subframes) when it enters DRX mode. When the configurable short DRX timer expires, the UE transitions to the long DRX cycle (10–2560 subframes). The optimal periodicity of the DRX cycle depends on the QoS (quality of service) requirements, especially on the maximum latency.

- **On-duration timer** – Specifies the number of consecutive PDCCH sub-frames (1–200 sub-frames) during which the UE should monitor the PDCCH for possible scheduling.
  - **Start condition** – At the first subframe of a DRX cycle. It is calculated as an offset from the long or short DRX cycles.
  - **Stop condition** – After it expires or the UE receives a DRX command MAC control element.
  - **Expiry** – The UE enters the sleep duty and no longer monitors the PDCCH.
- **The DRX inactivity timer** – specifies the number of consecutive PDCCH sub-frames (1–2560 subframes). It provides means for the network to keep a UE awake beyond the on-duration period when data are buffered. In the case of the VoIP (voice over Internet protocol) application this parameter is less useful since the buffer is typically emptied in one TTI (transmission time interval), that is, no need to stay awake.
  - **Start condition** – When the UE successfully decodes a PDCCH indicating an initial UL grant or DL user data.
  - **Stop condition** – after it expires or the UE receives a DRX command MAC control element.
  - **Expiry** – The UE applies a short DRX cycle (if configured), and the DRX short cycle timer starts or restarts. Or, the UE applies the long DRX cycle if the short DRX cycle is not configured.
- **The DRX retransmission timer** – If the UE decodes data with CRC (cyclic redundancy check) errors, and this was not associated with a broadcast HARQ (hybrid automatic repeat request) process, then following the expiration of the HARQ RTT (round trip time) timer, which specifies the earliest time at which a UE could expect a retransmission, the DRX retransmission timer is started (1–33 subframes). The UE must wait for this timer to expire and continue to monitor the DL before it can enter the DRX sleep state.

As discussed in Chapter 2, PDCCH is also used to schedule the UE on the UL. Hence, C-DRX controls the UL transmissions as well. The discontinues transmission (DTX) of UL channels applies, depending on the commands being received on the PDCCH during active duty. Figure 4.4 illustrates the UL mechanism during C-DRX. The UE sends an SR (scheduling request)[4] when data is available in the UE buffer. If the UE is in C-DRX sleep duty, it immediately switches to active duty and begins monitoring the PDCCH for a possible UL grant. In this case, the HARQ transmissions take priority over sleep duty in order to receive the HARQ feedback (ACK (acknowledgment) or NACK (negative acknowledgment)) within the HARQ timelines, regardless of the on-duration timer.

There are several other parameters that could be taken into account to further enhance the overall performance, such as the UE's reported channel quality indicator (CQI) and sounding reference signal (SRS). These connected mode parameters are used to regulate the data transmission on the DL or UL. They establish some indirect relation to the C-DRX, especially after the UE becomes in active duty. Those parameters are generally expected to be optimized as part of the overall LTE performance, but can be taken into account to reduce the overheads while operating in C-DRX.

---

[4] Refer to Chapter 2 for detailed explanation of SR reporting.

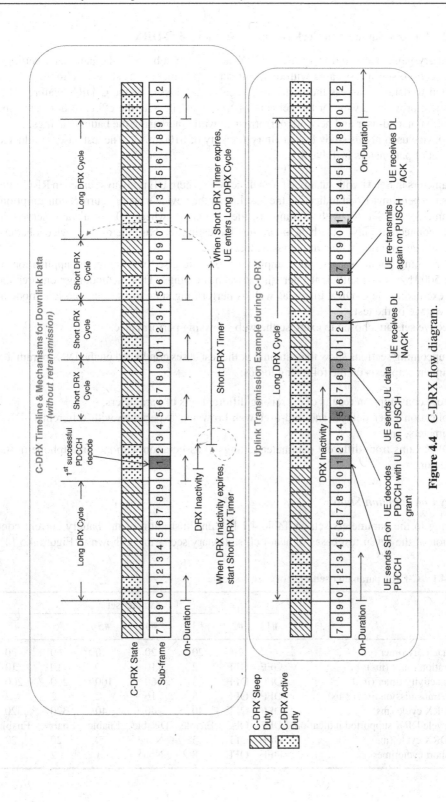

Figure 4.4   C-DRX flow diagram.

4.1.2.2 Battery Saving and Performance Overview of C-DRX

The battery gains estimated from the C-DRX feature are subject to the parameter settings. In this section, seven test cases with different set of parameters, as shown in Table 4.1, are evaluated in detail to assess the performance gains and losses of the C-DRX feature. This assessment provides a way for the network operators to alleviate the effect of non-real-time application on end-user and network performance while increasing the battery savings.

The device used in this study has a battery capacity of 6100 mA h. The test cases conducted on the set of parameters in Table 4.1 are:

- Parameters in Set #1 without any UL or DL activity where UE is always served in RRC-idle. This is a benchmark test indicating the baseline of the device's battery current consumption.
- Parameters in Set #1 with continuous file download where UE is always served in RRC-connected. This is the benchmark test indicating the baseline of the device's battery current consumption with continuous data activity.
- The reaming parameter sets are comprehensively exercised with ping test application of size 5000 bytes and sufficient intervals between the pings. This methodology ensures the UE exercising the C-DRX timers as well as performing state transitions into idle mode at some stage of the test.
- Selective sets are chosen to exercise the web-browsing performance.

The upcoming sections show the outcome of the test cases conducted on the DRX parameter sets with an emphasis on the following topics:

- Battery consumption results of exercising different sets of parameters,
- Application layer delays, and end-user web-browsing experience with different C-DRX parameters, and
- C-DRX gains from different parameter settings in association with the Smartphone traffic behavior.

Battery Consumption Study

Referring to the parameters sets in Table 4.1 [4], an illustration of the battery current consumption of the seven test cases in near-cell stationary scenario, is shown in Figure 4.5 [4].

Table 4.1 C-DRX parameter settings for each test case

	Parameter set						
	#1	#2	#3	#4	#5	#6	#7
User inactivity timer (s)	60	20	20	20	20	60	20
On-duration timer (ms)	OFF	OFF	2	10	3	10	10
DRX inactivity timer (ms)	OFF	OFF	3	60	100	200	200
DRX retransmission timer (ms)	OFF	OFF	8	16	2	2	2
Long DRX cycle (ms)	OFF	OFF	40	40	40	320	320
Short-cycle DRX supported indication	OFF	OFF	Enable	Disable	Enable	Enable	Enable
Short DRX cycle (ms)	OFF	OFF	5	None	20	20	20
DRX short cycletimer	OFF	OFF	8	None	1	2	2

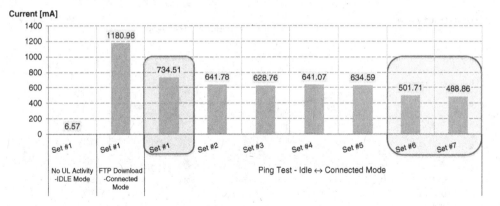

Figure 4.5   Current consumption measurements in near cell stationary.

The baseline battery current consumption without any user activity in idle mode is estimated to be 6.57 mA. The test cases running with sets #6 and 7 (i.e., C-DRX feature activated) have consumed the lowest current of ~ 501 and 488 mA, respectively. The test case in set #1 (i.e., C-DRX feature de-activated with long time in RRC-connected) have provided the highest current consumption for the ping applications. Finally, test case of set #2 (i.e., C-DRX de-activated with shorter time in RRC-connected) have provided similar current consumptions as in test cases of sets #3, #4, and #5 (i.e., C-DRX activated).

In the far cell scenario, Figure 4.6 [4] shows that the test case in set #7 provides the lowest current consumption. Sets #1 and #2 (C-DRX feature de-activated) still provide the highest current consumption for ping applications. This is mainly because the UE stays in RRC-connected for an extended amount of time. It is evident that when C-DRX is deactivated, the overall negative impact on the battery standby time is crucial.

This is a strong reason why a user-inactivity timer alone is not sufficient for battery gains with applications that occur in bursts. Consequently, implementing C-DRX combined with a shorter user-inactivity timer substantially improves the battery current consumption; the case

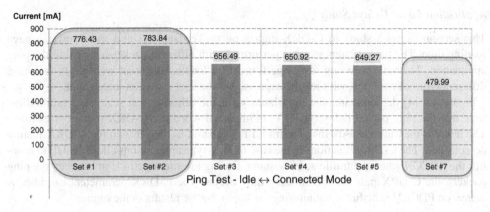

Figure 4.6   Current consumption measurements in far cell stationary.

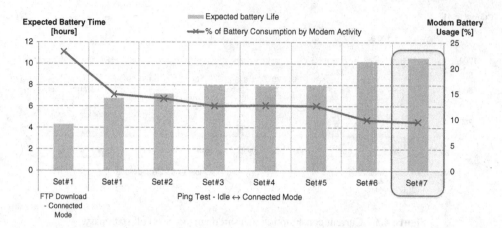

Figure 4.7   Expected battery life time for ping and FTP applications.

represented in Set #7. Additionally, during the wake-up cycles in RRC-idle, the UE is required to perform intra-frequency neighbor cell measurements (for cell reselection purposes) in far-ther RF conditions when compared to near-cell RF conditions. This also explains the higher current consumptions for sets #1 and #2 in far-cell compared to near-cell.

On the other hand, Figure 4.7 [4] explains that set #7 provides the best battery lifetime and lowest battery usage by modem activity. For continuous file download activities, the battery lifetime can sustain roughly 4 hours duration. At the same time, set #7 shows ~ 8% reduction of battery consumption produced by modem activities (i.e., PDCCH monitoring and measure-ments) solely from the frequent C-DRX sleep duty cycles. Set #1 shows a 20% increase in battery consumption coming from continuously monitoring the PDCCH and measurements in the absence of the C-DRX sleep duty.

As shown in the study, the C-DRX benefits for the battery lifetime are tied to the parameter settings and optimizations. However, the associated performance drawbacks of putting the device in deep sleep cycles arestudied in the next section.

Application Layer Delays Study

This section shows a study of C-DRX impact on packet latencies and web-browsing page loading time. The packet latency test is performed with different packet sizes using the ping application. The web-browsing test is performed by loading different web pages that have different types of contents, with results expressed in the user's session throughput.

Figure 4.8 [4] demonstrates the ping latency for the different test sets. The results illus-trate that the ping packet delay (or RTT) is impacted when the C-DRX feature is enabled. The delay is measured as ~10 ms of extra RTT compared to the set with the C-DRX feature de-activated. The C-DRX functions in the case of ping-type applications allow the UE to mon-itor the PDCCH when the traffic activity starts. During the silent periods in between the ping packets, the C-DRX puts the UE in sleep duty cycles. The C-DRX parameters can impose delays on PDCCH scheduling/monitoring, as shown by the results in the figure.

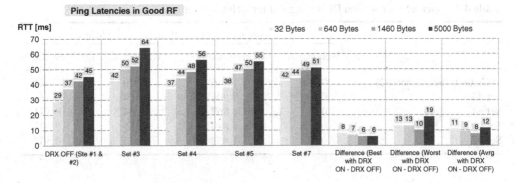

Figure 4.8   Ping round trip time (RTT) measurements.

In this study, set #7 provides a slight degradation in the ping RTT. However, considering battery and latency trade-off, set #7 can still provide acceptable latencies and minimum current consumptions with small or large ping packets. With the small increase in average RTT, the impact on the end user is expected to be negligible, especially in the case of non-real-time applications.

On the other hand, the web page loading time is described in Figure 4.9 [4] for different web pages. The results of this study show that there are no considerable performance drawbacks on web browsing page loading time when comparing the cases in which C-DRX feature is activated or de-activated. A maximum of 1 s web page loading time increase with an average of 0.5 s loading time are observed. Additionally, the web page loading time is clearly dependent on the contents and web page server locations. As shown in Figure 4.9 and Table 4.2 [4], the user's session throughput is maintained the same for all test cases without any sensible negative impact.

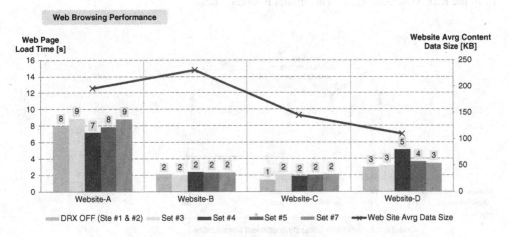

Figure 4.9   Web page load time measurements.

Table 4.2 Average user session DL throughput for web browsing

	Average user session DL throughput (Mbps)			
	Web site-A	Web site-B	Web site-C	Web site-D
DRX OFF (Set #1 and #2)	0.386	1.946	1.110	0.484
Set #3	0.358	1.894	1.154	0.571
Set #4	0.431	1.867	1.254	0.374
Set #5	0.385	1.849	1.155	0.503
Set #7	0.373	1.852	1.114	0.508

These results demonstrate that the C-DRX feature with a good battery consumption configuration, such as set #7, provide the projected end-user experience with an extended battery lifetime.

C-DRX Parameter Settings in Association with the Smartphone Traffic Behavior

It is imperative to evaluate the C-DRX parameters on the performance of smartphone/tablet applications. This section provides a C-DRX study of some applications' behavior in these devices, and their impacts on battery life.

Figure 4.10 [4] represents a client sync pattern of the Android operating system. Some Android client sync sessions cannot be finished within one data transaction and when the intervals between packets are larger than the user-inactivity timer, a multiple sync session (consisting of one to five data transactions) may occur, triggering RRC state transitions. A data transaction may take the UE from RRC idle to connected, and back to idle, depending on the configured user-inactivity timer. This type of application pattern can impact the battery lifetime if C-DRX is not enabled. Additionally, it can cause higher signaling overhead coming from the RRC state transitions. This model is studied next.

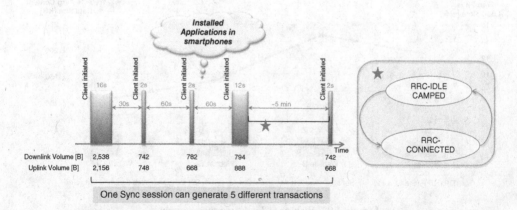

Figure 4.10 Example of Android application client sync pattern.

Table 4.3 [4] shows a Facebook client sync pattern collected through application profiling tools over 22 hours. Figure 4.11 shows the effect on the battery of applying this type of pattern on the DRX parameter sets explained before. It shows that set #7 significantly maximizes battery lifetime. The parameters in set #7 provide a battery lifetime of ~343 hours with only Facebook sync activities progressing on the device.

Similarly, in the case of multiple applications requiring parallel sync on Android OS, the nature of the patterns becomes different, as seen in Table 4.4 [4] which shows the sync patterns of 18 active applications and widgets in Android OS for 1-hour duration.

Table 4.3 Summary of Facebook client sync pattern

Application	Sync interval	Transaction duration	DL payload size (B)	UL payload size (B)	Who initiates the call?	Comments
Facebook	10–35 min; 61 sync in 22 h	Mostly 2 s (up to 11 s) due to packet delay	835–18 388	595–1 744	Client (DNS query)	Six syncs consists of two transactions separated by <1 min, all initiated by client

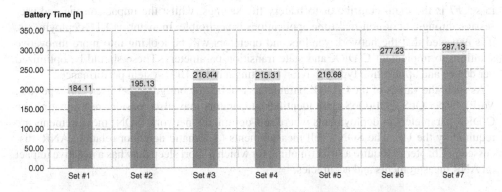

Figure 4.11 Expected battery lifetime for Facebook application.

Table 4.4 Summary of loaded Android application sync pattern

Application (widget)	Sync interval	Transaction duration (s)	DL payload size (B)	UL payload size (B)	Who initiates the call?	Comments
Fully loaded	6–79 s (144 data transactions in 1 h)	1–14	0–5486	0–2588	Client or network	More data transactions with smaller payload per transaction

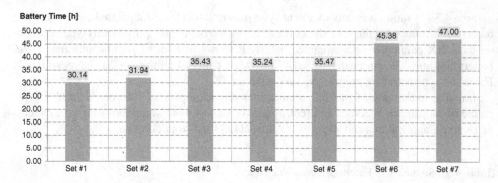

Figure 4.12 Expected battery lifetime with loaded application sync patterns.

Figure 4.12 shows the effect on the battery of applying this type of pattern on the DRX parameter sets explained before. It demonstrates that set #7 still maximizes battery savings compared to other sets of parameters. Hence, this C-DRX set benefits the smartphone users by providing the best battery savings for any kind of sync pattern produced by the applications. In the case where C-DRX is de-activated, the battery standby time is negatively impacted with any type of application.

From the results in this section, some basic conclusions can be drawn regarding the C-DRX parameters in Table 4.1. The higher "long DRX cycle" value with lower user-inactivity timer in set #7 is the main contributor to battery life savings, whilst the impact on the end-user session throughput of non-real-time applications is negligible. In maturing LTE deployments, it is expected that the network vendors and operators will be looking into more innovative solutions to implement C-DRX and state transition parameters. These should be optimized per device and application type/behavior, targeting a better LTE system performance.

As such, the values of the C-DRX parameters can be preconfigured per user, call type (i.e., VoIP), or per QoS QCI (QoS class identifier). Operators may also consider re-evaluating the C-DRX parameters in deployments where self-optimizing network (SON) measurements are required by the UE. The SON's cell measurements (automatic neighbor relation, ANR, discussed in Chapter 8) require a strict timeline, for which a short sleep duty has a negative impact on ANR measurements and accuracies.

4.2 Circuit Switch Fallback (CSFB) for LTE Voice Calls

An all-IP based network in LTE offers richer data services to the end user. However, voice service over IP multimedia architecture was a challenge for early LTE deployment. The IP multimedia subsystem (IMS) is a framework for delivering all-IP based services. Voice via IMS has been defined as a possible solution from 3GPP Release 5, prior to LTE. However, the cost and immediate need for IMS services prevented operators from migrating, especially with the wide presence of CS domain services offered in today's GSM and UMTS networks. Hence, operators set a migration pathway allowing them to start from LTE as data-centric, and reuse the CS domain services in 2G and 3G systems until the point when voice via IMS articulates.

CSFB, introduced in 3GPP Release 8, enables the support of voice service without IMS. This was a possible scenario for many operators when LTE was initially rolled out. CSFB and IMS can also be deployed simultaneously, meaning that an operator can gradually roll out an

IMS system while still supporting a fallback mechanism, as necessary. Therefore, 3GPP in TS 23.221 describes how a UE can be configured to be either "voice centric" or "data centric" to allow coexistence between different devices in one LTE solution [5].

Furthermore, 3GPP TS 23.221 allows mechanisms for the UE's preferences, whether it is capable or prefers CS voice and/or IMS PS voice. Four settings are possible: "CS voice only," "IMS PS voice only," "prefer CS voice with IMS PS voice secondary," and "prefer IMS PS voice with CS voice Secondary." A specific combinations of settings and EPS attach results enforces the UE to either remain in E-UTRAN (evolved universal terrestrial radio access network) or reselect to another RAT (radio access technology), depending on the network support and device capabilities.

For example, in an extreme case where the device is defined as "voice centric" and its preference setting is "IMS PS voice only," the UE could end up reselecting to the UMTS network to be able to make voice calls. This may occur if the UE roams in an LTE network that does not support IMS voice services. In another case, the device could be capable of a "CS voice preferred" setting but, while roaming in an LTE network not supporting the CSFB feature, the device would initiate a RAT change to either GSM or UMTS networks for voice calls.

Hence, the standards generally cover all possible combinations of UE and network capabilities to ensure voice continuity to the best level of support. Such bridging between device capability and network support is negotiated at the time the device is initially attached to the EPS network.

The CSFB architecture requires new interfaces to ensure voice continuity and interaction throughout the fallback to other RATs. The interfaces of CSFB architecture are described in Figure 4.13, as per 3GPP in [6, 7]. The SGs (serving grants) interface between the MSC (mobile services switching center) sever and the MME (mobility management entity) is based on the Gs protocols. It is utilized to enable mobility management and paging procedures between the CS domain and the LTE EPC (evolved packet core). As a result of a combined attached procedure, a terminating voice call arriving at the MSC will cause a paging notification to be sent to the MME. Then, the MME is responsible for paging the UE of the incoming voice call.

Additionally, the S3 interface between the MME and the SGSN (serving GPRS support node) enables idle mode signaling reduction (ISR). This is a feature enabling the UE and the network to share the UE's context between an MME and a SGSN. Doing so extends the "paging area"

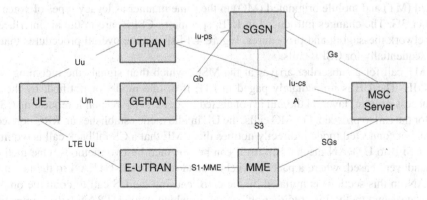

Figure 4.13   3GPP architecture for CSFB interfaces.

for the UE to simultaneously cover all the tracking areas (TAs) served by the MME and the routing area (RA) served by the SGSN [7]. Thereby, ISR allows the UE to roam across these TAs and the RA without having to inform the network of a change in its location. As a result, when the UE transitions between an E-UTRAN eNB serving one of the UE's registered TAs and the UTRAN RNC that serves the UE's registered RA, it need not perform the TA or RA updating procedures. This effectively reduces signaling in idle mode, a much needed gain in already signaling-overwhelmed networks. It additionally reduces the CSFB call setup time, as will be discussed next.

On the other hand, the PS data continuity is another case to consider during the fallback from LTE to the legacy RAT. The S4 interface between the S-GW (serving gateway) and the SGSN enables data forwarding from E-UTRAN to other 3GPP RATs. In summary, it provides related control and mobility support between the GPRS (general packet radio system) core and the 3GPP anchor function of S-GW.

Another important service relevant to these mechanisms is the short messaging system (SMS). If the home PLMN (public land mobile network) selects the CS domain for the delivery of SMS messages, the delivery utilizing the SGs interface between the MSC/VLR (visitor location register) and the MME is supported without the need to perform CS fallback. All other existing protocols and entities for SMS are reused. In either user's originating or terminating SMS scenarios, the message is encapsulated as part of a packet that is exchanged between the UE and the MME in EPC. Once the UE acknowledges the delivery report, the MME and MSC (alongside Service Center) interacts to complete the message delivery status. Hence, SMS over the LTE network can be done on a level separate from CSFB. The UE can request SMS-only service but not CSFB, as in the case of data centric devices. Thus, fallback to another RAT is not necessary, and the UE can register in the CN and receive SMS messages via the SG's interface, while still on LTE.

4.2.1 CSFB to UTRAN Call Flow and Signaling

The CSFB procedures start from the instant the UE initially attaches to the LTE EPC. The voice-centric UE begins the registration procedure with the MME indicating "combined attached". In practice, this means that the device requests to also register its presence in the 2G/3G CS network.

After a successful combined attach, voice calls initiated from LTE involve both mobile terminated (MT) and mobile originated (MO) in the same manner as legacy types of voice calls in 2G or 3G. The changes introduced in LTE, specific to CSFB, are in the air interface and core network messaging and procedures. Figure 4.14 shows the overall procedures that take place sequentially for CSFB calls.

An MT call for a subscriber arrives at the MSC, which then signals the incoming call to the MME. The UE is immediately paged in LTE if in idle mode, or notified, by the NAS (non-access stratum) layer, of the call in connected mode. The UE responds requesting CS fallback for the call to proceed. For MO calls, the UE in idle mode establishes an RRC connection first, and in connected mode it directly notifies the MME that a CS fallback call is required.

The CSFB to UTRAN mechanism itself can be executed by two methods. One method is "PS handover" based, where a packet switched handover from E-UTRAN to the target RAT (UTRAN, in this section) is initiated by the eNB, causing the CS call to continue on 3G as well. The second method is "redirection" based, in which the E-UTRAN RRC connection is

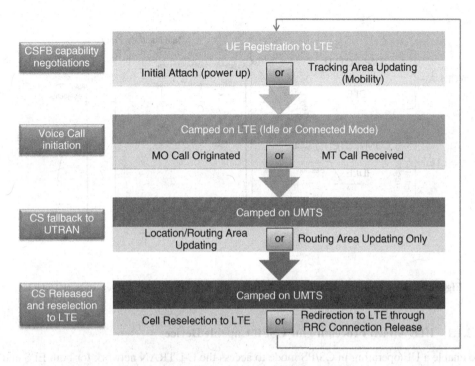

Figure 4.14   General layout of CSFB call flow.

released, including redirection information, to the other RAT with the PS being interrupted. Both mechanisms are applicable for MT or MO CS calls while the UE is in RRC idle or connected mode. There are a number of improvements in the "redirection" mechanism in Release 9 in order to reduce the CS call setup delays and PS interruption time.

The last step as shown in Figure 4.14, involving the cell reselection from UTRAN to E-UTRAN, is exactly similar to what has been comprehensively covered in Chapter 3. Alternatively, the network and UE could support a fast redirection back to LTE through the RRC connection release message received in UMTS at the voice call release. In this case, the UE camps back to LTE immediately and without a cell reselection procedure, enhancing the end-user experience by being on the high-speed network faster. This type of redirection to LTE may be done as a blind handover (i.e. referred to as fast return to LTE) or with LTE measurements (i.e. referred to as service based redirection). The blind redirection is initiated to the LTE frequency deployed and done without any UE measurements on the LTE cells. On the other hand, the measurement-based redirection to LTE requires compressed mode activation in WCDMA followed by event triggering (i.e. event 3C in UMTS). Once the UE reports event 3C, the RNC redirects it to LTE through RRC Connection Release message.

This section is organized to present the CSFB to UTRAN related procedures. As depicted in Figure 4.15, it discusses topics from when the CSFB capable device attaches to the LTE network, making voice calls, until the call is released and it camps back on the LTE. During each of the stages in Figure 4.15, optimizations are required to improve the end-user voice call setup delays, success rate, and perceived experience of camping back quickly to the LTE network after the call is released. The CSFB optimization aspects are also discussed in this section.

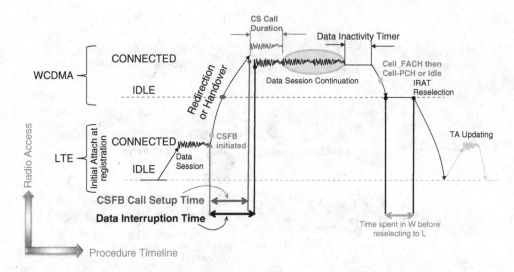

Figure 4.15 A timeline illustration of CSFB to UTRAN transition and return back to LTE.

4.2.1.1 LTE Attach Procedure for CSFB Capable Devices

To enable a UE operating in CS/PS mode to access the E-UTRAN network for both EPS and non-EPS services (which comprise CS SMS over NAS and/or CSFB), a different type of attach procedure is introduced in 3GPP [8], referred to as combined attach. This procedure results in the creation of the SGs context for the UE in the MSC/VLR and the MME. Figure 4.16 illustrates the full steps of the combined attach.

At step 1 in the flow diagram, the MME uses the IMSI (international mobile subscriber identity) in the "initial EPS attach" message to determine if it has an existing UE context. During initial attach, the MME is not expected to have the UE context. At this point, the MME begins to create the UE context by storing the UE network capability information, PDN (packet

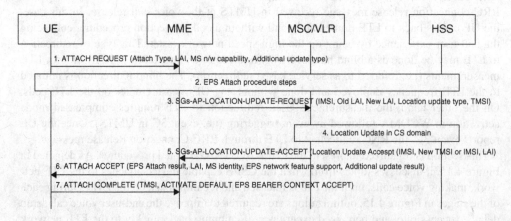

Figure 4.16 Successful combined attach procedure flow diagram (CSFB).

data network) connectivity request, and so on. These are used later during security activation and bearer establishment.

For voice-centric devices, a required device setting for CSFB, "EPS attach type" information element, is set to "combined EPS/IMSI attach" and sent by UE to MME in the attach request message shown in step 1 in Figure 4.16. This signals to the MME that the UE requires registration in the CS domain and is CSFB capable. Table 4.5 shows the "combined EPS/IMSI attach" message information sent by the UE (note that a device can signal SMS-only in this procedure which, in return, triggers a different attach response from the network).

After MME and MSC/VLR negotiates the UE's CS domain context and location areas (LAs), the MME sends an attach accept, as in step 6, containing the result of the combined attach request. Once the combined attach is successful, the core network includes the location area identifier (LAI) and TMSI (temporary mobile subscriber identity) as supplied by the MSC/VLR (conveyed at the previous steps). Table 4.6 lists the main values sent in the attach accept message from the MME to the UE.

The optional field "additional update result" in the attach accept can take the values "SMS only" or "CSFB not preferred." The action of the UE when either of these is present is controlled by the UE and the network configuration. The contents of this field may indicate that the UE will not be able to access CS services, such as SMS over SG and CSFB to UTRAN, within the registered TA. This ultimately implies that the UE should disable E-UTRAN and reselect to a 2G/3G in order to be able to get the necessary CS services.

Moreover, the inclusion of the UMTS LAI in the attach accept helps the UE to virtually consider this LAI when a CSFB call is initiated. Once the UE falls back to UMTS during CSFB call setup, it first reads the 3G cell's LAI sent as part of the SIB Type 1. If SIB-1 LAI indicates the same LAI as in the attach accept message, the UE would not need to immediately perform location area updating (LAU). Later, when the CS call is released, the UE could perform the LAU procedure, depending on the Network Mode of Operation (NMO). This virtual LA mapping eliminates LAU signaling at CSFB setup in UMTS,[5] saving a few seconds of the call setup time. This procedure is described in later sections as part of the CSFB call setup latency improvements.

Table 4.5 Main values of combined attach procedure in attach request

Information sent to EPC	Value	Description
EPS attach type	Combined EPS/IMSI attach	Indicates that the UE is performing combined attach for voice centric devices
Old UMTS location area identification	Old LAI	Include if available
MS network capability	CSFB capability bit = 1	Indicates that the UE is CSFB capable
Supported codecs	Codecs supported by the UE	This is an encoded list of codecs for GSM and UMTS
Additional update type	SMS only	Value shall be set to *SMS only* if this procedure is being executed for SMS services only

---

[5] In legacy UMTS-only CS calls, the LAU procedure takes place before the CS call is attempted, that is, when the UE first camps on a 3G cell of a different LAI than the serving one, in idle mode.

Table 4.6 Main values of combined attach procedure in attach accept

Information sent to UE	Value	Description
EPS attach result	EPS only	Indicates attach **not** successful for CS services such as CSFB and SMS over SG
	Combined EPS/IMSI	Indicates attach successful for SMS over SG. The attach is considered successful for CSFB if additional update result does not contain *SMS only*
UMTS location area identification	LAI	Indicates the LAI where the UE is registered for CS services. The UE stores the received LAI for CSFB procedures (referred to in later sections as virtual LA)
EPS network feature support	IMS voice over PS supported	Indicates support for voice over IMS in this TA
	IMS voice over PS not supported	
Emergency number list	List of local emergency numbers	Used for the various emergency services
Additional update result	SMS only	Indicates that the UE is CS domain attached for SMS over NAS services and is not allowed to perform CSFB
	CSFB not preferred	Indicates that the UE is CS domain attached for SMS over NAS services and for CSFB. However, voice centric devices shall not use CSFB in this TA

4.2.1.2 LTE Tracking Area Updating Procedure for CSFB-Capable Devices

A CSFB-capable UE that is already EPS-attached (according to the procedure previously described) can be moving between LTE cells of different TAs. Different TAs consisting of a group of cells could be serving the UEs in several geographical locations. The LTE device in idle mode is basically known to the MME at a level of the last registered TAs. The MME notifies the different TAs, in which the UE registered, of a user's incoming page. The paged UE is then able to continue with call establishment. This procedure is illustrated in Figure 4.17.

To connect to the 3GPP CS domain via E-UTRAN during mobility in different TAs, the procedure "combined tracking area updating (combined TAU)" is used [8]. The signaling of the combined TAU for CSFB-capable devices is highlighted in Figure 4.18.

In general, the contents of the tracking area update request, shown in step 1 of Figure 4.18, are the same as in the case of the normal EPS-only tracking area update procedure. Additionally, the CSFB related values sent by the UE in this message are similar to those discussed in the combined attach procedure in Table 4.5. The MME prepares the TAU accept message, in step 6, based on the outcome of the EPS updating and location update procedure carried out between the MME and MSC/VLR.

Similar to the combined attach procedure, the TAU accept message is comprised of CSFB related values such as those in Table 4.6. Additionally, the TAU accept would determine the

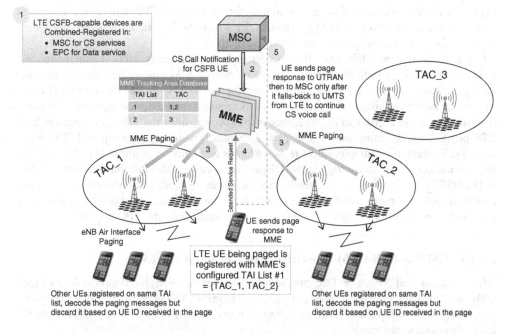

Figure 4.17   Non-EPS paging procedure in LTE (CSFB).

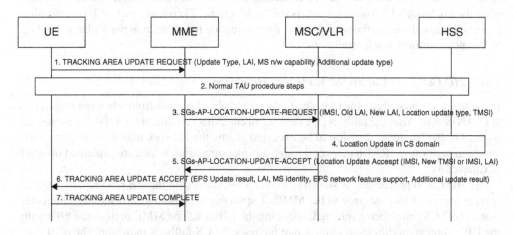

Figure 4.18   Successful combined TAU procedure flow diagram (CSFB).

virtual LA for the UE to consider during the fallback into UTRAN for CS call setup and avoid an extra LAU procedure.

Some failure scenarios could arise in the procedure of the combined attach or TAU. The handling of such failure scenarios by the UE must be efficient to be able to access the services accordingly. Several failure cases could occur especially in early EPS deployments. Thus, knowledge of how to identify the issue and redesign the EPC parameters is important in order to provide the user with a level of service, at least for voice calls. 3GPP in [8] gives all types of abnormal combined attach or TAU procedures, by which the UE should proceed, depending on the value of the reject/failure cause conveyed by the EPC.

Let us take one real-life case that occurred in several early LTE deployments. An LTE-capable device trying to register to the EPC with a SIM (subscriber identity module) card provisioned without LTE (i.e., a SIM card for use only with UMTS and GSM) could experience an attach failure from the EPC. In this scenario, as the device supports LTE radio, it would first try to register to the LTE network. However, since the SIM card does not have LTE provisioning to camp on the EPC, the EPS attach would end up rejected. The reject reason other than, for example, reject reason #15 (as explained in 3GPP in [8]), "no suitable cells in tracking area," may trigger the UE to lose service altogether on the entire PLMN (LTE, UMTS, or GSM) after the attach attempt. Hence, a modification to the attach reject reason in such a scenario could provide a fix to the issue. Accordingly, it is essential to cover all possible attach failure scenarios and how they are handled by the EPC and UE.

4.2.1.3   CSFB to UTRAN Mobile Originated Call Setup Procedure

After the combined Attach or TAU procedure completes successfully, the procedure of handling the CSFB voice calls becomes dependent on being an MO or MT call. Additionally, the procedure can be different in terms of delays or messaging, if the call is originated from the LTE idle or connected mode.

The CSFB from LTE to UMTS employs one of two procedures: handover or redirection. Both of these procedures are described in this section. The most common CSFB mechanism used in today's LTE deployments is the "redirection." Therefore, the CSFB optimization aspects discussed later in the section cover the redirection mechanism in particular, according to the most common implementations.

CSFB MO Call from Connected Mode with Handover

In the handover procedure, the target cell is prepared ahead of time from when the fallback to UTRAN occurs. After the fallback, the device can enter that cell directly in UTRAN connected mode. For the handover decision to be executed stably, the network may have to trigger the device to perform an IRAT handover through gap measurements, which are explained in detail in Chapter 3.

The MO CSFB procedure, as shown in Figure 4.19, starts with the UE sending an extended service request (ESR) message to the MME. The service type in this NAS message indicates that an MO CS call is being initiated, requiring the fallback. The MME notifies the eNB with the UE's context modification request that includes the CS fallback indicator. The eNB then starts the PS handover process to UTRAN. The handover could be blind (without prior knowledge of the UTRAN cell RF quality) or non-blind. For non-blind, the eNB configures the target RAT measurements, as explained in detail in Chapter 3.

Initially, as shown in steps 4–7 in Figure 4.19, the UE needs to perform IRAT measurements on the configured UTRAN cells from the LTE connected mode, a process enabling the eNB to execute the IRAT handover. Then, the eNB instructs the UE to execute the handover by sending the mobility from EUTRA command message. The CS-fallback indicator in this RRC message informs the UE that this procedure is being initiated for CSFB. Once the handover to UTRAN is completed, the UE tunes to the target UTRAN cell, as instructed in the handover message.

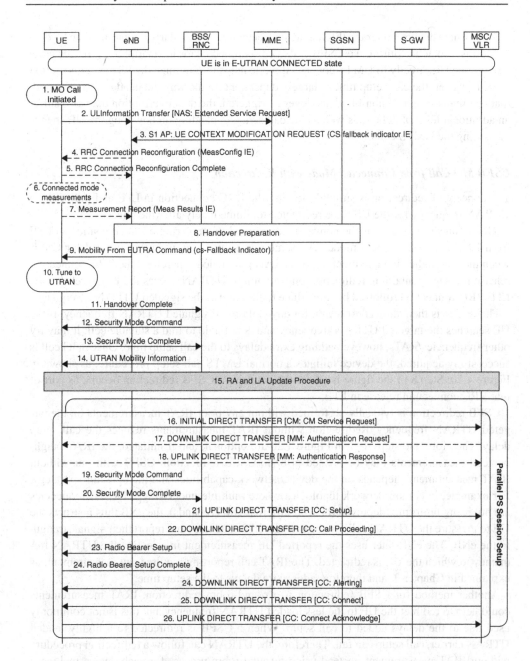

Figure 4.19   MO CSFB from E-UTRAN connected to UTRAN connected using handover.

The UE then initiates the setup of the CS call in UTRAN connected mode, by sending an initial direct transfer message to the RNC containing a CM (connection management) service request message. The RNC forwards the CM service request to the MSC/VLR which then executes a CS call setup.

As shown in the flow diagram, this handover procedure is particularly important if the PS data session is also active during the CSFB call, a common behavior in smartphones. In this case, moving the UE directly to UMTS connected mode helps to minimize the PS data interruption time. However, the call setup time is largely dependent on the way the handover is executed, that is, blindly or not. If non-blind handover is triggered, the data interruption may also suffer an additional level of delays, as well as the call setup, depending on how long the UE spends measuring the UMTS cells.

CSFB MO Call from Connected Mode with Redirection

The concept of redirection is simply releasing the RRC connection in LTE and indicating a UTRAN frequency for the UE to be redirected to, immediately after the release.

RRC connection release is the conventional method of triggering a state transition in LTE from RRC connected to idle mode. The same release message is further utilized for CSFB when the field indicating a redirection to UTRAN is signaled. Therefore, the RRC connection release message containing redirection information to UTRAN, forces the UE to release the LTE RRC connection followed by immediate redirection to the signaled UTRAN's frequency.

The device is then allowed to search for any cell on the signaled UTRAN frequency. If the UE searches the targeted UTRAN frequency and is not able to find a suitable cell, it may try other frequencies/RATs, however, adding extra delays to the call setup time. Once a 3G cell is successfully acquired, the device initiates a normal UMTS call setup procedure, as shown in Figure 4.20. Step 8 in the figure is where the concept of CSFB redirection occurs, as part of that RRC connection release in LTE.

CSFB redirection is typically performed without any prior IRAT measurements on the targeted UTRAN frequency. Redirection without IRAT measurements reduces the call setup delay. Therefore, the steps 4–7 in Figure 4.20 are optional, and if initiated, the IRAT neighbor cell measurements may be needed prior to CSFB redirection. CSFB with our without IRAT measurements depends on the device/network capabilities and the operator's strategy. For example, in certain network topologies where multiple underlying UTRAN frequencies are not being uniformly deployed (or with different UMTS bands), the eNB thus instructs the UE to measure the UTRAN cells on the configured frequencies and report their signal strength to the eNB. The eNB later uses the reported UE measurement in selecting the UTRAN frequency to which the UE is redirected. The IRAT cell reporting requires gap measurement, as explained in Chapter 3, and therefore adds extra delays to the setup time.

Another method for CSFB redirection is a round-robin redirection. IRAT measurements could help to redirect the UE to the less loaded UTRAN frequency, but this is not commonly used due to the delays added to call setup. When a CSFB is redirected to a highly loaded UTRAN carrier, call setup can fail. Therefore, the UTRAN can follow a redirection procedure without IRAT measurement, instead using a round-robin process for each device making a CSFB call. One device is redirected to one of the UTRAN carriers, and the other to the second UTRAN carrier in a round-robin manner. This may offload the UTRAN carriers, expediting the call setup time without compromising the call setup success rate.

The main call setup delay occurring during the redirection procedure (with or without IRAT gap measurements) is caused by the structure of the redirection information in 3GPP Release 8. This revision of the standard allows only UTRAN frequencies (without any particular cell

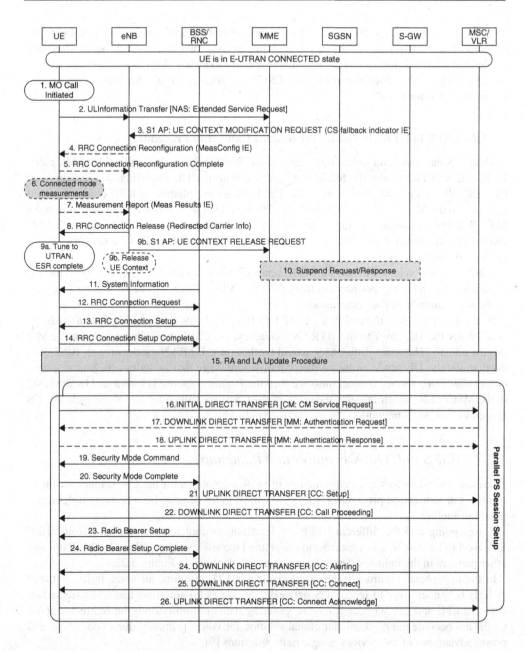

Figure 4.20 MO CSFB from E-UTRAN connected mode to UTRAN using redirection.

information) to be signaled in the LTE RRC release message executing the redirection. Consequently, the UE performs a full search in UMTS to acquire one of the 512 PSCs (primary scrambling codes) on the redirected frequency, as well as the SIB reading requirements.

So far, the MO CSFB from LTE connected mode has been discussed in Sections 4.2.1.3.1 and 4.2.1.3.2. The scenario of MO CSFB from idle mode is still possible if the device does not have PS data connectivity. However, the procedure of setting up the voice call is not entirely different. The only addition to the call flow shown in these two sections will be the signaling needed for the transition from LTE idle to connected mode. The MO call setup time will be increased due to this state transition. The CSFB call then continues from connected mode normally, as shown before.

4.2.1.4   CSFB to UTRAN Mobile Terminated Call Setup Procedure

Both the PS handover and redirection mechanisms for CSFB are also applicable to MT calls. If the UE is in idle mode, the MME initiates paging to the UE following notification from the MSC/VLR of an incoming voice call. The UE then establishes an RRC connection and responds to the MME with an ESR message. The service type in this message indicates that a MT CS call being initiated, requires the fallback. The MME notifies the eNB, and E-UTRAN can then initiate either the PS handover procedure or redirection based on the network configuration, as discussed earlier.

If the UE is in connected mode, the MME sends a CS service notification message to the UE on reception of a paging notification from the MSC/VLR. There will be no paging message, as the UE is already in connected mode.

Figure 4.21 describes the call flow of MT CSFB from idle or connected based on "redirection". After the UE tunes to the UTRAN frequency, the call flow is the same as in the MO originated call. The main differences would be that once UTRAN establishes an RRC connection and the UE sends an initial direct transfer message that includes a paging response encapsulated in a service request message then the Paging response is forwarded by the RNC to the MSC/VLR, which initiates the setup of the CS voice call, similar to the legacy MT CS procedure in UMTS networks.

4.2.2   CSFB to UTRAN Features and Roadmap

There are several factors and combinations of implementations for the CSFB mechanisms in handover or redirection procedures. Each one can be more suitable for a certain implementation than another.

Before going into the different CSFB implementations and features evolving from 3GPP Release 8 to Release 9, it is important to understand the voice evolution branded by 3GPP and other partners in the industry. The voice evolution is depicted in Figure 4.22.

In the first phase, commonly deployed in today's LTE systems, all voice traffic initiated in LTE is handed over to legacy CS networks. Single radio solutions use CSFB to switch between LTE and 2G/3G access modes, ensuring data continuity onto the redirected RAT. CSFB has become the predominant global solution for voice, primarily due to cost, size, and power advantages of the devices' single radio solutions [9].

The second phase introduces VoIP on LTE, and is known as VoLTE (voice over long term evolution). In this stage, single radio voice call continuity (SRVCC) is introduced. SRVCC enables handover to WCDMA/GSM (wideband code division multiple access) CS voice outside VoLTE coverage to ensure voice call continuity.

The third phase introduces enhanced capacity and services of all-IP networks (VoIP and video over IP) for continuous coverage across the broader range of network access methods, including LTE, and 3G/HSPA+ (high speed packet access), with interoperability across operators (i.e., roaming) and legacy telephony domains.

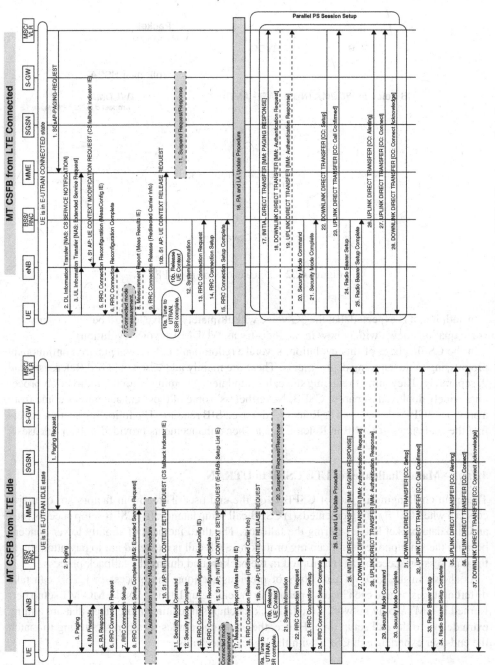

Figure 4.21 MT CSFB from E-UTRAN idle or connected mode to UTRAN using redirection.

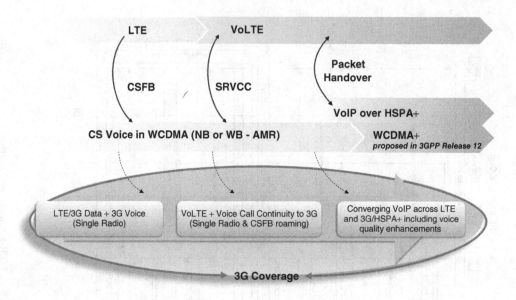

Figure 4.22 Voice evolution in 3GPP.

In addition, it is expected that the 3G voice will ultimately evolve toward better capacity and voice quality, known widely now in the industry as WCDMA+ voice evolution.

In the CSFB phase of this evolution, several predominant enhancements are continuously progressing in the ongoing deployments. These are mainly addressed in 3GPP Releases 8 and 9, separately. They aim at reducing the call setup latency coming from the redirection procedure widely deployed as part of CSFB. Nevertheless, some of these enhancements come from a previous 3GPP release, such as Release 7 deferred SIB reading. The following section covers first the challenges of CSFB in Release 8, and then enhancements provided in later releases.

4.2.2.1 Main Challenges of LTE CSFB to UTRAN

The main challenging impact of CSFB is on the end user. Three main factors are of major concern when CSFB is implemented: (i) CSFB call setup latency, (ii) CSFB setup success rate, and (iii) data packet latency during the fallback. The third factor is of slightly lower concern because the smartphone user receiving or making a CS call is expected not to pay attention to the data session speed being interrupted in the background during the fallback process.

As shown in the MO/MT call flows of previous sections, the process of CSFB can take different stages, where each incurs a penalty in the call setup latency. Unlike MO/MT setup performed in UMTS-only networks, the fallback mechanism requires more steps that are adding extra delays to the call setup time. Hence, the challenge comes for operators trying to match the end-user experience of CSFB with the legacy voice calls.

CSFB Call Setup Latency Challenges

Before going into the details of those steps impacting the call setup, let us first evaluate the call setup latency in the legacy UMTS network for MO calls initiated by UMTS-only devices,

Table 4.7 Mobile-originated (outgoing) call setup times comparisons

Call setup delays (ms)	Device in stationary conditions near cell			Device in mobility conditions	
	UMTS only	CSFB from LTE idle	CSFB from LTE connected	UMTS only	CSFB from LTE idle
Average	1670	3456	3142	1890	5240
Minimum	1590	2535	2400	1600	3499
Maximum	1760	5264	4118	3790	9495

see a breakdown in Table 4.7. In the same table, the setup delays are compared with a CSFB device implementing the basic features of 3GPP Release 8.

The table shows two cases: a device in stationary conditions near a single cell, and one in different mobility RF conditions. The CSFB case in the table is using "redirection" procedures. Also note that the values shown are measured in one of the deployed networks, and may change from one architecture to another, however, the main point is to show the substantial difference in the call setup latency which tends to vary within the same percentages as in the trial illustrated in [9].

The CS call setup values in Table 4.7 show that in the best case scenario, the average CSFB setup latency lags UMTS-only latencies by ~1.5 s. In the worst case scenario, that is, the mobility case, the CSFB latency can increase by 5.7 s, which impacts the end-user experience. The CSFB calls in this table are made from a mobile to a landline. The CSFB delays can be magnified if a LTE-to-LTE voice call is made which would add delays from both directions, specifically noticeable to the originating user.

As a result, the CSFB call setup delays need improvements to match the UMTS-only user's experience, at best. To carry out any improvement, one should first consider where the delays are coming from. Figure 4.23 explains the CSFB call setup latency stages with "redirection" and compares it to the UMTS-only procedure.

It is evident from the CSFB setup delay stages that extra steps (i.e., 1, 2, 3, and parts of 5) are required. With reference to the values in Table 4.7, the delay breakdown of each of the setup stages is illustrated in Table 4.8.

Hence, several optimization processes are required in each of these delay stages. 3GPP improvements basically deal with minimizing the SIB reading stage impacting the setup delays. In mobility conditions, from Table 4.8, the SIB reading takes 1.3 s, a process that does not necessarily exist in the UMTS-only call setup. In UMTS-only calls, the device has already read all SIBs, prior to the CS call setup, when it camps on the serving UTRAN cell. As a result, a delay from SIB reading requires additional improvements, addressed in 3GPP Release 9.

The Release 9 optimization in SIB readings comes from improving the redirection process which is part of step 1. Additionally, some improvements in Release 7 have been utilized to reduce SIB reading time.[6] The improvements are discussed in Section 4.2.2.2.

Both steps 4 and 5 are common in CSFB and UMTS-only calls. The only difference can come from the fact that, in CSFB, the UE may newly register to a LA or RA and, therefore, requires an updating process prior to the call setup. This process occurs in UMTS-only calls prior to the call setup, in idle mode, when the UE crosses into different LA/RA areas.

---

[6] SIB reading has been initially a concern in UMTS networks even before CSFB and Release 7 have addressed several ways of improving SIB reading time in RRC idle mode.

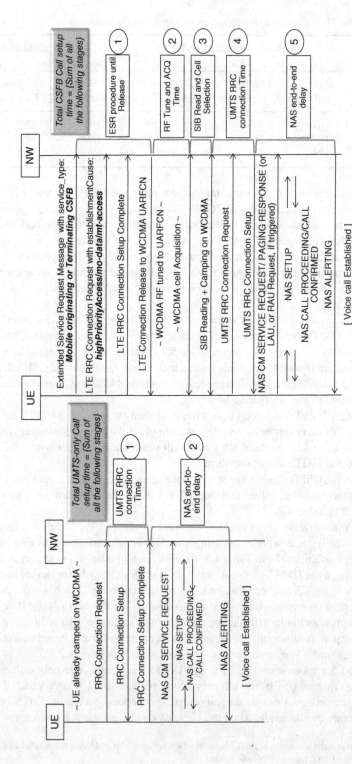

Figure 4.23 CSFB call setup delay definitions as compared to UMTS-only voice calls.

Table 4.8   Mobile-originated CSFB call setup times breakdowns

Call setup stage (average values in ms)	Device in stationary conditions near cell		Device in mobility conditions
	CSFB from LTE idle	*CSFB from LTE connected*	*CSFB from LTE idle*
Step 1: ESR to release	284	173	325
Step 2 + 3: tune to WCDMA and SIB read	381	342	1326
Step 4: WCDMA RRC time	551	237	823
Step 5: NAS end-to-end	1768	1927	2055

Therefore, as shown in the call flow in previous sections, the LTE combined attach/TAU procedure could pass to the UE a virtual LA identity that avoids LA updating in step 5, a solution that seems reasonable to reduce the call setup delays. However, it only works in cases where the TA is collocated on the LA at the time the CSFB is made. Hence, in mobility conditions, it is unlikely that this collocation would occur, inevitably triggering the LAU.

As for the RA updating in step 5, ISR (explained earlier in the section) can play a role in minimizing the occurrence of such signaling that adds delays to the setup time. However, ISR is not widely deployed due to requirements complexities.

CSFB Setup Success Rate Challenges

In today's UMTS-only deployment, the CS call setup success rate has been made robust by rounds of optimization processes over the years. The CS call setup success rate is usually firm at >98% in mature UMTS networks. Adding the CSFB factor requires a new round of optimization or additional features.

As the CSFB redirection in Release 8 is a blind-handover process, it requires an additional level of optimization.

(i) **Case Study 1**

In Figure 4.24 deployment, there are three UMTS carriers (i.e., frequencies) referred to as F1, F2, and F3 which are collected with LTE in cell A. On the other hand, cell B is deployed without LTE and with UMTS F1 and F2 deployed in one-to-one coexistence over the entire network sites. F3 is typically added as another capacity-relief carrier, and hence deployed in hot-spots where F1 and F2 could be highly loaded.

If a CSFB call is initiated from LTE cell A, and the eNB redirects the UE to camp on F3, a troubling scenario of a CS call failure can occur. If the LTE site itself is a hot-spot layer, and a UE is camped on LTE such that its coverage is dragged to the area of cell B, then, UTRAN F3 cells may no longer be suitable to handle a CS call setup in that location. The UE in the shown coverage area may try another carrier on UTRAN (either F1 or F2 within the Cell B coverage location) through a frequency search operation. By the time UE camps on F1 or F2, the call setup timers might expire in the core network causing a setup failure.

(ii) **Case Study 2**

From the same figure, in another possibility where the redirection to F3 succeeds, but the camped cell is weak, an extra setup delay occurs while exchanging the extensive CSFB messages. While F3 may be weak, call setup itself may fail in degrading RF conditions or while the UE is trying to perform inter-frequency handover from F3 to F1/F2 in the shown location in the figure.

(iii) **Case Study 3**

If in a similar situation in Figure 4.24, the UE camps on a weak F3, it may immediately trigger an IRAT handover from UMTS into GSM. This has a negative impact on the data speed of the end user. In a scenario where a network does not support GSM to LTE transitions, the device ending the CS call would need to first camp back into UMTS before being able to camp on the LTE network for high-speed data. A 3G to 2G IRAT reselection process typically takes a long time and is highly visible to the end user.

This scenario does not normally occur in the UMTS-only case, as hot-spot F3 would be set as a barred layer for idle mode camping. This means that the network does not prefer F3 for camping, to avoid CS call setup from hot-spot layers that most likely end up with setup failures. The multi-carrier strategies are crucial in achieving a balance between CSFB setup failure/delays and the desired 3G carriers load sharing.

(iv) **Case Study 4**

The CSFB call setup success rate can be improved if the redirection is executed as non-blind one. This is only possible in 3GPP Release 9, where the RRC connection release message is expanded to include additional information elements reducing redirection delay. A list of PSCs (UTRAN scrambling codes, distinguishing the cell ID) is sent in the LTE RRC connection release message for the redirection process. For the network to redirect the UE into a cell-level and not frequency-level (Release 8), the UE may need to perform measurements on the UTRAN cells prior to the redirection so the eNB is made aware of which PSC to chose. In the example in Figure 4.24, a Release 9 network may request the UE to measure F1, F2, and F3 cells and report their signal levels before the redirection process. Hence, the redirection may trigger the UE to move to F1/F2, non-blindly, avoiding all failures discussed in previous case studies.

This theoretically leads to a more robust CSFB success rate. However, the trade-off ,as can be seen, is the extra CSFB setup delays associated with the measurements needed.

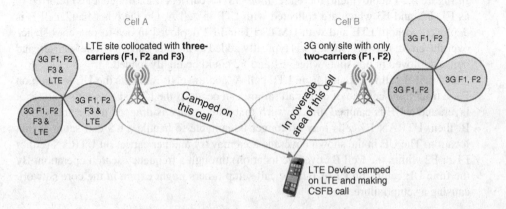

Figure 4.24   Example of call setup failure.

The trade-off is subject to an optimization process.Figure 4.25 shows the results of one trial in mobility conditions relating the CSFB calls setup time to the IRAT gap measurement time. During this trial, the CSFB call setup success rate has been improved, but the delay has been increased. This does not imply that the Release 9 features should be less recommended, but rather that careful planning and optimization should be carried out to ensure the feature is working as intended. Measurement optimization, including IRAT neighbor set relations and the number of UTRAN carriers to measure simultaneously can be areas to consider. More details are in Chapter 3, as part of IRAT connected mode handovers.

Moreover, to avoid call setup failure, the link budget (LB) difference between LTE and UMTS should be taken into account. For example, a UE camping on the LTE 1800 MHz band and falling back into the UMTS 2100 MHz band requires an additional step of LB verification to ensure CSFB stability at the coverage difference between the two systems. The LB details are explained in later chapters. In the mean time, the cell reselection parameters (IRAT specifically) can play a role in the CSFB success rate. Keeping a UE in LTE at extreme RSRP (reference signal received power) levels may result in CSFB setup failure if the UE is unable to process the CSFB messaging while in weak LTE coverage. This parameterization aspect is discussed in detail in IRAT cell reselection optimization in Chapter 3.

(v) **Case Study 5**

Lastly, some of the real-life abnormal scenarios observed in CSFB calls are summarized in Table 4.9. Any case outside the expected MO or MT call flow is usually considered as abnormal. The handling of such cases should be tested and verified in each device, as not all cases are comprehensively covered in the standards.

However, the standard in [8] requires that for lower layer failures or abnormal behavior during CSFB call initiation, the UE should autonomously select GERAN (GSM/EDGE radio access network) or UTRAN RATs. In this way, the device implementation can treat

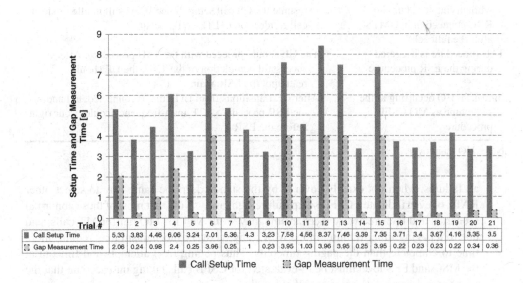

Figure 4.25 A trade-off between measurements prior to CSFB setup and CSFB call setup delays.

Table 4.9 Abnormal scenario leading to possible CSFB call setup failures

Abnormal condition during CSFB call origination	Description
Access barred for LTE cell	UE is not able to perform RACH procedure on the camped E-UTRAN eNB due to barring for MO access. UE will not be able to enter LTE connected mode to send ESR
RACH failures or LTE RRC setup timer expires	The UE is not able to perform the LTE RRC connection establishment procedure on the camped eNB due to RACH failures, or if the UE does not receive RRC connection setup
RRC connection establishment rejection	UE is not able to transition to RRC-connected to perform the ESR procedure, including the reception of the RRC connection reject message from the camped E-UTRAN eNB
ESR message transmission failure	The transmission failure reported by the serving eNB. Spec allows the UE to repeat ESR again
Service reject during ESR procedure	UE receives service reject from EPC with various cause codes in [8]
Radio link failure during CSFB	Lower layer failure and UE may be allowed to select to UTRAN/GERAN to initiate the CS call as in [8]
Redirection failure	UE receives RRC connection release from the eNB but unable to acquire a suitable UTRAN cell on the target frequency
UE performs LTE cell reselection to another LTE cell in different tracking area while ESR is being sent	When UE makes a call and attempts to move into RRC-connected to send ESR, cell reselection can happen after sending RRC connection request to another LTE cell in different TA. Spec usually allows UE to move to UTRAN to continue the call
UE makes IRAT cell reselection from UTRAN to E-UTRAN while trying to set up the RRC connection in UMTS after the fallback	If the UE successfully camps on a target UTRAN cell, it should try to establish an RRC connection in order to resume the CS call setup. Since UE is still in idle mode, a cell reselection to LTE could occur
TAU procedure is triggered during the ESR procedure	The UE should proceed with the TAU procedure. After the successful completion of the TAU; the UE should reattempt the ESR again
Intra-LTE HO occurring at the same time as MO CSFB procedure	After UE transitions from LTE idle to connected, to handle the ESR procedure, LTE intra-frequency could occur right before the ESR

Note: HO, handover.

all failures, when not exactly covered by the standard, in the same way to select other RATs, or even indicate a CSFB setup failure right away. Note that most of these abnormal scenarios and their recovery mechanisms may work well for MO but for MT calls, and due to a paging timer running in the MSC, any recovery mechanism may require more time to complete than the paging timeouts. Thus paging (re-) attempts and timeout at the MSC and EPS level must be re-evaluated for CSFB calls taking into account that the worst case scenario observed from field testing.

Data Packet Latency Challenges during CSFB

Based on what is explained in Figure 4.15, CSFB voice calls may happen alongside PS data activity, especially for smartphones. Once the fallback occurs, MME and SGSN communicate to handle the transfer of the UE's data contexts to ensure PS data continuity while the CS call is being established in parallel. Such a situation usually happens for background data activities, as the user mostly attends to the voice call on the user interface when originating or receiving a call.

However, the operator should evaluate the impact of the CS fallback on the data continuity, in terms of the data packet delays during the context transfer from 4G to 3G systems, known as the data interruption time. Data interruption can also occur during a normal LTE to UMTS IRAT handover situation, even without a CSFB call, when the LTE coverage reaches its edge.

The DL data interruption time during the fallback to UTRAN is typically calculated at the RLC (radio link control) level from the last received data packet on the LTE to the first received data packet on the WCDMA. This time interval includes the redirection delays, SIB reading, and CS/PS call establishment on UTRAN. Table 4.10 explains one example of the data interruption time measured from a Release 8 device.

The data interruption time shown is calculated for the condition where DL packets are flowing continuously, such as in the case of an FTP (file transfer protocol) download. In this trial from a live-network, Table 4.10 shows three cases:

- Data interruption during HSDPA (high speed downlink packet access) serving cell change,
- Data interruption time in the non-CSFB case of LTE intra-frequency handover, and
- Data interruption time of active data during CSFB calls.

The measured data interruption time of the PS data during CSFB is shown to be much higher, reaching 5 s on average, than the other cases. This high delay is mainly due to the fact that the data packets are paused until the LTE to 3G/HSDPA fallback is completed at the air interface level first, and then between the different entities in the core network. Additionally, the UE is required to perform location updating or RA updating after the fallback, increasing the interruption further until the procedure is completed.

The CS call establishment between the UE and core network usually takes higher priority than the other PS domain related procedures. For example, if a CM service request and a GMM (GPRS mobility management) service request come in parallel, the network may choose to handle the CS CM service request for the CS call first before the PS call. Additionally, to improve the CS call setup performance, some devices can deploy delaying the PS domain

Table 4.10 DL data interruption time comparison

Data interruption time (s)	HSDPA during serving cell change without HSPA+	LTE during intra-frequency handover	LTE to HSDPA during CSFB or coverage based IRAT handover
Average	0.128	0.050	4.99
Minimum	0.039	0.039	4.49
Maximum	0.300	0.080	5.77

related requests until the CS call is established. In the later case, the 3GPP standard does not specify when the PS call should be established. Thus devices can act as if there are no data activities, and delay the PS call establishment (including GMM service request and, in extreme cases, RA updating) until the CS call is fully established at the air interface and in the core network.

This procedure saves a few milliseconds on the call setup delays, but increases the data interruption time, as shown in Table 4.11. The trade-off shown in this trial is clear. When the device chooses to delay the PS call establishment to expedite the CS call setup time, the CSFB call setup time in this case decreases by 0.340 s, and the data interruption time increases by 2.5 s. It is worth mentioning that the values are calculated for a device in near cell stationary conditions. In mobility conditions, the values can increase.

4.2.2.2   List of CSFB to UTRAN Features

What has been discussed so far in this section shows the immediate need to optimize the CSFB to closely match the performance of UMTS-only calls. Hence, there are several rounds of optimizations introduced in devices and on the network side in Release 9, mainly. The whitepaper in [9] discusses several of these enhancements, and they are summarized in Table 4.12.

Table 4.11   CSFB versus DL data interruption time in various implementations

	Delayed PS establishment until after CSFB	Normal PS establishment in parallel with CSFB
Average CSFB setup delay (s)	2.80	3.14
Average data interruption time (s)	7.50	4.99

Table 4.12   CSFB deployment variants

CSFB deployment type	CSFB impact due to network/device support	Notation to CSFB call setup delay
3GPP Release 8 with basic functions	Redirection-based CSFB, and devices required to read UMTS SIBs	Longest
3GPP Release 8 with optimized functions	Redirection-based CSFB, with SIB skipping feature: deferred measurement control reading (DMCR)	Shorter
3GPP Release 9	Redirection-based CSFB, with RRC release including list of 3G cells and tunneled SIB information	Shortest
CSFB LTE to UTRAN handover-based with measurements	Target cell is prepared in advance and device can camp on the cell directly in connected mode	Flash[a]

[a]UMTS cell measurements may be required while on LTE in this procedure, prior to making the handover (~300–600 ms delay), depending on time/reliability needed for measurements.

CSFB with 3GPP Release 8 Basic Functions

Once CSFB is deployed, it is expected that the network operator is at this stage of deployment. The impact on the call setup time delay is the largest. The expected values are expressed in Tables 4.7 and 4.8 from live-network trials and are also evaluated in [9].

CSFB with 3GPP Release 8 Optimized Functions

The main feature introduced in this enhanced version is the usage of 3GPP's SIB deferment feature, referred to as deferred measurement control reading (DMCR) and highlighted in [9]. DMCR can be implemented in two ways:

- **UE-based** – The UE will only read UMTS SIBs 1, 3, 5, 7, ignoring all other SIBs broadcast from the UTRAN cell (e.g., SIB11, 11bis, 12). Helpful in mobility conditions.
- **Network-based** – The DMCR indication is sent in UMTS SIB 3, and the UE can skip SIBs 11, 11bis, and 12.

The UMTS System Information Type 11 and 12 broadcasts the network's neighbor cells for the UE to measure in idle mode or cell/URA-PCH (paging channel) states. According to 3GPP in [10], the SIB 11 or 12 size is generally divided as follows

- The maximum SIB size is 16 segments (each with max. size 222 bits) $= 16 \times 222 = 3552$ bits
- Due to the physical size of the SIB-11 data, it has the capacity of 50–75 intra-frequency, inter-frequency, and IRAT cells
  - Each intra-frequency neighbor requires 2 bytes
  - Each inter-frequency and IRAT neighbor requires 6 and 5 bytes, respectively
  - Each neighbor $Q_{qualmin}$ parameter if deviating from the serving cell is 1 byte
  - Each neighbor $Q_{rxlevmin}$ parameter if deviating from the serving cell is 1 byte
  - The use of $Q_{offset,1,n}$ is 1 byte
  - The use of $Q_{offset,2,n}$ is 1 byte
  - The HCS (hierarchy cell structure) consumes 15 bits more, for each neighbor.

Given the large size of SIB 11 or 12, then skipping them adds extra gains on the CSFB call setup latency. Additionally, skipping these SIBs does not expose any risks on the CSFB performance because it is expected that the network would send a measurement control message to the UE once entering the cell_DCH state; containing the neighbor lists the UE uses for handover measurements. Therefore, this feature has potentially a major gain for reducing setup delays.

On the other hand, in demanding environments, the current size of SIB 11 may not be enough. If the network is optimized with a high number of intra- and inter-frequency, and inter-RAT cells, then 50–75 neighbors per cell type is not enough. Also, when relevant neighbors are missing from SIB 11, the UE cannot make the optimal cell reselection, which may negatively impact the call success rate. Therefore 3GPP Release 6 introduces SIB 11bis, an extension to the SIB 11 Neighbor List. SIB 11bis allows defining of 32 intra-frequency, 32 inter-frequency, and 32 inter-system neighboring cells for idle mode UEs.

The advantage of SIB 11bis is better inter-frequency planning, especially with more than two carriers implemented. However, the disadvantage is that the feature is still optional in 3GPP Pre-Release 7 devices. In this case, if the network sends important neighbor list information in SIB 11bis to Pre-Release 7 UEs, SIB 11bis will be discarded by the UE, thus it would not have any information about the neighbors of SIB 11bis. As a result, it would cause missing neighbor relations, preventing the UE from making optimal cell reselections. Since LTE devices are Release 8 and beyond, and if SIB 11bis is broadcast by the UMTS cell, then skipping it will improve the CSFB call setup latency without any impact on the system performance as explained for SIB 11/12 case.

To evaluate the gains of the DMCR feature, a trial was conducted with the results shown in Figure 4.26. The trial was performed in mobility conditions where deferred SIB reading is important in terms of skipping the large size SIB 11/11bis/12. Note that the spikes seen in the non-DMCR case in the figure are due to the high errors observed while reading SIB 11 (discussed later). The average SIB reading time with DMCR = 0.82 s, while it is 1.3 s when DMCR is disabled. Therefore, DMCR has a good gain when skipping SIB11/12, with the precondition that the network sends the neighbor list during ell_DCH call setup.

Now, let us evaluate the gains of DMCR on the CSFB call setup delays, listed in Table 4.13. Comparing these values with Release 8 basic functions in Table 4.7, the gains are visible. In DMCR-based implementation, the average call setup is reduced by ~0.500 s in stationary conditions. In mobility conditions, it is reduced by ~1.8 s. The 1.8 s gain on average is expected, as all SIBs periodicities as configured by UTRAN usually set to 1.28 s. The extra time difference here is due to mobility RF conditions causing some BLER (block error rate) during reading some of the mandatory SIBs. Hence, the UEs may wait for the next round of SIB scheduling periodicity to decode them.

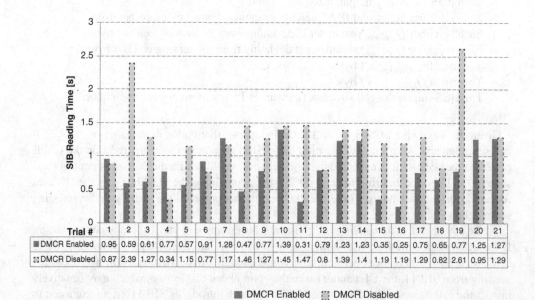

Trial #	1	2	3	4	5	6	7	8	9	10	11	12	13	14	15	16	17	18	19	20	21
■ DMCR Enabled	0.95	0.59	0.61	0.77	0.57	0.91	1.28	0.47	0.77	1.39	0.31	0.79	1.23	1.23	0.35	0.25	0.75	0.65	0.77	1.25	1.27
▒ DMCR Disabled	0.87	2.39	1.27	0.34	1.15	0.77	1.17	1.46	1.27	1.45	1.47	0.8	1.39	1.4	1.19	1.19	1.29	0.82	2.61	0.95	1.29

■ DMCR Enabled   ▒ DMCR Disabled

Figure 4.26 SIB reading time for DMCR enabled versus disabled.

Table 4.13 Mobile-originated (outgoing) call setup times baseline with Release 8 optimized

Call setup delays (ms)	Device in stationary conditions near cell			Device in mobility conditions	
	UMTS only	CSFB from LTE idle	CSFB from LTE connected	UMTS only	CSFB from LTE idle
Average	1670	2997	2789	1890	3395
Minimum	1590	2561	2326	1600	2678
Maximum	1760	4755	4886	3790	5500

CSFB with 3GPP Release 9

3GPP Release 8 optimized functions have shown encouraging improvements in bringing the CSFB call setup time to similar values to those in UMTS-only CS calls. Release 9 is promising more optimization in that regard. This is potentially possible with the introduction of the "UMTS SIB tunneling" concept during the redirection process from LTE to UMTS [1]. In Release 9, for the redirection to UTRAN, a list of PSCs, each with a container that includes the associated SIBs, can be defined. Similarly for GERAN, a list of carrier frequencies can be defined, each with its associated Network/BSS (base station subsystem) color codes and system information message.

For CSFB to UTRAN, the call flows shown in previous sections remain the same in Release 9. The change introduced in the SIB tunneling case is to have the eNB sending a different content in the LTE RRC connection release. For example, in Figure 4.20 showing an MO call flow with redirection, the message in Step 8 is renewed in 3GPP Release 9. The contents of this message are listed in Table 4.14.

With the UMTS SIBs being tunneled through the LTE RRC Release, step 11 in figure 4.20 consisting of UMTS SIB reading is no more required. The gains of this feature are expected to outperform those in DMCR because DMCR skips some of the non-mandatory SIBs for CSFB UTRAN camping. SIB tunneling allows the UE to get the SIBs in LTE prior to camping on UMTS in octets embedded inside the LTE RRC connection releasing.

Table 4.14 3GPP Release-9 LTE RRC connection release

Information element	Value	Comments
redirected carrier info	Utra-FDD utra-TDD	Carries ARFCN of the UTRAN indicating the target UTRA frequency to which the UE shall tune
Idle mode mobility control info	Refer to [1] for actual values	Network should use this value to control the UE's behavior after the UE transitions to the target UTRAN NB. It provides dedicated cell reselection priorities, also discussed in Chapter 3
cell info list-r9		This carries lists containing a sequence of physical cell IDs and the corresponding UTRAN BCCH containers (SIBs)

TDD, time division duplex and ARFCN, absolute radio frequency channel number.

SIB tunneling is studied in [9]. The gains in setup delays are estimated to be from 500 ms to 1 s, when compared to the DMCR. However, they are still not on a par with the UMTS-only CS call setup time, as the results showeded 13% extra delays in CSFB.

One of the challenges that may be envisaged in the SIB tunneling is at the level of how to communicate the SIBs from the UMTS cells into LTE cells. This requires upgrades into several nodes in the UMTS and LTE networks. Additionally, there is a problem with the neighbor relation between UMTS cells (having different SIB parameters) and the LTE cell during CSFB. This means there will be a many-to-one SIB neighbor relation in the LTE cells, requiring extra planning and optimization.

Handover-based CSFB with UTRAN Measurements

The handover-based CSFB procedure, as fully explained in Section 4.2.1.3.1, is expected to be the flash CSFB, in terms of setup time gains. When compared to the redirection CSFB, the UE initiating handover-based CSFB can skip several steps during the fallback to UTRAN, including SIB reading and RRC connection establishment in UMTS. This is evident from the call flows in Figure 4.19 (handover-based) and Figure 4.20 (redirection-based). Accordingly, the performance of the setup delays is expected to be on a par with SIB tunneling, and better than DMCR.

However, this gain is highly dependent on the ways the handover is triggered, whether blindly or not. The performance of non-blind handover with gap measurement is shown in Figure 4.25. The other consideration would be the CSFB call setup success rate if the handover is executed as a blind one. This is an area for further optimization. However, the same concepts of IRAT handover optimizations discussed in Chapter 3 can still apply to this CSFB case, and can be used as a reference. From [9], handover-based CSFB provides the best CSFB setup delays, a gain that still underperforms the UMTS-only by an estimated 9%.

Closing Notes

The eNB's decision to select any of the CSFB features discussed so far depends on the device support and network availability. The eNB initially retrieves the device capability during the LTE RRC capability message exchange at attach or during the time the RRC connection is being set up. The UE informs the eNB of its capability list through the feature group indicators (FGIs). The FGI contents are detailed in [1] and also explained in Chapter 3.

All in all, CSFB is an important feature during migrations toward VoIP, and it remains the only way to allow the networks to make voice calls until VoLTE is widely deployed. CSFB operations in Release 9 are fully optimized to minimize call setup delays at the radio interface, however, they are still below UMTS-only CS call performance, but are generally acceptable in the industry.

• To close on the CSFB roadmap, 3GPP Release 10 introduces a new feature known as "CSFB-specific access barring." In this feature, the E-UTRAN is made capable of providing access class barring parameters indicative of the congestion situation in UTRAN/GERAN for LTE-capable UEs. The objective behind this enhancement is to prevent Release 10 UEs from originating normal CSFB calls that not only require accessing the E-UTRAN, but also subsequently the UTRAN/GERAN. This prevents the congestion situation from getting worse as a result of CSFB calls. These new requirements apply to MO CSFB calls only. The requirements for handling MT CSFB and emergency MO CSFB remain unchanged, as in previous 3GPP releases.

4.2.3 Optimizing CSFB to UTRAN

4.2.3.1 Network Optimization and Configuration

Several optimization aspects are required for CSFB to achieve the targeted performance. CSFB mobile-to-mobile performance represents the most challenging case in terms of call setup delays. Extra delays due to the CSFB procedure on either side, originating or terminating parties, impact the end-user experience.

Moreover, mobility conditions of both CSFB devices can introduce additional factors to consider in the delay performance. RF variations, handovers, or loading conditions, for instance, can introduce additional CSFB call setup delays. Hence, CSFB should be evaluated in different RF conditions.

Some of the cases used in verifying and benchmarking the CSFB performance are highlighted in Table 4.15. During the benchmark testing, the related key performance indicators (KPIs) should be noted down. The test cases are to be conducted with the current status of the network CSFB features. Additionally, several devices should be used for the benchmark because the devices may have different CSFB performance or implementations. Both CSFB

Table 4.15 CSFB to UTRAN verification test scenarios

CSFB test scenario	Type of calls	RF conditions	Measurement KPIs
CSFB from LTE idle mode CSFB from LTE connected mode	MO to landline MT from landline Mobile-to-mobile	Near cell: good LTE and UMTS RF Far cell: bad RF in LTE or UMTS (once in each case) Mobility conditions: good coverage route and marginal coverage route Indoor vs. outdoor	CSFB call setup delays Setup failures Call drop rate Voice quality
Data interruption time during CSFB	Active data session with DL/UL packet activities during CSFB		User-plane delays Data resumption rate
UTRAN ↔ E-UTRAN handover	UE in LTE/UMTS connected mode	Mobility in routes with edge of coverage in LTE and UMTS	Handover success rate Call drop rate
UTRAN ↔ E-UTRAN Cell reselection	UE in LTE idle mode, or UMTS idle, Cell_PCH or URA_PCH	Mobility in routes with edge of coverage in LTE and UMTS	Reselection success rate UE reachability Paging performance

and UMTS-only CS calls should be running in parallel to benchmark the related KPIs. CSFB and UMTS-only CS call setup delays can be calculated as shown in Figure 4.23.

According to the planning of those test cases, once a failure case is detected, troubleshooting should be conducted to investigate the source of the issue. The key issues encountered that contribute to the CSFB call setup failure rate can be further categorized into four areas: (i) network optimization, (ii) network implementation, (iii) network configuration, and (iv) device failures. Figure 4.27 depicts a troubleshooting mechanism that can be used during CSFB failure investigations.

CSFB calls involve interactions between the E-UTRAN and UTRAN access parts as well as the core network of both EPS and UMTS systems. As a result, the lack of appropriate RF optimization/planning, interference management or integration between both networks leads to various issues. These performance culprits negatively impact CSFB call setup in multiple ways. Each of the areas shown in Figure 4.27, either in LTE or UMTS, can be investigated thoroughly upon detecting a failure.

On the LTE side and during CSFB calls, suboptimal E-UTRAN RF coverage can lead to RACH procedure failures, delaying LTE RRC connection setup and jeopardizing subsequent redirection to UTRAN. It can also be manifested through missed pages for MT CSFB calls. Pilot pollution in E-UTRAN leads to frequent handover executions which, if it occurs during CSFB call establishments, causes a delay in the reception of RRC connection release to redirect the UE to UTRAN, and eventually can result in CSFB failures.

Similarly, improper network planning can cause unnecessary frequent TAU procedures in the middle of the CSFB procedure execution. A failed TAU procedure may cause the MME to

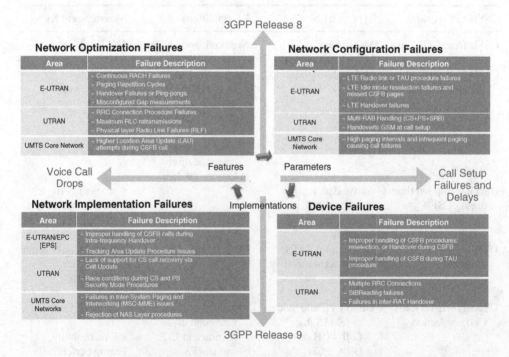

Figure 4.27 CSFB failure troubleshooting mechanism.

dislocate the UE, leading to missed pages for MT CSFB calls. Several other anomalies may occur during the CSFB setup attempts, as explained in Table 4.9.

On the UMTS side, during or after the fallback, and in areas of improper UTRAN RF coverage suffering high load/interference, either MO or MT CSFB call setup may encounter challenges. UMTS physical layer radio link failures (RLFs), RLC maximum retransmissions or failed RRC connection procedures can occur at any stage of the setup or during the call. The failures can also occur due to establishing multiple concurrent UTRAN radio access bearers (RABs), one for CS voice and one for PS data in addition to the signaling RAB. This type of call is known as multi-radio access bearer (mRAB). The next section covers the UMTS readiness for topics that should be considered in enhancing the CSFB performance.

4.2.3.2   UMTS Network Readiness for CSFB

The UMTS radio interface has undergone extensive optimization over the years by network operators, vendors, and third party consultant companies. In the mean time, HSPA+, the evolution of UMTS starting from 3GPP Release 7, has introduced add-on features to continuously improve the radio access performance. Some of the HSPA+ features, other than data-speed enhancements, have not gained solid attention because of device capabilities, commercially available or yet to come.

With regard to CS call setup delay performance, the UMTS radio interface was already in good shape as the setup time is normally low. Therefore, when CSFB to UTRAN is introduced, it is generally expected that UMTS radio re-optimizations will not be extensively required. However, this is not entirely true. As a matter of fact, it is time for operators to revisit some of the UMTS/HSPA+ features since CSFB is already in much need of improvement at the radio interface level throughout the call flow. Hence, UMTS networks should be equipped with some of the UMTS/HSPA+ unused features to improve not only the CSFB setup delays, but also the UMTS-only CS calls, whenever possible by device capabilities.

This section reviews some of the UMTS/HSPA+ features, and improvements that can directly relate to the CSFB call setup delays, setup success rate, or call retainability rates.

Signaling Radio Bearer over HSPA(+)

Signaling radio bearer over HSPA, also known as SRB over HSPA (or SRBoHSPA) is a feature introduced in 3GPP and further enhanced in Release 7.

In the very first 3GPP release, Release 99, the SRB (signaling radio bearer) is carried on the dedicated physical layer channels.[7] A dedicated channel consumes power and resources for each user making any connection with the network – voice, data, or even signaling-only calls as in location updating procedures. Even though the speed of the SRB is low, being a control plane related process, the consumption of spreading factor (SF) codes and power impact the capacity of the network.

Since the HSPA system was introduced in Release 5/6, the shared channel concept has been introduced to maximize the capacity of the networks, by serving more users within the same time slot or by sharing the same code tree. HSPA was initially intended or deployed for PS data services, however, with the resource constraints, operators needed to create other sources for improved capacity.

---

[7] Refer to Chapter 2 for a complete UMTS channel mapping and structure.

Some of the capacity improvements can come from deploying multiple UMTS frequencies (i.e., carriers) within the same band or in a different one. This process of multi-carrier deployment adds an overhead on optimization and design around the coverage-edge in between the carriers. Having said that, multi-carrier is still the main method of improving capacity constrains in UMTS networks nowadays.

Reaching the maximum number of possible carriers per design and cost implies that network operators need to find other ways to improve capacity. One of the ideas is to utilize the shared channels in the HSPA system to carry signaling messages and hence relieve some of the SFs and power to serve other users. Signaling is one of the main exhausting sources of capacity and resources. This signaling phenomenon comes from the increasing number of smartphones making a large number of PS data calls with a small volume of packets.

In HSPA+, SRB can be carried on the shared download channels by introducing another physical channel, the fractional dedicated physical channel (F-DPCH). It replaces the legacy DPCH (dedicated physical channel) dedicated for each user. F-DPCH can be used among a number of users in the system, saving power and resources in the cell. In Release 7, F-DPCH has been improved to enhanced F-DPCH to allow multiplexing of more users into one physical channel.

With signaling being carried on HSPA channels, a feature referred to in 3GPP as SRB over HSPA, the system can benefit from a higher capacity, and the user can observe a faster call setup, and possibly lower drop rate. This comes from the fact that HSPA channels are transmitted in the time domain every 2 ms, referred to as the transmission time interval. In Release 99 DPCH, the signaling messages can be transmitted on UL or DL over 10–40 ms TTI (a common network side configuration).

In CSFB, SRBoHSPA (signaling radio bearer over high speed packet access) is one of the areas where this feature becomes handy. To estimate the gains on call setup delays, the Release 99 SRB over DPCH can be used as a comparison. Taking the CS call setup call flow in Figure 4.20, the UMTS call setup starts from step 13 onwards. In step 13 where the UE receives the UMTS RRC connection setup, the RNC establishes the SRBs with the options to either map them into HSPA or Release 99 DPCH. If mapped into HSPA, then the following configuration options can be available.

1. UL SRB mapped into the HSUPA (high speed uplink packet access) transport channel, E-DCH (enhanced dedicated channel) with 10 ms TTI,
2. UL SRB mapped into the HSUPA transport thannel, E-DCH with 2 ms TTI,
3. DL SRB mapped into the HSDPA transport channel, HS-DSCH (high speed physical downlink shared channel) (always 2 ms TTI), and
4. If DL SRB is mapped into HSDPA, then configure enhanced F-DPCH.

On the other hand, if SRBs are mapped into Release 99 DPCH, then multiple options can be configured:

1. UL and DL SRB mapped into Release 99 DPCH with 10 ms TTI, referred to as SRB with rate 13.6 kbps
2. UL and DL SRB mapped into Release 99 DPCH with 40 ms TTI, referred to as SRB with rate 3.4 kbps.

A mix of combinations can also be configured by the UMTS RNC. For example, UL SRB can be mapped on E-DCH and DL SRB on Release 99 DPCH, and so on. Additionally, the network may establish SRB on Release 99 DPCH at the RRC setup case (i.e., step 13 in Figure 4.20), and later reconfigure the SRBs to HSPA when the user-plane RABs are established in the radio bearer (RB) setup (i.e., step 23 in Figure 4.20), or vice versa. Also, the network can reconfigure the SRB to any combination during the call. This can happen if the UE moves from a cell that supports only DPCH to another cell supporting HSPA, and vice versa. The choice is based on the device support and the network planning and requirements.

For the DPCH side, 3.4 kbps improves coverage at the expense of longer setup times. SRB with 3.4 kbps is mainly used during the registration process and in procedures where call setup time is not a concern. On the other hand, 13.6 kbps provides for short setup times. SRB with 13.6 kbps can be used during CS/PS call setup and then the UE can be reconfigured later to 3.4 after the CS/PS RAB setup since fewer signaling messages are exchanged across the air interface after RAB is set up. Note that the power control characteristics are different between these two configurations, and hence 3.4 kbps is the one recommended to use after the call has been already set up. This is an important factor also when multiple concurrent services are present (mRAB scenario where CS and PS in addition to signaling). Lower speed SRB can lead to lower call drops in cell-edge coverage but it may lead to higher drops during a handover situation due to delayed messaging exchange. This is a subject of RF optimization, which is also covered in the mRAB section later.

Related to this, for CS voice calls, the user-plane (actual voice packets) are still mapped into the DPCH, unless VoIP is supported in the HSPA+ network. In the case of CS calls, and because DL DPCH channels are always present for a user in dedicated mode, then the presence of SRBoHSDPA (i.e., DL SRB mapped on HS-DSCH) is not possible. Hence, from Figure 4.20, the RNC may choose to set up SRBoHSPA only at the RRC phase (control plane signaling from steps 13 to 22) and then disable it when the CS user-plane RAB is activated in RB setup (from step 23 onwards). The use-case of this scenario is only to expedite the CS call setup time. After Step 24, where the user-plane RABs are established, the voice (or data packet) will flow as the call setup has already been established.

At the CS call setup, where the PS call can also be established concurrently for a smartphone, the amount of signaling is immense. In steps 13 to 24, the UE, RNC, and core networks are all exchanging signaling messages on the control plane. Comparing the setup options explained above, Table 4.16 shows the amount of delays contributing to each stage of the call setup. The

Table 4.16 Comparison of call setup delays between SRBoHSPA and SRB over DPCH with 3.4 kbps rate

Call setup stage (values in ms)	SRB over HSPA		SRB over Release 99 DPCH with 3.4 kbps	
	CS call	PS call	CS call	PS call
Step 1: WCDMA RRC time	292	290	256	247
Step 2: Radio bearer setup	708	469	582	346
Step 3: NAS end-to-end	2531	1107	3420	1612
Total average	*2827*	*1402*	*3680*	*1864*

trial is conducted with SRBoHSPA (DL SRB mapped into HSDPA with F-DPCH, and UL into HSUPA 10 ms TTI), and SRB over Release 99 DPCH (both UL and DL mapped into DPCH with 40 ms). The trial shows a reduction in CS and PS call setup delays on average of ~25% (see Figure 4.23 for the call setup stages in UMTS-only calls).

From this live-network trial, the gains of SRBoHSPA are visible on the SRB packet delays transmitted in the control plane. Furthermore, Figure 4.28 shows the distribution of control-plane signaling delays in the case of SRBoHSPA and SRB over Release 99 DPCH in mobility conditions. The control-plane signaling delay depicted in the figure is defined in terms of handovers. It is the time from when the device sends the UMTS measurement report to add a cell into the active set until the RNC replies with the active set update message to complete the handover. The same delay sample can represent any signaling process, including call setup delays.

With the SRBoHSPA (DL 2 ms, and UL 10 ms TTI), the average SRB delay in this process is 199 ms. In the case of SRB mapped into the DL DPCH and UL HSUPA (with 10 ms TTI), the average SRB delay is 423 ms. The SRB time delay reduction from using SRBoHSPA compared to SRBoDCH/HSUPA is 53%. This delay reduction in SRB transmission improves the handover and data interruption time and thus the call stability. Applying the gains shown in these trials into CSFB, the call setup delays can be reduced significantly, once the SRBoHSPA feature operates properly.

Lastly, while optimizing the SRBoHSPA for better CSFB call setup delays, one of the legacy concepts in UMTS is required to be revisited. The concept of activation time (defined as the absolute "frame number" at which the physical channel operation should take effect) is important for optimizing the SRBoHSPA. For example, looking back at Table 4.16, when the network sends RB setup to establish the user-plane RAB, the RNC first needs some time to configure these RABs and their attributes. Hence, the calculation of activation time in the RNC takes into account: (i) the signaling message size, (ii) RLC retransmission (for SRB on DPCH) and HARQ retransmission (for SRBoHSPA), and (iii) DL and UL processing delay in the RNC and UMTS cells.

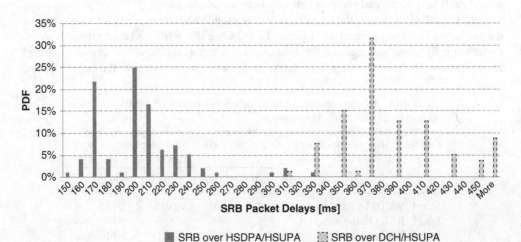

Figure 4.28 Control plane signaling delays of SRBoHSPA versuss SRB over DPCH 3.4 kbps in mobility conditions.

The relative activation time (the activation time would be obtained by adding the relative activation time to the time the RNC sends the radio setup message) should be re-optimized in the case of SRBoHSPA when compared to SRB over DPCH. If the activation time selected by the RNC is actually the same for the SRBoHSPA and SRBoDPCH, then it effectively means that the time from the UE receiving RB setup to the activation time is longer for SRBoHSPA since the RB setup arrives at the UE faster in the case of SRBoHSPA. This is visible in step 2 in Table 4.16 where the activation time selected by the RNC is the same for SRBoHSPA and SRBoDPCH, hence impacting the overall delays in this call setup stage.

The point here is that the RNC should use a smaller relative activation time for SRBoHSPA than for SRB over the DPCH to make use of the feature's benefits, as the transmission time is less for SRBoHSPA. However, the activation time should not be reduced dramatically, in order to account for the retransmissions on the air interface in challenging RF conditions. The rule of thumb is that if the activation time is set too small, the time of retransmitting signaling packets with error bits may be reduced accordingly, which increases the possibility of call drops in adverse environments. Also, the larger the size of the SRB message, then the higher the activation time should be, accounting for RLC (HARQ) retransmission and SRB RTT.

UMTS SIB Periodicity and Scheduling

In legacy UMTS operations, the SIBs are needed for the UE to gather the parameters broadcast from the network while in idle mode or at the time the RRC is being established. Once the UE enters connected mode, the SIBs parameters are overridden by the connected mode messages.

In CSFB call setup scenario with redirection, the SIBs are needed since the UE comes from the LTE into UMTS idle mode first before setting up the RRC connection. The SIB reading is required as shown in the call flow of Figure 4.20, steps 9a–11.

As discussed in Section 4.2.2.2.2, DMCR is introduced to skip some of the non-mandatory SIBs related to CSFB call setup. However, some SIBs are required even with DMCR enabled, such as SIBs 1, 3, 5, and 7. In Release 9 with the SIB tunneling concept, the SIBs are all read ahead of time, and hence the UE does not even go through this stage. Thus, this section covers the SIB periodicity enhancements needed for Release 8 CSFB calls with redirection.

Based on the UMTS SIB scheduling facts in 3GPP in [10], the SIBs are all broadcast unitarily from each of the UMTS cells and from each frequency. Depending on the size of each SIB scheduled in its time period, the SIB is segmented over the duration of the BCCH (broadcast control channel) time bandwidth of 20 ms. The master information block (MIB) is the block responsible for conveying to the UE which of the SIBs are being scheduled, along with their periodicities. With DMCR, the UE is required to read the MIB, and SIBs 1, 3, 5, and 7. Therefore, the scheduling and periodicity of broadcasting those SIBs are now important for Release 8 CSFB. At the scheduled period, if the UE misses reading one of these mandatory SIBs in mobility conditions due to deteriorating RF, the UE needs to wait for the next period of time where the same SIB is scheduled again.

Table 4.17 shows an example of SIBs periods scheduled in the MIB. For the mandatory SIBs 1, 3, 5, and 7, it is important to set the periodicity to a low value to avoid the UE waiting for the next scheduling period if the first instance is missed due to RF conditions. With a total SIB periodicity being typically configured as 1.28 s in UTRAN cells, a significant increase in CSFB call setup latency can occur when combined with bad RF.

Table 4.17 Example of SIB
scheduling and periods

SIB type	Periodicity (ms)
SIB Type 1	640
SIB Type 3	320
SIB Type 5	320 (3 segments)
SIB Type 7	80
SIB Type 11	1280 (12 segments)

In this case, if SIB 3 is missed by the UE in its first transmission instance during the CSFB procedure, the UE waits for the next scheduling period after 320 ms. This effectively adds an extra 320 ms to the CSFB call setup time.

In addition, the size of the SIB is important. In Table 4.17, SIB Type 5 periods are scheduled every 320 ms with three segments. As the size of the SIB is too big to fit into one 20 ms BCCH period, it is segmented into three parts. If one of the segments is lost, the UE needs to wait to complete the decoding of the entire SIB. Note that each segment is not transmitted every 320 ms, but can be scheduled in MIB in a way to avoid SIB scheduling collisions with other broadcast SIBs.

For illustrative purposes, the SIB scheduling used in the example shown in Table 4.17 is characterized in Figure 4.29 from a live-network CSFB call failure case. In this network, the DMCR is not implemented, and hence the UE is required to read all SIBs according to the periods in Table 4.17. First, let us look at the ways SIBs are scheduled according to this figure:

- SFN-prime (system frame number) is set to the period of the BCCH channel of 20 ms. It is a frame number used by the UE and incremented every 20 ms, as defined in [10]. UE usually gets to know the frame number after acquiring the cell from the PCCPCH (primary common control physical channel) to synchronize to the cell.
- SIB Type 7 received at SFN-prime numbers 1409 and 1413. The distance between these two SFNs is four frames, each of 20 ms length, making the periodicity of SIB 7 to be $4 \times 20 = 80$ ms, in line with the configuration of the network.
- Similarly, SIB 3 in the first instance is decoded by the UE at SFN 1415 and then repeated at 1431. The difference is 16×20 ms $= 320$ ms.

In this scenario, the UE is in challenging RF conditions, and a high number of SIBs missed by the UE. In the entire round of SIB scheduling, the UE receives all of the SIBs multiple times, but is still waiting for SIB 11 segment 6. Due to the UE missing this segment and then the serving cell's RF conditions going very poor a cell selection failure results. Hence, the CSFB call setup fails while the UE is trying to decode the SIBs. The total SIB reading BLER observed in this duration (i.e., missed SIB segments) is calculated as 38 missed segments out of $127 = 30\%$ BLER.

Based on this example, with the DMCR implemented, and if SIB 3 is missed in its first broadcasting period at SFN 1415, then the UE will wait to the next round of SIB 3 scheduling at SFN 1431. If eventually received there, this effectively means an extra delay of 320 ms to the CSFB call setup time. The same applies to SIB 1 or SIB 5.

SFN-Prime	1407	1408	1409	1410	1411	1412	1413	1414	1415	1416	1417	1418	1419	1420	1421	1422
SIB Type	11 indx11	MIB	7	5 indx0	5 indx1	MIB	7	5 indx2	3	MIB	7	1	11 indx12	MIB	7	11 indx1
SFN-Prime	1423	1424	1425	1426	1427	1428	1429	1430	1431	1432	1433	1434	1435	1436	1437	1438
SIB Type	11 Indx2	MIB	7	MISSED	5 indx1	MIB	7	5 index2	3	MIB	7	MISSED	MISSED	MIB	7	MISSED
SFN-Prime	1439	1440	1441	1442	1443	1444	1445	1446	1447	1448	1449	1450	1451	1452	1453	1454
SIB Type	MISSED	MIB	MISSED	5 indx0	MISSED	MISSED	MISSED	MISSED	MISSED	MIB	7	MISSED	MISSED	MISSED	7	MISSED
SFN-Prime	1455	1456	1457	1458	1459	1460	1461	1462	1463	1464	1465	1466	1467	1468	1469	1470
SIB Type	11 indx8	MIB	7	5 indx0	5 indx1	MIB	7	5 indx2	3	MIB	7	Empty	11 indx9	MIB	7	11 indx10
SFN-Prime	1471	1472	1473	1474	1475	1476	1477	1478	1479	1480	1481	1482	1483	1484	1485	1486
SIB Type	11 Indx11	MIB	7	5 indx0	5 indx1	MIB	7	5 indx2	3	MIB	7	1	11 indx12	MIB	7	11 indx1
SFN-Prime	1487	1488	1489	1490	1491	1492	1493	1494	1495	1496	1497	1498	1499	1500	1501	1502
SIB Type	MISSED	MISSED	7	MISSED	MISSED	MIB	7	5 indx2	3	MIB	7	Empty	11 Indx3	MIB	7	11 indx4
SFN-Prime	1503	1504	1505	1506	1507	1508	1509	1510	1511	1512	1513	1514	1515	1516	1517	1518
SIB Type	11 indx5	MIB	7	5 indx0	5 indx1	MIB	MISSED	MISSED	MISSED	MISSED	MISSED	MISSED	MISSED	MISSED	7	11 indx7
SFN-Prime	1519	1520	1521	1522	1523	1524	1525	1526	1527	1528	1529	1530	1531	1532	1533	1534
SIB Type	11 indx8	MIB	7	MISSED	MISSED	MISSED	MISSED	MISSED	MISSED	MISSED	MISSED	MISSED	MISSED	MISSED	7	Call Setup Fails

Figure 4.29   UMTS SIB scheduling and periodicity example from live-network.

In conclusion, in challenging UMTS RF conditions while the CSFB call is being set up, the SIB reading is an important factor. Skipping some of the SIBs allowed by DMCR is one way to improve the call setup delay. Furthermore, optimizing the SIB periodicity and the number of segments for each of the mandatory ones is another important factor with or without DMCR. Keeping the mandatory SIB Types 1 and 3 within periods of 160 ms is a recommended configuration for CSFB. Adding any new SIB to the cell, like SIB 19, SIB 4, and SIB 11bis, also requires additional scheduling optimizations. Any change made in this regard requires the assurance that non-CSFB users are not impacted. Ultimately, SIB scheduling optimization overhead is avoided once Release 9 SIB tunneling for CSFB is deployed in a network.

Tracking Area and Location Area Mapping as Part of Combined Attach or Combined TAU Procedures

The use of virtual LA during LTE combined attach/TAU is important. The main purpose of defining LAI inside the LTE attach/re-attach procedure is to eliminate the time for the UE to perform LAU in every CSFB to UTRAN. This mapping usually happens in step 6 of Figures 4.16 and 4.18, respectively.

The proper mapping between LA and TA in this case is significant. Let us first look at the impact on setup delays by performing LAU during CSFB calls. Table 4.18 shows a case from a live network for CSFB MO call setup delays from a mobile to a landline where in one run LAU is needed and in another run LAU is not required. As shown, 600 ms of extra CSFB setup time was added to complete LAU (in stationary conditions). In mobility conditions, up to 3 s of delays due to the LA updating procedure were observed.

LAU is usually performed in UMTS after the CSFB if the LA of the UE in the camped serving cell is different from that LA identity conveyed to the UE in the last combined attach/TAU procedure. This is the case shown in step 15 in Figure 4.20. To alleviate the delay impact of LAU, a proper mapping of LAU is required of the physical cell during the combined attach or combined TAU procedures.

Figure 4.30 illustrates an example of this potential mismatch. In this example, both the UMTS cell ID 46 and the LTE cell ID 58 are collocated physically in the same cell site. The LA ID of the 3G cell is 3329, while the TA ID of the LTE cell is 200. So in the combined TAU procedure explained in Section 4.2.1.2, it is expected that the TAU accept message will send

Table 4.18   Example of SIB scheduling and periods

RF conditions	Near LTE cell in stationary condition	Near LTE cell in stationary condition
CSFB stage (values in seconds)	*CSFB without LAC procedure*	*CSFB with LAC procedure*
LTE ESR to real time	0.18	0.22
Tune to WCDMA and ACQ time	0.31	0.34
SIB read time including cell selection	0.83	0.81
RRC latency	0.19	0.22
NAS E2E time	1.84	2.40
Total average (s)	*3.50*	*4.13*

Note: LAC, location area code.

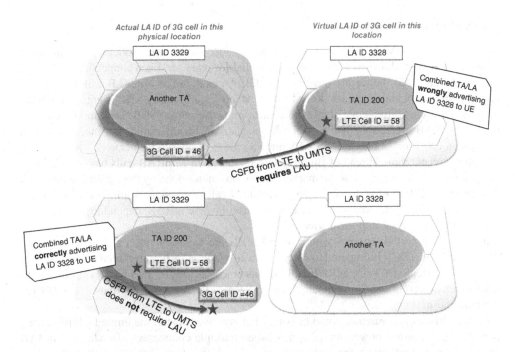

Figure 4.30 Examples of mismatch between TA/LA in combined TAU procedure, compared to the correct configuration.

the LA ID 3329 to the UE. Once the UE in stationary conditions is under the coverage of the LTE cell with ID 58, the CSFB is performed and the UE is blindly redirected to the collocated UMTS cell with ID 46. The UE then reads SIB 1 from cell ID 46 and finds its LA ID to be 3329, which is similar to the one saved from the combined TAU procedure. Hence, the UE does not perform LAU at the CSFB setup, saving on the call setup delays.

However, the troubling scenario (upper part of Figure 4.30) occurs if the TAU accept message misconfigures the LA identity and sends a different ID from that which the UMTS cell is physically configured with. In this example, TAU accept sends the LA identity 3328 and not 3329. Therefore, when the UE falls back to UMTS during the CSFB call, and reads the LA ID configured correctly in the SIB 1 as 3329, the UE is required to perform the LAU procedure, adding an extra delay to the setup time. The solution is to replan the configuration setting of cells in TA ID 200 to match LA ID 3329, in this example (as shown in the lower part of Figure 4.30). A manual planning of such a process is extensive and requires careful settings. The case can be most noticeable in stationary conditions where a fallback to the UMTS cell is expected to happen from a collocated LTE cell within the coverage area.

On the UMTS side, an additional network planning aspect is with regards to LA and MSC boundaries. These are not always optimized and if the LA and the serving MSC change during a CSFB call setup, some failures could occur. A failure symptom is typically a missed page when the new MSC (after LA change) does not page the UE over the correct interface nor send the required messages to proceed with the call after getting the UE profile. This case is usually mitigated by deploying MSC pooling, as explained in [9]. With MSC pool architecture, all MSC servers within a pooled area serve all belonging cells, eliminating MSC borders' impact on paging failures or time delay of inter-MSC LAUs within the pool.

UMTS Multi-RAB Improvements

The definition of UMTS mRAB is when two or more concurrent services from the same domain or different domains are ongoing in parallel for one user. Each connection is mapped into a separate RAB, hence the term multi-RAB. An example of a mRAB call is when the CS-domain user-plane voice (also known as the adaptive multi rate, AMR) is going in parallel with the PS-domain user-plane data. The control-plane signaling (SRB) is common with one or multiple RABs. Additionally, PS + PS calls are also considered as mRAB. Since this section focuses on CSFB calls, only the CS + PS mRAB case is discussed further.

With the recent rapid spread of smartphones, the usage of MmRAB calls has risen accordingly. It is estimated in some of the markets, based on studies done by the authors of this book that, on average, 16% of the CS calls have at least one mRAB session. Additionally, with some smartphones 47% of the total calls made by one user involve mRAB, depending on the brand and the operating system of the device.

It has been reported worldwide that mRAB calls are experiencing a significantly higher dropped call rate (DCR) compared to voice-only calls. Operators have reported anywhere from double to quintuple DCR for MRAB calls versus voice-only calls. This has caught the attention of both operators and handset manufacturers because the mRAB performance is associated with the end-user experience.

Inferior mRAB performance could be somewhat intuitive because the limited UE resources, primarily UL transmit power, are shared between multiple connections. In addition, mRAB calls were not prevalent in the early days of 3G and, therefore, not much effort has been spent in optimizing and configuring the network and mobile devices for better mRAB performance. Nowadays, optimizations aimed at mitigating the mRAB performance are needed across the spectrum, including network configuration, network features, UE features, and signaling protocol enhancements.

The main factor in the high drop rate in the first place is that the UMTS air interface in 3GPP is designed so that if any of the RABs experience protocol or transmission issue (CS, PS, or SRB), the entire connection could drop. Therefore, in mRAB calls, if the PS call is the source of issues, the CS in parallel could also drop. Thus, the rate of drop can be much higher than single RAB calls. In the same way, PS call drops have not received much attention because, if the connection is lost in PS-only data calls, the re-establishment that takes place on the air interface will revive the call, a process that is transparent to the user. But in CS-only calls, once the call drops, reviving the call in the real-time conversation is going to be visible to the user.

Recently, 3GPP has recognized the need to address mRAB performance issues and the efforts in this area are ongoing. Some of the 3GPP enhancement efforts are highlighted in the 3GPP contributions in [11, 12]. As operators are looking for immediate solutions, additional effort is needed to mitigate the mRAB performance in a shorter time frame.

Moreover, with the introduction of LTE and its backward compatibility with other radio access networks, including UMTS, the inferior mRAB drops can impact the overall user experience, in particular, those with CSFB support. As already explained, CSFB requires the fallback from LTE into the UMTS network. Also, while on LTE and a data session is active, the CS call, originated or received by the user, drags the device into UMTS, creating a higher possibility of an mRAB situation. Hence, addressing the mRAB performance is crucial not only to the UMTS network users, but also to the LTE users.

For a CSFB call, the mRAB session usually occurs immediately after the call setup if the PS data are successfully transferred from LTE into UMTS during the ongoing CS call. This is illustrated in steps 23 and beyond, in Figure 4.20. There are several options and configurations of an mRAB call. These options are shown in Figure 4.31. Mainly, a CS call can concurrently coexist with PS data mapped onto UL/DL Release 99 DPCH channels, or on UL/DL HSPA(+) channels. The configuration choice is essential to the call performance and end-user experience.

From extensive studies by the authors in this book in multiple networks, the causes of the inferior mRAB versus voice-only drops are embedded in the system design or parameters. Figure 4.32 illustrates some of the most common reasons for mRAB drops.

Some predictable causes are sometimes quantifiable, like the LB impact, whereas other aspects of the inferior mRAB performance are typically tracked via network KPIs and other means of statistical performance monitoring.

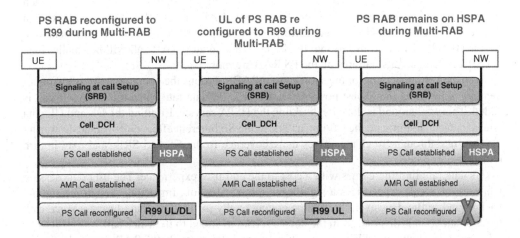

Figure 4.31   PS RAB rates in multi-RAB.

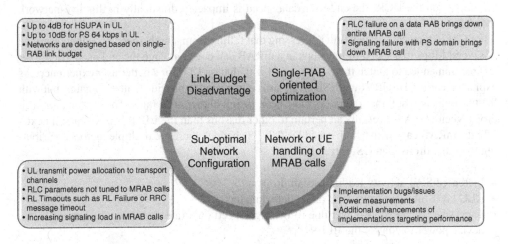

Figure 4.32   Common reasons for suboptimal performance of MRAB calls.

In the LB area, some mRAB combinations use a lower SF than voice-only calls. For example, voice-only calls typically use UL SF 64, whereas mRAB combinations AMR12.2 + 16 kbps use UL SF 32. There is an associated LB loss of 3 dB for mRAB in this example. The LB loss for mRAB with UL PS 64 kbps (SF 16) would be 6 dB. Meanwhile, in mRAB combinations involving HSUPA in the UL, AMR and SRB will use the same SF as voice-only calls because EUL (enhanced uplink) is mapped on a separate E-DPDCH (enhanced dedicated physical data channel). Therefore, mRAB combinations with EUL would not have an LB disadvantage due to lower SF with respect to voice-only calls.

Additionally, in an mRAB call at the cell edge (max UE transmit power) scenario, UL data transmission (Release 99 DPCH or HSUPA) will be allowed only when there is no AMR or SRB activity. So the nominal data rate of UL data RABs in mRAB at the cell edge must be scaled down with respect to the voice activity factor (VAF) and SRB AF. For example, assuming VAF of 50% and SRB AF of 10%, the maximum possible PS AF is 40%, which means that the nominal data rate of the UL 16 kbps RAB in mRAB call at the cell edge in this case is $16 \times 0.4 = 6.4$ kbps. This corresponds to an LB loss of around 4 dB applicable to data RAB only. This also means that the PS coverage in an mRAB call will be smaller than CS/SRB coverage, thus, failures on the PS RAB are more likely.

Given the LB impact, one of the short-term mRAB solutions the operators have been implementing is by using the lowest possible DL and UL PS data rate when the CS AMR call is added. This workaround is done by disabling the HSPA service. Limiting the DL and UL data rate is intended to mitigate the LB impact due to the SF inherent effect on coverage. Lowering the data rate through SF change gives more coverage to the CS and SRB, and may in turn reduce the drop rate.

This data rate limiting comes with a cost to the end-user experience. The user data session while in a CS call experiences a significant speed reduction, impacting web-browsing and background traffic. However, this can be commercially justified, in that, during a voice call, the user attention is diverted from the device screen to attend to the voice call.

Figure 4.33 shows the SF comparison between mRAB (with PS RAB mapped into the Release 99 DPCH channel) and single-RAB (CS only or PS only). The presence of PS RAB in mRAB calls is intuitively using a lower SF than CS-only. With this workaround in order to save LB impact on the DCR, the end-user data speed is impacted drastically in this live-network example.

For CSFB-type calls, it is worth addressing that falling from a high-speed broadband service of several Mbps to a speed of few kbps in a mRAB call can challenge the end user. Therefore, it is recommended to match the UMTS speed with that on LTE for a better user experience. As explained earlier, HSUPA (and HSDPA in turn) is expected to provide better LB gains, but with the unfortunate result that operators have been facing higher drop rates. The reasons for such drops when HSPA is present can pertain to other reasons than the LB, this is discussed next.

In the mRAB case with a high speed DL/UL PS channel, there are multiple factors contributing to higher drops than CS-only calls:

- Increased SRB activity leading to call drops
- RLC layer time-out from SRB or PS data domains
- Excessive UL transmit power due to the high speed data rate requiring an increase in the device power to control the BLER.

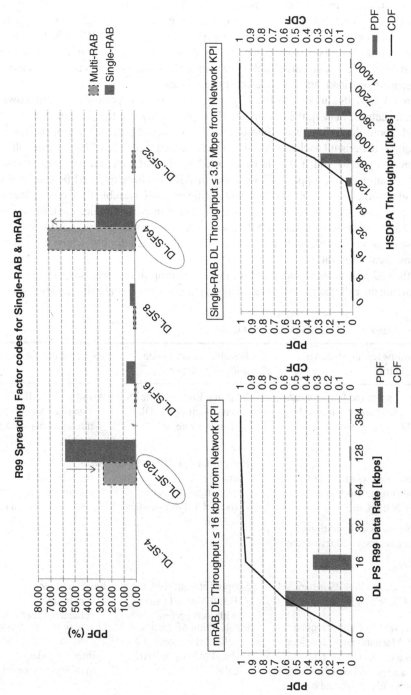

Figure 4.33   Spreading factor impact and data rate comparison of mRAB versus single CS RAB.

mRAB control-plane signaling interacts with two core network domains, PS and CS. In CS-only calls, the amount of signaling is expected to be limited to handover messaging overheads. Conversely, in mRAB, the amount of signaling is expected to increase with the presence of the PS domain extra signaling. In challenging RF conditions, once SRB fails, the voice and PS data calls could both drop. Hence, the increasing amount of signaling here comes with a higher probability of dropped calls. Table 4.19 summarizes some of the reasons for increasing mRAB signaling overload, and possible short-term solutions.

When an acknowledge mode (AM) entity in the RLC layer experiences an unrecoverable RLC error, as defined in [13], the cell update (CU) procedure is initiated by the UE to inform the network about the failure, according to [10]. An unrecoverable RLC error indicates that further attempts to transmit packets to the network are abandoned. The CU procedure following RLC failure is optional and if it is not supported by the network, the UE will release the connection, resulting in a dropped call. Because mRAB calls have two AM RLC entities (one for SRB and the other for data RAB) the risk of hitting an unrecoverable RLC error is increased with respect to voice-only calls (AMR is configured as transparent mode, TM). Moreover, as discussed earlier in the section, data RAB may have smaller coverage than SRB/CS, meaning that the risk of unrecoverable RLC error in mRAB is more than double, which directly maps onto increased call drop rate.

Even if the CU procedure is supported it may still end up in call release or in a significant delay impacting the call quality, potentially causing the user to end the call. The same logic can

Table 4.19   Signaling overload in multi-RAB calls

Possible reasons for high mRAB signaling activities	Possible network-side mitigation actions	Expected outcomes
The need to support the frequent data call (re)establishment initiated by user/applications in smartphones	Enabling low rates for PS data, such as UL/DL 0/0 kbps rate	Avoid extra SRB signaling during the transition between PS activity and no-activity
PS RAB configurations/re-configurations and additional core network signaling from PS CN domain For example, PS reconfiguration of HSPA parameters, or adaptive rate change	Prevent DL/UL R99 rate change reconfigurations by fixing UL/DL rates for the UE	Reduction of PS signaling, and minimizing signaling conflicts between calls
Mobility in mRAB CS + PS is handled either by CS or by PS criteria For example, IRAT or inter-frequency handover. Mechanisms to consider mRAB as a distinct entity adding challenges to parameter optimization, hence not widely supported by network vendors	Support of both R99 and HSDPA on all carriers with load preference Establishing of specific cell/carrier mobility thresholds for mRAB calls	Easier and more synchronized cell/carrier mobility Decrease of compressed mode activations and inter-frequency mobility/handovers

be applied to DL, even though the behavior of the RNC with respect to DL RLC unrecoverable errors is less regulated in the specifications.

The conventional network implementation in single PS RAB is that the downside to delaying RLC resets and RLC recovery through connection re-establishment from data RABs is in the reduced long-term throughput that could be achieved on a fresh connection. This does not apply to mRAB calls: here the main objective is not to increase the probability of higher throughput during connection re-establishment but rather to preserve the connection as long as possible. In a typical network deployment, RLC parameters for mRAB calls are not necessarily configured to follow this objective. Figure 4.34 explains the flow of the RLC packets and the contribution of the UL/DL parameters and implementations. The RLC parameters to optimize are too extensive to name one by one in this section. The point is to review such mechanisms with the aim to improve the mRAB drops (more RLC parameters and mechanisms are given in [13]).

Finally, let us address the UL transmit power factor and its impact on the mRAB with HSPA configured. The nominal UE transmit power (AKA (authentication and key agreement) DPCH power gain) is the transmit power before the UL transmit power commands (TPC) and initial DPCCH (dedicated physical control channel) power offset are applied. It is given by 3GPP [14]:

$$\text{UE\_tx\_pwr} = 1 + \frac{\beta_d^2}{\beta_c^2} + A_{hs} = 1 + \frac{\beta_d^2}{\beta_c^2} + \frac{\beta_{hs}^2}{\beta_c^2}$$

The gain factors β_c, β_d, β_{hs} are configured by the Network and they determine the fraction of the nominal UE power assigned to the DPCCH, DPDCH (dedicated physical data channel), and HS-DPCCH, respectively. If the HSUPA is also transmitted in the UL, there are additional terms for HSUPA-specific gain factors β_{ed} and β_{ec}. Beta gain factors are configured for each transmit format combination (TFC) separately. Successful reception of the UL DPCCH is needed to maintain radio link synchronization and detection of TFCI (transport format combination indicator) bits. Poor reception of the DPCCH will result in increased DCR due to loss of synchronization in the UL (UL RLF). Successful reception of DPDCH is needed to maintain the target packet loss rate (DTCH (dedicated traffic channel), DCCH (dedicated control channel) BLER) and avoid RLC errors. Successful reception of the HS-DPCCH is needed for providing the physical layer acknowledgments (ACK/NACK) and CQI for DL data transmission on the HS-PDSCH (high speed-physical downlink shared channel).

In mRAB calls in poor RF conditions, priority is given to maintaining the call and its quality over maximizing data throughput on the HS-PDSCH. This philosophy needs to be reflected in the configuration of the beta gain factors. In mRAB calls in poor RF conditions, the transmit power should be allocated to the DPDCH and DPCCH with higher priority than to HS-DPCCH. In current deployments, this may not always be the case. Additionally, the power consumed by the HS-DPCCH will effectively reduce the power available to other UL channels, including the UL DPCCH, UL DPDCH, and/or HSUPA. Depending on the configuration of the beta factors and the HS-DPCCH AF, the reduction in transmit power available to the DPCCH and DPDCH due to the HS-DPCCH transmission ranges from 1.5 to 2.5 dB. This disadvantage is applicable to all mRAB combinations across the board.

In conclusion, the setting of CS calls concurrently with the PS call and signaling control plane has an impact on the user experience. Each PS configuration of low or high data speed on

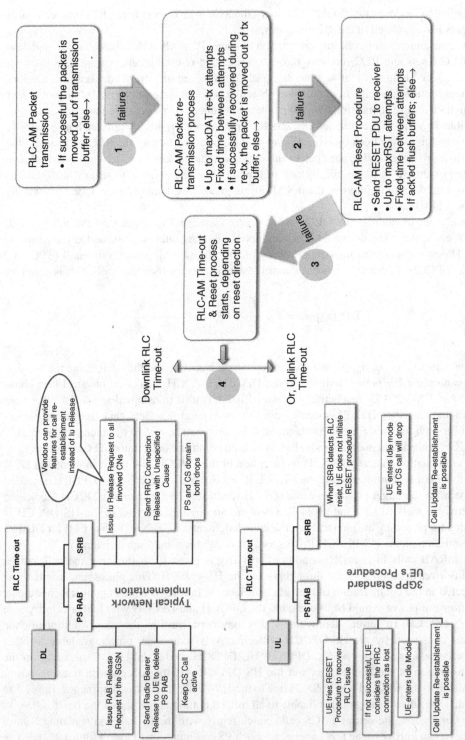

Figure 4.34 Optimizing oPS and SRB RLC parameters for mRAB calls.

Release 99 or HSPA channels has its own pros and cons on the system KPIs and end-user experience. For CSFB calls from LTE to UMTS, keeping the mRAB drop rate as low as possible requires some compromise in the user experience. The optimization processes should account for this trade-off and a review of UMTS network design and parameters is thus required when CSFB and LTE are deployed.

Fast Return to LTE from UMTS

So far in this section, different stages during the CSFB have been discussed. In Figure 4.15, the main aspects of the CSFB call are covered: the LTE attach/re-attach, the transition to UMTS during the CSFB setup, CSFB call setup performance and delays, data interruption performance, and the CS call stability while in UMTS. Once the user with an LTE device releases the CS call, the system needs to ensure the user is pushed back again to LTE as soon as possible to gain from the high-speed network. This section covers the fast return to LTE (FR2L) aspect which concludes the last step in the stages of the CSFB timeline in Figure 4.15.

FR2L could utilize several techniques depending on the layers configured: multi-UMTS layers, GSM, and multi-LTE layers. Figure 4.35 and Table 4.20 show the possible multi-RAT transitions between the different systems.

When the interactions between LTE and GSM are not directly supported by the operators (e.g., scenarios 5 and 6 in Table 4.20), and while inter-system transitions between GSM and UMTS are commonly used, some delays in camping back to LTE occur. Additionally, delays to camp back to LTE from UMTS can occur if the network is designed to depend only on cell reselection methods. The following case studies highlight possible IRAT scenarios in some deployments.

(i) **Case Study 1**

In the case of CSFB to UTRAN, and after the call setup, and with aggressive thresholds, the voice call can experience a quick iRAT handover from UMTS to GSM. This is a typical case as network operators usually push the CS voice traffic quickly to GSM to maintain a better call drop rate, and minimize the load on the UMTS layer to be able to serve more data users. When the CS call is released by the user in GSM, the UE has to wait to reselect back to UMTS. From UMTS, a reselection back to LTE becomes possible, in this example of a common network deployment. Therefore, the user is not served on LTE for a long time after the CSFB call release, depending on the transition time from GSM to UMTS, back to LTE.

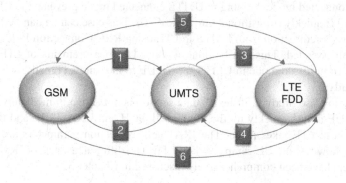

Figure 4.35   Inter-system transitions.

Table 4.20   Inter-system transitions explanation

Scenario	During PS call	During CS call	During CS connection release
1	IRAT cell re-selection	CS handover (usually not supported by operators)	Fast return from GSM to UMTS, or as cell reselection right after CS release
2	Handover (usually disabled by operators)	Handover, depending on coverage-based compressed mode thresholds	Fast return from UMTS to GSM is not needed. Cell reselection from UMTS is possible, right after CS release
3	Handover (usually not supported by operators)	Not possible	Fast redirection from UMTS to LTE, or cell reselection from UMTS Idle, Cell_PCH or URA_PCH right after release
4	Supported as re-direction (common) or as IRAT handover	CSFB Call (redirection, or handover, using Release 8 or Release 9 techniques)	Not possible (no CS calls in LTE), but possible in normal LTE RRC Release for load balancing (dedicated priority at RRC connection release)
5	Supported as re-direction or as IRAT handover (not widely supported by operators)	CSFB call (not widely supported by operators)	Not possible (no CS calls in LTE)
6	IRAT cell re-selection (usually disabled by operators)	Not possible	Fast return from GSM to LTE, or as cell reselection, right after release

(ii) **Case Study 2**

Another typical scenario is where a CSFB is made and the UE falls into the UMTS network and continues with a call. Immediately after the CS call is released, the UE is expected only to be pushed back to LTE through IRAT idle mode cell reselection parameters. However, if PS data are still active after the CS call release, the UE remains in UMTS connected mode. Staying in UMTS connected mode prevents the UE from going back to LTE quickly (transition #3 in Table 4.20). This case can be handled if the operator deploys a redirection from UMTS to LTE using RRC Connection Release right after CSFB call is ended. This can be done without UE measurements of LTE cells, or by enabling compressed mode in UMTS for the UE to measure LTE cells.

(iii) **Case Study 3**

Similarly to Case Study 2, if the PS data session is not active immediately after the CS call, the UE moves quickly to idle mode, cell_PCH and URA_PCH, and the IRAT cell reselection to LTE takes place. The IRAT cell reselection parameters are the decisive factor for how fast the UE camps back to LTE. The different aspects of IRAT reselection parameters have been comprehensively discussed in Chapter 3.

To improve the camping time back to LTE, for any of the case studies mentioned earlier, 3GPP allows the concept of fast return to WCDMA (FR2W) or FR2L. The two features are different, but for CSFB calls they serve the same purpose of allowing the UE to quickly camp back to LTE.

FR2W is introduced as a feature in GSM. The network directs the UE straight to a WCDMA cell after a call release in GSM. In GSM's channel release message sent to the UE while releasing the CS call, the GSM cell can indicate a list of UTRAN frequencies to which the UE camps directly. This reduces the time needed for the UE to reselect back to UMTS. Figure 4.36 illustrates the concept, and explains the delays observed with and without FR2W.

FR2W is particularly beneficial to a network deployment that does not support CSFB or reselections between LTE and GSM directly. For LTE-capable devices experiencing a transition from UMTS to GSM during the CS call, then they could camp back quicker into LTE right after the call release (through a quick transition from GSM back to UMTS back to LTE). This feature can resolve the delays occurring in Case Study 1.

On the other hand, FR2L is introduced in 3GPP to redirect the UE from UMTS to LTE after the CSFB call is released. The support for idle mode IRAT cell reselection alone does not guarantee the UE returning to LTE quickly. Instead, FR2L provides a mechanism for CSFB when, after the UE releases the call in UMTS, the RNC sends the RRC connection release message with redirection information to LTE. If the "E-UTRA target information" is present, UE attempts to camp on any of the LTE frequencies included in the RRC connection release message.[8] Figure 4.37 illustrates the concept, and explains the delays observed with and without FR2L.

The FR2L feature is beneficial to any CSFB network deployment. For LTE-capable devices falling back to UMTS during a CSFB call, then they could camp back quicker into LTE right after the call is released. This feature resolves the delays occurring in Case Study 3, specifically.

Lastly, some device-related features also lead to FR2L. In the situation mentioned in Case Study 2, there is a potential need for the UE itself to be able to transition to LTE from the UTRAN connected state. This device-based FR2L thus provides a mechanism enabling the UE to autonomously return to LTE from UMTS when a CSFB call ends. This is usually done by means of "fast dormancy" device-based solutions, where the UE sends the signaling connection indicator (SCI) to the RNC, which in turn releases the RRC connection and transitions the devices quickly to idle mode. In idle mode, the device reselects to LTE. UE-based

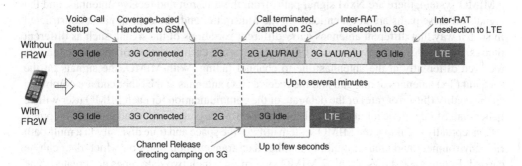

Figure 4.36   Fast return to WCDMA from GSM, and its time delay interaction with LTE camping.

---

[8] Excluding any cell indicated in the list of "not allowed" cells for that RAT, if present.

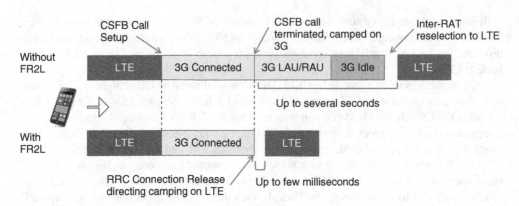

Figure 4.37   Fast return to LTE from UMTS, and its time delay association.

"fast dormancy implementation" is widely used in smartphones [15], and the same concept can be extended to CSFB, expediting the return to LTE after the call release, hence enhancing the user experience and battery consumption.

However, there are known negative impacts of the UE-based fast dormancy in terms of increasing signaling load discussed in [15]. This could imply that CSFB UE-based FR2L through fast dormancy may impose risks to the network KPIs. Hence, a standardized solution is preferred for CSFB calls. 3GPP has been introducing more features to control CSFB by adding a concrete realization between the UE and the network for the call handled being a CSFB or a legacy CS call. 3GPP solutions would eventually enforce the network-side FR2L to be a common solution in returning to LTE after the CSFB call is released.

4.3   Multiple-Input, Multiple-Output (MIMO) Techniques

4.3.1   Introduction to MIMO Concepts

A wireless communication system utilizing multiple transmit antennas (inputs) and multiple receive antennas (outputs) over the wireless channel is often referred to as a MIMO system. In a MIMO system, there are NxM signal paths from the transmit and receive antennas, and the signals on these paths are not identical. On the transmitter end, the data signal is constructed in such a way that different antennas carry different variations of the signal, such as different phases, amplitudes, or waveforms. At the receiver end, each variation of the data signal is received differently at the antennas due to channel fading. With MIMO, the signals on the transmit (Tx) antennas at one end and the receive (Rx) antennas at the other end are combined so the quality (bit-error rate) or the data rate of the communication for each MIMO user will be better than SISO (single input and single output) or SIMOs (single input and multiple outputs).

Conceptually speaking, the MIMO system utilizes the space and time diversity in a multipath rich environment and creates multiple parallel data transmission pipes on which data can be carried. Figure 4.38 shows a rank-2 MIMO system, in which two data pipes are created. The data pipes are realized with proper digital signal processing by combining signals on the NxM paths. A transmission pipe does not correspond to an antenna transmission chain or any one particular signal path. These transmission pipes are orthogonal in the space–time domain and thus create little interference to each other. The pipes are not equal, that is, the interference

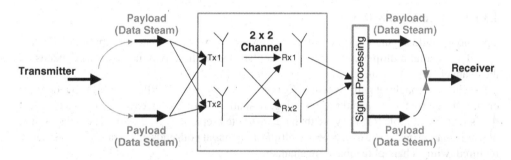

Figure 4.38 2×2 MIMO mechanism.

on the pipes is different. A pipe that has low interference can support high rate data streams, while a pipe with high interference supports lower rate data streams.

The rank of the MIMO system is limited by the number of transmitting or receiving antennas, whichever is lower. The wireless channel condition also affects the rank of the system, but does not disable the antennas. A multipath rich channel with high SNR (signal to noise ratio) typically results in a full-rank MIMO system, while in a line-of-sight (LOS) channel that lacks multipath components or has low SNR, the rank of a MIMO system can become 1.

MIMO builds on SIMO, also called receive diversity (RxD), as well as multiple-input single-output (MISO), also called transmit diversity (TxD). Both of these techniques seek to boost the SNR in order to compensate for signal degradation. As a signal passes from Tx to Rx, it gradually weakens, while interference from other RF signals also reduces the SNR. In addition, in dense urban environments, the RF signal frequently encounters objects which alter its path or degrade the signal. Multiple-antenna systems can compensate for some of the loss of SNR due to multipath conditions by combining signals that have different fading characteristics, as the path from each antenna will be slightly different [16].

SIMO and MISO systems achieve SNR gain by combining signals that take multiple paths to the Tx and Rx in a constructive manner, taking the best piece of each signal. Because different antennas receive or transmit the same signal, these systems can achieve SNR gains even in LOS situations. The boost in SNR can then be used to increase the range of the connection or boost data rates by using a modulation scheme such as 16QAM (quadrature amplitude modulation) or 64QAM rather than QPSK (quadrature phase shift keying).

However, SIMO or MISO systems may not be fully suitable for the high-speed data rates promised in 3GPP next generation cellular systems. Therefore, the full flavor of the MIMO version can achieve benefits for both SNR increase and throughput gains. MIMO in 3GPP exploits several concepts, as highlighted in Table 4.21. All these techniques mentioned in the table fall under two main MIMO categories: open loop or closed loop.

Table 4.21   MIMO techniques in 3GPP

MIMO concept	Purpose
Spatial multiplexing	Maximize user throughput in high SNR
Transmit diversity or beamforming	Improves SNR in cell edge
Multi-user MIMO	Increases the overall system capacity

4.3.1.1 Open-Loop MIMO

Open-loop multi-input, multiple-output (OL-MIMO) mode in general supports TxD and open-loop spatial multiplexing. The use for either one of these two mechanisms depends on the channel conditions and rank.

To achieve throughput gains where SNR is already very high, 3GPP uses a MIMO technique on the DL referred to as spatial multiplexing. In spatial multiplexing, each Tx sends a different data stream to multiple Rx. They are then reconstructed separately by the UE. The transmission power is shared among the data pipes. Multiple antennas at both ends of the radio link are hence required, with a channel feedback mechanism.

In general, spatial multiplexing gain does not require additional spectral bandwidth or power. The result is that spatial multiplexing can theoretically multiply throughput by the transmission rank. Under rich scattering conditions, signals from different Tx take multiple paths to reach the UE at different times. In order to achieve the promised throughputs in 3GPP MIMO systems, networks deployment must target multipath conditions for MIMO to achieve both rich scattering conditions and high SNR. This type of environment ensures that the data are spatially carried across good channel conditions so that all data can be successfully decoded by the receiver.

In 2×2 MIMO systems, the Rank-2 multiplexing gain of increasing the throughput is mainly achieved with good channel conditions experiencing little interference with rich multi-path environment. With good channel conditions, the data streams are transmitted simultaneously where the total transmission power is shared among multiple data streams. Hence, the total SNR is also shared among multiple data streams, resulting in a lower SNR on each individual data stream. If the total SNR is low, then the SNR on each individual stream will be small and the throughput on each data stream will suffer. This indicates that the spatial multiplexing gain for MIMO is mostly achieved in the high SNR region, where good throughput can be achieved on each of the independent data streams.

In open-loop spatial multiplexing operations, the network receives minimal information from the UE: a rank indicator (RI), and a CQI. The RI indicates the number of streams (stream is used for now, but in a later subsection it will be changed to layers) that can be supported under the current channel conditions and modulation scheme. CQI indicates the channel conditions under the current transmission scheme, roughly indicative of the corresponding SNR. Hence, only one CQI is reported by the UE, which is the spatial average of all the streams. The network scheduler then uses the CQI to select the corresponding modulation and coding scheme for the channel conditions (MCS). The network adjusts its transmission scheme and resources for the UE to match the reported CQI and RI with an acceptable BLER.

On the other hand, at the cell edge or in other low SNR or poor multipath conditions, instead of increasing the data rate or capacity, MIMO is used to exploit diversity and increase the robustness of data transmission. In TxD mode, MIMO functions much like a MISO system. Each antenna transmits essentially the same stream of data, so the receiver gets replicas of the same signal. This increases the SNR at the receiver side and thus the robustness of data transmission, especially in fading scenarios. Typically, an additional antenna-specific coding is applied to the signals before transmission to increase the diversity effect and to minimize co-channel interference. The UE receives the signals from both Tx at both Rx and reconstructs a single data stream from all multipath signals.

The most popular open loop TxD scheme is space/time coding, where a code known to the receiver is applied at the transmitter. Of the many types of space/time codes, the most popular is orthogonal space/time block codes (OSTBCs), or Alamouti code. This code has become the most popular TxD scheme, with its ease of implementation and linearity at both the transmitter and the receiver.

4.3.1.2   Closed-Loop MIMO

Similar to open-loop, closed-loop also supports TxD and closed-loop spatial multiplexing. The use of either one of these two mechanisms depends on the channel conditions and rank. In 3G, closed- loop transmit diversity is referred to as CLTD, while in LTE it is referred to as Rank-1 spatial multiplexing.

In closed-loop operations, the UE analyzes the channel conditions of each Tx, including the multipath conditions. The UE provides an RI as well as a precoding matrix indicator (PMI), which determines the optimum precoding for the current channel conditions. This is unlike open loop spatial multiplexing that uses a fixed set of precoding matrices to enable multiple-layer spatial multiplexing for fast-moving UEs. In a closed loop, each data stream is processed with precoding before being sent over all the transmission antennas. It is the weighting and phase shifts needed for signal processing. The UE provides a CQI given the RI and PMI, with a CQI for each stream. This allows the network scheduler to quickly and effectively adapt the transmission to channel conditions.

In the case of UEs with high velocity, the quality of the feedback may deteriorate. Thus, an open loop spatial multiplexing becomes more robust, especially in LTE systems. The open loop is based on predefined settings for spatial multiplexing and precoding. The network and operators usually select to use a closed-loop or open-loop mechanism, depending on the load of the network, the RF conditions, and targeted throughput.

On the other hand, for a LOS channel, or in areas with low SNR, MIMO spatial multiplexing gains are still exploited but in a different mechanism. The intent of spatial multiplexing in degrading RF conditions is to boost the SNR and user throughput at the cell edge. For this to happen, lower rank MIMO spatial multiplexing can be utilized. This is a process usually referred to as Rank-1 spatial multiplexing (or closed-loop transmit diversity).

In closed-loop Rank-1 spatial multiplexing, the transmitter sends only one set of data for both Tx, and hence the receiver decodes the same data which may not increase the throughput but increases the signal decoding reliability and hence SNR. Closed-loop Rank-1 spatial multiplexing uses a linear precoding matrix to improve multipath conditions for spatially multiplexed signals, similar to the idea in closed-loop Rank-2. However, depending on the rank of the channel, the precoding on the transmitter side is usually different. The optimum precoding matrix is selected from a predefined set which is known at the network and UE side. The optimum precoding matrix is the one which offers maximum capacity. The UE estimates the radio channel and selects the optimum precoding matrix and feedback to the network alongside the rank of the channel preferred.

In general, in the closed-loop mechanism (either Rank-2 or Rank-1 spatial multiplexing), the network scheduler is able to receive detailed information on channel conditions from the UE and choose the best among the precoding matrices. With this mechanism, matching the precoding matrix to the UE's channel conditions produces an additional increase in the overall SNR. Compared to more conventional TxD techniques, closed loop Rank-1 spatial multiplexing increases the likelihood that MIMO antenna schemes will result in significant SNR increases.

This can further extend the range and throughput potential at cell edges. This is discussed in later sections.

4.3.1.3 Multi-User MIMO

Spatial multiplexing allows transmission of different streams of data simultaneously by exploiting the spatial dimension of the radio channel. The different data streams can be arranged by the network scheduler (i.e., transmitter) to be sent to a single user (known as single user multi-input, multiple-output, SU-MIMO) or to different users (known as multi-user multi-input, multiple-output, MU-MIMO). While SU-MIMO increases the data rate of one user, MU-MIMO allows an increase in the overall system capacity. Like SU-MIMO, MU-MIMO is dependent on rich scattering conditions for each UE to decode the data stream meant for that UE. It is worth mentioning that MU-MIMO is introduced in LTE mainly but is still rarely deployed due to complexity in design and challenges [16].

4.3.1.4 Beamforming

Similar to the Rank-1 spatial multiplexing (i.e., closed-loop multi-input, multiple-output (CL-MIMO)), the beamforming gain is realized because both antenna paths carry the same information in low SNR or in less multipath conditions. This increases the effective SNR and provides a gain compared to what a UE with a single antenna system would achieve. Beamforming also requires precoding that helps to equalize the SNR between the layers.

Additionally, beamforming requires a different antenna configuration from other MIMO operations. Generally speaking, beamforming and spatial multiplexing have conflicting antenna configuration requirements. Beamforming requires antenna to be correlated with the same polarization, while spatial multiplexing requires transmit antennas to be de-correlated with cross-polarization. Hence, the beamforming in Release 8 (for LTE, specifically) is limited to a single antenna port and categorized in a different transmission mode than that of spatial multiplexing, as will be seen later in this chapter.

Dual-layer beamforming combines beamforming with 2×2 MIMO spatial multiplexing capabilities and is planned in more advanced 3GPP versions (starting in Release 9 as shown in the next section) for both MU-MIMO and SU-MIMO. These beamforming techniques require deployment of beamforming antenna arrays as well as special configuration of the network and UEs. As a result, beamforming techniques require additional capital expenditure and, therefore, are still not widely used in LTE or UMTS systems [16].

4.3.2 3GPP MIMO Evolution

MIMO has been introduced in the HSPA+ system part of 3GPP Release 7 as an optional feature depending on the device and network support. In the HSPA+ system, MIMO is only deployed as closed-loop spatial multiplexing and limited to SU-MIMO. Open-loop TxD has been introduced in UMTS system since Release 99, and is known as space time transmit diversity, STTD. STTD's main goal is to provide power balancing between the streams. In general, the use of STTD is quite flexible and is initially used for common channels to improve the coverage. The need of multiple antennas limited the initial deployment of STTD, and hence lacked the proper support from the terminal side, even though being a mandatory feature since

Release 99. However, when MIMO was introduced, the power balancing needs achieved by STTD faced some legacy issues on the terminal side. This forced the network operators and vendors to provide several workarounds to the MIMO design. Therefore, due to the challenges for network operators, HSPA+ MIMO has not yet gained the expected deployment traction. Several workarounds have been provided in HSPA+ MIMO systems, as addressed in [17], which could limit the expected gains of MIMO for different RF conditions.

On the other hand, MIMO has been a standard feature of next-generation LTE networks, and a major technique of LTE's high-speed data rates and overall increased system capacity. MIMO deployment in LTE is less challenging than in HSPA+ given the different and flexible deployment options that operators can use. LTE from the very first 3GPP Release 8 has been supporting closed- and open-loop spatial multiplexing, TxD, beamforming, SU-MIMO, and MU-MIMO.

Figure 4.39 shows the 3GPP DL MIMO evaluations for both HSPA+ and LTE systems. LTE MIMO techniques have been made more flexible than in HSPA+, depending on the network conditions and UE capabilities. Hence, with the various MIMO options and modes in LTE, it has realized significant deployment and operators are mostly interested in understanding how and when to deploy these different techniques. The next sections give several examples of the LTE MIMO transmission modes and performance comparisons.

On the UL side, MIMO is also possible in LTE and HSPA+ systems. UL MIMO in HSPA+ is not of major interest as UL dual-cell operation can provide the same spectral efficiency and throughput increase without major complexities. On the other hand, in LTE, in order to be fully IMT-Advanced compliant for UL peak spectral efficiency, the LTE UL must be extended with the support for UL MIMO. Up to 4×4 UL SU-MIMO is introduced in LTE-Advanced starting in 3GPP Release 10, while Release 8/9 supported only switched UL TxD.

As shown in the MIMO evolution roadmap, several different options are available to the network, and hence choice of the option that improves the system capacity and user throughput becomes important. As of today, the interest of operators is still on optimizing the available

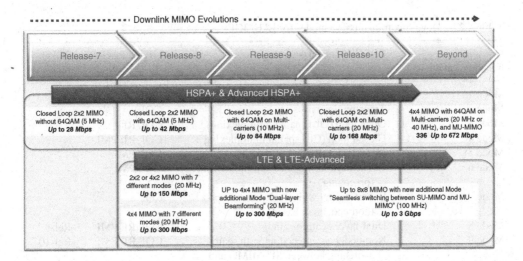

Figure 4.39   MIMO evolutions in 3GPP.

3GPP Release 8 MIMO schemes. Hence, the following sections focus on LTE Release 8 DL SU-MIMO with comparisons between different LTE MIMO transmission modes in order to provide ways to maintain and optimize the commercial LTE networks.

4.3.3   MIMO in LTE

As shown in previous sections, several MIMO techniques are allowed in LTE's Release 8. Since each eNB can be configured differently in terms of how it adapts transmissions in real time, it is important to understand the key transmission modes available in LTE, as well as the conditions under which they are most useful.

As illustrated in Table 4.22, LTE Release 8 supports seven different transmission modes, with an eighth available in Release 9 and ninth in Release 10, as part of LTE-Advanced. These modes are designed to take the best advantage of different channel and multipath conditions and eNB antenna configurations, as well as differences in UE capabilities and mobility.

Modes 2, 3, 4, and 6 are SU-MIMO modes. They form the main concepts of LTE's MIMO operations, in which more than one antenna at the eNB communicates with more than one antenna at a single UE. Selection of the desired SU-MIMO mode depends on factors such as mobility, SNR, and channel correlation, where low correlation indicates signal orthogonality. Modes 4 and 6 are considered as CL-MIMO, while modes 2 and 3 are based on OL-MIMO operations.

Transmission modes 3 and 4 are the most commonly used in the deployed LTE networks. Various antenna combinations are possible with these modes, such as 2×2, 4×2, or 4×4. However, the 2×2 MIMO is still the common configuration due to cost and performance.

Reconfigurations between different transmission modes within the same LTE call for the same user are possible. This is referred to as "adaptive mode selection" and is commonly used between modes 2, 3, 4, and 6. This, however, requires the eNB to reconfigure the mode through RRC messages which can produce increasing singling load, and additional delay in adapting

Table 4.22   Downlink transmission modes for LTE, 3GPP

Downlink transmission mode	PDSCH transmission on	UE feedback	3GPP release
Mode 1	Single antenna port (SISO, SIMO)	CQI	Rel-8
Mode 2	Transmit diversity	CQI	
Mode 3	Open loop (OL) spatial multiplexing	CQI, RI	
Mode 4	Closed loop (CL) spatial multiplexing	CQI, RI, PMI	
Mode 5	MU-MIMO (Rank-1 to the UE)	CQI, PMI	
Mode 6	CL, with Rank 1 spatial multiplexing (precoding)	CQI, PMI	
Mode 7	Beamforming with single antenna port (non-codebook based)	CQI	
Mode 8	Dual-layer beamforming	CQI, RI, PMI	Rel-9
Mode 9	Non-codebook precoding with seamless switching between SU-MIMO and MU-MIMO up to Rank 8	CQI, RI, PMI	Rel-10

to the best RF conditions suitable for the selected mode. Even though there is a potential gain in using such an adaptive mechanism, it is not widely used for increasing the throughput or capacity.

However, one common adaptive mode selection mechanism is used during LTE handover. In this scenario, if the eNB and UE are using transmission mode 3 on the serving LTE cell, and when the UE reports a request to change the serving cell, the eNB may choose to switch the UE to transmission mode 2 during the handover process. Later, the eNB reconfigures the UE to mode 3 after the handover is successfully completed. The MIMO adaptation during the handover process may provide more robust handover performance and less data interruption time. If the channel conditions are not really suitable for a higher rank channel on transmission mode 3, higher BLER could be observed, but then controlled over time by rank switch within the same transmission mode. However, since the handover process is time-sensitive and the needed control of the BLER over time may cause higher handover failures or longer data interruption, switching to single codeword transmission on mode 2 potentially provides better overall performance. This is one of the cases of the use of adaptive mode selection at the RRC stage by the eNB scheduler. It is also worth mentioning that Release 10 supports SU and MU MIMO switching without RRC reconfiguration.

The remaining modes are less relevant to current LTE MIMO techniques. Modes 1 and 7 represent non-MIMO-based antenna techniques. Mode 1 with a single antenna port is mainly a SISO or SIMO operation that gives LTE networks an option when the UE or eNB or UE does not support MIMO operations (likely in indoor solutions or small cells). The mode 7 single antenna port with beamforming requires a different antenna configuration from MIMO operations, and is discussed in a later section. Mode 5 supports MU-MIMO, discussed earlier. Dual-layer beamforming in mode 8 is introduced in LTE Release 9, which combines beamforming with 2×2 MIMO spatial multiplexing capabilities. Mode 8 can be used for either MU-MIMO or SU-MIMO. Lastly, Mode 9 is added in LTE-A and used in transparent SU-MIMO and MU-MIMO switch with up to 8×8 MIMO configuration.

Figure 4.40 from [18] gives an overview of LTE DL baseband signal generation, including the steps relevant for MIMO transmission (layer mapper and precoding). Several important principles in the DL MIMO structure are: codewords, layers, transmit antenna, and precoding.

The codeword represents an output from the channel coder. With multiple-layer transmissions, data arrives from higher level processes in one or more codewords. In Release 8/9, one codeword is used for Rank-1 transmission, and two codewords for Rank 2/3/4 transmissions. The switch between the number of codewords is done in the OL-MIMO or CL-MIMO, depending on the channel conditions. Practically, a codeword corresponds to a TB (transport block), discussed in Chapter 2. Each TB size is determined by the reported CQI. Each TB (i.e., codeword) is then an input to the modulation mapper shown in Figure 4.40. The modulation

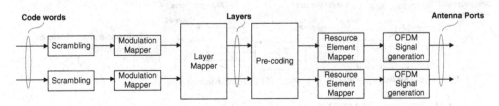

Figure 4.40   Overview of downlink baseband signal generation, 3GPP.

scheme (MCS, a function of TBS (transport block size) and 16/64-QAM or QPSK) is selected based on the CQI reported. The codeword term in LTE is similar to that in HSPA+ MIMO, referred to as a "stream" [17].

Each codeword is then mapped onto one or more layers. The number of layers depends on the number of transmit antenna ports and the channel rank report by the UE (RI). There is a fixed mapping scheme of codewords to layers, depending on the transmission mode used.

Furthermore, each layer is then mapped onto one or more antennas using a precoding matrix. In Release 8/9, there is a maximum of four antenna ports which potentially form up to four layers. Precoding is used to support spatial multiplexing. When the UE detects a similar SNR from both Tx, the precoding matrix will map each layer onto a single antenna. However, when one Tx has a high SNR and another has a low SNR, the precoding matrix will divide the layers between the Tx antennas in an effort to equalize the SNR between the layers. The layer mapping and precoding in LTE are similar to the HSPA+ MIMO concept of "virtual antenna mapping" [17].

Figure 4.41 explains the codewords, layer, antenna, and precoding concepts for transmission modes 1, 2, 3, 4, and 5.

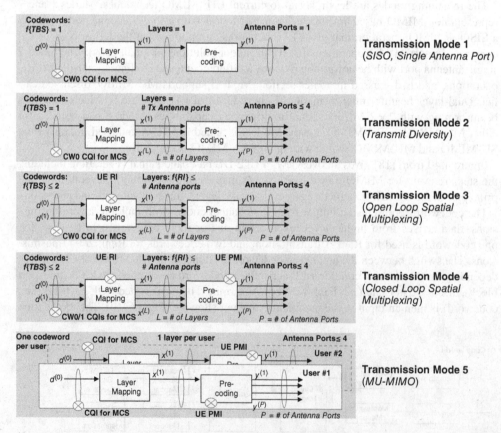

Figure 4.41 Concepts of LTE MIMO codewords, layers, antennas, and precoding for different PDSCH (physical downlink shared channel) transmission modes.

The number of Tx antenna ports depends on the eNB configuration (i.e., 1, 2, or 4 antennas) and the information is conveyed to the UE in the RRC layer (i.e., MIB or other RRC messages in connected mode). Then, the layer mapping rules from [18] are:

- **For a single antenna port** – In transmission mode 1, for example, a single layer is used.
- **For spatial multiplexing** – In transmission modes 3 and 4, for example, the number of layers must be less than or equal to the number of antenna ports used for transmission of the physical channel. There is a maximum of two codewords where the case of a single codeword mapped to two layers is only applicable when the number of antenna ports is 4.
- **For TxD** – In transmission modes 2, for example, there is only one codeword and the number of layers must be equal to the number of antenna ports used for transmission of the physical channel.

4.3.4   Closed-Loop MIMO (TM4) versus Open-Loop MIMO (TM3)

The goal of optimizing a MIMO system is to achieve the highest throughput and system capacity in different RF conditions by leveraging the multipath potential of the environment. With the various MIMO options in different transmission modes, operators should optimize the algorithms the eNB uses to select the best MIMO mode given the UE capabilities and multipath conditions.

In 2×2 MIMO deployments, the main question typically raised is whether to deploy OL-MIMO in Mode 3 or CL-MIMO in Mode 4. While each mode has its own advantages and disadvantages, the network vendors and operators have different views on the choice of the transmission mode to use.

Conceptually, CL-MIMO is better suited for low speed scenarios when the PMI feedback is accurate, while OL-MIMO provides robustness in high speed scenarios when the feedback may be less accurate. The advantage of CL-MIMO over OL-MIMO is limited due to the small number of PMI choices for 2×2 configurations in the current LTE standard.

With reference to Table 4.22 and Figure 4.41, transmission mode 3 will depend only on CQI and RI feedbacks from the UE in an open loop manner. For 2×2 MIMO, this means that when the channel is suitable for Rank-2, two codewords will be used by the eNB, mapped into two layers, and then precoded with a predefined precoding matrix. When the channel conditions allow only a Rank-1 channel and the UE reports $RI = 1$, the TxD technique is used, where one codeword is mapped into two layers, exactly similar to layer mapping and precoding in mode 2, shown in Figure 4.41. Briefly, in mode 3, the number of layers is equal to the number of antenna ports (2 in 2×2 MIMO). There is only one CQI reported by the UE and it is calculated as the special average of all streams. The DCI 2A (downlink control signal) is usually used for scheduling the UE.[9]

Transmission mode 4 on the other hand depends on CQI, RI, and PMI feedbacks from the UE in a closed loop manner. For 2×2 MIMO, this means that when the channel is suitable for Rank-2, two codewords will be used by the eNB mapped onto two layers and then precoded with a precoding matrix matching the current channel conditions. When the channel conditions allow only a Rank-1 channel and the UE reports $RI = 1$, the Rank-1 spatial multiplexing technique is used where one codeword is mapped into one layer. Alternatively, the option of

---

[9] Refer to Chapter 2, Table 2.10 for PDCCH DCI used with C-RNTI for each transmission mode.

using TxD in mode 4 is possible in the same RF conditions, by mapping one codeword into two layers. The determination of using either case (i.e., Rank-1 spatial multiplexing or TxD) in the RI = 1 channel is based on the precoding information configured in the PDCCH DCI. The precoding information determines the PMI to use as well as the number of layers, which could possibly be less than the number of antenna ports. Hence, CL-MIMO provides different options within the same transmission mode. The concepts discussed here are summarized in Table 4.23.

Typically, two CQI are reported by the UE given the RI and PMI, rather than basing the CQI on the current operation mode. In this case, transmission mode 4 supports Rank-1, Rank-2 spatial multiplexing as well as TxD. TxD in mode 4 is supported when eNB regards the UE

Table 4.23 Release 8 transmission mode 3 versus 4 codewords, layer and 2 × 2 MIMO techniques

Transmission mode 3			Transmission mode 4		
Number of antenna ports	2		Number of antenna ports	2	
PDCCH DCI format	2A Precoding information not sent		PDCCH DCI format	2 Precoding information decides on the number of layers	
Number of codewords	Number of layers	MIMO technique	Number of codewords	Number of layers	MIMO technique
One codeword when UE reports RI = 1 and no PMI (fixed precoding)	2	Transmit diversity	One codeword when UE reports RI = 1 with unreliable PMI	2	Transmit diversity
Two codewords when UE reports RI = 2 and no PMI (fixed precoding)	2	OL spatial multiplex-ing	Two codewords when UE reports RI = 2 with reliable PMI (three precoding choices)	2	Rank-2 CL spatial multiplex-ing
			One codeword when UE reports RI = 1 with reliable PMI (four precoding choices)	1	Rank-1 CL spatial multiplex-ing

PMI report as unreliable. For Rank-2, the LTE standard allows three precoding choices, and for Rank-1, it allows four precoding choices. DCI 2 is usually used for scheduling the UE.

A practical performance comparison between modes 3 and 4 has been conducted. The performance comparison was conducted in a multipath-rich scattering environment in urban morphology with 2×2 MIMO in the 20 MHz channel in band 3, and unloaded conditions with low to medium speed mobility. The main areas of comparison are:

1. SNR gain,
2. Throughput, and
3. BLER.

From the SNR point of view, and as shown in Figure 4.42, the average values are very comparable between both MIMO modes. These values of SNR are plotted as a function of RSRP. In bad RF conditions, represented by a low range of RSRP, mode 4 in CL-MIMO shows a gain in SNR over mode 3 in OL-MIMO. With the CL-MIMO mechanism, the scheduler has a better input on the channel conditions and precoding matrix, hence the SNR is boosted more in mode 4 in low ranges of RSRP. The gain observed in SNR in CL-MIMO is expected to reflect on the CQI reported by the UE and thus the network scheduler allocates the UE with more data in degrading RF conditions when compared to OL-MIMO. There is also a slight increase in SNR in the higher range of RSRP (i.e., in good RF conditions). However, the increase in SNR in very good RF conditions may not reflect much on the throughput since the UE in both modes will support up to 100 Mbps as it is CAT 3 modem.

After analyzing the SNR gains, the next step is to validate if the gain of SNR actually reflects on the overall DL throughput. Figure 4.43 shows the DL throughput as a function of SNR for CL-MIMO mode 4 and OL-MIMO mode 3. The throughput with CL-MIMO gains an extra 2 Mbps in average. As discussed before, since the SNR is higher for a lower range of RF conditions, the medium range of DL throughput distributions (in the range 35–55 Mbps) is

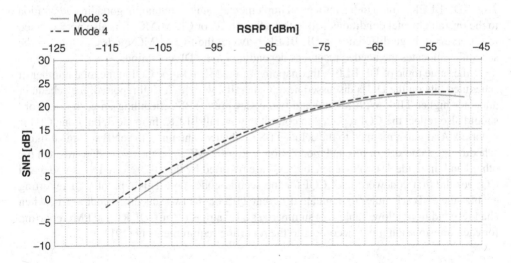

Figure 4.42   Mobility SNR versuss RSRP for transmission mode 3 and 4 in same drive route cluster.

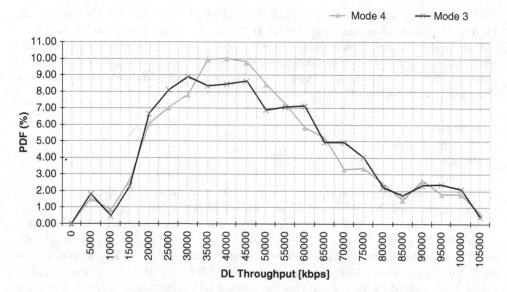

Figure 4.43 Mobility DL throughput distribution for transmission modes 3 and 4 in the same drive route cluster.

showing stronger usage in mode 4 compared to mode 3. In the other ranges of DL throughput, both showed the same distribution. Hence, CL-MIMO is benefiting in terms of higher DL throughput in medium-to-low range SNR (i.e., around cell edge coverage).

In the last step of this study, the performance comparison between the two modes is represented in terms of BLER. It is expected that with the slight gain in SNR and throughput in CL-MIMO, the DL BLER will be maintained similar to OL-MIMO and not impact the overall upper layer throughput of the UE. As illustrated in Figure 4.44, both modes show an average of 7–8% DL BLER, which indicates a good implementation in maintaining good link adaptation to the overall channel conditions with either OL-MIMO or CL-MIMO. It is also not expected to observe any huge difference in DL BLER between the two MIMO modes, as long as the scheduler allocates the needed resources depending on the DL conditions.

On the same topic of DL BLER, the mapping of the MIMO-type CQI to the total number of scheduled bits on the DL is discussed next. Generally, with high CQI reported, the scheduler allocates higher TBS. With MIMO and two codewords, TBS scheduling on each codeword should also reflect the CQI reported to control the overall BLER. In mode 3, only one CQI is reported. And when mapped to the actual TBS on the two antennas, the eNB will most likely schedule the same size of bits and modulation (i.e., same MCS) on the two codewords. On the other hand, in mode 4, two independent CQI are reported by the UE, referred to as wideband CQI for the first codeword, and CQI for the second codeword. This type of CQI reporting enables the eNB scheduler to maintain different TBS on the two scheduled codewords, when channel conditions allow. Table 4.24 summarizes an example of the CQI, RI, and PMI reporting for each transmission mode according to the reporting requirements in [19].

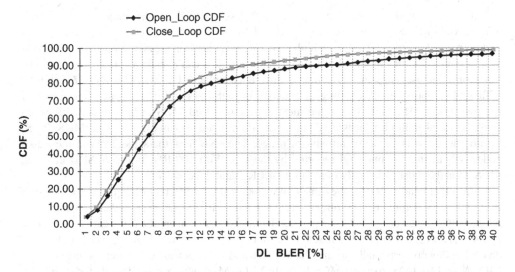

Figure 4.44 Mobility DL BLER distribution for transmission modes 3 and 4 in the same drive route cluster.

Table 4.24 CQI, RI, PMI reporting rules for transmission modes 3 and 4

CQI reporting	Transmission mode 3	Transmission mode 4
Periodic CQI, RI, PMI reporting	MODE_1_0 Type 3 report supports RI feedback	MODE_1_1 Type 2 report supports wideband CQI and PMI feedback
	Type 4 report supports wideband CQI	Type 3 report supports RI feedback
	Wideband CQI assuming transmission over all subbands	Wideband CQI/PMI conditioned on last RI
Aperiodic CQI,RI,PMI reporting	RM3-0 No PMI reporting One CQI for each subband	RM3-1 Single PMI One CQI per Codeword per each subband
	One wideband CQI	One wideband CQI per each codeword
Periodic reporting example from a network configuration	CQI reported = 40 ms RI reported = 160 ms	CQI/PMI reported = 20 ms RI reported = 80 ms

Transmission mode 4 therefore adds an extra level of CQI to TBS mapping that finally aims at yielding an acceptable BLER (within 10%, normally). However, independent MCS on the two codewords may produce higher BLER, especially on areas of 64QAM to 16-QAM switch-points. BLER alignment of mode 4 versus mode 3 can also be an area of further optimization when assessing the performance difference between CL-MIMO and OL-MIMO.

Table 4.25 shows the scheduling mechanism in mobility conditions in the same route for the two modes. The table shows the most commonly used MCS in both codewords when Rank-2 spatial multiplexing is used. In mode 3, both codewords use the same MCS (TBS) symmetrically with the same amount of usage. The eNB uses the same TBS on these two codewords as there is one CQI reported by the UE. With 100 RBs in the 20 MHz channel, MCS 23 is the most commonly used, with BLER maintained < 7%.

On the other hand, and from the same table, mode 4 shows different MCS on each codeword. In the table, the second most commonly used MCS is 17 in the first codeword and 16 in the second codeword. The decision to use independent MCS comes from the different CQI reported by the UE for each codeword. The scheduler in this case observes a higher variation of BLER on each codeword, especially around the MCS 16 and 17, which are the switch-points between 64QAM and 16QAM. As seen, MCS 16 (with 16-QAM) produces much less BLER than MCS 17 (with 64-QAM). Also, since the same number of RBs is utilized among the two codewords, it is important to ensure the power is distributed similarly on both, so as the BLER is kept within an acceptable range. Overall, the BLER is maintained < 10% for the most commonly used MCS, in this case indicating a good scheduler performance, also as shown in Figure 4.44.

One last performance aspect to consider is the comparison of CL-MIMO and OL-MIMO in a cell loading scenario. With higher cell loading the SNR (translated into signal-to-interference-noise ratio, SINR) will be different between the two modes. It is

Table 4.25 MCS usage and BLER for each codeword in rank-2 OL-MIMO *vs.* CL-MIMO

| | | Transmission mode 3 (OL-MIMO) | | | | | |
| | | *Codeword0* | | | | *Codeword1* | |
#RB	MCS1	MCS usage (%)	DL BLER (%)	#RB	MCS2	MCS usage (%)	DL BLER (%)
100	23	18.32	6.21	100	23	18.59	3.72
100	7	8.53	0.00	100	7	8.43	0.00
92	7	6.93	0.00	92	7	6.85	0.00
84	12	4.21	0.00	84	12	4.16	0.00
100	15	4.19	0.00	100	15	4.12	0.39
100	20	4.17	6.52	100	20	4.10	7.07
92	14	3.69	0.00	92	14	3.65	0.00
		Transmission mode 4 (CL-MIMO)					
		Codeword0				*Codeword1*	
#RB	MCS1	MCS usage (%)	DL BLER (%)	#RB	MCS2	MCS usage (%)	DL BLER (%)
80	27	12.26	2.33	80	27	14.01	1.55
100	17	11.27	10.17	100	16	11.01	4.95
100	16	8.73	5.37	100	18	6.67	11.41
100	15	6.48	5.06	100	20	6.42	13.03
100	20	3.95	15.02	100	15	4.44	3.67
100	18	3.87	14.43	100	21	4.21	17.44
84	25	3.80	2.69	84	27	3.09	1.64

expected that the gain of CL-MIMO in loaded conditions is higher than OL-MIMO due to the improved SINR, especially in deteriorating RF conditions where the interference tends to increase even further. In [20], it has been provided that for 2×2 MIMO, an ideal configuration of CL-MIMO provides 2 dB link gains, resulting in approximately 20% higher spectrum efficiency on a system level when compared to OL-MIMO. Practically, the relevant wideband PMI in CL-MIMO with moderate error results in the 0.9–1.0 dB link and approximately 10% capacity gain assuming a fully loaded network compared to OL-MIMO.

4.3.5   MIMO Optimization Case Study

An LTE MIMO systems optimization process and troubleshooting typically considers the algorithms and parameters the eNB uses to ensure the increase in throughput, as long as the channel conditions (in terms of SNR) allow. It is important to ensure that the use of Rank-2 MIMO is possible in a channel with a reasonable SNR in a way that still produces good BLER. This can be controlled in the deployed location with suitable MIMO parameters configured by the eNB vendor.

One case study tested in a 20 MHz network in band-3 with 2×2 MIMO is summarized in Table 4.26.

The case is analyzed based on the troubleshooting flow diagrams discussed in Chapter 3. A troubleshooting on DL related channels is necessary once a low throughput case is observed. As a first step, the optimizer can look at the general statistics in a route taking some benchmarking data (from a different eNB implementation or vendor) into account to list the steps

Table 4.26   Low LTE downlink throughput

Drive test conditions	Mobility test in unloaded conditions
Measurements	Downlink data transmissions
Issue symptom	Low two-codewords MIMO utilization
Expected impact	Low physical layer throughput
Debugging guidelines	RF and PDSCH

Table 4.27   Downlink throughput performance KPIs

KPI (average values)	Benchmarked route	Troubled route
Serving cell RSRP (dBm)	−78.0	−82.0
Serving cell RSRQ (dBm)	−7.9	−8.3
SINR (dB)	19.9	16.1
Wideband CQI (periodic) (−)	11.6	10.6
RI request by UE/two-codewords MIMO scheduled by eNBs (%)	1.9/85.0%	1.8/47.9
DL physical layer throughput (Mbps)	49.9	36.9
Peak instantaneous throughput (Mbps)	102.0	102.0
DL scheduling rate in time domain (%)	88.7	84.4
DL BLER on all transmissions (%)	5.6	10.0
RBs scheduled on DL (−)	91.7	93.0
DL MCS (−)	15.4 (Max = 28)	15.9 (Max = 28)

of debugging needed. Table 4.27 shows the final statistics of areas impacting the throughput, where the troubled area and the benchmarked statistics are compared.

As listed in Table 4.27, the two routes show very similar RF conditions in terms of RSRP, RSRQ (reference signal received quality), and CQI. The DL throughput in the troubled area is 13 Mbps less than the benchmarked area. Therefore, the troubleshooting is focused on the reasons behind the lower throughput.

Since the RF conditions are very close, it is expected to see the network scheduler utilizing the same MCS (i.e., 64QAM and TBSs) in the two areas, which is in fact the case, as shown in the table. However, the main difference is in the utilization of the two-codewords MIMO. In both routes, and based on the similar RF conditions, the UE requests the same amount of Rank-2 channel (i.e., two-codewords) but the scheduler is very conservative in the case of the troubled area.

As a first step in troubleshooting, one should look at the main factors impacting the MIMO utilization. One of these factors is the SINR estimation of the DL channel. Figure 4.45 shows the SINR values as a function of RSRP. As shown, in the troubled route, the SINR tends to go

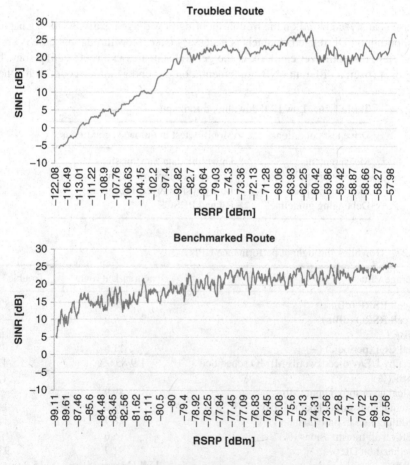

Figure 4.45 SINR comparison between the two routes.

to 5 dB at RSRP < −105 dBm. In the mean time, in the benchmarked route, the SINR tends to go to 5 dB at RSRP < −99 dBm. This generally indicates that the troubled area should allow even better DL throughput, especially at the cell-edge because the SINR is sustained at high levels with lower RSRP.

It is hence expected to observe better 64QAM and MIMO utilization in the troubled area, especially for a low range of SINR, but the actual results in the statistics in Table 4.27 have shown only 64QAM benefiting from this advantage whilst MIMO utilization is not. 64QAM is proven to show better utilization, as in Figure 4.46, with 12% higher usage in the troubled area due to the higher SINR values observed at lower RSRP.

Therefore, a deeper look at the MIMO functionality is required. After post processing the data, the two-codewords MIMO utilization as a function of SINR is depicted in Figure 4.47. From the results, it is seen that, at high SINR values, the two areas perform almost the same way with 100% utilization of two-codewords MIMO, which in turn would give a similar DL throughput in the two routes. However, at low SINR where the RF conditions tend to degrade, the scheduler in the troubled area refrains from using more of the two-codewords MIMO starting at an SINR of 10 dB. While in the benchmarked route, the scheduler starts to reduce the two-codewords MIMO utilization at an SINR of 2 dB or less. This indicates a difference between the two scheduler mechanisms at low SINR values causing the observed difference in the DL throughput.

It is generally expected that with transmission mode 3, the scheduler would sustain a good utilization of two-codewords MIMO at low SINR. This is proven in the level of DL BLER observed. In the benchmarked area, even when the scheduler utilizes the two-codewords MIMO at an SINR level up to 2 dB, the average DL BLER stays at 5%. In the troubled area, the scheduler is conservative in using more codewords (i.e., doubling the DL throughput) to avoid increasing the DL BLER, which is already at a high level of 10%, on average. The solution to this issue is to adjust the MIMO utilization thresholds to a lower value and to

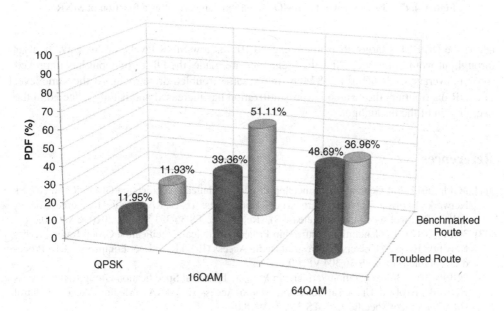

Figure 4.46   MCS and 64QAM utilization comparison between the two routes.

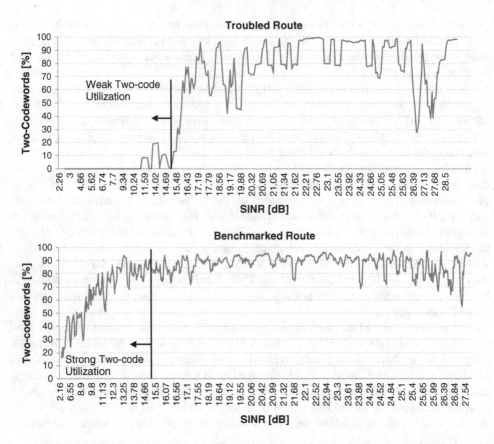

Figure 4.47   Two-codewords MIMO utilization comparison as a function of SINR.

adjust the DL BLER target so as not to exceed 10% at low SINR levels. In this way, the user throughput would improve. After these changes are made, the DL throughput has improved from an average of 36 Mbps to 48 Mbps in the same troubled drive route. At the same level of SINR distribution, the two-codewords utilization has increased and is thus reflected in the overall gain of the throughput.

References

[1] 3GPP (2011) 3rd Generation Partnership Project; Technical Specification Group Radio Access Network; Evolved Universal Terrestrial Radio Access (E-UTRA); Radio Resource Control (RRC); Protocol specification (Release 8), (Release 9). TS 36.331 V8.9.0, 2010 & V9.9.0.
[2] 3GPP (2012) 3rd Generation Partnership Project; Technical Specification Group Radio Access Network; Evolved Universal Terrestrial Radio Access (E-UTRA); User Equipment (UE) Procedures in Idle Mode. TS 36.304 V8.8.0.
[3] 3GPP (2012) 3rd Generation Partnership Project; Technical Specification Group Radio Access Network; Evolved Universal Terrestrial Radio Access (E-UTRA); Medium Access Control (MAC) Protocol Specification. TS 36.321 V8.8.0.

[4] Elnashar A. and El-Saidny M.A., Practical Performance Analysis and Evaluation of Discontinuous Reception (DRX) Mechanism for Power Saving in LTE System, submitted for publication at IEEE Vehicular Technology Magazine.

[5] 3GPP (2011) 3rd Generation Partnership Project; Technical Specification Group Services and System Aspects; Architectural Requirements. TS 23.221 V8.8.0.

[6] 3GPP 2010 3rd Generation Partnership Project; Technical Specification Group Services and System Aspects; Circuit Switched (CS) Fallback in Evolved Packet System (EPS). TS 23.272 V8.8.0.

[7] 3GPP (2011) 3rd Generation Partnership Project; Technical Specification Group Services and System Aspects; General Packet Radio Service (GPRS) Enhancements for Evolved Universal Terrestrial Radio Access Network (E-UTRAN) Access. TS 23.401 V8.10.0.

[8] 3GPP (2011) 3rd Generation Partnership Project; Technical Specification Group Core Network and Terminals; Non-Access-Stratum (NAS) protocol for Evolved Packet System (EPS). TS 24.301 V8.10.0.

[9] Qualcomm and Ericsson Whitepaper (2012) Circuit-switched Fallback; The First Phase of Voice Evolution for Mobile LTE Devices.

[10] 3GPP (2010) 3rd Generation Partnership Project; Technical Specification Group Radio Access Network; Radio Resource Control (RRC); Protocol Specification. TS 25.331 V8.10.0.

[11] 3GPP (2010) Cell Update-Less RLC/PDCP Unrecoverable Error Recovery. R2-113178.

[12] 3GPP (2011) RLC Reset on a Signalling Radio Bearer. R2-105835.

[13] 3GPP (2006) 3rd Generation Partnership Project; Technical Specification Group Radio Access Network; Radio Link Control (RLC) Protocol Specification. TS 25.322 V6.9.0.

[14] 3GPP (2006) 3rd Generation Partnership Project; Technical Specification Group Radio Access Network; Spreading and Modulation (FDD). TS 25.213 V6.5.0.

[15] GSMA Whitepaper (2011) Fast Dormancy Best Practices.

[16] Rev, B. (2011) Maximizing LTE Performance Through MIMO Optimization, PCTEL Whitepaper, April 2011.

[17] GSMA Whitepaper (2011) MIMO in HSPA: The Real-World Impact.

[18] 3GPP (2009) 3rd Generation Partnership Project; Technical Specification Group Radio Access Network; Evolved Universal Terrestrial Radio Access (E-UTRA); Physical Channels and Modulation. TS 36.211 V8.9.0.

[19] 3GPP (2009) 3rd Generation Partnership Project; Technical Specification Group Radio Access Network; Evolved Universal Terrestrial Radio Access (E-UTRA); Physical Layer Procedures. TS 36.213 V8.8.0.

[20] BallI, C., Mullner, R., Lienhart, J., and Winkler, H. (2009) Performance analysis of closed and open loop MIMO in LTE. IEEE European Wireless.

5

Deployment Strategy of LTE Network

Ayman Elnashar

5.1 Summary and Objective

This chapter presents the practical deployment scenarios for the LTE (long term evolution) system, including access network (AN) and core network (CN). In addition, it will cover the LTE CN elements and network architecture. Moreover, the high level design (HLD) of the LTE network will be presented in detail to address key design aspects. Practical HLD scenarios will be presented along with a pros and cons analysis. The AN will be presented in detail, including macro sites deployment scenarios and indoor building solution (IBS) (sometimes termed as DAS (distributed antenna system)) deployment scenarios. Also, this chapter will cover infrastructure-sharing scenarios for capital expenditure (CAPEX) and operational expenditure (OPEX) saving, including antenna/feeder sharing scenarios, single RAN (radio access network) solutions, and coexisting with 2G/3G networks. LTE spectrum options as well as refarming and guard band requirement are also discussed. An LTE business case (BC) will be illustrated to analyze the gain from introducing LTE on top of the existing 2G/3G network. Also, key relevant financial factors will be presented and analyzed. Finally, we will present a case study on a deployment scenario of an LTE network operator inter-operating with another 2G/3G network operator with different PLMNs (public land mobile networks) in the same country to offer voice/data services in both directions based on a national roaming agreement.

5.2 LTE Network Topology

The LTE system brings a flat and all IP (internet protocol) architecture [1–4], as shown in Figure 5.1. This flat architecture offers savings in CAPEX and OPEX thanks to eliminating the radio network controller (RNC) and the circuit switch (CS) core while introducing an IP multimedia subsystem (IMS) [3, 4]. Figure 5.2 demonstrates the evolution from 2G/3G to LTE topology. As shown in Figure 5.2, the RNC functionality has been moved to the eNB (eNodeB) and the mobility management entity (MME). Therefore, the eNB has become more intelligent

Design, Deployment and Performance of 4G-LTE Networks: A Practical Approach, First Edition. Ayman Elnashar, Mohamed A. El-saidny and Mahmoud R. Sherif.
© 2014 John Wiley & Sons, Ltd. Published 2014 by John Wiley & Sons, Ltd.

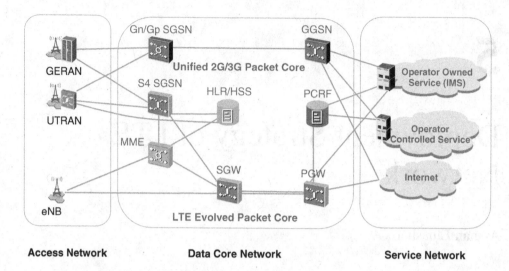

Figure 5.1   High-level topology of LTE evolution.

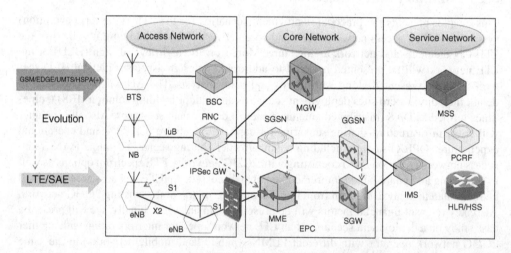

Figure 5.2   Evolution from 2G/3G/HSPA(+) to LTE.

and more sophisticated. This will lead to faster decisions within the eNB as compared to the NB in the 3G, where major functionalities are processed in the RNC. This topology acts as an evolution from a centralized network topology to a distributed network topology, there-fore eliminating the single point of failure (SPOF) in the RNC. The RNC is the main network element in the UMTS (universal mobile telecommunications system) that is deployed with-out redundancy such as 1 + 1 hot standby or a load-sharing mechanism. Another advantage of eliminating the RNC is the offload of the signaling load to the eNB. It has been estab-lished that the 3G networks suffer from signaling storms due to smartphone behavior, that is, a frequent state change from idle mode connected mode. The RNC is the first network ele-ment that is impacted by this signaling load and as a result the signaling processors of the RNC hit their maximum load before the data processors. It is not expected to face this issue with LTE system as each eNB will manage the subscribers connected to this eNB and the signaling load can be managed in this case. In addition, LTE supports very short inactivity timer such as 10 s with a transition duration from idle state to active state that is less than 100 ms. The analysis in [5] illustrates that the requirement for the state transition from idle state to active state can be achieved within the 100 ms requirement. Moreover, LTE-Advanced (LTE-A) targets reduced C-plane latency (e.g., including the possibility to exchange user-plane data starting from camped-state with a transition time of less than 50 ms (excluding S1 transfer delay)) [6]. The LTE system offers higher network performance and increased efficiency. This is achieved by reducing the latency, since the eNBs are directly connected via the S1 interface to the evolved packet core (EPC), and also to faster handover thanks to the direct connectivity between eNBs via the X2 interface. The EPC consists of the serving gateway (S-GW or SGW, or the SAE GW (system architecture evolution gateway)), the MME, and the packet data network (PDN) gateway (PGW or P-GW) [3, 4]. The SGW is responsible for handovers with neighboring eNB, data transfer in terms of all packets across the user plane, and is the mobility anchor to other 3GPP (third generation partnership project) systems (i.e., 2G and 3G). In addition, SGW performs replication of the user traffic in the case of lawful interception (LI). The MME is the centralized control unit for key operations on the AN and CN. The roles of the MME include radio resource management (RRM) and network access control. The PGW acts as an anchor of mobility between 3GPP and non-3GPP technologies [3, 4]. The PGW provides connectivity from the UE (user equipment) to external packet data networks by being the point of entry/exit of traffic for the UE. Finally, the LTE system offers increased throughput with 100 Mbps/50 Mbps DL/UL (downlink/uplink) peak throughputs with a category 3 (i.e., CAT3) modem and with seamless evolution to 150 Mbps/50 Mbps peak DL/UL throughputs with the introduction of the category 4 (CAT 4) modem [1, 7]. Moreover, the introduction of MIMO (multiple input multiple output) 4×4 with major HW upgrade in the eNB will boost the peak throughput to 300 Mbps/75 Mbps using CAT 5 modems. The list of LTE modem categories is shown in Figure 5.3. The introduction of the CAT 4 modem is just a software upgrade in the eNB that will increase the maximum number of DL-SCHs (downlink-shared channels) transport block within the LTE transmission time interval (TTI) from 102 048 to 150 752, as shown in Table 5.1 [7]. Therefore, this upgrade can be considered as a minor upgrade from a network perspective but is more challenging from a terminal perspective due to the processing complexity in the UE modem. On the other hand, the introduction of MIMO 4×4 needs major hardware and software upgrades in the eNB where either two RRU (remote radio unit) (two TX/RX) modules will be deployed per sector (i.e., a total of six RRUs per site with a three sector configuration) or a new RRU with 4TX/4RX will replace the existing 2TX/2RX RRU. Additionally, a cross-polarized antenna per sector will be added or the existing two ports cross-polarized antenna can be replaced with a four ports antenna using two pairs of cross-polarized antennas, designed for MIMO 4×4 operation. Moreover, the baseband module will be upgraded by adding additional baseband units or by replacing the existing baseband unit with a higher capacity one according to the vendor roadmap. It is worth mentioning that, the MIMO 4×4 will need a rich multipath environment to be able to increase the rank of the channel and therefore achieve the maximum gain of the MIMO 4×4. Therefore, the evolution to MIMO 4×4 will not be a systematic upgrade to every site but rather it will be for urban or dense urban morphologies. The anticipated gain from CAT 4 and MIMO 4×4 introduction is shown in Table 5.1. A detailed performance analysis for CAT 4 UE is provided in Chapter 8. The UE categories 1−5 are defined in 3GPP Release 8 and 9. Release 10 introduced three more

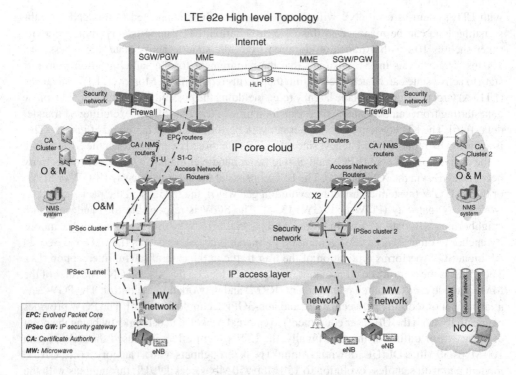

Figure 5.3 Typical high-level LTE network end-to-end topology.

categories Cat 6 to Cat 8 as shown in Table 5.1 [22]. From deployment perspective, it seems that the CAT6 UE will be launched before CAT5, tentatively Q4 of 2014. This is due to the complexity of upgrade to MIMO 4 × 4 from network and terminal perspectives. Addtionally, the LTE spectrum is available in small chunks less than 20 MHz and therefore, CA is the next major evolution in LTE roadmap.

The deployed LTE network topology will be completely different from the simplified topology in Figure 5.1. A typical high-level LTE network topology is depicted in Figure 5.3. The basic system architecture configuration and its functionality are provided in [3]. The network consists of two main domains: the CN domain and the AN domain that will be explained in detail in this chapter.

5.3 Core Network Domain

This is the main domain that includes the IP CN cloud that carries the AN traffic to the CN elements. Note that the EPC nodes can have local redundancy and/or geo-redundancy for high reliability and high availability. Also, the home subscriber server (HSS) is the combination of the HLR (home location register) and the AuC (authentication center). The HSS is responsible for storing and updating the database containing all the user subscription information, including the IMSI (international mobile subscriber identity) and MSISDN (mobile subscriber

Table 5.1 LTE-Release 10 UE categories

Category*		1	2	3	4	5	6	7	8
Peak rate Mbps**	DL	10	50	100	150	300	300	300	3000
	UL	5	25	50	50	75	50	100	1500
Maximum number of bits of a DL-SCH TB within TTI 1ms		10296	51024	102048	150752	299552	301504	301504	2998560
Maximum number of bits of a UL-SCH TB within TTI 1ms		5160	25456	51024	51024	75376	51024	102048	1497760
DL spatial multiplexing layers & Carrier Aggregation capability: #CCs/BW(MHz)(#L) CC: Carrier Component L: Spatial Layers		(1)	(2)	(2)	(2)	(4)	• 1/20MHz (4) • 2/10MHz + 10MHz (4) • 2/20MHz + 20MHz (2) • 2/10MHz (4) + 20MHz (2)	• 1/20MHz (4) • 2/10MHz +10MHz (4) • 2/20MHz + 20MHz (2) • 2/10MHz (4) + 20MHz (2)	• 2/20MHz + 20MHz (8)
UL spatial multiplexing layers & Carrier Aggregation capability: #CCs/BW(MHz)(#L) CC: Carrier Component L: Spatial Layers		(1)	(2)	(1)	(1)	(1)	• 1/20MHz (1) • 2/10 + 10MHz (1) • 1/10MHz (2)	• 2/20MHz +20MHz (1) • 2/10 (2) + 20MHz (1) • 1/20MHz (2)	• 2/20MHz + 20MHz (4)
Capabilities for physical functionalities									
RF Bandwidth		20MHz	20MHz	20MHz	20MHz	20MHz	40MHz	40MHz	Upto 100MHz
Modulation	DL	QPSK, 16 QAM	QPSK, 16 QAM	QPSK, 16 QAM	QPSK, 16 QAM	QPSK, 16 QAM, 64 QAM	QPSK, 16 QAM, 64 QAM	QPSK, 16 QAM, 64 QAM	QPSK, 16 QAM, 64 QAM
	UL	QPSK, 16 QAM	QPSK, 16 QAM	QPSK, 16 QAM	QPSK, 16 QAM	QPSK, 16 QAM, 64 QAM	QPSK, 16 QAM	QPSK, 16 QAM	QPSK, 16 QAM, 64 QAM
Multi-antenna									
2x2 MIMO		Not supported	Mandatory	Mandatory	Mandatory	Mandatory	Mandatory	Mandatory	Mandatory
4x4 MIMO		Not supported	Not supported	Not supported	Not supported	Mandatory	Supported	Supported	Supported
8x8 MIMO		Not supported	Not supported	Not supported	Not supported	Not supported	Not supported	Not supported	Supported

* The UE categories 1–5 are defined in Release 8 &9. Release 10 added three more categories Cat 6 to Cat 8.

** The peak throughput rates are theoretical and cannot be achieved in practice.

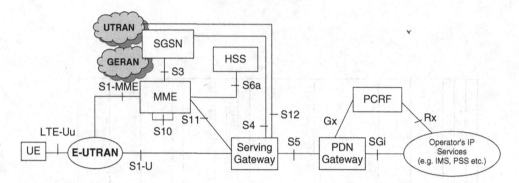

Figure 5.4 EPS reference model architecture. (Source: [3] 3GPP TS. Reproduced with permission of ETSI.)

integrated services digital network-number), user profile information such as service subscription states and user-subscribed quality of service (QoS), network-terminal authentication, ciphering, and integrity protection as defined in it [3, 4]. Therefore, HSS is the subscription data repository for all permanent LTE user data. It also keeps the location of the user, that is, the MME. The HSS stores the master copy of the subscriber profile, which contains information about the services that are applicable to the user and whether roaming to a particular visited network is allowed or not [8]. The HSS is connected to all MMEs in the home network and keeps the MME of each UE in the network and updates this information when the UE is moving from one MME to another. A typical 3GPP non-roaming architecture model network with main reference points is demonstrated in Figure 5.4 [3]. Also, the functional split between E-UTRAN (evolved universal terrestrial radio access network) and EPC is explained in Figure 5.5 [1].

The EPS (evolved packet system) reference points are defined as follows [3]:

- **S1-MME:** Reference point for the control plane protocol between E-UTRAN and MME.
- **S1-U:** Reference point between E-UTRAN and SGW for the per bearer user plane tunneling and inter eNB path switching during handover.
- **S3:** Enables user and bearer information exchange for inter 3GPP AN mobility in idle and/or active states. This reference point can be used intra-PLMN or inter-PLMN (e.g., in the case of inter-PLMN handover).
- **S4:** Provides related control and mobility support between the GPRS (general packet radio service) core and the 3GPP anchor function of the SGW. In addition, if direct tunnel is not established, it provides user plane tunneling.
- **S5:** Provides user plane tunneling and tunnel management between SGW and PGW. It is used for SGW relocation due to UE mobility and if the SGW needs to connect to a non-collocated PGW for the required PDN connectivity. Some vendors have adopted the same platform for SGW and PGW and therefore this reference point will be internal connectivity within same platform.
- **S6a:** Enables transfer of subscription and authentication data for authenticating/authorizing user access to the evolved system (AAA (authorization and accounting) interface) between the MME and the HSS.

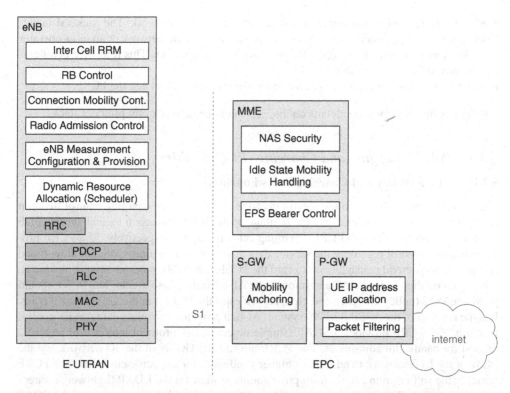

Figure 5.5 Functional split between E-UTRAN and EPC. (Source: [1] 3GPP TS 2009. Reproduced with permission of ETSI.)

- **Gx:** Provides transfer of (QoS) policy and charging rules from the PCRF (policy and charging rules function) to the policy and charging enforcement function (PCEF) in the PDN GW.
- **S8:** Inter-PLMN reference point providing user and control plane between the SGW in the visited PLMN and the PGW in the home PLMN. S8 is the inter-PLMN variant of S5.
- **S9:** Provides transfer of (QoS) policy and charging control information between the home PCRF and the visited PCRF in order to support the local breakout function.
- **S10:** Reference point between MMEs for MME relocation and MME to MME information transfer. This reference point can be used intra-PLMN or inter-PLMN (e.g., in the case of inter-PLMN handover).
- **S11:** Reference point between MME and SGW.
- **S12:** Reference point between UTRAN and SGW for user plane tunneling when direct tunnel is established. It is based on the Iu-u/Gn-u reference point using the GTP-U (GPRS tunneling protocol) as defined between the SGSN (serving GPRS support node) and UTRAN, or between the SGSN and GGSN (gateway GPRS support node). Use of S12 is an operator configuration option.
- **S13:** Enables a UE identity check procedure between the MME and the EIR (equipment identity register).

- **SGi:** Reference point between the PGW and the packet data network. The packet data network may be an operator external public or private packet data network or an intra-operator packet data network, for example, for provision of IMS services. This reference point corresponds to Gi for 3GPP accesses.
- **Rx:** This reference point resides between the AF (activity factor) and the PCRF.

Refer to Chapter 1 for more details on the EPS interfaces and relevant protocol stacks.

5.3.1   Policy Charging and Charging (PCC) Entities

5.3.1.1   PCRF (Policy and Charging Rules Function)

The PCRF function [9] is a policy control decision and flow-based charging control function entity. These two functionalities are the heritage of the 3GPP Release 6 logical entities PDF (policy decision function) and CRF (charging rules function), respectively. The PCRF provides network control regarding the service data flow detection, gating, QoS, and flow-based charging (except credit management) toward the PCEF. The PCRF receives session and media related information from the AF and informs the AF of traffic plane events. The PCRF should provision PCC (policy charging and charging) rules to the PCEF via the Gx reference point. It is the key part of the 3GPP R7 IMS system. Also, it provides a wireless data service control function. The PCRF provides a SOAP (simple object access protocol interface to receive a request for bandwidth adjustment. The PCRF informs the GGSN of the 3G network and the PGW of the LTE network to adjust to a higher bandwidth for a specific customer. The PCRF retrieves the subscription profile from provisioning system via the LDAP (lightweight directory access)/ SOAP protocol and integrates with the portal/application system via the SOAP interface. Figure 5.6 provides the PCRF/PCEF connectivity diagrams with other network elements and the 3GPP interfaces [9]. The PCRF architecture works on a service data flow level. The PCRF architecture provides the functions for policy and charging control as well as event reporting for service data flows.

PCRF functions are described in more detail in [9]. In a non-roaming scenario, there is only a single PCRF in the home PLMN associated with one UE's IP-CAN (Internet protocol connectivity access network) session. The PCRF terminates the Rx interface and the Gx interface. In a roaming scenario with local breakout of traffic there may be two PCRFs associated with one UE's IP-CAN session: the home PCRF that resides within the home PLMN and the visited PCRF that resides within the visited PLMN.

The purpose of the PCC rule is to [10]:

- Detect a packet belonging to a service data flow.
- The service data flow filters within the PCC rule are used for selection of the DL IP CAN bearers.
- The service data flow filters within the PCC rule are used to ensure that UL IP flows are transported in the correct IP CAN bearer.
- Identify the service that the service data flow contributes to.
- Provide applicable charging parameters for a service data flow.
- Provide policy control for a service data flow.

The PCEF selects a PCC rule for each received packet by evaluating received packets against service data flow filters of PCC rules in the order of precedence of the PCC rules. When a

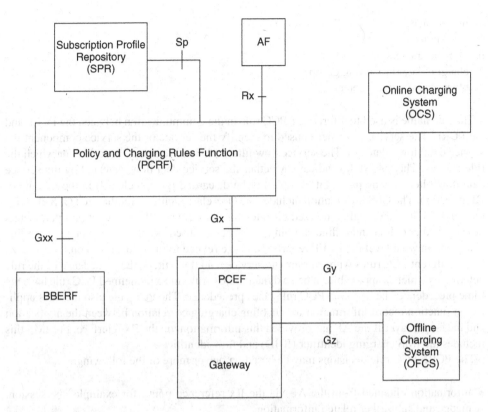

Figure 5.6 Overall PCC logical architecture (non-roaming) when SPR is used. (Source: [9] 3GPP TS. Reproduced with permission of ETSI.)

packet matches a service data flow filter, the packet matching process for that packet is completed, and the PCC rule for that filter is applied.

There are two different types of PCC rules as defined in [10]:

- **Dynamic PCC rules.** Dynamically provisioned by the PCRF to the PCEF via the Gx interface. These PCC rules may be either predefined or dynamically generated in the PCRF. Dynamic PCC rules can be activated, modified, and deactivated at any time.
- **Predefined PCC rules.** Preconfigured in the PCEF. Predefined PCC rules can be activated or deactivated by the PCRF at any time. Predefined PCC rules within the PCEF may be grouped allowing the PCRF to dynamically activate a set of PCC rules over the Gx reference point.

The operator may define a predefined PCC rule, to be activated by the PCEF. Such a predefined rule is not explicitly known in the PCRF and not under the control of the PCRF.

A PCC rule consists of:

1. a rule name
2. a service identifier
3. service data flow filter(s)

4. precedence
5. gate status
6. QoS parameters
7. charging key (i.e., rating group)
8. other charging parameters.

The rule name is used to reference a PCC rule in the communication between the PCEF and the PCRF. The service identifier is used to identify the service or the service component the service data flow relates to. The service flow filter(s) is used to select the traffic for which the rule applies. The gate status indicates whether the service data flow, detected by the service data flow filter(s), may pass (gate is open) or be discarded (gate is closed) in the UL and/or DL direction. The QoS information includes the QoS class identifier (authorized QoS class for the service data flow) and authorized bit-rates for the UL and DL. The charging parameters define whether online and offline charging interfaces are used, what is to be metered in offline charging, on what level the PCEF reports the usage related to the rule, and so on.

For different PCC rules with overlapping service data flow filters, the precedence of the rule determines which is applicable. When a dynamic PCC rule and a predefined PCC rule have the same precedence, the dynamic PCC rule takes precedence. The PCC rule also includes application function record information for enabling charging correlation between the application and the bearer layer if the AF has provided this information via the Rx interface. For IMS this includes the IMS charging identifier (ICID) and flow identifiers.

The PCRF PCC rule decisions may be based on one or more of the following:

- Information obtained from the AF via the Rx reference point, for example, the session, media, and subscriber related information.
- Information obtained from the PCEF via the Gx reference point, for example, IP-CAN bearer attributes, request type, and subscriber related information.
- Information obtained from the SPR (subscription profile repository) via the Sp reference point, for example, subscriber and service related data.
- Own PCRF pre-configured information.

The PCRF reports events to the AF via the Rx reference point. The PCRF informs the PCEF through the use of PCC rules on the treatment of each service data flow that is under PCC control, in accordance with the PCRF policy decision(s). For GPRS it is possible to support policy control, that is, access control and QoS control, on a per PDP (packet data protocol) context basis for the UE initiated case.

The PCRF is able to select the bearer control mode that applies for the IP-CAN session and provide it to the PCEF via the Gx reference point. Upon subscription to loss of AF signaling bearer notifications by the AF, the PCRF shall request the PCEF to be notified of the loss of resources associated with the PCC rules corresponding to AF signaling IP flows, if this has not been requested previously.

5.3.1.2   PCEF (Policy and Charging Enforcement Function)

The PCEF [9–11] is the functional element that encompasses policy enforcement and flow-based charging functionalities. These two functionalities are the heritage of the

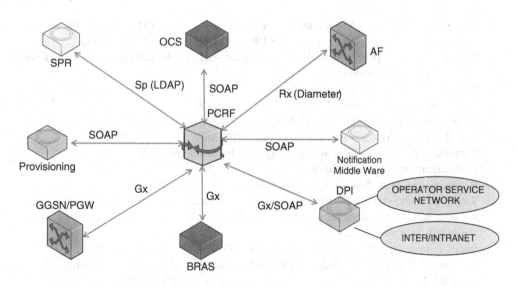

Figure 5.7   Connectivity diagram of the PCRF.

Release 6 logical entities PEP and TPF, respectively. This functional entity is located at the gateway (e.g., GGSN in the GPRS case, PGW in the EPS case, and PDG (packet data gateway) in the WLAN (wireless local area network) case). It provides control over the user plane traffic handling at the gateway and its QoS, and provides service data flow detection and counting, as well as online and offline charging interactions. For a service data flow that is under policy control the PCEF allows the service data flow to pass through the gateway if and only if the corresponding gate is open. For a service data flow that is under charging control the PCEF allows the service data flow to pass through the gateway if and only if there is a corresponding active PCC rule and, for online charging, the OCS (online charging system) has authorized the applicable credit with that charging key. The PCEF may let a service data flow pass through the gateway during the course of the credit re-authorization procedure. This is a critical scenario as if the response time of the re-authorization procedure is higher the network speed based on the volume being authorized, that is, 20 MB, then the user speed will be impacted. The PCEF will hold the authorization if the volume is consumed while the authorization response is not received. This issue can be resolved by increasing the volume, however, this increase may impact the revenue and the optimal volume in this case needs to be practically estimated. If requested by the PCRF, the PCEF reports to the PCRF when the status of the related service data flow changes. This procedure can be used to monitor an IP-CAN bearer dedicated for AF signaling traffic.

A typical connectivity diagram for PCRF is provided in Figure 5.7.

5.3.2   Mobility Management Entity (MME)

MME [3] is the key control-node in the EPC for the LTE access-network. In accordance with the 3GPP standard, the MME provides the following functions and procedures:

1. Non-access stratum (NAS) signaling and signaling security (The NAS is the highest stratum of the control plane between the UE and MME at the radio interface)
2. UE access in ECM (EPS connection management)-IDLE state (including control and execution of paging retransmission)
3. Tracking area (TA) list management
4. PGW and SGW selection
5. MME selection for handovers with MME change
6. SGSN selection for handovers to 2G or 3G 3GPP ANs
7. Terminates interface to HSS and roaming (S6a toward home HSS)
8. Authentication and authorization
9. Bearer management functions including dedicated bearer establishment
10. Transparent transfer of HRPD (high rate packet data) signaling messages and transfer of status information between E-UTRAN and HRPD access, as specified in the pre-registration and handover flows
11. Inter CN node signaling for mobility between 3GPP ANs (terminating S3)
12. UE reachability in ECM-IDLE state (including control and execution of paging retransmission)
13. Mapping from UE location (e.g., TAI (tracking area identity)) to time zone, and signaling a UE time zone change associated with mobility
14. SGSN selection for handover to 2G or 3G 3GPP ANs
15. Lawful interception of signaling traffic
16. Warning message transfer function (including selection of appropriate eNB)
17. UE reachability procedures
18. Interfaces with MSC (mobile services switching centre) for voice paging
19. Interfaces with Gn/Gp SGSN for interconnecting to legacy network
20. MAP (mobile application part) -based Gr interface to legacy HLR
21. Support relaying function (RN (relay node) attach/detach).

The MME works in conjunction with the eNB, the SGW within the EPC, or the LTE/SAE CN as follows:

1. Involved in the bearer activation/deactivation process and also responsible for choosing the SGW for a UE at the initial attach and at the time of intra-LTE handover involving CN node relocation.
2. Provides PGW selection for subscriber to connect to PDN.
3. Provides idle mode UE tracking and paging procedure, including retransmissions.
4. Responsible for authenticating the user (by interacting with the HSS).
5. Works as a termination point for NAS signaling.
6. Responsible for generation and allocation of temporary identities to UEs.
7. Checks the authorization of the UE to camp on the service provider's PLMN and enforces UE roaming restrictions.
8. The MME is the termination point in the network for ciphering/integrity protection for NAS signaling and handles the security key management.
9. Communicates with the MME in the same PLMN or on different PLMNs. The S10 interface is used for MME relocation and MME to MME information transfer or handoff.

Besides the above-mentioned functions, the lawful interception of signaling is also supported by the MME. The MME also provides the control plane function for mobility between LTE and 2G/3G ANs with the S3 interface terminating at the MME from the SGSN. In addition, the MME interfaces with Gn/Gp SGSN for interconnecting to the legacy network. The MME also terminates the S6a interface toward the home HSS for roaming UEs. Figure 5.4 provides MME connectivity and integration with other network elements. The SGW and the MME may be implemented in one physical node or separated physical nodes, based on the vendor's topology.

5.3.2.1   MME Load balancing

The MME load balancing functionality permits UEs that are entering into an MME pool area to be directed to an appropriate MME in a manner that achieves load balancing between MMEs. This is achieved by setting a weight factor for each MME, such that the probability of the eNB selecting an MME is proportional to its weight factor. The weight factor is typically set according to the capacity of an MME node relative to other MME nodes. The weight factor is sent from the MME to the eNB via S1-AP (S1 application protocol) messages [12]. The MME load rebalancing functionality permits UEs that are registered on an MME (within an MME pool area) to be moved to another MME.

- **MME selection:** The MME selection function selects an available MME for serving a UE. The selection is based on network topology, that is, the selected MME serves the UE's location and for overlapping MME service areas, the selection may prefer MMEs with service areas that reduce the probability of changing the MME. When an MME/SGSN selects a target MME, the selection function performs a simple load balancing between the possible target MMEs. When an eNB selects an MME, the selection achieves load balancing. Each MME has a weight factor. It is typically set according to the capacity of an MME node relative to other MME nodes.
- **MME pooling:** The MME pooling function was designed to be efficient from the UE movement point of view and has been developed to reduce MME change when serving within certain operation boundaries. It provides support to configure the MME pool area consisting of multiple MMEs within which a UE may be served without any need to change the serving MME. The benefits of MME pooling are:
  - Enables geographical redundancy, as a pool can be distributed across sites.
  - Increases overall capacity, as load sharing across the MMEs in a pool is possible, as explained above.
  - Converts inter-MME tracking area updates (TAUs) to intra-MME TAUs for moves between the MMEs of the same pool. This substantially reduces the signaling load as well as the data transfer delays.
  - Eases introduction of new nodes and replacement of old nodes as subscribers can be moved in a planned manner to the new node.
  - Eliminates a SPOF between an eNB and MME.
  - Enables service downtime free maintenance scheduling.
- **MME pool area:** An area within which a UE may be served without need to change the serving MME. An MME pool area is served by one or more MMEs ("pool of MMEs") in

parallel. MME pool areas are a collection of complete TAs. MME pool areas may overlap each other.

5.3.2.2   Tracking Area List Management

TA list management comprises the functions to allocate and reallocate a TA identity list to the UE. The TAI is the identity used to identify TAs [3]. The TAI is constructed from the MCC (mobile country code), the MNC (mobile network code), and the TAC (tracking area code). A TAI should be associated with a single time zone. All TAIs served by one eNB should be in the same time zone. All the TAs in a tracking area list to which a UE is registered are served by the same serving MME. The "TA list concept" is used with E-UTRAN. With this concept, when the UE registers with the network, the MME allocates a set (a "list") of TAs to the UE. By making the center of this set of TAs close to the UE's current location, the chance of a UE rapidly making another TAU can be reduced. The MME assigns the TAI list to a UE so as to minimize the TAUs that are sent by the UE. The TAI list should be kept to a minimum in order to maintain a lower paging load. To avoid a ping-pong scenario, the MME includes the last visited TAI (provided that the TA is managed by the MME) in the TAI list assigned to the UE. TA lists assigned to different UEs moving in from the same TA should be different to avoid TAU message overflow.

5.3.2.3   Connection of eNodeBs to Multiple MMEs

An eNB may connect to several MMEs. This implies that an eNB must be able to determine which of the MMEs, covering the area where an UE is located, should receive the signaling sent from a UE. To avoid unnecessary signaling in the CN, a UE that has attached to one MME should generally continue to be served by this MME as long as the UE is in the radio coverage of the pool area to which the MME is associated. The concept of pool area is a RAN-based definition that comprises one or more TA(s) that, from a RAN perspective, are served by a certain group of MMEs. This does not exclude that one or more of the MMEs in this group serve TAs outside the pool area. This group of MMEs is also referred to as an MME pool.

To enable the eNB to determine which MME to select when forwarding messages from a UE, this functionality defines a routing mechanism. A routing mechanism is defined for the MMEs. The routing mechanism is required to find the correct old MME (from the multiple MMEs that are associated with a pool area). When a UE roams out of the pool area and into the area of one or more MMEs that do not know about the internal structure of the pool area where the UE roamed from, the new MME will send the identification request message or the context request message to the old MME using the GUTI (globally unique temporary identifier). The routing mechanism in both the MMEs and the eNB utilizes the fact that every MME that serves a pool area must have its own unique value range of the GUTI parameter within the pool area.

5.3.3   Serving Gateway (SGW)

The SGW [3] is the gateway which terminates the interface toward E-UTRAN. For each UE associated with the EPS, at a given point of time, there is a single SGW. The functions of the SGW, for both the GTP-based and the PMIP-based (proxy mobile IP) S5/S8, include:

- The local mobility anchor point for inter-eNB handover.
- Functions (for both the GTP-based and the PMIP-based S5/S8) include packet routing and forwarding.
- Sending of one or more "end marker" to the source eNB, source SGSN, or source RNC immediately after switching the path during inter-eNB and inter-RAT (radio access technology) handover, especially to assist the reordering function in eNB.
- Providing the local mobility anchor point for inter-eNB handover and assisting the eNB reordering function by sending one or more end marker packets to the source eNB immediately after switching the path.
- Mobility anchoring for inter-3GPP mobility (terminating the S4 interface from an SGSN and relaying the traffic between a 2G/3G system and a PGW.
- ECM-idle mode DL packet buffering and initiation of network triggered service request procedure.
- Replicating user traffic in the event that lawful interception is required.
- Transport level packet marking in the UL and the DL, for example, setting the DSCP (DiffServ code point) DiffServ code point, based on the QCI (quality of service Class indicator) of the associated EPS bearer.
- User accounting and QCI granularity for charging.
- UL and DL charging per UE, PDN, and QCI.
- Accounting for inter-operator charging. For GTP-based S5/S8, the serving GW generates accounting data per UE and bearer.
- Handling of router solicitation and router advertisement messages if PMIP-based S5 and S8 are used.
- Interfacing the OFCS (offline charging system) according to charging principles and through reference points specified in [11].
- MAG (mobile access gateway) for PMIP-based S5 and S8.

The SGW routes and forwards data packets from the UE and acts as the mobility anchor during inter-eNB handovers. Signals controlling the data traffic are received on the SGW from the MME, which determines the SGW that will best serve the UE for the session. Every UE accessing the EPC is associated with a single SGW. The SGW is also involved in mobility by forwarding DL data during a handover from the E-UTRAN to the eHRPD network. An interface from the eAN/ePCF to an MME provides signaling that creates a GRE (generic routing encapsulation) tunnel between the SGW and the eHRPD SGW.

5.3.4 PDN Gateway (PGW)

The PGW [3] is the gateway, which terminates the SGi interface toward the PDN. The PGW provides connectivity to the UE to external packet data networks by being the point of exit and entry of traffic for the UE. If a UE is accessing multiple PDNs, there may be more than one PDN GW for that UE, however, a mix of S5/S8 connectivity and Gn/Gp connectivity is not supported for that UE simultaneously. The PGW performs policy enforcement, packet filtering for each user, charging support, lawful interception, packet screening, and other functions including the following for both the GTP-based and the PMIP-based S5/S8:

- Terminates the interface toward the PDN (SGi).
- Per-user based packet filtering (by e.g., deep packet inspection (DPI)).

- Lawful interception.
- Policy enforcement (gating and rate enforcement).
- Mobility anchor for mobility between 3GPP access systems and non-3GPP access systems. This is sometimes referred to as the SAE anchor function.
- UE IP address allocation.
- Per-user based packet filtering (DPI).
- Transport level packet marking in the UL and DL, for example, setting the DiffServ code point, based on the QCI of the associated EPS bearer.
- Accounting for inter-operator charging.
- UL and DL service level charging as defined in [9] (e.g., based on SDFs (service data flows) defined by the PCRF, or based on DPI defined by local policy).
- Interfacing OFCS according to charging principles and through reference points specified in [11].
- UL and DL service level gating control as defined in [9].
- UL and DL service level rate enforcement as defined in [9] (e.g., by rate policing/shaping per SDF).
- UL and DL rate enforcement based on APN-AMBR (access point name-aggregate maximum bit rate) (e.g., by rate policing/shaping per aggregate of traffic of all SDFs of the same APN that are associated with non-GBR QCIs).
- DL rate enforcement based on the accumulated MBRs (maximum bit rates) of the aggregate of SDFs with the same GBR QCI (e.g., by rate policing/shaping).
- DHCPv4 (dynamic host configuration protocol) (server and client) and DHCPv6 (client and server) functions.
- The network does not support PPP point-to-point bearer type in this version of the specification. Pre-Release 8 PPP functionality of a GGSN may be implemented in the PDN GW packet screening. Additionally, the PDN GW includes the following functions for the GTP-based S5/S8:
  - UL and DL bearer binding as defined in [9]
  - UL bearer binding verification as defined in [9]
  - Accounting per UE and bearer. ·

The PGW provides PDN connectivity to both GERAN/UTRAN-only UEs and E-UTRAN-capable UEs using any of E-UTRAN, GERAN, or UTRAN. The PGW provides PDN connectivity to E-UTRAN-capable UEs using E-UTRAN only over the S5/S8 interface. The PGW may interact with an AAA server over the SGi interface. This AAA server may maintain information associated with UE access to the EPC and provide authorization and other network services. This AAA server could be a RADIUS (remote authentication dial in user service) or Diameter server in an external PDN network, as defined in [13]. This AAA server is logically separate from the HSS and the 3GPP AAA server. Another key role of the PGW is to act as the anchor for mobility between 3GPP and non-3GPP technologies such as WiMAX (worldwide interoperability for microwave access) and 3GPP2 (CDMA (code division multiple access) 1X and EvDO).

Within the EPC there are two mobility protocols defined to use on the S5/S8 interface between the SGW and the PGW. The architecture and functionalities of the two gateways are different either of GTP or PMIP is deployed as the mobility management protocol on the S5/S8 interface. The PMIP provides mobility to IP device without its involvement. This is achieved by moving the mobility management functions from the device to the network. If

PMIP is adopted on the S5/S8 interface, the EPS bearer lasts only from the UE over the eNB to the SGW. The PGW is not anymore part of the EPS bearer.

5.3.5   Interworking with PDN (DHCP)

In current LAN environments the most commonly used configuration protocol is DHCP and DHCPv6 (DHCP for IPv6) [13]. It provides a mechanism for passing a large set of configuration parameters to hosts connected to a TCP/IP (transmission control protocol) network (IP address, sub-net mask, domain name, MTU (maximum transmission unit), etc.) in an automatic manner. Moreover, DHCP may assign IP addresses to clients for a finite lease time, allowing sequential reassignment of addresses to different users. The lease time is chosen by the administrator of the DHCP server in the external network.

The packet domain may obtain the IP address via an external DHCP server during the packet bearer establishment procedures (e.g., PDP context activation in 3G, default bearer establishment in 4G). In this case, the GGSN/PGW acts as a DHCP client toward the external DHCP server. The packet domain offers the end user the possibility to run DHCP end-to-end in the same way as he does when connected directly to a LAN (e.g., an enterprise intranet). No modifications should be required in common implementations of DHCP clients and servers. However in a non-EPC-based packet domain, a DHCP relay agent function is needed in the GGSN so as to allow correct routing of DHCP requests and replies between the TE and the DHCP servers. In an EPC-based packet domain, the PGW acts a DHCP server toward the UE and it acts as a DHCP client toward the external DHCP server. In a non-EPC-based packet domain, at PDP context activation no IP address is allocated, this is done afterwards through DHCP. After the TE's configuration has been completed by DHCP, the PDP context is updated by means of the GGSN-initiated PDP context modification procedure in order to reflect the newly assigned IP address. In the following cases the bearer associated with the allocated IP address (i.e., IPv4 address or IPv6 prefix) is released:

- If the DHCP lease expires
- If the DHCP renewal is rejected by the DHCP server
- If the IP address is changed during the renewal process. Usually when the lease is renewed, the IP address remains unchanged. However, if for any reason (e.g., poor configuration of the DHCP server) a different IP address is allocated during the lease renewal process the associated bearer is released.

5.3.5.1   PDN Interworking Model of GGSN for DHCP

A DHCP relay agent is located in the GGSN used for interworking with the IP network, as illustrated in Figure 5.8 [13].

The DHCP relay agent relays the requests received from the DHCP client to the DHCP server(s), and the replies received from the server(s) to the corresponding client. The DHCP relay agent allows the replies from DHCP servers to be delivered to the correct terminal, as the logical connection from the MT terminates in the GGSN, and consequently only the GGSN holds enough information to locate the DHCP client. DHCP provides mechanisms for user authentication and integrity protection, but does not offer any message confidentiality, therefore additional mechanisms (e.g., IPsec tunnel) may be provided if the link toward the

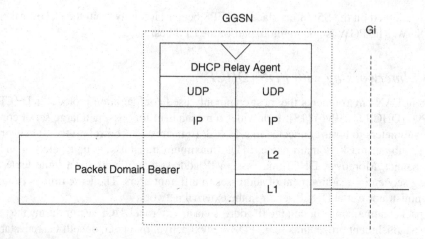

Figure 5.8   The protocol stacks for the Gi IP reference point for DHCP. (Source: [13] 3GPP TS. Reproduced with permission of ETSI.)

external network is not secure. This model is basically the same as that for interworking with IP networks. Using DHCP corresponds to the transparent access case as the GGSN does not take part in the functions of authentication, authorization, address allocation, and so on.

5.3.5.2   Address Allocation by the Intranet or ISP

The MS is given an address belonging to the intranet/ISP addressing space. The address is given dynamically immediately after the PDP context activation. This address is used for packet forwarding between the intranet/ISP and the GGSN and within the GGSN. The MS may authenticate itself to the intranet/ISP by means of the relevant DHCP procedures. The protocol configuration options are retrieved from the DHCP server belonging to the intranet/ISP. The end-to-end protocol stack for access with the DHCP is shown in Figure 5.9 [13].

5.3.5.3   PDN Interworking Model of PGW for DHCP

A DHCP client is located in the PGW used for interworking with the IP network, as illustrated in Figure 5.10 [13].

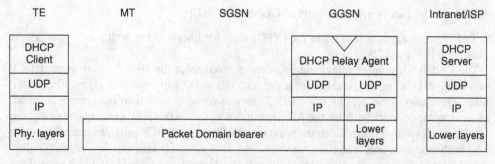

Figure 5.9   Protocol stack for access with DHCP end-to-end. (Source: [13] 3GPP TS. Reproduced with permission of ETSI.)

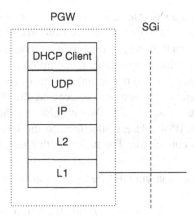

Figure 5.10 The protocol stacks for the SGi IP reference point for DHCP. (Source: [13] 3GPP TS. Reproduced with permission of ETSI.)

The DHCP client function in PGW is used to allocate IP address(es) to the UE and/or to configure associated parameters via external DHCP servers in PDN(s). As both IPv4 and IPv6 address allocation is supported in EPS, the PGW has both DHCPv4 and DHCPv6 client functions. The procedures where the DHCP client function in the PGW is used are further described in [3, 14, 15]. The procedures are IPv4 address allocation and IPv4 parameter configuration via an external DHCPv4 server in a PDN; IPv6 prefix allocation via stateless address autoconfiguration; and IPv6 parameter configuration via stateless DHCPv6.

5.3.6 Usage of RADIUS on the Gi/SGi Interface

GGSN/PGW may, on a per APN basis, use RADIUS authentication to authenticate a user and RADIUS accounting to provide information to an AAA server [13].

5.3.6.1 RADIUS Authentication and Authorization

RADIUS authentication and authorization is used according to [16, 17]. The RADIUS client function may reside in a GGSN/PGW. When the GGSN receives a create PDP context request message or the PGW receives an initial access request (e.g., create session request) the RADIUS client function may send the authentication information to an authentication server, which is identified during the APN provisioning. The authentication server checks that the user can be accepted. The response (when positive) may contain network information, such as an IPv4 address and/or IPv6 prefix for the user when the GGSN or PGW is interworking with the AAA server.

The information delivered during the RADIUS authentication can be used to automatically correlate the user's identity (the MSISDN or IMSI) to the IPv4 address and/or IPv6 prefix, if applicable, assigned/confirmed by the GGSN/PGW or the authentication server, respectively. The same procedure applies when sending the authentication to a "proxy" authentication server.

RADIUS authentication is only applicable to the primary PDP context. When the GGSN receives an access accept message from the authentication server it completes the PDP context activation procedure. If access reject or no response is received, the GGSN rejects the PDP context activation attempt with a suitable cause code, for example, user authentication failed. The GGSN may also use the RADIUS re-authorization procedure for the purpose of IPv4 address allocation to the MS for a PDP type of IPv4v6 after establishment of a PDN connection.

For EPS, RADIUS authentication is applicable to the initial access request. When the PGW receives an access accept message from the authentication server it completes the initial access procedure. If access reject or no response is received, the PGW rejects the initial access procedure with a suitable cause code. The PGW may also use the RADIUS re-authorization procedure for the purpose of IPv4 address allocation to the UE for a PDN type of IPv4v6 after establishment of a PDN connection. The use cases that may lead to this procedure are:

- Deferred IPv4 address allocation via DHCPv4 procedure after successful attach on 3GPP accesses.
- Deferred IPv4 address allocation after successful attach in trusted non-3GPP IP access on S2a.
- Deferred IPv4 home address allocation via DSMIPv6 re-registration procedure via S2c.

5.3.6.2 RADIUS Accounting

RADIUS accounting is used according to [17, 18]. The RADIUS accounting client function may reside in a GGSN/PGW. The RADIUS accounting client may send information to an accounting server, which is identified during the APN provisioning. The accounting server may store this information and use it to automatically identify the user. This information can be trusted because the packet domain network has authenticated the subscriber (i.e., SIM (subscriber identity module) card and possibly other authentication methods). The GGSN/PGW may use the RADIUS accounting request start and stop messages during IP-CAN bearer (e.g., primary and secondary PDP context, default and dedicated bearer) activation and deactivation procedures, respectively. For an EPC-based packet domain, if the PGW is not aware of the IP-CAN bearers, for example, in the case of a PMIP-based S5/S8, the PGW may use the RADIUS accounting request start and stop messages per IP-CAN session as it would be one IP-CAN bearer. Accounting request stop, and in addition accounting on and accounting off messages may be used to ensure that information stored in the accounting server is synchronized with the GGSN/PGW information. If the AAA server is used for IPv4 address and/or IPv6 prefix assignment, then, upon receipt of a RADIUS accounting request stop message for all IP-CAN bearers associated with an IP-CAN session defined by APN and IMSI or MSISDN, the AAA server may make the associated IPv4 address and/or IPv6 prefix available for assignment.

For PDN/PDP type IPv4v6 and deferred IPv4 address allocation, when the IPv4 address is allocated or re-allocated, the accounting session that was established for the IPv6 prefix allocation is used to inform the accounting server about the allocated IPv4 address by sending a RADIUS accounting request interim update with framed-IP-address attribute and its value field containing the allocated IPv4 address. Similarly, the release of the IPv4 address is indicated to the accounting server by sending the RADIUS accounting Request interim update without the framed-IP address attribute.

In order to avoid race conditions, the GGSN/PGW includes a 3GPP vendor-specific sub-attribute "session stop indicator" when it sends the accounting request stop for the last IP-CAN bearer of an IP-CAN session and the IP-CAN session is terminated (i.e., the IPv4 address and/or IPv6 prefix and any associated GTP tunnels or PMIP tunnel can be released). The AAA server does not assume the IP-CAN session terminated until an accounting request stop with the session stop indicator is received.

5.3.7   IPv6 EPC Transition Strategy

According to current projections, no new IPv4 address blocks will be delivered by 2012 and all public IPv4 addresses will be in use for 2015–2016. The depletion of public IPv4 addresses becomes a reality and is approaching rapidly. On the other hand, the demand for IP connectivity is expanding with the introduction of new technologies, such as HSPA (high speed packet access), LTE, and IMS. Mobile networks are also evolving toward always-on connectivity, where an IP address is assigned to the UE from its attachment to its detachment from the network and, therefore, the lease time from DHCP servers will be increased and, accordingly, the efficiency of IP pool assignment will be significantly decreased. Furthermore, the proliferation of devices foreseen in the coming years leads to a significant need for IP addresses.

IPv6 has several advantages over its predecessor, including a larger and more diverse address space, built-in extensibility, and the power to support a more robust security paradigm. As such, it serves as a powerful foundation for the creation of a new and improved net-centric set of products and services [19]. Network providers are analyzing the best migration strategy and the optimum time to introduce IPv6 in their mobile/fixed networks. Several large providers have even begun IPv6 deployment or will begin to do so within the next years.

While waiting for a permanent address space solution, there have been numerous optional "fixes" (such as network address translation – aka "NAT") and extensions to IPv4 to try to overcome the address space limitations. NAT allows multiple devices to be "hidden" behind one or more real IPv4 address. Such mechanisms restrict the end-to-end transparency of the Internet. While NAT has to some extent delayed the pressure on IPv4 address space in the short term, it places severe restrictions on application bi-directional communication.

IPv4 allows as many as 232 addresses, which is 4 394 967 296. The most important feature of IPv6 is a much larger address space than that of IPv4: addresses in IPv6 are 128 bits long (up to 3.4×10^{38} hosts), compared to 32-bit addresses in IPv4.

The ongoing transition of the global Internet from IPv4 to IPv6 will span many years, and is projected by many to last longer than a decade. Many organizations introducing IPv6 into their infrastructure will operate in a dual-stack environment, supporting IPv4 and IPv6 concurrently for the foreseeable future. For organizations that are beginning a transition to IPv6, it is very hard to recommend any one IPv6 transition strategy. Though deploying dual stacks to support native IPv6 capability is the preferred solution, one size does not fit all. The transition strategy depends on the individual BC and whether new infrastructure is being installed and can be "born IPv6" or IPv6 is going to be integrated into an existing IPv4 infrastructure.

The three main transition strategies are [19]:

1. **Dual stack** – Network routers, switches, applications, services, management, and security infrastructure are upgraded to operate on both IPv4 and IPv6. Application level gateways or proxy servers built on dual stack servers can also be used to transition legacy IPv4-based client–server applications to IPv6. Since Release 8, the 3GPP EPS architecture (i.e., when interworking with a S4-SGSN) supports the coexistence of IPv4 and IPv6 with dual-stack operation. Dual-stack operation means that native IPv4 and native IPv6

packets are transported in parallel by tunneling them from the UE to the PGW within a single EPS bearer/PDP context. This dual-stack EPS bearer/PDP context is associated with both an IPv4 address and an IPv6 prefix.

The dual-stack PDP context is supported since Release 9 in GPRS networks (i.e., when interworking with Gn/Gp-SGSN). Prior to Release 9, dual-stack connectivity to a given PDN in GPRS networks required the activation of two parallel PDP contexts, one for IPv4 traffic and one for IPv6 traffic. Dual-stack support in pre-Release 9 GPRS networks introduces scaling issues since two PDP contexts are required per dual-stack connectivity.

For LTE networks that operate with Gn/Gp-SGSN, a network operator that is facing a depletion of public IPv4 addresses will then have the following options:

(a) Activating a single IPv6 PDP context for LTE-capable UE.
(b) Activating two PDP contexts (one in IPv4 and another in IPv6) for LTE-capable UE.
(c) Upgrading their GPRS packet CN to support dual-stack PDP contexts for LTE-capable UE.

2. **Tunneling** – encapsulating IPv6 packets within IPv4 packets for transmission over an IPv4-only network infrastructure. A network infrastructure may also be born "IPv6-only" and tunnel IPv4. Tunneling may be through manually set up tunnels, brokered tunnels, or numerous automated host-to-router tunneling solutions. If tunneling is used, an enterprise's security and network management infrastructure still needs to be upgraded for IPv6. The use of these techniques can significantly degrade the network performances and should then be very limited in the networks. These techniques are not recommended for mobile networks because LTE-capable terminals will support a dual stack.

3. **Protocol translation** – Full translation of IPv4 packets to IPv6 and vice versa, but only as a last resort because translation interferes with end-to-end network communications and security. Since most new IPv6 equipment is deployed with a dual stack, the IPv4 side is compatible with legacy IPv4 devices without translation.

An incremental phase-in transition approach allows a significant period where IPv4 and IPv6 will co-exist using some or all existing transition mechanisms. Depending on an organization's policy, transition to IPv6 could occur during a regular "technology refresh" program, where IPv6-capable products are introduced during regular network upgrades, or may be "mandated" to occur out of cycle. It is critical that planning is done prior to introducing IPv6 into a network. Successful IPv6 transition should be designed to have almost no impact on existing IPv4 infrastructure and allow co-existence of IPv4 and IPv6 networks. Figure 5.11 and Table 5.2 illustrate a IPv6 phase migration strategy for an LTE network.

5.4   IPSec Gateway (IPSec GW)

Internet protocol security gateway (IPSec GW) is a new network element introduced as part of the EPS architecture. IPSec GW is responsible for encryption and for terminating the IPSec tunnels with eNBs. All traffic and signals are encapsulated inside IPSec tunnel for security purposes and to prevent any attack on the EPC nodes. The deployment of IPSec GW can be in cluster fashion for reliability and simplicity. Therefore, the network can be divided into several clusters (i.e., regions) and eNBs traffic within a certain cluster is encapsulated toward an IPSec GW pair with $1+1$ hot standby. Furthermore, a feature such as S1-flex can allow the eNB to create two IPSec tunnels with two different EPCs nodes for reliability, and in this

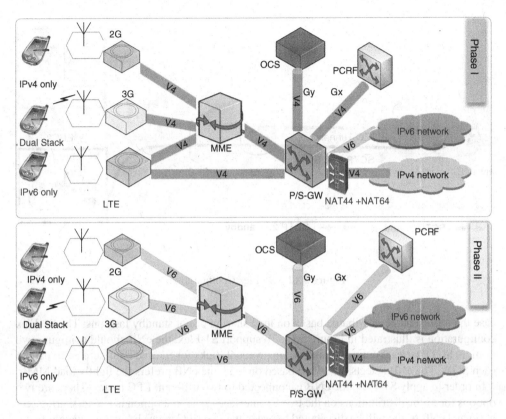

Figure 5.11   IPv6 migration strategy.

Table 5.2   IPv6 phase approach explanation

	Transition assumption	PS IPv6 transition strategy
Phase I – Step 1	Dual mode terminal introduction IPv6 service island introduction	Introduce dual stack bearer and PDP context Support IPv4 and IPv6 on Gi/SGi
Phase I – Step 2	IPv6 only terminal introduction Migrate charging and PCC control for IPv6 service	Enable IPv6 DPI Support IPv6 Gx/Gy External NAT64 for V6 terminal access to V4 service 2/3/4G interworking
Phase II	Migrate PS IP backbone and backhaul to native IPv6 network	Introduce IPv6 networking capability for IP backbone or backhaul Introduce IPv6 networking for Gx and Gy interface

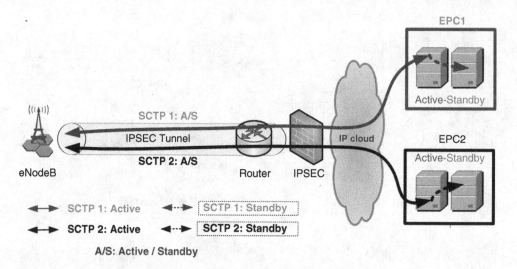

Figure 5.12 S1-Flex Configuration.

case the traffic can be distributed based on load sharing or hot standby fashions. The S1-Flex configuration is illustrated in Figure 5.12. To support S1-Flex, the eNB should configure two or more SCTP (stream control transmission protocol) links for S1, and configure an IP path to each SGW. The eNB selects the MME based on load, the eNB prefers the lightly load MME.

In order to apply S1-Flex, the eNB is connected to two different EPC nodes. There are two options for traffic distributions on the two S1 interfaces. The first option is a load-sharing scenario and eNB traffic will be distributed between the two EPC nodes based on a weight factor that is received from the MME each time the UE accesses the network. The second option is an 1 + 1 active standby operation where all traffic will be shifted to another S1 interface in case one EPC goes out of service. This option is not currently available with all vendors. This option is important as traffic from the same region can be connected to the EPC serving this region and only transferred to another EPC in another region if the EPC of this region is out of service for any reason. By maintaining this topology the latency will be managed while with the load sharing approach the user may have a different experience in terms of latency according to the EPC location, especially if the two EPCs are far from each other.

The IPSec protocol family is a series of protocols defined by the IETF (Internet engineering task force). It provides IP data packets with cryptology-based security, featuring high quality, and interoperability. The two sides of a communication perform encryption and data source authentication on the IP layer to assure confidentiality, data integrity, data origin authentication, and anti-replay for packets when they are transmitted on networks. The LTE system adopts the IPSec tunneling to ensure the data security and integrity when the data is transmitting in the non-trusted zone. The eNBs are connected to the EPC via IPSec GW using IPSec tunneling and the IPSec GW is connected to the MME and SGW as shown in Figure 5.13.

The eNB provides multiple IPSec subtunnels for data transition, different kinds of traffic can be mapped to different logical IPSec subtunnels and all subtunnels are encapsulated inside the main physical IPSec tunnel. Figure 5.14 provides an example for eNB with three IPSec subtunnels for separating O&M, user signaling (S1-MME), and user traffic (S1-U).

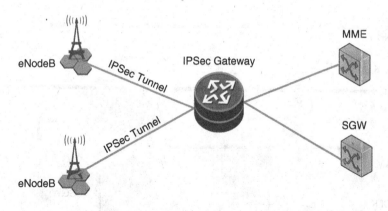

Figure 5.13 eNB high-level connectivity to EPC via IPSec GW.

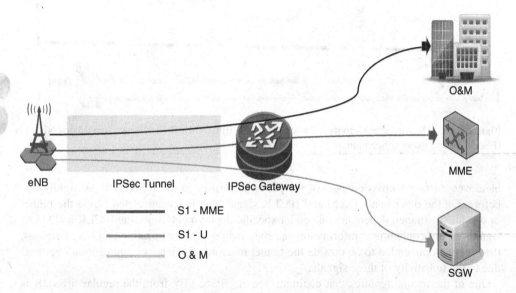

Figure 5.14 Connectivity between eNB and different NEs via IPSec subtunnels.

Although the eNB and SGW/MME in the EPC network can support an IPSec tunnel, the amount of tunnels will be large in the case of establishing these tunnels directly from the eNB to the EPC. Therefore, an IPSec gateway is introduced to tackle this problem. The benefit of this solution is that only three subtunnels need to be established between the eNB and the EPC network, which are S1-MME, S1-U, and O&M traffic. Figure 5.15 illustrates the connectivity between the eNB, the IPSec GW, the EPC, the O&M, and the IPCLK source. In this scenario, the user traffic (S1-U) and the user control plane (S1-C) are encapsulated inside the main IPSec tunnel between the eNB and the IPSec GW, and then from the IPSec GW they are terminated at the SGW and MME, respectively. Moreover, the X2 traffic between neighbor eNBs goes through the IPSec tunnels of the relevant eNBs as it contains user traffic during

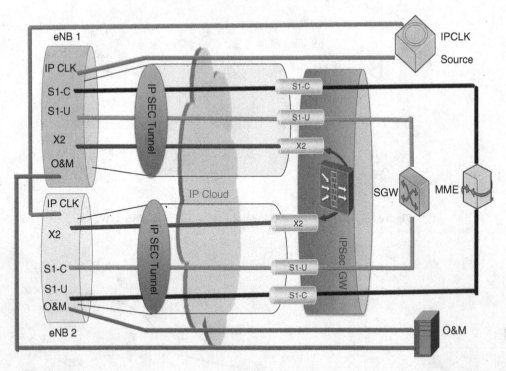

Figure 5.15 Detailed connectivity between eNB and EPC/NMS/IPCLK server/neighbor eNB via IPSec GW and using IPSec tunnel.

handover. Also, the cross connect of the X2 is performed in the IPSec GW, as highlighted before. On the other hand, O&M and IPCLK signals can be encapsulated inside the tunnel or outside the tunnel, depending on vendor specific deployment. Since the IPCLK and O&M signals do not contain user information and they include synchronization and O&M signals, then it is recommended to be outside the tunnel to avoid any delay or aggregation overhead due to the sensitivity of these signals.

One of the unique features that distinguishes the IPSec GW from the regular firewalls is the capability of intrusion prevention systems (IPSs) and intrusion detection systems (IDSs). The IPS function monitors the status of traffic on the network, accurately and comprehensively detects intrusions by means of DPI, and responds to the intrusions according to related policies.

The traditional firewall does not perform deep and comprehensive inspection of data at the application layer. For example, the HTTP traffic is widely used on the network. Therefore, HTTP dataflow accounts for a large proportion of network data. The data flows, however, also bring many hidden security threats. Thus a traditional firewall function fails to perform a deep inspection of the data flows. In this case, you can use of the IPSec GW to inspect them. IPS can block intrusions in a timely manner to secure the intranet. It also supports after-the-event auditing so that the information about the intrusions is recorded in real time. The IPS supports IP fragment reassembly and TCP flow reassembly, to prevent attacks from escaping the IPS inspection. It can effectively identify application protocols running on non-well-known ports

Table 5.3 IPSec GW dimensioning figures and corresponding units

Capacity figure	IPSec GW
Max throughput (Max)	xx Gbps
Max throughput (IMIX)	xx Gbps
Max throughput (HTTP)	xx Gbps
Max packets per second (64 bytes)	xx Mpps
VPN (virtual private network) performance	xx Gbps
Maximum IPS performance	xx Gbps
New sessions per second	x Million
Maximum concurrent sessions	xx Million
Maximum security policies	xxx K
Maximum users supported	xx Million

and inspect the protocol data to improve the rate of inspecting intrusions. It also supports protocol anomaly analysis and feature inspection to inspect attacks such as worm viruses, scanning, and spyware. Table 5.3 summarizes the key dimensioning figures for the IPSec GW node.

5.4.1 IPSec GW Deployment Strategy and Redundancy Options

The IPSec GW can be deployed in two scenarios, as illustrated in Figure 5.16. Option one is by using one IPSec GW node that is connected to two EPC nodes. In this case the EPC nodes will feature a geo-redundancy and the IPSec GW will feature a local 1 + 1 redundancy.

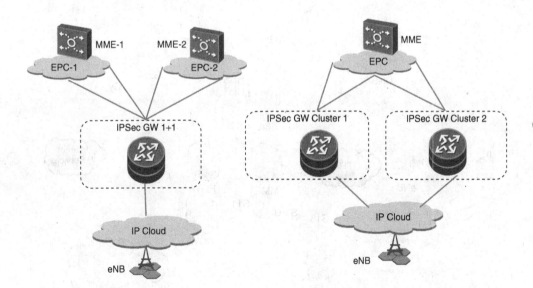

Figure 5.16 IPSec GW deployment strategies.

The second option is by using a geo-redundancy in IPSec GW by deploying two pairs of the IPSec GWs in different locations with local $1+1$ redundancy and then the two nodes will be connected to same EPC node. Also, the two nodes may be connected to two EPC nodes with geo-redundancy. However, the second option will add several complexities to the network design and traffic routing. As indicated before, a cluster approach would be perfect by adopting option one with several IPSec GWs nodes distributed based on regions with $1+1$ local redundancy.

5.5 EPC Deployment and Evolution Strategy

A practical EPC system architecture is shown in Figure 5.17.

In the case where the operator already deploys a 3G/PS (packet switched) core and needs to evolve to EPS there are two options to introduce EPC:

Option 1: By introducing a new EPC In this option, new EPC nodes will be introduced and integrated with the existing PS, as shown in Figure 5.17. The EPC will serve the LTE users and the existing PS will serve the 2G/3G users. The SWOT(strengths, weaknesses, opportunities and threats) of this option are provided in Figure 5.18. The integration with the existing PS core can be via two solutions

Scenario 1: Inter-working with legacy PS by Gn/Gp In this scenario, the EPC is integrated with the legacy PS core based on Gn/Gp interfaces, as shown in Figure 5.19. MME translates the security context of EPC into the legacy PS format, the R8 QoS profile into

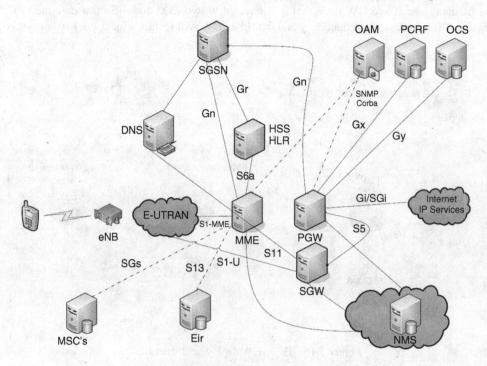

Figure 5.17 Practical LTE-EPC topology.

<table>
<tr><td>

Strengths

- Vendors' solutions are available and widely deployed and tested.

- No major impact on the existing live nodes (overlay network)

- More flexibility by having a dedicated EPC core which can integrate with multiple access vendors.

</td><td>

Weakness

- Extra cost due to the introduction of the new platform and integration with existing PS core.

- Longer deployment due to IOTs with existing PS core, IN, OCS, PCRF and other backend systems.

</td></tr>
<tr><td>

Opportunities

- Introduce new features and services

- Encourage competition by having potentially a new platform for EPC and then either to upgrade the existing PS to support LTE or to upgrade the EPC to support the PS functionality as part ·of convergence strategy.

</td><td>

Threats

- Extra complexity due to the introduction of new platform.

- Limited references for EPC deployment may complicate the IOT with other platforms.

- Potential instability on the EPC nodes since they are new platforms till fully optimized.

</td></tr>
</table>

Figure 5.18   SWOT of Option 1 – introducing a complete new EPC.

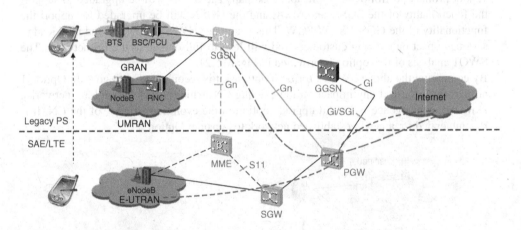

Figure 5.19   Inter-working with legacy PS by Gn/Gp.

an R99/R97 QoS profile and just acts as a SGSN to interconnect with the legacy PS core. The gateway also supports the GGSN function at the same time and can act as the common anchoring point for the multi-mode UE wherever the UE is in the EPC or in the legacy PS core. As to the multimode UE, the common anchoring point should be selected to ensure there is no PGW/GGSN relocation when the UE moves from the legacy PS core to LTE. In this solution there is no impact on legacy or roaming networks. However, SGSN is able to select the GGSN or GGSN/PGW based on the type of subscribers.

Scenario 2: Inter-working with legacy PS by S3/S4 interfaces In this scenario, the EPC is integrated with the legacy PS core based on S3/S4 interfaces, as shown in Figure 5.20. The EPC core is interconnected with the legacy 2G/3G PS core by the S3/S4 interface. The existing SGSN should be upgraded to become S4 SGSN and the existing GGSN should be upgraded to become SAE-GW. The SGW becomes the common anchoring point between LTE and 2G/3G. In this case, the legacy PS core can enjoy some enhancement of R8, such as the label QoS profile, the idle signaling reduction, and so on. Optimization of the signaling load for all terminals in idle mode leads to ISR, idle mode signaling reduction. The ISR function provides a mechanism to limit signaling during inter-RAT cell reselection in idle mode. ISR is a feature that allows the UE to move between LTE and 2G/3G without performing a TA or routing area (RA) update. ISR can be used to limit the signaling between the UE and the network as well as signaling within the network. The idea behind the ISR feature is that the UE can be registered in a GERAN/UTRAN RA at the same time as it is registered in an E-UTRAN TA (or list of TAs). The UE keeps the two registrations in parallel and runs periodic timers for both registrations independently. Similarly the network keeps both registrations in parallel and also ensures that the UE can be paged in both the RA and the TA(s) in which it is registered. The MME/SGSN activates ISR only if the SGW supports the ISR. How MME/SGSN determines a SGW supports ISR is implementation dependent.

Option 2: By fully upgrading the existing PS core to support 2G/3G/LTE In this option the existing PS core will be upgraded to support 2G/3G/LTE. This upgrade depends on the PS vendor product portfolio and applicable roadmap. The SGSN will be upgraded to support the functionality of the SGSN and MME and the GGSN will be upgraded to support the functionality of the GGSN/SGW/PGW. This is a major HW/SW upgrade and may lead to serious impact on existing customers and will lead to full outage to the PS network. The SWOT analysis of this option is provided in Figure 5.21.

By comparing the above SWOTs for both options, it is recommended to go with Option 1 at this stage to avoid any impact on the existing CN, which serves 2G and 3G networks. However, convergence is a valid upgrade option in the evolution strategy of the CN. It is recommended for a newly established network to consider Option 2.

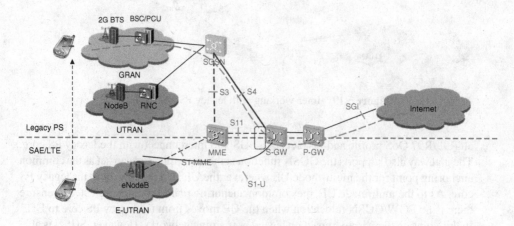

Figure 5.20   Inter-working with legacy PS by S3/S4.

Strengths	Weakness
• Re-use of the existing PS nodes. • Re-use of the current interfaces and therefore no new IOT is needed with existing nodes. • Reduced OPEX/CAPEX	• Limited capacity if we upgrade the existing GGSN to support PGW functionality. • The SGSN upgrade to support MME will lead to reduced capacity as well.

Opportunities	Threats
• Faster deployment due to reuse of the existing interfaces and existing PS platform. • Leverage the operator knowledge and experience of the existing network.	• The need for IOT with new RAN vendors for LTE. • The upgrade on live nodes may brings major operational risks. • No references for previous upgrades.

Figure 5.21 SWOT of Option 2 – upgrade of the existing PS core.

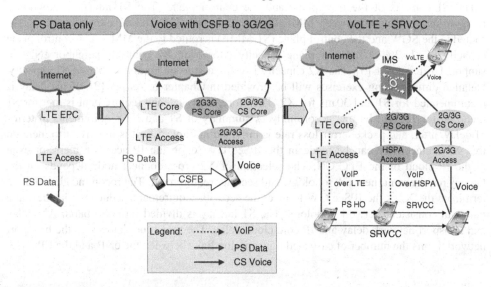

Figure 5.22 Voice roadmap with LTE network.

Voice roadmap The voice roadmap with LTE is shown in Figure 5.22. The phase 1 of LTE started with data only and in phase 2 the voice is introduced using circuit switch fall back (CSFB) and then IMS with VOIP (voice over Internet protocol) and SR-VCC (single radio voice continuity call) will be implemented in phase 3. Finally, full multimedia services will be offered in phase4. The detailed LTE voice evolution is discussed in Chapter 7.

5.6   Access Network Domain

5.6.1   E-UTRAN Overall Description

The AN consists of eNBs that provide the E-UTRA user plane (PDCP/RLC/MAC/PHY) (packet data convergence protocol/radio link control/medium access control/Physical layer) and control plane protocol (RRC, radio resource control) terminations toward the UE and IP access layer that carry the traffic of eNB toward the CN domain [1]. Functionally, eNB acts as a layer 2 bridge between the UE and the EPC, by being the termination point of all the radio protocols toward the UE, and relaying data between the radio connection and the corresponding IP-based connectivity toward the EPC [8]. In this function, the eNB performs ciphering/deciphering of the user data plane, and also IP header compression/decompression, to avoid sending the same or sequential data in an IP header several times [8, 20]. Each eNB may have one S1 connectivity with one EPC node and it can have two S1 connectivities with two EPC nodes if geo-redundancy is adopted, similar to the scenario in Figure 5.3 [20]. More specifically, both MMEs and SGWs of the EPC may be pooled, which means that a set of those nodes is assigned to serve a particular set of eNBs. From a single eNB perspective this means that it may need to connect to many MMEs and SGWs. However, each UE will be served by only one MME and one SGW at a given time, and the eNB has to keep track of this association. This association will never change from a single eNB point of view, because MME or SGW can only change in association with inter-eNB handover [8].

The S1 consists of the user plane and the control plane. The S1 interface supports a many-to-many relation between MMEs/SGWs and eNBs [1]. The S1 user plane (S1-U) is routed to the SGW and the control plane (S1-MME) is routed to the MME. The control plane capacity is ~1–3% of the user plane capacity. Moreover, X2 connects neighbor eNBs to support UE handover [3]. The X2 capacity is estimated to be around 3–5% of S1 capacity. Detailed dimensioning exercises will be provided in Chapter 6. As per [2], 15 ms delay is recommended for S1 and 30 ms for X2. Therefore, if the latency requirement is met for S1 then that for X2 will be met accordingly. A summary of S1 and X2 requirements in terms of delay, jitter, and packet error loss rate is provided in Table 5.4. It is important to mention that X2 connectivity can be made in the IPSec GW, or in the IP access routers, or even in the IP cloud of the EPC CN. The selection of X2 cross connect node depends on the latency requirement, network topology, and security requirement. The recommended node to terminate the X2 is the IPSec GW to meet the security requirement, reduce the latency, and maintain consistent network topology. The S1 latency is divided into two parts: AN delay and transport network delay and IP core cloud delay. The maximum latency of the transport network limits the number of cascaded transmission links between the eNB and the EPC.

Table 5.4 S1 and X2 requirements

Type		Delay (ms)	Jitter (ms)	Packet error loss rate (%)
S1 interface	Best	5	2	0.0001
	Recommended	15	4	0.001
	Tolerable	20	8	0.5
X2 interface	Best	10	4	0.0001
	Recommended	30	7	0.001
	Tolerable	40	10	0.5

The eNB main functions include, but are not limited to [3, 2, 8, 20]

1. RRM, radio bearer control, and scheduling
2. Prioritizing and scheduling traffic according to the required QoS
3. IP header compression and encryption of user data streams
4. Selection of the MME at the UE attachment (if not determined by information sent from the UE)
5. Scheduling and transmission of paging messages (originated from the MME)
6. Scheduling and transmission of broadcast information (originated from the MME or O&M)
7. Measurement and measurement reporting configuration for mobility and scheduling
8. Controls and analyses radio signal level measurements sent by the UE, makes similar measurements, and, based on those, makes decisions to handover UEs between cells.

5.6.2 Home eNB

The E-UTRAN architecture adopts a home eNodeB gateway (HeNB GW) to allow the S1 interface between the HeNB and the EPC to support a large number of HeNBs in a scalable manner. The HeNB GW serves as a concentrator for the C-Plane, specifically the S1-MME interface. The S1-U interface from the HeNB may be terminated at the HeNB GW, or a direct logical U-Plane connection between HeNB and S-GW may be used, as shown in Figure 5.23. The HeNB improves the indoor coverage and delivers network savings through data offload over FTTH/DSL networks and, therefore, enables fixed/mobile convergence.

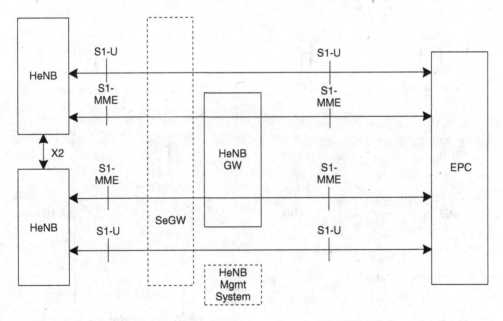

Figure 5.23 E-UTRAN HeNB logical architecture. (Source: [1] 3GPP TS. Reproduced with permission of ETSI.)

The HeNB GW appears to the MME as an eNB. The HeNB GW appears to the HeNB as an MME. The S1 interface between the HeNB and the EPC is the same, regardless of whether the HeNB is connected to the EPC via a HeNB GW or not. The HeNB GW connects to the EPC in such a way that inbound and outbound mobility to cells served by the HeNB GW does not necessarily require inter-MME handovers. One HeNB serves only one cell.

The functions supported by the HeNB are the same as those supported by an eNB and the procedures run between an HeNB and the EPC are the same as those between an eNB and the EPC. X2-based HO between HeNBs is allowed if no access control at the MME is needed, that is, when the intra-PLMN handover is between closed/hybrid access HeNBs having the same CSG ID or when the target HeNB is an open access HeNB. Direct X2-connectivity between HeNBs, is supported independently of whether any of the involved HeNBs is connected to a HeNB GW. The overall E-UTRAN architecture with deployed HeNB GW is shown in Figure 5.24.

In Figure 5.24, a HeNB operating in LIPA (local Internet protocol access) mode is represented with its S5 interface. Only if the HeNB supports the LIPA function, will it support an S5 interface toward the S-GW and an SGi interface toward the residential/IP network. For an LIPA PDN connection, the HeNB sets up and maintains an S5 connection to the EPC. The S5 interface does not go via the HeNB GW, even when present. For an LIPA PDN connection, the HeNB sets up and maintains an S5 connection to the EPC. The S5 interface does not go via the HeNB GW, even when present. The mobility of the LIPA PDN connection is not supported up to 3GPP Release 10 [1, 3]. The LIPA connection is always released at outgoing handover, as described in [3]. The L-GW function in the HeNB triggers this release over the S5 interface.

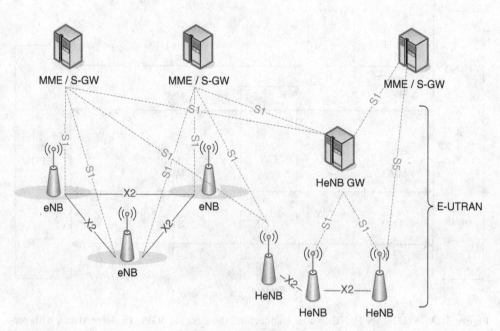

Figure 5.24 Overall E-UTRAN architecture with deployed HeNB GW. (Source: [1] 3GPP TS. Reproduced with permission of ETSI.)

Therefore, we can summarize the HeNB and its GW as follows:

1. The HeNB is a customer-premises equipment (CPE) that uses the operator's licensed spectrum and it is used to enhance network coverage and/or capacity similar to the femto concept.
2. The HeNB includes the functions of an eNB as well as some additional HeNB-specific configuration/security functions.
3. The HeNB GW is an optional and transparent gateway through which the HeNB accesses the CN. Therefore, it addresses the issue of supporting a large number of S1 interfaces in the CN and also provides security function similar to the IPSec GW.
4. There are three different operation modes for the HeNBs:
   (a) **Closed mode:** HeNB provides services only to its associated closed subscriber group (CSG) members. This mode is suitable for HeNB installed inside customer premises and services could be offered at reduced rates.
   (b) **Hybrid mode:** HeNB provides services to its associated CSG members and to non-members. However the CSG members are prioritized over non-members.
   (c) **Open mode:** HeNB appears as a normal eNB and provides services to all customers. This HeNB is used as an offloading layer with capacity boosting or convergence gap filling.

5.6.3   Relaying

E-UTRAN supports relaying by having a RN wirelessly connected to an eNB serving the RN, called a donor eNodeB (DeNB), via a modified version of the E-UTRA radio interface, the modified version being called the Un interface. The RN supports the eNB functionality, meaning it terminates the radio protocols of the E-UTRA radio interface, and the S1 and X2 interfaces. From a specification point of view, functionality defined for eNBs, for example, the RNL (radio network layer) and the TNL (transport network layer, also applies to RNs unless explicitly specified. In addition to the eNB functionality, the RN also supports a subset of the UE functionality, for example, the physical layer, layer-2, RRC, and NAS functionality, in order to wirelessly connect to the DeNB. Inter-cell handover of the RN is not supported. It is up to implementation when the RN starts or stops serving UEs. An RN may not use another RN as its DeNB.

The architecture for supporting RNs is shown in Figure 5.25. The RN terminates the S1, X2, and Un interfaces. The DeNB provides S1 and X2 proxy functionality between the RN and other network nodes (other eNBs, MMEs, and S-GWs). The S1 and X2 proxy functionality includes passing UE-dedicated S1 and X2 signaling messages as well as GTP data packets between the S1 and X2 interfaces associated with the RN and the S1 and X2 interfaces associated with other network nodes. Due to the proxy functionality, the DeNB appears as an MME (for S1-MME), an eNB (for X2), and an S-GW (for S1-U) to the RN. In phase II of RN operation, the DeNB also embeds and provides the S-GW/P-GW-like functions needed for the RN operation. This includes creating a session for the RN and managing EPS bearers for the RN, as well as terminating the S11 interface toward the MME serving the RN. The RN and DeNB also perform mapping of signaling and data packets onto EPS bearers that are set up for the RN. The mapping is based on existing QoS mechanisms defined for the UE and the PGW. In phase II of RN operation, the PGW functions in the DeNB allocate an IP address for the RN for

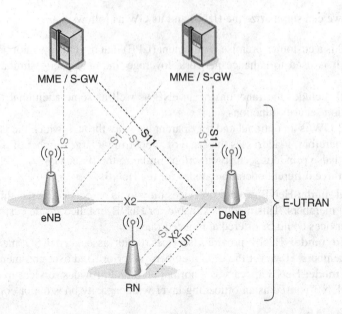

Figure 5.25 Overall E-UTRAN architecture supporting RNs. (Source: [1] 3GPP TS. Reproduced with permission of ETSI.)

the O&M which may be different than the S1 IP address of the DeNB. If the RN address is not routable to the RN O&M domain, it is reachable from the RN O&M domain (e.g., via NAT).

5.6.4 End-to-End Routing of the eNB

The routings for S1 user traffic and signaling are shown in Figure 5.26. The user traffic is routed to the SGW while the signaling is routed to the MME. The X2 traffic is routed through the IPSec GW, as shown in Figure 5.27. X2 contains user traffic and signaling to support fast inter-eNB handover, replacing the RNC in the legacy 3G network. As indicated before, the cross connect of X2 for adjacent eNBs is conducted in the IPSec GW to establish a consistent network topology and to avoid loading on the other IP network elements in the IP cloud of the EUTRAN network or the EPC CN that is designed to serve the transport network. This will also allow protection of X2 traffic as it goes inside the IPSec tunnel that is established between the eNB and the IPSec GW.

The clock signaling (IPCLK, Internet protocol clock) routing is shown in Figure 5.28. The clock signaling is routed toward the IPCLK server. With the introduction of the LTE network the exiting 3G clock server can be used or a new clock server may be introduced according to the adopted synchronization method. The IPCLK signaling can be routed inside the main IPSec tunnel or outside the tunnel but it is recommended to be outside the tunnel as there are no users traffic in this signaling and it is merely used for synchronization. The IPSec GW and tunneling may also introduce delay and, therefore, it is recommended to avoid additional delay with the IPCLK signaling. If the IEEE 1588V2 clock is adopted for the synchronization then Table 5.5 provides the IP requirement for this synchronization scheme. The O&M signaling routing is demonstrated in Figure 5.29. The O&M signaling is routed outside the IPSec GW tunnel or it

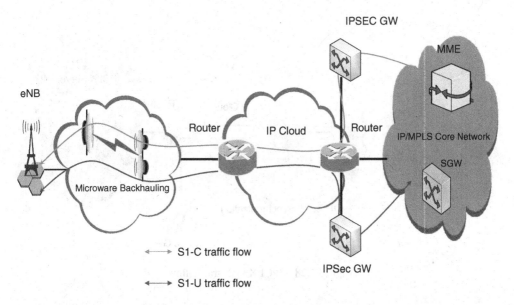

Figure 5.26 S1 user traffic and signaling routing.

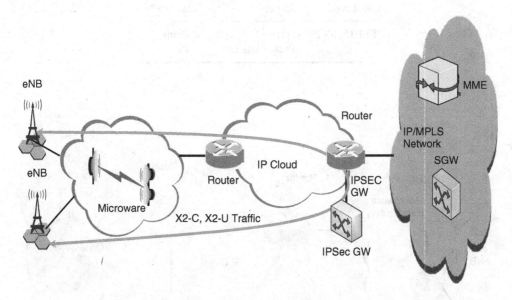

Figure 5.27 X2 routing through IPSec GW.

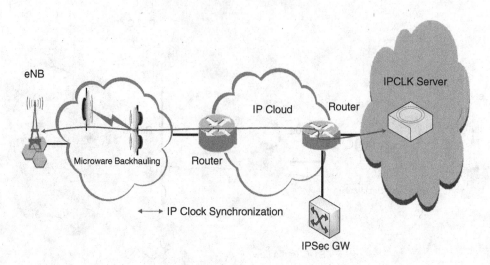

Figure 5.28   IPCLK sync traffic flow.

Table 5.5   IEEE 1588 V2 requirement

Clock type	Target	Restriction
IEEE1588V2	Jitter	< 20 ms
	Packet loss rate	< 1%

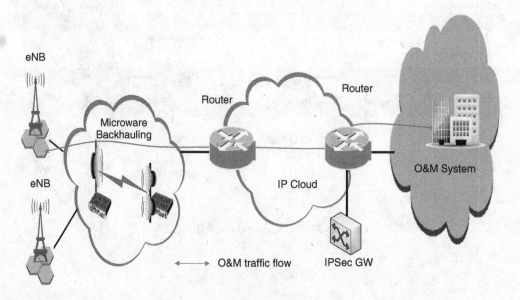

Figure 5.29   O&M signaling traffic flow.

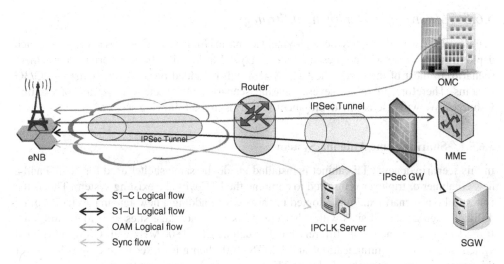

Figure 5.30   End to end traffic flow for one eNB.

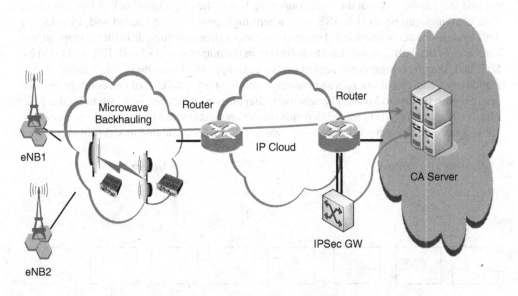

Figure 5.31   eNB and IPSec GW connectivity to CA server.

can be encapsulated inside the IPSec tunnel to protect the NMS (network management system) from any possible attack. The complete end-to-end routing is illustrated in Figure 5.30.

The last connectivity toward the eNB is the CA server connectivity. The eNB and IPSec GW need to be connected to the CA server for authentication, as shown in Figure 5.31.

5.6.5   Macro Sites Deployment Strategy

In this section, we will discuss the deployment scenarios for macro sites. Pros and cons for each deployment scenario will be presented and analyzed to identify the appropriate deployment scenario. In most of the cases, the LTE system will be added on top of the existing 2G/3G systems. Therefore, there are several sharing scenarios for efficient utilization of existing resources. We will discuss the deployment scenarios in the following sections.

5.6.5.1   Sharing Same Antenna System

In this scenario, a new LTE cabinet is installed inside the same shelter used for 2G/3G and a new combiner or triplexer is inserted to combine the LTE with the existing system. This is the fastest rollout scenario and it is adopted in cases where adding a new antenna system is quite difficult, such as shared sites with other operators or camouflage sites with low wind load. If the LTE is combined with different bands, such as LTE 1800 MHz with GSM900 (global system for mobile communication) and UMTS2100, then a triplexer is an easy solution for this scenario. On the other hand, if the LTE system is in the same band as the existing system, then the combination is not an easy solution. There are two types of combiner that can be used to combine the same band systems. The well-known hybrid combiner is the easiest method that can be used but the 3 dB combining loss is the major drawback of this combiner. This combiner can be used in IBS sites where high power is not needed and, therefore, a 3 dB loss can be accommodated. However for macro sites it is quite difficult to adopt such a combiner as the 3 dB loss will lead to reduction in the output power of both LTE and GSM by 50% and, therefore, significant impact on the coverage will be experienced. Therefore, it is mandatory to develop a low loss same antenna sharing unit (SASU) that can combine the two systems with minimum losses. The schematic diagram for such a unit is shown in Figure 5.32. The selection of an SASU model depends on the application and the required specification. Table 5.6 provides a comparison between two SASU models. The first model provides higher

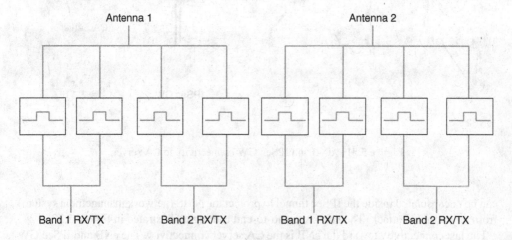

Figure 5.32   Schematic diagram for SASU.

Table 5.6 Comparison between two SASU models

Specifications	Model A	Model B
Frequency range; ports 1,3	1730–1750 MHz/ 1825–1845 MHz (GSM)	
Frequency range; ports 2,4	1765–1785 MHz/ 1860–1880 MHz (LTE)	
Insertion loss	0.5 dB	0.5 dB
Isolation port 1 < – > port 2	40 dB	30 dB
Isolation port 3 < – > port 4		
VSWR	< 1.2	< 1.2
Intermodulation (3rd order)	≤ −140 dBc	< −100 dBm

Table 5.7 Field testing results for SASU models

Model 1								
Without combiner			With combiner			Δ		
LTE1800		GSM1800	LTE1800		GSM1800	LTE		GSM1800
RSRP	RSRQ	Rx_Lev	RSRP	RSRQ	Rx_Lev	RSRP	RSRQ	Rx_Lev
−51.5	−10.7	−37	−51.88	−11	−37	−0.38	−0.3	0
−40	−9	−37	−40	−8.06	−37	0	0.94	0
−40.1	−9	−38	−41	−10.8	−37	−0.9	−1.8	1
−73.9	−10.6	−40	−71	−10.9	−38	2.9	−0.3	2
−94.3	−10.88	−50	−88.9	−10.81	−55	5.4	0.07	−5
−98.3	−10.88	−57	−99.25	−10.75	−64	−0.95	0.13	−7
−103.1	−10.81	−74	−98.5	−10.94	−76	4.6	−0.13	−2

Model 2								
Without combiner			With combiner			Δ		
LTE		GSM1800	LTE		GSM1800	LTE		GSM1800
RSRP	RSRQ	Rx_Lev	RSRP	RSRQ	Rx_Lev	RSRP	RSRQ	Rx_Lev
−48.13	−10.83	−49	−50.56	−10.81	−47	−2.43	0.02	2
−40	−8.75	−39	−40	−8.75	−36	0	0	3
−40.13	−10.44	−42	−40.19	−10.75	−37	−0.06	−0.31	5
−64.56	−10.88	−53	−70.09	−10.94	−63	−5.53	−0.06	−10
−90.19	−10.81	−79	−92.89	−10.56	−81	−2.7	0.25	−2
−90.81	−10.81	−90	−92.38	−10.88	−86	−1.57	−0.07	4
−100.88	−10.81	−92	−102.88	−10.94	−94	−2	−0.13	−2

isolation between GSM and LTE and lower intermodulation. In order to validate the SASU, it is recommended to have a quick field testing at live GSM/LTE sites by testing the Rx level for GSM and the RSRP/RSRQ (reference signal received power/reference signal received quality) for LTE with and without the combiner. Table 5.7 demonstrates the field test results for two models demonstrated at Table 5.6. By comparing the overall average of the values we find that the first model introduced an average of 0.08 dB losses while the second model introduced an average of 0.7 dB losses. This is a quick test and the most important factors are the isolation and intermodulation rejection. These two criteria will need a RF lab for detailed testing. These factors are critical factors, especially for IBS sites where the passive intermodulation (PIM) is a key issue due to aging of the DAS. Therefore the first model is the preferred model due to the better isolation and lower intermodulation capability. The drawback of the SASU is the need for a high guardband >10 MHz to guarantee the isolation between GSM and LTE. Also, this combiner type is not tunable in the field and therefore once installed it cannot be tuned to support other frequencies in the same band. This combiner will therefore limit the operator choice in terms of re-farming, or changing the frequency bands, or adding new frequency bands, that is, carrier aggregation with additional carrier in same band.

The antenna system sharing scenario is shown in Figure 5.33. The GSM1800 and LTE1800 are combined (using SASU) on top of the cabinet then the output of the combiner can go directly to the antenna if no other systems exist. In the case of other systems in different bands,

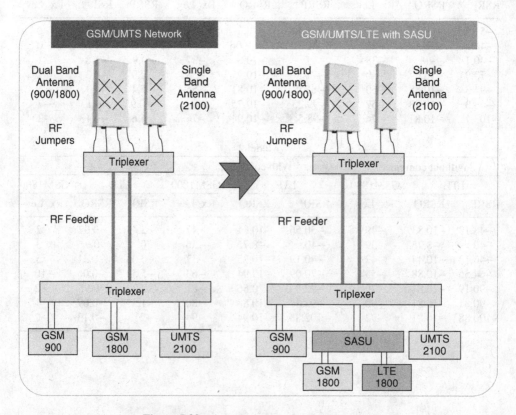

Figure 5.33 Antenna system sharing exercise.

such as UMTS at 2.1 GHz and GSM900 then triplexers are mandatory to combine all bands (900 MHz (GSM), 1800 MHz (LTE + GSM), 2.1 GHz (UMTS)) into one pair of RF feeders. Another triplexer is needed at the top of the tower to separate the three bands and the GSM1800 and LTE1800 will go to same antenna port as they will not be separated. The pros and cons of this scenario are summarized as follows:

1. Reuse of existing antenna system
2. Faster rollout with minimum installation time.
3. Less CAPEX as only an additional three combiners are needed per site. The cost of the combiner is ~1000 USD.
4. Customization of the combiner to the specific bands of the operator is needed.
5. Guardband requirement to isolate GSM from LTE in the same band.
6. Limits operator flexibility in terms of re-farming and frequency change
7. LTE and GSM will have same azimuth and same tilt and therefore optimization is quite complicated.

5.6.5.2 LTE/UMTS Co-Antenna with Indoor Cabinet

In this scenario the UMTS and LTE can share a dual band antenna with four ports. A complete indoor cabinet is installed inside the shelter and a new set of RF feeders is installed for LTE. Figure 5.34 illustrates this scenario. In this scenario, the dual port UMTS antenna will be

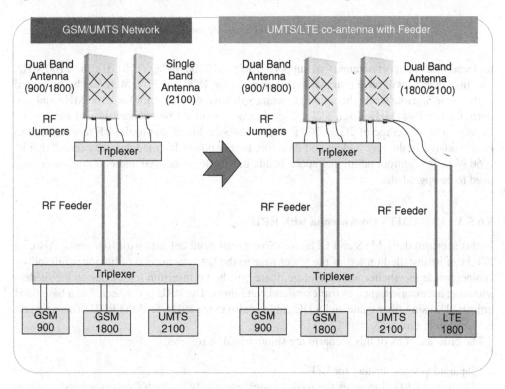

Figure 5.34   LTE/UMTS co-antenna with indoor cabinet scenario.

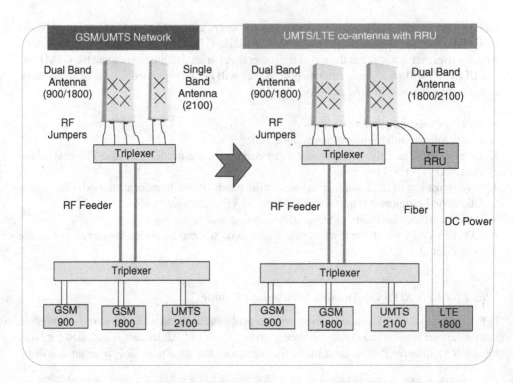

Figure 5.35 LTE/UMTS co-antenna with RRU.

replaced by a four port antenna that supports 1800 and 2100 MHz or 2.6 GHz and 2100 MHz. The first antenna will have four symmetric ports as the 1800 and 2100 MHz can be supported in the same antenna while the second antenna will have two ports for the 2100 MHz and two ports for the 1800 MHz. Each technology can be optimized independently and there is no impact on the coverage of 2G or 3G. However, 2–3 dB losses due to the RF feeders need to be considered in the link budget and therefore higher power RF modules are needed, that is, ≥ 60 Watt. In addition, additional space inside the shelter is needed and cooling power may need to be upgraded.

5.6.5.3 LTE/UMTS Co-Antenna with RRU

In this scenario the UMTS and LTE can share a dual band antenna with four ports. Also, an RRU can be installed on top of the tower near to the antenna instead of having a full indoor cabinet inside the shelter and, therefore, there will be no insertion losses due to the feeders and other accessories, such as triplexers and combiners. The RRU is connected to a base band unit (BBU) using fiber cable and a DC cable is also extended to feed the RRU. This scenario is shown in Figure 5.35.

The pros and cons of this scenario are summarized as follows:

1. Separate antenna system for LTE.
2. The use of RRU eliminates the feeder losses, especially for high frequency bands such as 2.6 GHz.

3. Less space inside the shelter as the BBU can be installed inside a 19 in rack and there is no need for a full indoor cabinet.
4. RRU installation on top of tower is needed.
5. LTE and UMTS have the same azimuth but different tilt and therefore optimization of both systems is almost independent.
6. Difficulty of maintenance, and tower climbing is need for the RRU change.
7. Installation of fiber and DC power cables is needed.
8. Less power consumption from A/C as the RRU is installed in an outdoor environment and therefore less cooling power is needed inside the shelter.
9. A sunshade for the RRUs is needed for high temperatures of 50 °C and above.

Figure 5.36 summarizes the pros and cons of the co-feeder and co-antenna scenarios.

5.6.5.4 Antenna Separation

To avoid interference and intermodulation between LTE and other systems in the same band or adjacent bands, such as GSM or UMTS, a safety distance should be maintained between LTE antenna and other technology. Figure 5.37 demonstrates the vertical and horizontal safety distance between LTE1800 and GSM1800. As shown, a 20 cm vertical separation is sufficient while a 0.5 m horizontal distance is mandatory for separation.

5.6.6 IBS Deployment Strategy

Since the legacy IBS sites were deployed prior to LTE commercialization MIMO was therefore not considered in the design of the IBS sites. In addition, the LTE deployment in IBS sites using

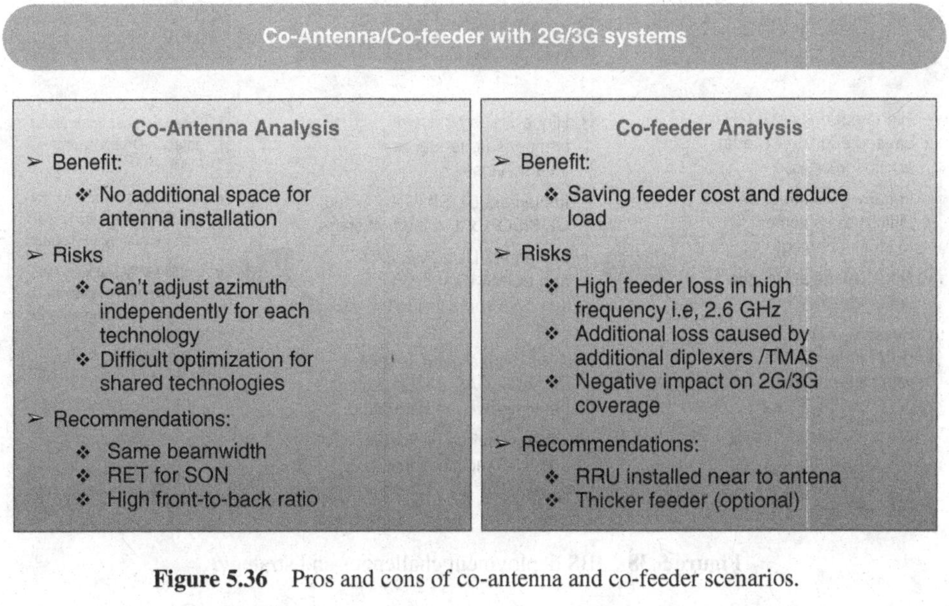

Co-Antenna/Co-feeder with 2G/3G systems	
Co-Antenna Analysis	**Co-feeder Analysis**
➢ Benefit:	➢ Benefit:
❖ No additional space for antenna installation	❖ Saving feeder cost and reduce load
➢ Risks	➢ Risks
❖ Can't adjust azimuth independently for each technology ❖ Difficult optimization for shared technologies	❖ High feeder loss in high frequency i.e, 2.6 GHz ❖ Additional loss caused by additional diplexers /TMAs ❖ Negative impact on 2G/3G coverage
➢ Recommendations:	➢ Recommendations:
❖ Same beamwidth ❖ RET for SON ❖ High front-to-back ratio	❖ RRU installed near to antena ❖ Thicker feeder (optional)

Figure 5.36 Pros and cons of co-antenna and co-feeder scenarios.

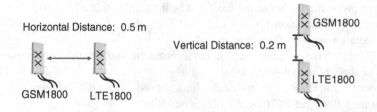

Figure 5.37   Safety distance between LTE and GSM in 1800 MHz band.

a high frequency band such as the 2.6 GHz band is challenging. The major challenges that may face the introduction of LTE in IBS sites are:

1. Additional feeder losses due to higher frequency. Therefore, coverage will be significantly impacted and the link budget and IBS design need to be revisited.
2. Existing DAS system does not support MIMO or RX diversity.
3. The legacy DAS systems are designed to work up to 2.5 GHz and therefore a major swap for some components is necessary to support high frequency bands such as 2.6 GHz.
4. The possibility of PIM, that may appear due to the injection of a high power LTE system into the existing DAS. Aging or existing minor issues in the DAS may became major issues with the injection of high power to the DAS.

Figure 5.38 summarizes the IBS challenges and strategy describing the deployment and hereunder are the main guidelines:

1. Full coverage with SISO (single input single output) can be provided where the maximum theoretical DL throughput is 75 Mbps.

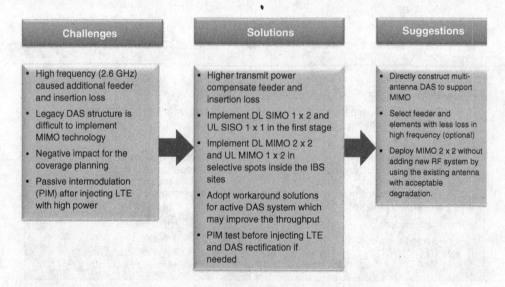

Figure 5.38   IBS deployment challenges and strategy.

2. Full peak theoretical throughput (~150 Mbps) with MIMO 2 × 2 can be provided at selected spots within the IBS coverage.
3. Workaround solutions for active DAS system to deploy MIMO 2 × 2 with less optimality can be adopted. These solutions will depend on the DAS supplier and the DAS design.

The main challenges for LTE introduction in IBS sites are the MIMO 2 × 2 deployment, high frequency bands such as 2.6 GHz, and the PIM that may appear with the injection of a high power LTE system to existing legacy DAS systems. The MIMO deployment strategy can be summarized in the following steps:

- Full coverage by implementing SIMO in the first stage to expedite the LTE rollout and the throughput in this case will be as shown in Figure 5.39. The DL/UL throughputs are obtained via simulation using the generalized protocol parameters specified in [7, 21, 22].
- Deploy MIMO 2 × 2 in selective spots inside the IBS sites such as high traffic areas or important locations. In this case extension of additional DAS to these locations is mandatory.
- Adopt workaround solution to deploy MIMO 2 × 2 without adding a new RF path (i.e., almost double the existing DAS) by using the existing antenna to form a MIMO setup with acceptable degradation. This will depend on the DAS design and whether the solution is passive or active. Also, MIMO performance may be impacted in this case as the antenna separation will not be optimal.
- Directly construct new multi-antenna DAS to support MIMO 2 × 2 from the beginning for new sites.

The other main challenge for the IBS system is deployment of LTE at a high frequency band such as 2.6 GHz, which is one of the main bands for LTE, especially for hotspot deployment. The legacy DAS systems were designed to work up to 2.5 GHz and, therefore, a significant degradation is expected if we insert LTE at 2.6 GHz to these legacy IBS systems. A complete revamping is needed to change the passive or active components that are designed to work up to 2.5 GHz with other components that support up to 2.7 GHz. The final major challenge for LTE deployment in IBS sites is the PIM which may appear due to the injection of the LTE system with high power into legacy IBS sites that were not designed to accommodate such high power. The nonlinearity of some components due to aging may also lead to PIM. The PIM will be explained in detail in the next section.

5.6.7   Passive Inter Modulation (PIM)

PIM [23–26] is unwanted signals being created by high power Tx signals in passive RF components, such as connectors, filters, antennas, splitters, combiners, and so on. PIM generates intermodulation products that can fall within the Rx band of the receiver of adjacent bands in the base station or in user equipment and, therefore, interfere with the real signals from/to the cellular users served by these systems. PIM will lead to reduced coverage, capacity, and data throughputs. PIM is a growing issue for cellular network operators, especially in IBS solutions, due to aging and difficulty of maintenance. PIM issues may occur as existing equipment ages, when co-locating new carriers, or when installing new equipment, like adding LTE to existing 2G/3G systems. PIM is a particular issue when adding new carriers to old antenna systems. PIM can create interference that will reduce a cell's receive sensitivity or even increase the call

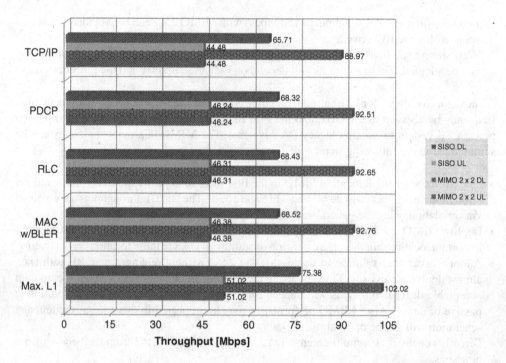

Figure 5.39   Theoretical LTE UL and DL throughputs for MIMO 2 × 2 and SIMO.

drop rate. This interference can affect both the cell that creates it and other nearby cells. PIM is created by high transmitter power so on-site PIM testing needs to be conducted at or above the original transmitter power levels to make sure that the test reveals any PIM issues. It is worth noting that PIM testing does not replace impedance-based line sweeps, that is, VSWR (voltage standing wave ratio) testing, rather, it complements line sweeping, which is now more important with the introduction of the LTE system with high output power and with MIMO, that is, 2 × 60 W output power.

High-speed LTE systems make PIM testing critical. As cell usage and throughput grows, the peak power produced by the new digital modulations increases dramatically, contributing heavily to PIM problems. Slight increases in PIM may lead to significant decreases in download speeds for broadband systems like 3G/LTE.

PIM testing versus VSWR testing [25] The PIM test is a measure of system linearity while a return loss (RL) measurement is concerned with impedance changes. It is important to remember that they are two independent tests, consisting of mostly unrelated parameters that are testing opposite performance conditions within a cellular system. It is possible to have a PIM test pass while RL fails, or PIM fail while RL passes. Essentially, PIM testing will not find high insertion loss and RL will not find high PIM. Line sweeps and PIM testing are both important and the latter is critical at the introduction of new systems or new carriers on existing IBS sites.

Defining PIM [23–26] PIM is a form of intermodulation distortion that occurs in components normally thought of as linear, such as cables, connectors, and antennas. However, when subject to the high RF powers found in cellular systems, these devices can generate intermodulation signals at −80 dBm or higher. Intermodulated signals are generated late in the signal path, they cannot be filtered out and may cause more harm than the stronger, but filtered, IM products from active components.

The PIM test is a measure of the linearity of the antenna system and thus the installation and material quality. PIM shows up as a set of unwanted signals created by the mixing of two or more strong RF signals in a nonlinear device, such as a loose or corroded connector, or nearby rust.

The following formulas can predict PIM frequencies for two carriers [23]:

$$nf1 - mf2 \qquad\qquad (5.1)$$

$$nf2 - mf1 \qquad\qquad (5.2)$$

where f1 and f2 are the carriers' center frequencies and the n and m are positive integers.

When referring to PIM products, the sum $n + m$ is called the product order, so if m is 2 and n is 1, the result is referred to as a third-order product (as shown in Figure 5.40) [23]. Typically, the third-order product is the strongest, causing the most harm, followed by the fifth- and seventh-order products. Since PIM amplitude becomes lower as the order increases, higher order products typically are not strong enough to cause direct frequency problems, but usually assist in raising the adjacent noise floor (as shown in Figure 5.41) [23]. It is unlikely that a third order product will fall directly into a designed cellular receive band. It is highly likely that energy from other external transmissions will mix within the nonlinear

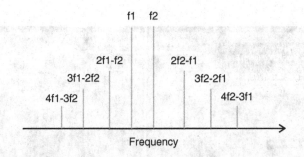

Figure 5.40   Frequency components generated by Intermodulation distortion.

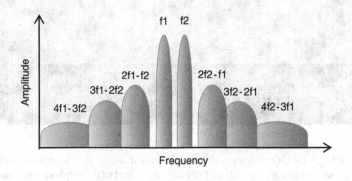

Figure 5.41   Increase of frequency components bandwidth by Intermodulation distortion order.

transmission line causing many smaller PIM levels to mix over and over, resulting in a wideband raised noise floor that usually spans all operators licensed spectra. Once this raised noise floor crosses into the Rx band, it then has an interference impact on the BTS receiver.

Causes of PIM [24–26] PIM comes from two or more strong signals and a nonlinear junction. The strong signals normally come from transmitters sharing an antenna run, transmitters using adjacent antennas, or nearby towers with conflicting antenna patterns. Damaged or poorly torqued RF connections, contamination, fatigue breaks, cold solder joints, and corrosion can create nonlinear junctions. Since the nonlinear junction may be outside the cell cabinet, eliminating PIM by filtering may not be possible. Often, it is necessary to identify and eliminate the root cause of the problem. Damaged connectors, cables, duplexers, circulators, and antennas can all contain nonlinear components. In addition, nearby corroded objects, such as fences, barn roofs, or rusty bolts can cause PIM, as long as the signals that reach them are strong enough. This effect is common enough to have its own name, which is the "rusty bolt effect". There are a number of different places to look for nonlinear junctions.

PIM testing in the field [23–26] There are many commercial devices that can measure the PIM in the field. With the introduction of an LTE system on top of existing 2G/3G technologies, and especially in the case of the antenna system, such as in the IBS solution and in the case of combining the use of SASU or hybrid combiners; it is mandatory to conduct a PIM testing before integrating the sites. From practical cases, the PIM issue may already exist but it became worst with the installation of a new system and this test will give the operator the opportunity to validate the antenna system and to rectify any fault that may impact the overall performance in terms of coverage, throughput, and even call drop. One of the common

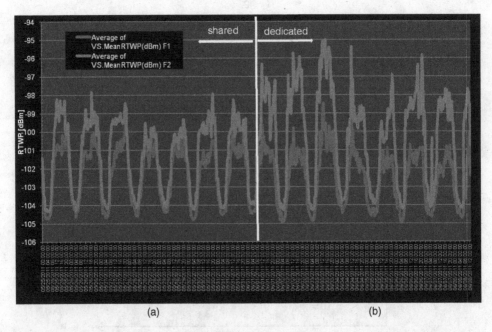

Figure 5.42 RTWP for 3G sectors with two cells and using two scenarios: (a) with shared (voice + data) on second carrier (i.e., f2) and (b) with dedicated (data only) on same carrier.

problems is with the introduction of a new LTE system the collocated 2G and 3G systems are impacted while they are using different bands. The maintenance team will blindly lock the LTE site and will return the cause to the new system, that is, LTE.

One of the parameters that can be used as an indication of the existence of the PIM is the received total wideband power (RTWP) which is average with the 3G system. The normal range for the average RTWP is less than −100 dBm and therefore an increase in the average RTWP indicates that there is an increase in the interference level and, therefore, a possible PIM or VSWR issue. Figure 5.42 provides the average RTWP for two 3G cells with different scenarios for the second carrier F2, that is, shared between data and voice and a dedicated carrier for data where the load is increased on the second carrier. As shown in the figure, it is normal that the average RTWP increases slightly with the loaded scenarios, as shown for the second carrier. This degradation is expected to happen only in loaded scenarios and during busy hours. In the case of PIM, the average RTWP is expected to increase consistently and with higher order, as shown in Figure 5.43. The figure indicates the RTWP for IBS sites with 2G/3G at 1800 MHz and 2.1 GHz, respectively, and when LTE at 1800 MHz with 20 MHz channel was plugged to the same IBS using an SASU combiner the average RTWP increased from −110 to −90 dbm.

In the 2G system, the average UL interference can be used as an indication of the existence of PIM, similar to the RTWP with 3G. The average UL interference for the same site

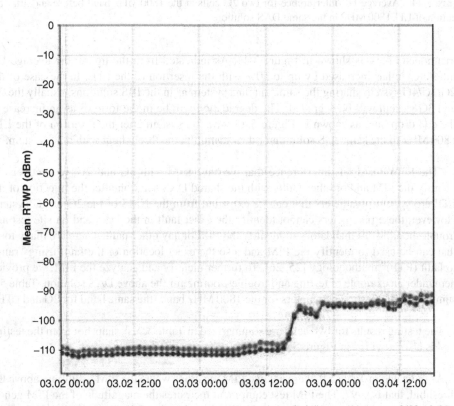

Figure 5.43   RTWP for two 3G cells before and after the injection (inserted between 3/03 12:00 and 3/04 00:00) of LTE1800 MHz at the same DAS solution.

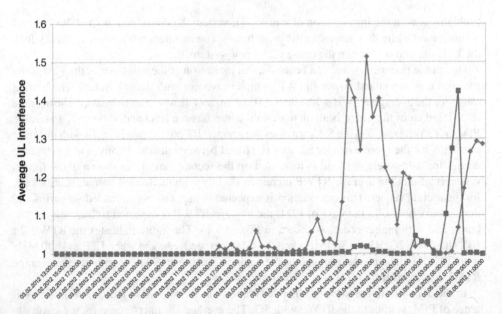

Figure 5.44 Average UL interference for two 2G cells at the 1800 MHz band before and after the injection of LTE1800 MHz in the same DAS solution.

mentioned above is shown in Figure 5.44. As indicated from the figure, the average UL interference has increased by up to 50% with the insertion of the LTE. In the case of the 2G/3G/4G system sharing the same antenna system as in the IBS solutions, usually the 2G and 3G system will be degraded. The degradation will be in the form of as an increase in the call drop rate, as shown in Figure 5.45 for a 2G system after the insertion of the LTE 1800 MHz to the same IBS solution or degradation in the throughput with a 3G system.

The RTWP and UL interference that we have used with 3G and 2G, respectively, to identify the PIM and/or other faults with the shared DAS solution after the injection of the LTE are system parameters that can be extracted from the NMS of the 2G or 3G system. However, these parameters cannot identify the exact fault in the DAS and on-site test and troubleshooting are mandatory to identify the PIM or any other fault. There are many tools that can be used to identify the PIM and also the exact location of the fault using "range to fault (RTF)" methodology [25, 26]. To further identify and analyze the PIM, we provide hereunder an example of testing and troubleshooting for the above DAS solution. Table 5.8 summarizes the testing parameters for the 1800 MHz band (the same band for 2G and LTE).

The testing results for two cells are summarized in Table 5.9. A snapshot from the testing tool is also provided in Figure 5.46 for one RF feeder of one cell.

As indicated in Table 5.9, there are two PIM locations where the PIM results are above the threshold, that is, −97. The PIM test equipment measures the magnitude of the PIM generated by the test signals and displays this information, as shown in Figure 5.46. As shown in

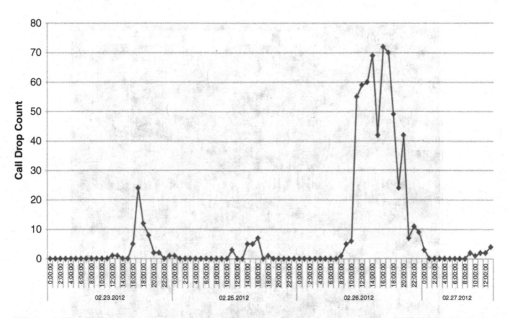

Figure 5.45 Call drop count for 2G cell before and after the insertion of the LTE1800 MHz in the same DAS solution.

Table 5.8 PIM Parameters for 1800 MHz band [25, 26]

Tone 1 frequency (MHz)	Tone 2 frequency (MHz)	IM3 frequency (MHz)	PIM threshold (dBm)	RF threshold (dB)
1805.0	1880.0	1730.0	−97.0	−15.0

Table 5.9 PIM testing results

Test of rack	P1 P2 (dBm)	RL (dB)	PIM (dBm)	Peak PIM (dBm)	Result
Top of cabinet	43.0 43.0	−30.1	−67.2	−59.1	Fail
Top of cabinet	43.0 43.0	−30.0	−71.7	−71.7	Fail

Figure 5.46, there are two peaks for PIM at 38.9 and 68 m using the RTF technology. Also, Figure 5.47 provides the RL (return losses) versus distance overlaid with the RTF. Using the relative distance of a PIM problem to a known RL peak allows users to locate faults on the line with more precision than with PIM versus distance alone. Therefore, as indicated by Figure 5.47, there is a fault at 38.9 m.

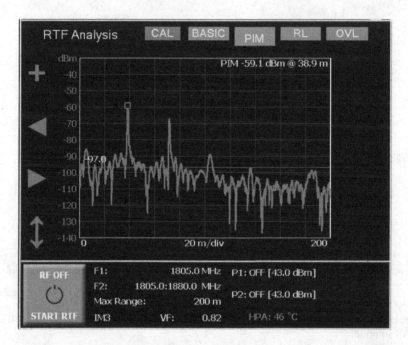

Figure 5.46 PIM testing results. (Reproduced with permission of KALUS.)

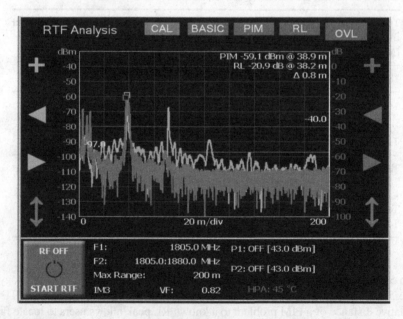

Figure 5.47 Return loss versus distance and PIM versus distance. (Reproduced with permission of KALUS.)

5.7   Spectrum Options and Guard Band

5.7.1   Guard Band Requirement

One of the important aspects when introducing the LTE system on top of existing systems like 2G or 3G is the guardband requirement. More specifically, when LTE is collocated with other technologies in the same site where nearby antennas or sharing scenarios are adopted then a guardband needs to be maintained to avoid interference and PIM between collocated systems. Table 5.10 summarizes the guardband requirement for LTE at different channel BW with different collocated technologies.

The guardband requirement can be summarized as follows:

- In the case of LTE with GSM collocated in the same band, that is, 1800 MHz, then a single carrier of GSM 200 kHz is enough to avoid interference between the two systems, as shown in Figure 5.48. Based on the simulation results in 3GPP Report 36.942, it can be concluded that 300 kHz offset from the LTE channel edge (separation between the nearest GSM carrier center frequency and LTE channel edge) is sufficient for the protection of GSM UL/DL against interferences from LTE (i.e., 200 kHz guard band edge-to-edge between the LTE1800 FDD (frequency division duplex) and GSM1800) [27, 28]. Some vendors can allow the use of an adjacent GSM carrier but not for the BCCH (broadcast control channel). In this case the RRU or the TX module deploys a sharp band pass filter on the LTE BW to avoid leakage to the adjacent GSM carrier [29].
- In the case of LTE collocated with another system in a different band (such as LTE 2.6 GHz and GSM 1800 MHz) then no guardband is needed.
- In the case of LTE and UMTS collocated in the same band no guardband is required, however, both systems need to deploy a strict filter on the RF power amplifiers to avoid leakage.
- In the case of LTE FDD collocated with another TDD (time division duplex) system, such as LTE or WiMAX, then half of the channel BW (the higher channel BW of either the FDD or the TDD system) is needed as the guardband.
- In the case where LTE is collocated with another operator's LTE in the same band, such as 2.6 GHz, digital dividend, or 1800 MHz, recent studies regarding LTE-FDD to LTE-FDD interference have shown that there is minimum to no interference without guardbands. Due to the "orthogonal" nature of the OFDMA/LTE (orthogonal frequency division multiple access), there is no need for any guardbands if all adjacent technologies are LTE-FDD. Further coordination is needed between collocated systems to avoid possible interference and also a safety distance between antennas is mandatory in this case.

5.7.2   Spectrum Options for LTE

Figure 5.49 summarizes the spectrum options and possible refarming. The selection of appropriate spectrum depends on many factors, such as the regulatory policy, spectrum fees, existing technologies, and so on. In this section we will summarize the major LTE bands and pros and cons for each band and how we can develop a spectrum strategy.

As per 3GPP TS 36.101 [7], Table 5.11 illustrates the supported LTE bands up to 3GPP Release 12.1.

Table 5.10 Guardband requirement for co-existing systems

Co-existing systems	LTE bandwidth			
	5 MHz	10 MHz	15 MHz	20 MHz
LTE1800 + GSM1800	0.2 MHz[a]	0.2/0[a] MHz	0.2/0[a] MHz	0.2/0[a] MHz
LTE band X + LTE band Y	0	0	0	0
LTE band X + LTE band X different operators sharing same sites	0	0	0	0
LTE band X and UMTS band X such as 2.1 GHz	0	0	0	0
LTE FDD LTE/WiMAX TDD (same band i.e., 2.5 GHz)	2.5 MHz	5 MHz	7.5 MHz	10 MHz

[a]Some vendors can allow the use of next GSM carrier to LTE for non-BCCH channel.

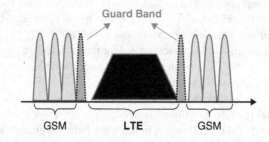

Figure 5.48 Guardband requirement for collocated LTE and GSM system in the same band.

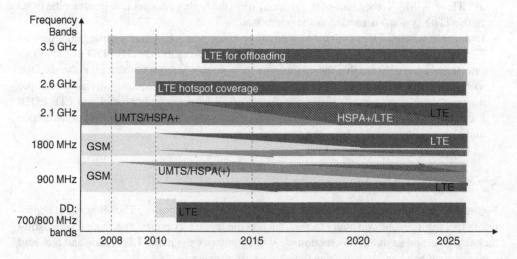

Figure 5.49 Spectrum options and refarming options.

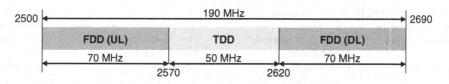

Figure 5.50 Band arrangement for 2.5 GHz.

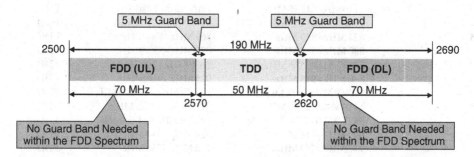

Figure 5.51 Band arrangement for 2.5 GHz with guardband.

5.7.2.1 LTE in 2.6 GHz

This is the first and widest available LTE band because this band was empty or deployed for TDD technologies such as WiMAX. Therefore, the early-deployed LTE networks have adopted this spectrum for fast rollout. Also, this band is well supported by terminal manufacturers. The 2.6 GHz band is arranged as shown in Figure 5.50. There is 70 MHz for LTE-FDD (band 7 in Figure 5.49) and 50 MHz for LTE-TDD (band 38 in Figure 5.49) or WiMAX.

In order to avoid interference between FDD and TDD systems a 5 MHz guardband is mandatory if the TDD channel BW 10 MHz. As a result, the band arrangement can be as shown in Figure 5.51. Therefore, this band can provide three LTE-FDD channels with a 20 MHz band and one LTE-FDD channel with 10 MHz and four WiMAX 16e channels with 10 MHz BW or two LTE-TDD channels with 20 MHz.

The interference between FDD and TDD in this band is quite challenging and the following need to be considered while designing LTE-TDD for coexistence with LTE-FDD:

- The interference from eNB to UE and UE to eNB has negligible impact on either the FDD or TDD system
- eNB to eNB: Additional filtering is needed, together with at least a 5 MHz guard band
- UE to UE:
  - *Worst case* (max power: 23 dBm): 10 MHz guard band and 1 m UE-UE separation (1 dB desensitization)
  - *Best case* (6 dBm power in 95% of the cases): 5 MHz guard band and 5 m UE-UE separation (3 dB desensitization).
- Good site planning is strongly required.
- Advanced interference cancelation schemes to mitigate interference between LTE FDD and LTE TDD is a valid option.

Table 5.11 3GPP LTE E-UTRA supported bands

EUTRA operating band	Uplink (UL) operating band BS receive UE transmit $F_{UL\_low} - F_{UL\_high}$	Downlink (DL) operating band BS transmit UE receive $F_{DL\_low} - F_{DL\_high}$	Duplex mode
1	1920 MHz–1980 MHz	2110 MHz–2170 MHz	FDD
2	1850 MHz–1910 MHz	1930 MHz–1990 MHz	FDD
3	1710 MHz–1785 MHz	1805 MHz–1880 MHz	FDD
4	1710 MHz–1755 MHz	2110 MHz–2155 MHz	FDD
5	824 MHz–849 MHz	869 MHz–894 MHz	FDD
6[a]	830 MHz–840 MHz	875 MHz–885 MHz	FDD
7	2500 MHz–2570 MHz	2620 MHz–2690 MHz	FDD
8	880 MHz–915 MHz	925 MHz–960 MHz	FDD
9	1749.9 MHz–1784.9 MHz	1844.9 MHz–1879.9 MHz	FDD
10	1710 MHz–1770 MHz	2110 MHz–2170 MHz	FDD
11	1427.9 MHz–1447.9 MHz	1475.9 MHz–1495.9 MHz	FDD
12	699 MHz–716 MHz	729 MHz–746 MHz	FDD
13	777 MHz–787 MHz	746 MHz–756 MHz	FDD
14	788 MHz–798 MHz	758 MHz–768 MHz	FDD
15	Reserved	Reserved	FDD
16	Reserved	Reserved	FDD
17	704 MHz–716 MHz	734 MHz–746 MHz	FDD
18	815 MHz–830 MHz	860 MHz–875 MHz	FDD
19	830 MHz–845 MHz	875 MHz–890 MHz	FDD
20	832 MHz–862 MHz	791 MHz–821 MHz	
21	1447.9 MHz–1462.9 MHz	1495.9 MHz–1510.9 MHz	FDD
22	3410 MHz–3490 MHz	3510 MHz–3590 MHz	FDD
23	2000 MHz–2020 MHz	2180 MHz–2200 MHz	FDD
24	1626.5 MHz–1600.5 MHz	1525 MHz–1559 MHz	FDD
25	1850 MHz–1915 MHz	1930 MHz–1995 MHz	FDD
26	814 MHz–849 MHz	859 MHz–894 MHz	FDD
27	807 MHz–824 MHz	852 MHz–869 MHz	FDD
28	703 MHz–748 MHz	758 MHz–803 MHz	FDD
29	NA	717 MHz–728 MHz	FDD[b]
30	2305 MHz–2315 MHz	2350 MHz–2360 MHz	FDD
31	452.5 MHz–457.5 MHz	462.5 MHz–467.5 MHz	FDD
...			
33	1900 MHz–1920 MHz	1900 MHz–1920 MHz	TDD
34	2010 MHz–2025 MHz	2010 MHz–2025 MHz	TDD
35	1850 MHz–1910 MHz	1850 MHz–1910 MHz	TDD
36	1930 MHz–1990 MHz	1930 MHz–1990 MHz	TDD
37	1910 MHz–1930 MHz	1910 MHz–1930 MHz	TDD
38	2570 MHz–2620 MHz	2570 MHz–2620 MHz	TDD
39	1880 MHz–1920 MHz	1880 MHz–1920 MHz	TDD
40	2300 MHz–2400 MHz	2300 MHz–2400 MHz	TDD
41	2496 MHz–2690 MHz	2496 MHz–2690 MHz	TDD
42	3400 MHz–3600 MHz	3400 MHz–3600 MHz	TDD
43	3600 MHz–3800 MHz	3600 MHz–3800 MHz	TDD
44	703 MHz–803 MHz	703 MHz–803 MHz	TDD

[a]Band 6 is not applicable.
[b]Restricted to E-UTRA operation when carrier aggregation is configured. The downlink operating band is paired with the uplink operating band (external) of the carrier aggregation configuration that is supporting the configured Pcell
(Source: [7] 3GPP TS. Reproduced with permission of ETSI.)

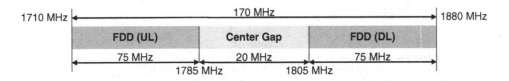

Figure 5.52   Band 1800 MHz arrangement for LTE.

In addition to the above, the LTE deployment in this band is quite challenging in terms of coverage and IBS deployment. Therefore, it is not recommended to use this band for LTE nationwide deployment as a huge number of sites will be needed and also a major overhaul to IBS will be needed to introduce LTE in this band. This spectrum can be used in hotspot areas as a capacity layer in a multicarrier LTE scenario.

5.7.2.2   LTE in 1800 MHz

This is the most promising LTE band as it can be used for nationwide coverage and for dense urban, urban, and suburban convergence. The band is arranged as per the B2 arrangement in ITU-R M.1036-3, as shown in Figure 5.52.

This band was widely used for GSM 1800 MHz and it can be refarmed for LTE1800 MHz. Many operators have deployed LTE in this band after refarming of 10 MHz. Other operators have acquired an additional 10 or 20 MHz in this band and rolled out a nationwide LTE network in this band.

The main advantages of this band for LTE deployment are:

- Coverage area is about 2× larger than LTE2.6 GHz with better indoor penetration.
- 35% improvement in cell edge throughput compared to LTE 2.6 GHz.
- Reduction of extra sites results in quick delivery of the LTE to market.
- Reuse of existing GSM1800 coverage polygons and possibility to share antenna system of GSM1800, as shown in this chapter.
- Reuse of existing IBS system without upgrade to support 2.6 GHz and without coverage degradation, as explained before in this chapter.
- Possibility of nationwide coverage using this band.
- Availability of LTE terminals in this band. Even some major smartphone manufacturers have supported this band before the 2.6 GHz band.

5.7.2.3   Digital Dividend Band (800 MHz: LTE Band 20)

The World Radio-communication conference identified the band 790–862 MHz (aka Digital Dividend) for mobile broadband services in Europe, the Middle East, and Africa "EMEA" (ITU Region 1). This will allow the development of a mass market for terminals using this band. International roaming and economies of scale will drive down terminal prices. As of May 2009, two proposals for frequency plans (which are not compatible) have been submitted. The FDD plan has been retained as the preferred plan for harmonization of the channel arrangement in the band, while the TDD plan is only offered as guidance for those administrations, which could not implement the FDD, harmonized plan. The band is arranged as shown in Figure 5.53.

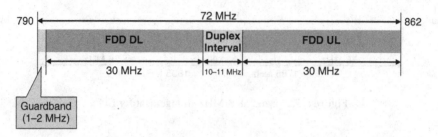

Figure 5.53   Digital dividend band arrangement.

This band is now LTE Band 20 in Table 5.11: 832:862 MHz for UL and 791:821 MHz for DL. It is worth to note that, the DL and UL order is reversed to avoid interference with broadcast at 790 MHz. This will increase the gap in the link budget between UL and DL since the LTE is mainly UL limited as explained in Chapter 6.

The pros and cons of this band are:

- Expected gain/saving in the number of sites is 40–50% for the 800 MHz band compared to the 2.5 GHz band in rural areas (where it is coverage-limited versus capacity-limited).
- Suited for the coverage rollout (rural and suburban) due to its excellent path loss characteristics.
- Major vendors are either planning to support this band or have already released an eNB that supports this band.
- The available 30 MHz can accommodate an "equally" divided spectrum.
- No need for guardbands within the 30 MHz FDD band if the deployed technology is LTE-FDD.
- Digital dividend band (790 –860 MHz) (band 20) will cater only for 3 × 10 MHz FDD carriers and, therefore, it is not possible to achieve the peak throughput of the LTE. Alternatively, a 10 MHz can be used as a second carrier to the 20 MHz carrier in LTE1800 MHz for rural areas convergence with LTE.
- Spectrum fees will be almost double the 1800 MHz band.

Spectrum strategy and refarming options can be summarized as follows:

- New bands well suited for LTE to avoid re-farming when introducing new technology
  - 2.6 GHz spectrum mainly for LTE and as hotspot deployment only.
  - "Digital Dividend" (e.g., 800 MHz/700 MHz bands): trend is to use it for LTE for rural area convergence or as a second carrier.
- 2.1 GHz spectrum: bandwidth mostly for UMTS/HSPA/HSPA+ and few countries for LTE such as DoCoMo in Japan
  - 900 and 1800 MHz, AWS refarming is needed before UMTS/HSPA or LTE usage
  - 900 MHz refarming already started for UMTS usage and may not be possible to use it for LTE in most countries.
  - 1800 MHz will be refarmed directly to LTE and will be the main LTE band.
  - Band 20 (800 MHz band) is currently one of the main candidate for rural broadband and 30 MHz is available in this band which can be divided between three operators to be used for rural breadboard and/or carrier aggregation with higher band such as 1800 MHz.

- Band 28 (700 MHz) is one of the promising bands for LTE which will be made available in many countries in the near future. This band can be used as well for rural broadband or for carrier aggregation. Alternatively, it is a good candidate for broadband critical communication applications.

Figure 5.54 summarizes the deployment and refarming options in terms of technology and clutter type for the most common bands [30].

5.8   LTE Business Case and Financial Analysis

This section provides a strategic overview of the LTE BC and is designed to assist a technology strategy team to develop a viable BC to deploy the LTE system [31–33]. A BC is typically a presentation or a proposal to an executive management or to the investors by the technology strategy team to seek funding, approval, or both, for LTE network deployment. A BC puts a proposed investment decision into a strategic context and provides the information necessary to make a decision about whether to proceed with network rollout or not and the plan to implemnt the network.

A BC is a justification for a project that needs investment. The BC requests a budget for a project from the decision-makers or from the investing committee. The BC must demonstrate that the benefits of the proposed project outweigh the requested budget/fund. A positive BC should offer an increase in revenue, reduce the cost, and provide strategic advantage over competitors. The BC should include financial analysis that provides and analyzes the key financial KPIs (key performance indicators) such as a cash-flow statement, return on investment (ROI), net present value (NPV), internal rate of return (IRR), and payback period. Moreover, the BC should provide a business justification to deploy the LTE network and show how LTE will help the company to meet its objectives, for example to be the best mobile broadband operator in the country or to use LTE to serve fixed residential customers and SME.

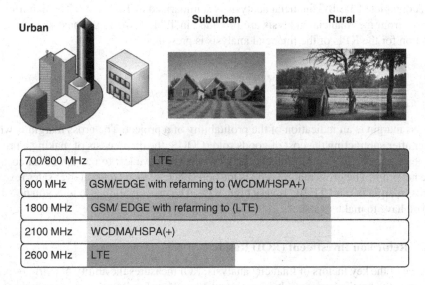

Figure 5.54   Deployment options (clutter type and technology) for different frequency bands.

Table 5.12   Business case contents for LTE

1.0 Executive summary
2.0 Business objective
3.0 Current situation, opportunity statement
4.0 Critical assumption and constraints
5.0 Market analysis
6.0 Analysis of options and recommendation
7.0 Preliminary project requirements
8.0 Budget estimate and financial analysis
9.0 Schedule estimate and implementation plan
10.0 Potential risks and sensitivity analysis
11.0 Appendixes: Exhibit A: Financial analysis
         Exhibit B: project plan

This BC provides the context for an investment decision, a description of viable options, analysis thereof, and a recommended decision. The recommendation describes the proposed investment and all its characteristics, such as benefits, costs, risks, time frame, change requirements, impact on stakeholders, and so forth.

The importance of the BC in the decision-making process continues throughout the entire life cycle of an investment: from the initial decision to proceed to the decisions made at scheduled project gates to continue, modify, or terminate the investment. The BC would be used to review and revalidate the investment at each scheduled project gate and whenever there is a significant change to the context, project, or business function. The BC would be revisited and considered again if the context changed significantly during the course of the project.

Note that a BC is used to identify and explore options and then develop recommendations for the proposed investment and propose a plan to implement the proposed investment. Table 5.12 summarizes the main items that need to be included in the BC.

In this section we will provide the detailed financial analysis, including all key financial KPIs. A complete LTE BC financial analysis is summarized in Table 5.13. The definitions for each item from the financial analysis are provided in Table 5.14. In the next section, a brief description for the KPIs of the financial analysis is presented.

5.8.1   Key Financial KPIs [31]

5.8.1.1   Gross Margin

The gross margin is an indication of the profitability of a project. The gross margin is what is left over after subtracting the cost of goods sold (COGS; the direct costs of making a product) from the revenues. Gross profit tells you how much you have left to pay overhead costs and make a net profit. The gross margin ratio is the percentage of gross margin over sales revenues. The gross margin in our LTE BC is very high, 97%. The gross margin also tells you how much room you have to make mistakes.

5.8.1.2   Return on Investment (ROI) Ratio

ROI is one of the key factors of financial analysis. ROI measures the ability of a firm or project to create profits for the investors. It represents the number of dollars of net income earned per

Table 5.13 Financial analysis for LTE business case

Profit and loss calculation	Unit	2011 / 0	2012 / 1	2013 / 2	2014 / 3	2015 / 4	2016 / 5
Customers forecast							
Mobile broadband customers: smartphone (voice/VoLTE + data)	- customers		10 000	100 000	400 000	700 000	1 000 000
Mobile broadband customers: USB dongles + embedded modems (data only)	customers		50 000	100 000	200 000	500 000	700 000
LTE routers with VoLTE for fixed services and SME (data + voice)	customers		25 000	50 000	100 000	200 000	300 000
ARPU (voice +broadband)							
Monthly ARPU of smartphone users From mobile broadband + VoLTE (2014)	USD		80	80	80	80	80
Monthly ARPU of USB/embedded modems from data only	USD		80	80	70	60	60
Monthly ARPU of routers (fixed broadband + VoLTE)	USD				75	100	110
Revenue		**2011**	**2012**	**2013**	**2014**	**2015**	**2016**
Smartphones	USD	0	9 600 000	96 000 000	384 000 000	672 000 000	960 000 000
Modems	USD	0	48 000 000	96 000 000	168 000 000	360 000 000	504 000 000
Routers	USD	0	0	0	90 000 000	240 000 000	396 000 000
Total revenue	**USD**	**0**	**57 600 000**	**192 000 000**	**642 000 000**	**1 272 000 000**	**1 860 000 000**
COGS							
LTE USB modem cost (subsidized)	USD	0	55	55	40	40	40
LTE router cost (subsidized)	USD	0	0	0	110	100	80
Total COGS	**USD**		**2 750 000**	**5 500 000**	**19 000 000**	**40 000 000**	**52 000 000**
Gross margin	**USD**		**54 850 000**	**186 500 000**	**623 000 000**	**1 232 000 000**	**1,808,000,000**
Gross margin (%)			**95**	**97**	**97**	**97**	**97**

(continued overleaf)

Table 5.13 (continued)

Profit and loss calculation	Year	2011	2012	2013	2014	2015	2016
	Unit	0	1	2	3	4	5
OPEX							
Accumulative number of sites			500	1000	1500	2000	2500
Operation and maintenance	USD		2 500 000	5 000 000	7 500 000	10 000 000	12 500 000
Sales/marketing	USD		3 000 000	2 000 000	1 500 000	1 300 000	1 000 000
Customer care	USD		500 000	1 000 000	1 500 000	2 000 000	3 000 000
Spectrum license	USD		200 000 000.00	200 000 000.00	200 000 000.00	200 000 000.00	200 000 000.00
Total OPEX	USD		206 000 000	208 000 000	210 500 000	213 300 000	216 500 000
EBITDA	USD		−151 150 000	−21 500 000	412 500 000	1 018 700 000	1 591 500 000
EBITDA margin (%)			−262	−11	64	80	86

Cash flow statement	Year	2011	2012	2013	2014	2015	2016
	Unit	0	1	2	3	4	5
EBITDA	USD		−151 150 000	−21 500 000	412 500 000	1 018 700 000	1 591 500 000
CAPEX (access network + core network)	USD	22 500 000	12 500 000	35 000 000	25 000 000	25 000 000	25 000 000
Free cash flow	USD	(22 500 000)	(163 650 000)	(56 500 000)	387 500 000	993 700 000	1 566 500 000
Cumulative cash flow	USD	(22 500 000)	(186 150 000)	(242 650 000)	144 850 000	1 138 550 000	2 705 050 000
Cumulative CF-break even		0	0	0	3	0	0

Key investment indicators

Discount factor (WACC)	12%
NPV in USD	1 412 987 325
IRR	120%
Pay-back period:	3
Five-years EBITDA margin	71%

Table 5.14 LTE BC formulas and definitions

Profit and loss Calculation	Formulas	Explanations
Customers forecast		
Mobile broadband customers: smartphone (voice/VoLTE + data)	A	Smartphone forecast (CSFB initially then VoLTE after two years)
Mobile broadband customers: USB dongles + embeddedmodems (data only)	B	USB modems and embedded modems (tablet, laptop, etc.) forecast
LTE routers with VoLTE for fixed services and SME (data + voice)	C	Routers for residential and SME with VoLTE after two years
ARPU (voice + broadband)		Average revenue per user
Monthly ARPU of smartphone users from mobile broadband + VoLTE (2014)	D	
Monthly ARPU of USB/embedded modems from data only	E	
Monthly ARPU of routers (fixed broadband + VoLTE)	F	
Revenues		Total income from the sale of goods and services to customers
Revenue of smartphones	$RS = A \times D \times 12$	Customers × Smartphone ARPU × 12
Revenue of modems	$RM = B \times E \times 12$	Customers × Modems ARPU × 12
Revenue of routers	$RR = C \times F \times 12$	Customers × Routers ARPU × 12
Total revenues	$TR = RS + RM + RR$	Sum of revenues
COGS		Cost of goods sold. The inventory costs of those goods a business has sold during a particular period
LTE USB modem cost (subsidized)	G	
LTE router cost (subsidized)	H	
Total COGS	$P = (G \times B) + (H \times C)$	Total cost of goods sold for LTE USB modem and routers
Gross margin	$Q = TR - P$	Difference between total revenue and total COGS
Gross margin %	$Q\% = Q / TR$	Percentage calculation of gross margin over revenue
OPEX		OPEX: operational expenditure
Accumulative number of sites	O	
Operation and maintenance	$R = O \times 5000$	OPEX per year for O&M (assuming $5000 per site)
Sales/marketing	S	OPEX per year (decreasing each year) for sales and marketing budget

(continued overleaf)

Table 5.14 *(continued)*

Profit and loss Calculation	Formulas	Explanations
Customer care	T	OPEX per year (increasing each year) for customer care budget
Spectrum license	U	Spectrum fees per year
Total OPEX	V = SUM(R : U)	Sum of all OPEX including O&M, sales and marketing, customer care, and spectrum fees
EBITDA	W = Q − V	Difference between gross margin and OPEX
EBITDA margin	X = W / TR	Earnings before interest, taxes, depreciation, and amortization. percentage calculation of EBITDA over revenue

Cash flow statement

EBITDA	W	Earnings before interest, taxes, depreciation, and amortization
CAPEX (access network + core network)	Y =(number of Sites per year) ∗ 50000	CAPEX: Capital expenditure: cost of sites (∼ $50K per site) deployed in the year + USD 10M for core network (2011) + USD 10 M for IMS (2013)
Free cash flow	Z = W − Y	Free cash flow is EBITDA minus CAPEX
Cumulative cash flow	Year on year added cash flow	Cumulative cash flow added from year to year
Cumulative CF-break even	When break even has been reached	Time calculated for break even; the year in which the cumulative cash flow became positive for first time

Key investment indicators

Discount factor (WACC)		WACC: Weighted average cost of capital
NPV in USD		Net present value = present value (PV) of the free cash flows discounted at WACC
IRR		Internal rate of return
Pay-back period:		The pay-back period is the number of years before the free cash flow became positive
Five-years EBITDA margin	EBITDA/total revenues	A measureme of a project's operating profitability

dollar of invested capital. Therefore, ROI is of great interest to investors, shareholders, and other people with a financial stake in the company.

$$\text{return on investment ratio} = \text{net income}/\text{owners'equity}$$

In our LTE BC we did not demonstrate this factor as the profit is calculated based on LTE only while the operator will usually have 2G/3G services and the net profit needs to consider all services offered by the firm. If we assume a native LTE operator that invested one billion dollars to establish the company, then after six years the ROI over six years will be 271%; this means an average of 46% ROI per year which is a very healthy indicator if we compare this with the ROI of investing the same amount in a fixed deposit where the ROI is 5–10%.

5.8.1.3 Net Present Value (NPV)

The net present value (NPV) is used in cost-benefit analysis to assess the potential profitability of an investment. NPV represents the difference between the total investment (costs) and the present-day value of anticipated future annual cash flows (benefits). More specifically, NPV is the anticipated profitability of a particular investment/project, considering projected cash flows discounted by a risk factor (i.e., discount factor) that takes into account inflation, level of risk, and returns required. In simpler terms, you compare the invested money today to projected money generated from the investment in the future, that is, the value of the money at that time in the future. The official formula for calculating net present value is

$$\text{NPV} = \sum_{t=1}^{n} \frac{C_t}{(1 + \text{WACC})^t} - C_0 \tag{5.3}$$

where t is the time of the cash flow, n is the total time of the project, C_t is the free cash flow (the amount of cash) at time t, C_0 is the capital outlay at the beginning of the investment time (i.e., $t = 0$) and WACC is the weighted average cost of capital, that is, discount rate.

5.8.1.4 WACC

WACC (weighted average cost of capital) is the average of the costs of these sources of financing (debt or equity), each of which is weighted by its respective use in the given situation. By taking a weighted average, we can see how much interest the company has to pay for every dollar it finances.

A firm's WACC is the overall required return on the firm as a whole and, as such, it is often used internally by the organization to determine the economic feasibility of expansionary opportunities and mergers. It is the appropriate discount rate to use for cash flows with risk that is similar to that of the overall firm. In our LTE BC, we assumed a 12% discount factor, that is, WACC.

5.8.1.5 Internal Rate of Return (IRR)

IRR is another approach to determining whether the return of a particular investment makes it worthwhile to pursue. IRR is simply the rate of return that the investment produces or the reward for making the investment. IRR is related to the NPV in that it represents the discount rate for which the NPV of a cash flow stream (inflows and outflows) equals zero. In fact, IRR and NPV are two sides of the same coin. With NPV you discount a future stream of cash flows by your minimum desired rate of return. With IRR you actually compute your break-even rate of return. Therefore, you use a discount rate above which you would have a negative NPV (and a poor investment) and below which you would have a positive NPV (and a great investment, all things being equal). IRR is often used to compare a potential investment against current rates of return in the securities market.

When calculating IRR, you make the present value of an investment's free cash flow equal to the cost of the project, then the NPV equals zero. Therefore, the IRR can be estimated as follows:

$$\text{NPV} = \sum_{t=1}^{n} \frac{C_t}{(1 + \text{IRR})^t} + C_0 = 0 \tag{5.4}$$

The IRR can be used to compare other options and then select the best option with the highest IRR value.

5.8.1.6 Payback Period

The payback period gives you a way to calculate how long it will take to earn back the money from a particular investment. The payback period is the number of years before the free cash flow became positive. In our LTE BC, it is three years. The lower the number – indicating a faster payback period – the better, because the investment will become profitable sooner.

5.8.1.7 Profitability Index

The profitability index gives a way to evaluate different investment options that have determined net present values. It is calculated by

$$\text{profitability index} = \text{NPV}/\text{initial investment}$$

In our LTE BC, the profitability index can be estimated by aggregating all investments before the payback period (2011–2013) and dividing the NPV by this value; then the profitability index = 1.948 which is very high.

5.8.1.8 EBITDA

EBITDA is earnings before interest, taxes, depreciation, and amortization, an indicator of a company's financial performance which is calculated as follows: EBITDA = revenue – expenses, excluding tax, interest, depreciation, and amortization. EBITDA is essentially the net income with interest, taxes, depreciation, and amortization added back to it, and can be used to analyze and compare profitability between companies and industries because it eliminates the effects of financing and accounting decisions. A common misconception is that EBITDA represents cash earnings. EBITDA is a good metric to evaluate profitability, but not free cash flow. EBITDA also leaves out the cash required to fund working capital and the replacement of old equipment, which can be significant. Consequently, EBITDA is often used as an accounting gimmick to dress up a company's earnings.

5.8.1.9 EBITDA Margin

A measurement of a project's operating profitability. It is equal to EBITDA divided by total revenue. Since EBITDA excludes depreciation and amortization, the EBITDA margin can provide a cleaner indicator of the project's core profitability. In our LTE BC, the EBITDA margin is 71%. The higher the EBITDA margin, the less operating expenses eat into a project bottom line, leading to a more profitable project.

5.8.1.10   Summary of the BC

It is preferable to present the financial analysis in the form of graphs to demonstrate and simplify the BC. Figure 5.55 presents the forecast of subscribers' growth for each category. As indicated in the figure, the LTE smartphone will grow exponentially from 2013 and with the introduction of VoLTE, rich applications can be offered. Figure 5.56 demonstrates the expected revenue from each category. The LTE routers will form a significant amount of revenue with the introduction of VoLTE where it can replace fixed services for residential customers with double play services (voice and data). Also, it will offer the data either using a WiFi or Ethernet interface for local WAN/LAN. Another typical application is by using the LTE network as a backhauling for WiFi to serve SME, as shown in Figure 5.57. In this case, the operator can offer the WiFi services as a hotspot with an operator-specific landing page. With this scenario, the WiFi controller needs to be connected to the PGW. The total revenues are presented in Figure 5.58. The free cash flow and cumulative cash flow are demonstrated in Figure 5.59. The EBITDA, OPEX, and COGS are illustrated in Figure 5.60 as a percentage of the revenue. The presented financial analysis clearly indicates that the LTE network is a viable investment option. This depends on the assumptions in this BC which are close to typical values in most of the market. The only factor that may impact this BC is the spectrum fees that vary from country to country and depend on the local telecom regulator.

5.9   Case Study: Inter-Operator Deployment Scenario

This case study demonstrates an inter-operator deployment scenario of an LTE network operator to another 2G/3G network operator with different PLMNs in the same country to offer

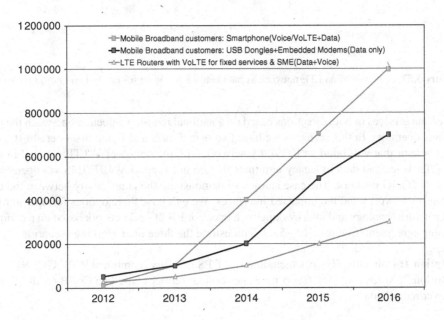

Figure 5.55   LTE subscribers forecast for different categories.

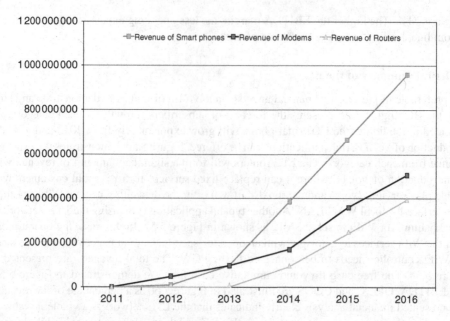

Figure 5.56   Revenue from each customer's category.

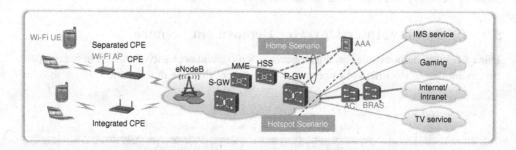

Figure 5.57   The use of an LTE network as backhauling for WiFi for hotspot and home scenario.

voice/data services in both directions based on a national roaming agreement between the two national operators. In this scenario, we have two operators C and B, the first operator (C) is a new operator that will build an LTE-only network with the option of VoLTE using IMS and will offer voice and data to legacy terminals that do not support VoLTE/IMS via operator B using its 2G/3G network. The case study will demonstrate the connectivity between the two operators' networks and the impacted interfaces. We will have three options for operator C's LTE network topology and interworking with operator B's 2G/3G network based on a national roaming agreement. Figures 5.61–5.63 demonstrate the three inter-working scenarios.

- **Option 1:** Data only Services including iRAT PS mobility (combined PGW/GGSN)
- **Option 2:** Voice and SMS (short message service) services based on CSFB (with PS handover/redirection)

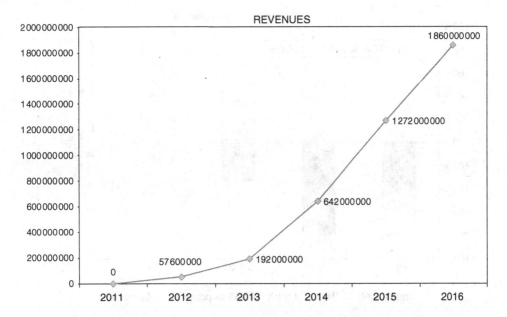

Figure 5.58 The total revenues from all customers' categories.

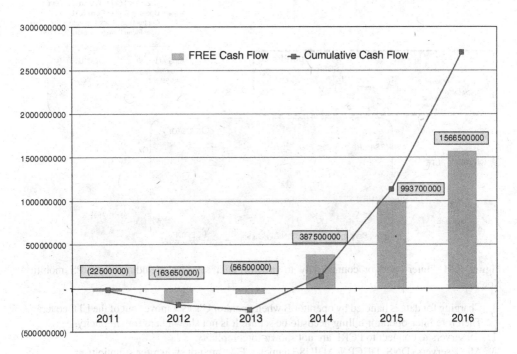

Figure 5.59 Free cash flow and cumulative cash flow.

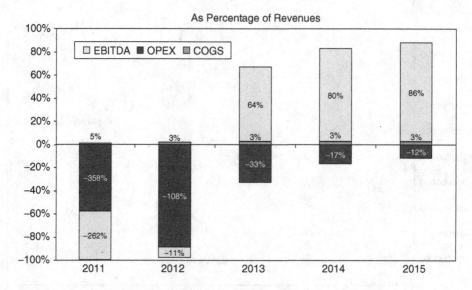

Figure 5.60 EBITDA, OPEX, COGS as percentage of revenues.

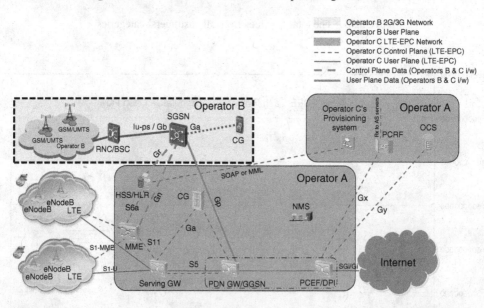

Figure 5.61 Inter-operator connectivity to provide data services including iRAT PS mobility. Notes –

1. Charging for data is handled by operator B when operator C's user moves out of the LTE coverage. Therefore inter-operator billing needs to be set up. It is not shown here for simplicity.
2. AS servers to connect to PCRF are not shown for simplicity.
3. All IT servers (DNS, DHCP, RADIUS/Diameter, FW) are not shown for simplicity.
4. If PGW/GGSN and PCEF/DPI are not collocated into the same platform, then there is a need to use a RADIUS/Diameter server to exchange the signaling information between both sets of equipment.

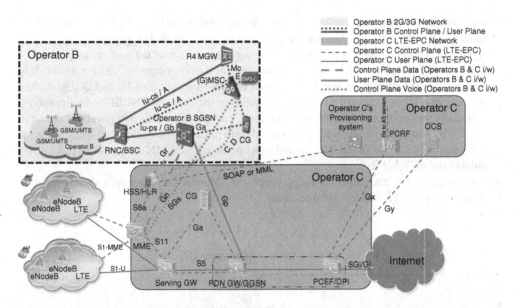

Figure 5.62 Inter-operator connectivity to provide voice using CSFB in addition to the data service.
Notes –

1. Charging for voice is fully handled by operator B. Inter-operator billing needs to be set up. It is not shown here for simplicity.
2. AS servers to connect to PCRF are not shown for simplicity.
3. All IT servers (DNS, DHCP, RADIUS/Diameter, FW) are not shown for simplicity.
4. If PGW/GGSN and PCEF/DPI are not collocated into the same platform, then there is a need to use a RADIUS/Diameter server to exchange the signaling information between both sets of equipment.
5. For this scenario we have two types of routing mechanism, the choice depending on operator B's GGSN capabilities as follows:
   (a) HR home routed. This can guarantee that operator C's subscribers can get the same data services set even when out of LTE coverage.
   (b) LBR local breakout routing, if this routing is adopted then it is not fully guaranteed that operator C's subscribers can get the same data services set when connecting to operator B's 2G/3G networks.

The 2G/RAN network is not shown, however, there are no changes as the LTE network will not be directly connected to the 2G of operator B and there is no direct handover between LTE and 2G. The recommended deployment is to have CSFB between LTE and 3G because the CSFB from LTE to 2G will cause extra call setup latencies experienced by the end user. However, for operator C LTE users once they have CSFB to the 3G of operator B, the user will be able to hand over to the 2G network of operator B for voice continuity, depending on the current operator B's 3G to 2G IRAT (inter-radio access technology) strategy without any core or air interface changes. After terminating the voice call, if an operator C user is on the 2G network, the user will reselect to the 3G network of operator B and then reselect to operator C's LTE network. The same scenario will be applied to an operator B user on 2G coming into 3G/LTE coverage as the user will reselect first to the 3G network of operator B then, once the user enters LTE coverage again, it can reselect to operator C's LTE network. Different air interface strategies can be applied on 2G to produce faster reselection to 3G and then back to LTE, and all are controlled through radio parameters and features without any interaction from the core at the stage of the reselection process.

It is possible to serve an LTE subscriber of operator B over operator C's LTE network. It is equivalent to the situation where two different PLMN-Id are set for LTE and 2G/3G within one operator. The selection of the network depends on the eNB and RNC/BSC configuration. Especially, there is a need to configure operator C's neighboring cells relations from operator B's 2G/3G networks. Gn/Gp interfaces should be in place in case of IRAT mobility during a data session (e.g., UMTS to LTE). In addition operator B's PLMN-Id should be configured into operator C's MME.

- **Option 3:** Voice and SMS services based on IMS-SRVCC (single radio voice call continuity) (with PS handover/redirection)

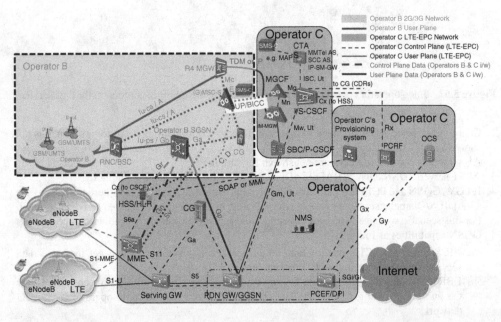

Figure 5.63   Inter-operator connectivity to provide voice and SMS services based on IMS-SRVCC (with PS handover/redirection). Notes –

- Inter-operator billing needs to be set up. It is not shown here for simplicity.
- Operator C's CTAS and MGCF (media gateway controller function) could generate CDRs which should be sent to operator C's CG.
- OCS could be connected to S-CSCF and CTAS for operator C's prepaid users. It is not shown here for simplicity.
- MRFC/MRFP (multimedia resource function controller/multimedia resource function processor) not shown for the simplicity.
- All IT servers (DNS, DHCP, RADIUS/Diameter, FW) not shown for simplicity.
- If PGW/GGSN and PCEF/DPI are not collocated into the same platform, then there is a need to use a RADIUS/Diameter server to exchange the signaling information between both sets of equipment.
- Depending on the LTE terminals capabilities (IMS-SRVCC or CSFB) it is recommended to add SGs interface for CSFB as a backup option in case LTE UEs do not support IMS.

The following summarizes the pre-conditions to have interworking between the two operators:

1. Operator B's MSCs provides (at least) SGs interface to operator C's MMEs. So, operator B's MSS (mobile services switching center server) needs to be upgraded to support SGs interfaces.
2. Operator B's GGSN needs to be upgraded to support an SGi interface toward operator C's P-GW.
3. Operator B's SGSN needs to be upgraded to support an S4 interface toward operator C's S-GW and S3 toward operator C's MME.
4. During call setup, operator B's CN needs to authenticate the UE of operator C. The security and authentication process may need rounds of testing and interoperability verification.
5. Defining cell reselection parameters from operator B's network to operator C's LTE network, and this is an air interface process
6. Definition of SIB-19 on operator B's Node-Bs is needed to allow operator C users to reselect to operator C's LTE after terminating the voice call and also for Operator B's users to reselect to operator C's LTE network if this scenario is adopted.
7. MME configuration needs to handle multiple PLMN-Id (operator C's PLMN-Id and in addition to be able to recognize operator B's PLMN-Id for CSFB). For operator C's subscribers, one has to configure the relation between the IMSI segment of operator C and the SGs interface. During the UE attach procedure; the MME has to select MSCs based on this relation.

The following interfaces need to be implemented:

1. **SGs interface** setup (to recognize a different PLMN Id for connecting the right MSC(s; MSS pooling) of operator B).
2. **D interface** between (visited network → operator B) MSC and (home network → operator C) 2G/3G HLR (assuming combined HSS/HLR). The D interface is necessary for normal roaming scenarios, also normally operator B's MSC has the address of operator C's HLR via the D interface. Note that operator C has to implement 2G/3G HLR as well (e.g., combined Ng-HLR for 2G/3G/LTE).
3. **Gr/S6d interface** between (visited network → operator B) SGSN and (home network → operator C) 2G/3G HLR (assuming combined HSS/HLR). Similar strategy as the D interface.
4. **Gn/Gp interface** (data services PS handover or redirection to the 3G PS domain) is needed. It is a classic requirement for operators B/C.
5. **Gn/Gp interface** (data services PS redirection to GPRS) is needed. It is a classic requirement for operators B/C.
6. **PCC architecture** for roaming support with S9 interface (between home PCRF and visited PCRF) is not necessary. It is only an optional solution. Another solution is Operator C P-GW direct address home PCRF by DRA (Diameter Routing Agent).

References

[1] 3GPP (2009) Evolved Universal Terrestrial Radio Access (E-UTRA) and Evolved Universal Terrestrial Radio Access Network (E-UTRAN); Overall Description. TS 36.300 V8.5.0.

[2] 3GPP (2010–2012) (Release 8) Network Architecture. TS 23.002 V8.7.0.

[3] 3GPP (2009) General Packet Radio Service (GPRS) Enhancements for Evolved Universal Terrestrial Radio Access Network (E-UTRAN) Access. TS 23.401, Version 8.6.0 Release 8.

[4] 3GPP (2007) Evolved UTRA and UTRAN Radio Access Architecture and Interfaces. TR R3.018 V.0.8.0.

[5] 3GPP TR 25.913 Requirements for Evolved UTRA (E-UTRA) and Evolved UTRAN (E-UTRAN).

[6] 3GPP TR 36.913 version 8.0.1 Release 8, Requirements for further advancements for Evolved Universal Terrestrial Radio Access (E-UTRA) (LTE-Advanced).

[7] 3GPP (2013) Evolved Universal Terrestrial Radio Access (E-UTRA); User Equipment (UE) Radio Transmission and Reception. TS 36.101 V12.1.0.

[8] Holma, H. and Toskala, A. (2009) *LTE for UMTS – OFDMA and SC-FDMA Based Radio Access*, John Wiley & Sons Ltd, Chichester.

[9] 3GPP (2010) Policy and Charging Control Architecture. TS 23.203 V8.9.0.

[10] 3GPP (2010) Policy and Charging Control Over Gx Reference Point. TS 29.212 V9.3.0.

[11] 3GPP (2010) Charging Management; Charging Architecture and Principles. TS 32.240 V8.6.0.

[12] 3GPP (2009) S1 Application Protocol (S1AP). TS 36.413 V8.7.0.

[13] 3GPP (2011) Interworking between the Public Land Mobile Network (PLMN) Supporting Packet based Services and Packet Data Networks (PDN). TS 29.061 V10.4.0.

[14] 3GPP (2010) General Packet Radio Service (GPRS); Service Description. TS 23.060 V8.10.0.

[15] 3GPP (2008) Architecture Enhancements for Non-3GPP Accesses. TS 23.402 V8.1.1.

[16] Rigney, C., Willens, S., Rubens, A., and Simpson, W., IETF (2000) Remote Authentication Dial in User Service (RADIUS). RFC 2865.

[17] Adoba, B., Zorn, G., and Mitton, D. IETF (2001) RADIUS and IPv6. RFC 3162.

[18] Rigney Livingston, C., IETF (2000) RADIUS Accounting. RFC 2866.

[19] Ladid, L., Bound, J., Pouffary, Y., *et al.* (2006) IPv6 Forum Roadmap and Vision Version 6.

[20] Lescuyer, P. and Lucidarme, T. (2008) *Evolved Packet System (EPS) The LTE and SAE Evolution OF 3G UMTS*, John Wiley & Sons Ltd, Chichester.

[21] 3GPP (2009) 3rd Generation Partnership Project; Technical Specification Group Radio Access Network; Evolved Universal Terrestrial Radio Access (E-UTRA); Physical Layer Procedures. TS 36.213 V8.8.0.

[22] 3GPP (2012) 3rd Generation Partnership Project; Technical Specification Group Radio Access Network; Evolved Universal Terrestrial Radio Access (E-UTRA); User Equipment (UE) Radio Access Capabilities. TS 36.306 V11.1.

[23] Anritsu (2012) Understanding PIM, White Paper.

[24] Cannon, N. (2010) Troubleshooting Passive Intermodulation Problems in the Field. Anritsu White Paper.

[25] Bell, T. and Nankivell, J. (2013) Range to Fault Technology. Kaelus White Paper.

[26] Kaelus Data Sheet (2012) Range to Fault Technology.

[27] 3GPP (2010–2012) Evolved Universal Terrestrial Radio Access (E-UTRA); Radio Frequency (RF) System Scenarios. TS 36.942 V10.2.0.

[28] CEPT (2010) Compatibility Study for LTE and WiMAX Operating within the Bands 880–915 MHz / 925–960 MHz and 1710–1785 MHz / 1805–1880 MHz (900/1800 MHz Bands). CEPT Report 40, Report from CEPT to the European Commission in response to the Mandate.

[29] Huawei White Paper (2011) LTE 1800 MHz Ecosystem Drivers, March 2011.

[30] Ericsson White Paper Entitled (2011) Mobile Broadband in 1800 MHZ Spectrum, July 2011.

[31] Allen, K. and Economy, P. (2008) *Complete MBA for Dummies*, 2nd edn, John Wiley & Sons Inc.

[32] Treasury Board of Canada Secretariat (2009) Business Case Guide, Her Majesty the Queen in Right of Canada, represented by the President of the Treasury Board, 2009.

[33] Kemp, A. (2006) *Business Case Primer*, Impact Technical Publications.

6

Coverage and Capacity Planning of 4G Networks

Ayman Elnashar

6.1 Summary and Objectives

This chapter presents the key practical aspects for capacity and coverage planning of a commercial LTE (long term evolution) network. The practical basics of coverage and capacity planning are introduced. The LTE coverage and link budget aspects are discussed in detail. The link budgets for different LTE channels are provided. The limiting channel and limiting link have been analyzed in different deployment scenarios and different frequency bands. Different propagation models and model tuning are analyzed in detail. Practical link budget examples are provided for data and VoLTE (voice over long term evolution) scenarios. The coverage aspects of the LTE system have been compared with the HSPA+ (high speed packet access) system. A case study on RF prediction for FDD-LTE (frequency division duplex) and TDD-LTE (time division duplex) is provided. The practical capacity planning aspects are presented. In addition, capacity dimensioning of LTE/HSPA+ systems is analyzed. Several dimensioning exercises are provided for data and voice scenarios. The dimensioning of the main LTE channels is discussed in detail along with practical examples. A comparative analysis with the HSPA+ network is provided to benchmark and evaluate the LTE system. This chapter can be used as a reference for best practices in LTE network design and planning.

6.2 LTE Network Planning and Rollout Phases

The target radio network planning should be a compromise between coverage, capacity, quality, and cost. The network designer should consider these factors during the planning phase of the network. The coverage objectives need to be selected in a smart way to meet the business requirement with minimum expenditure. On the other side, the network should be dimensioned properly to meet current and future capacity requirement without underestimation or overestimation of the traffic growth. The objective of this chapter is to provide the network

Design, Deployment and Performance of 4G-LTE Networks: A Practical Approach, First Edition. Ayman Elnashar,
Mohamed A. El-saidny and Mahmoud R. Sherif.
© 2014 John Wiley & Sons, Ltd. Published 2014 by John Wiley & Sons, Ltd.

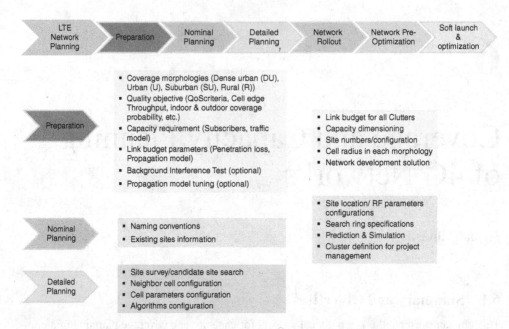

Figure 6.1   End-to-end network deployment phases.

designers and planning engineers with a framework to design and dimension a commercial LTE network from scratch. In this context, Figure 6.1 demonstrates the end-to-end planning and rollout phases of a commercial LTE network. The planning and rollout of an LTE network is divided into the following phases:

- **Preparation phase** In this phase coverage and capacity requirements are identified. This includes traffic profile, cell-edge throughput (CET), indoor and outdoor converge probability, QoS (quality of service) requirement, and so on. The clutter types need to be identified along with relevant indoor penetration loss. The propagation model selection and tuning for link budget calculation is conducted in this phase.
- **Nominal planning phase** In this phase, link budget, capacity dimensioning, and RF prediction are conducted. The outcomes of these exercises include cell radius for different clutters, supported number of subscribers, number of required sites for each clutter, and coverage maps for target areas. The converge maps include RSRP (reference signal received power), RSSI (received signal strength indicator), DL SINR (signal-to-interference-noise ratio), DL MAC (medium access control) Throughput, UL SINR, and UL MAC throughput.
- **Detailed planning phase** In this phase, we need to verify the nominal planning by identifying the site coordinates, conducting site surveys, and selecting the proper candidates that meet the coverage targets. Also, neighbor list preparation and cell parameters are defined as per vendor recommendation in this stage. Finally, the antenna type selection, antenna height, azimuth, and tilt need to be finalized in this phase to meet the required converge and to minimize the interference. Since the OFDM (orthogonal frequency division multiplexing) is based on overlapped orthogonal carriers, then there is no potential inter-carrier interference similar to legacy FDM (frequency division multiplexing) systems. Having said that, and unlike the UMTS (universal mobile telecommunications system), intra-cell interference

can be neglected. The inter-cell interference is the main concern that should be managed to boost the throughput at cell edge. Antenna tilt, height, and azimuth, and inter-site distance need to be optimized to reduce the inter-cell interference. Also, the antenna type needs to be selected to meet the converge target and clutter type. More details on antenna selection and antenna type will be provided later in this chapter.

- **Network rollout phase** In this phase the network rollout and site construction is conducted, based on the detailed planning phases. Based on the rollout model, the network acceptance may be conducted in cluster fashion or site by site, or a complete city.
- **Network pre-optimization** In this phase, the network is pre-optimized by validating the cell parameters, coverge target, and throughput. In this phase, we compare the detailed and nominal planning results with the actual network performance and tune the network parameters to meet the agreed KPIs (key performance indicators) before commercial launch. At this stage, the network remains in the supplier's custody till the operator accepts the network. Once the operator is satisfied with the network performance then it can be offered for soft launch as the last stage before full commercial launch.
- **Soft launch** This is the final phase when the network has passed all required KPIs and the SLA (service-level agreement). As a result, it can be launched in a soft launch mode or as a friendly user trial (FUT). A limited number of the users are allowed to access the network. The customers' feedbacks are combined to validate the network KPIs reported by the supplier's NMS (network management system). If the network performance is up to expectation and meets the agreed KPIs, then the operator will offer the network for commercial users.

6.3 LTE System Foundation

6.3.1 LTE FDD Frame Structure

In this section, we summarize the basic concepts of the LTE system that form the foundation for the LTE network planning. The LTE FDD frame structure is demonstrated in Figure 6.2 [1, 2, 22] for normal cyclic prefix (CP). As shown in Figure 6.2, each LTE FDD radio frame is $T_f = 307\,200 \cdot T_s = 10$ ms long and consists of 20 slots of length $T_{slot} = 15\,360 \cdot T_s = 0.5$ ms, numbered from 0 to 19. A subframe is defined as two consecutive slots where subframe i consists of slots $2i$ and $2i + 1$.

For LTE FDD, 10 subframes are available for downlink transmission and 10 for uplink transmissions in each 10 ms interval. Uplink and downlink transmissions are separated in the frequency domain.

The number of UL SC-FDMA (single-carrier frequency division multiple access) symbols in a slot depends on the CP length configured by the higher layer parameter and is given in Table 6.1 [1, 22]. The number of DL OFDM symbols in a slot depends on the CP length and subcarrier spacing configured and is given in Table 6.2. The CP length $N_{CP,l}$ that is used is provided in Table 6.2. Note that, in some cases, different OFDM symbols within a slot have different CP lengths.

An antenna port is defined such that the channel over which a symbol on the antenna port is conveyed can be inferred from the channel over which another symbol on the same antenna port is conveyed. There is one resource grid (RG) per antenna port. The antenna ports used for transmission of a physical channel or signal depend on the number of antenna ports configured for the physical channel or signal.

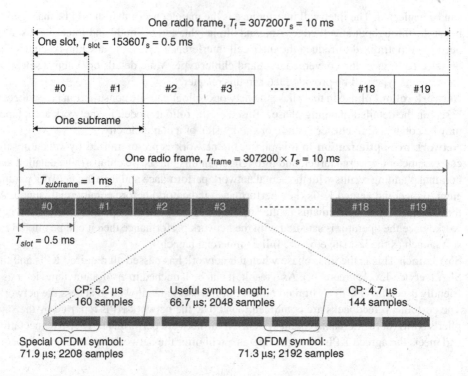

Figure 6.2 LTE FDD frame and slot structure for normal cyclic prefix. (Source: [2] Telesystem. Reproduced with permission of Telesystem.)

Table 6.1 CP type and relevant RB parameters in LTE

Configuration	N_{sc}^{RB}	$N_{symb}^{UL/DL}$
Normal cyclic prefix	12	7
Extended cyclic prefix	12	6

Table 6.2 CP and physical RB parameters in LTE DL with OFDM

Configuration	Δf (kHz)	OFDM symbol N_{symb}^{DL}	Sub-carrier N_{sc}^{RB}	Cyclic prefix length in samples $N_{CP,l}$	Cyclic prefix length (μs)
Normal CP	15	7	12	160 for $l = 0$ 144 for $l = 1, 2, \ldots, 6$	5.2 for first symbol and 4.7 for other symbols
Extended CP	15	6		512 for $l = 0, 1, \ldots, 5$	16.7
Extended CP	7.5	3	24	1024 for $l = 0, 1, 2$	33.3

6.3.2   Slot Structure and Physical Resources

This section summarizes the slot structure and physical resources in the LTE system that forms the LTE system capacity. Unlike UMTS and FDM systems, the LTE system resources consist of frequency domain and time domain resources. The UMTS system is based on spreading in the time domain with orthogonal spreading codes and, therefore, the resources of the UMTS system are the available codes in the time domain. The legacy FDM system consists of frequency domain resources (i.e., channels) with each resource occupying a chunk of spectrum without overlap and without adjacent channels usage in the same cell/site to avoid co-channel interference.

6.3.2.1   Resource Grid (RG)

The transmitted signal in each slot in DL or UL is described by one or several RGs of $N_{RB}^{UL/DL} N_{sc}^{RB}$ subcarriers and $N_{symb}^{UL/DL}$ SC-FDMA/OFDM symbols. The RG for UL/DL and the structure of the RB (resource block) and the RE (resource element) are illustrated in Figure 6.3. The quantity $N_{RB}^{UL/DL}$ depends on the UL/DL transmission bandwidth configured in the cell and fulfills the following inequality:

$$N_{RB}^{min,UL/DL} \leq N_{RB}^{UL/DL} \leq N_{RB}^{max,UL/DL} \tag{6.1}$$

where $N_{RB}^{min,UL/DL} = 6$ and $N_{RB}^{max,UL/DL} = 110$ are the smallest and largest UL/DL bandwidths, respectively [1, 22].

6.3.2.2   Resource Elements

Each element in the RG is called a resource element and is uniquely defined by the index pair (k, l) in a slot where $k = 0, \ldots, N_{RB}^{UL/DL} N_{sc}^{RB} - 1$ and $l = 0, \ldots, N_{symb}^{UL/DL} - 1$ are the indices in the frequency and time domains, respectively. RE(k, l) on antenna port p corresponds to the complex value $\lceil a_{k,l}^{(p)} \rceil$. The index p is dropped for simplicity. Quantities $a_{k,l}^{(p)}$ corresponding to REs not used for transmission of a physical channel or a physical signal in a slot are set to zero. The RE is the smallest defined unit, which consists of one OFDM subcarrier during one OFDM symbol interval.

6.3.2.3   Resource Blocks (RB)

A physical resource block (PRB) is defined as $N_{symb}^{UL/DL}$ consecutive SC-FDMA/OFDM symbols in the time domain and N_{sc}^{RB} consecutive subcarriers in the frequency domain for UL and DL, respectively. A PRB in the UL/DL thus consists of $N_{symb}^{UL/DL} \times N_{sc}^{RB}$ REs, corresponding to one slot in the time domain and 180 kHz in the frequency domain. Each RB consists of $12 \times 7 = 84$ REs in the case of normal CP and 72 REs for extended CP. The LTE DL physical layer parameters for different LTE channel bandwidths are shown in Table 6.3. As shown in Figure 6.4, and also Table 6.3, the maximum RB is 100. However, for some specific implementation and with reduced guardband, the number of RBs can be extended to 110, as explained above. Careful attention to interference with adjacent technology needs to be considered in such a case. The 100 RB corresponds to the transmission BW while 20 MHz is the channel BW.

As shown in Table 6.3, in the frequency domain, the number of sub-carriers N_{sc} ranges from 128 to 2048, where N_{sc} depends on the channel BW with 1024 and 2048 for 10 and

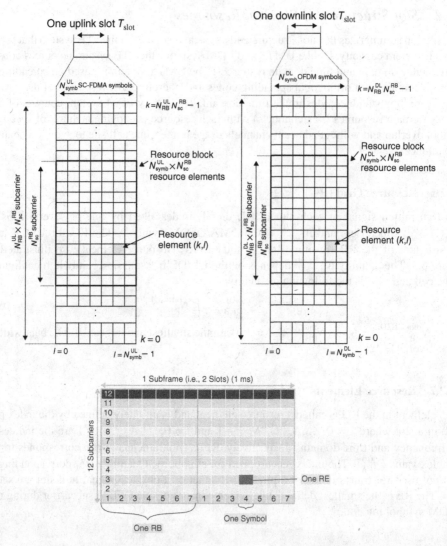

Figure 6.3 LTE FDD uplink and downlink resource grid and RB/RE structure. (Source: [1] 3GPP. Reproduced with permission of ETSI.)

20 MHz, respectively, being most commonly deployed in practice. The subcarrier spacing is $\Delta f = 15$ kHz. The sampling rate is $f_s = \Delta f \times N = 15\,000 \times N$. The LTE parameters have been chosen such that FFT (fast Fourier transform) lengths and sampling rates are easily obtained for all operation modes while at the same time ensuring the easy implementation of dual-mode devices with a common clock reference. Not all the subcarriers are modulated. The DC subcarrier is not used as well as subcarriers on either side of the channel band: approximately 10% of subcarriers are used as guard carriers in the case of only 100 RB being used in 20 MHz channel. In a macro cell, the coherence bandwidth of the signal is on the order of 1 MHz. Within the LTE carrier bandwidth of up to 20 MHz there are some subcarriers that are

Table 6.3 LTE downlink physical layer parameters

Channel bandwidth (MHz)	1.25	2.5	5	10	15	20
Frame duration (ms)			10			
Subframe duration (ms)			1			
Subcarrier spacing (kHz)			15			
Sampling frequency (MHz)	1.92	3.84	7.68	15.36	23.04	30.72
FFT size	128	256	512	1024	1536	2048
Occupied subcarriers (including DC subcarrier)	76	151	301	601	901	1201
Guard subcarriers	52	105	211	423	635	847
Number of resource blocks	6	12	25	50	75	100
Occupied channel bandwidth (MHz)	1.140	2.265	4.515	9.015	13.515	18.015
DL bandwidth efficiency (%)	77.1	90	90	90	90	90
OFDM symbols/subframe			7/6 (short/long cyclic prefix)			
CP length (short CP) (µs)			5.2 (first symbol)/4.69 (six following symbols)			
CP length (long CP) (µs)			16.67			

(Source [2] Telesystem. Reproduced with permission of Telesystem.)

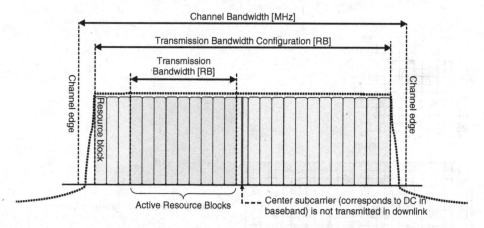

Figure 6.4 Definition of channel bandwidth and transmission bandwidth configuration for one E-UTRA carrier. (Source: [20] 3GPP. Reproduced with permission of ETSI.)

faded and other that are not faded. Transmission is done using those frequencies that are not faded. The transmission can be scheduled by RB, each of which consists of 12 consecutive subcarriers, or 180 kHz, for the duration of one slot (0.5 ms). This granularity is selected to limit signaling overhead [2].

During radio network planning, it is important to identify the CP. The radio network planning is usually performed considering the normal CP (4.7 µs), as it is able to sufficiently combat the multipath fading delay spread. In the 3GPP (third generation partnership project) LTE standard fading channel models, the maximum delay spread is 2.5 µs, which is much shorter than the normal CP. Moreover, the normal CP consumes fewer overheads in the total OFDM symbol than the extended CP and, therefore, it provides higher throughput performances. The extended

CP is used for two applications either for extended cell radius where the maximum delay spread is larger than 4.7 µs or with an eMBMS (evolved multimedia broadcast multicast service) feature that is needed for broadcasting TV channels on the LTE network.

6.3.3 Reference Signal Structure

6.3.3.1 Downlink Reference Signal

To allow for a robust detection a coherent demodulation is adopted at the UE (user equipment) receiver by sending pilot symbols (i.e., a reference signal (RS)) in the OFDM time–frequency grid to estimate the RF channel characteristics [1, 2, 22]. The coherent detection wastes some symbols used for pilot transmission to allow accurate channel estimation instead of blind channel estimation. The RS is similar to the "UMTS pilot" and it is used by the UE to predict the channel characteristics. The DL pilot symbols are transmitted within the first and third last OFDM symbol of each slot with a frequency domain spacing of six subcarriers (this corresponds to the fifth and fourth OFDM symbols of the slot in the case of normal and extended CP, respectively). Figures 6.5 and 6.6 demonstrate the RS for an LTE system with one, two, and four antennas and for normal and extended CP, respectively [1, 22]. The figures show the locations of the RS within each subframe when transmit antennae are used

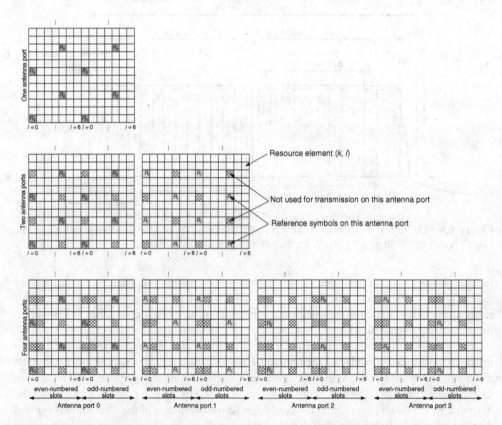

Figure 6.5 Mapping of downlink reference signals (normal cyclic prefix). (Source: [1] 3GPP TS. Reproduced with permission of ETSI.)

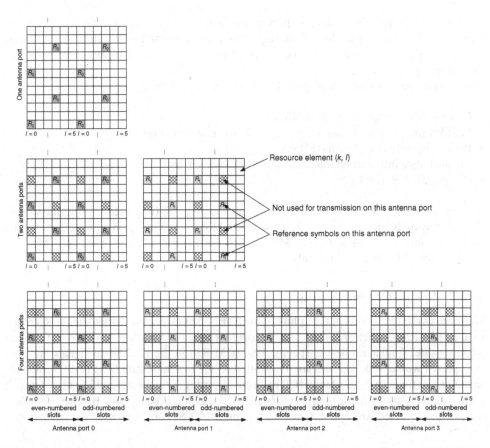

Figure 6.6 Mapping of downlink reference signals (extended cyclic prefix). (Source: [1] 3GPP TS. Reproduced with permission of ETSI.)

by the cell. Furthermore, there is a frequency domain staggering of three subcarriers between the first and second reference symbols. Therefore, there are four reference symbols within each RB. The UE will interpolate over multiple reference symbols to estimate the channel. As LTE is a MIMO (multiple input multiple output)-based technology, it can have more than two transmit antennas and in order to avoid RSs from the same cell interfering with each other, different antennas will be transmitting RS at different times and frequencies and how these are allocated are shown in Figures 6.5 and 6.6. Figures 6.5 and 6.6 illustrate the REs used for RS transmission according to the above definition for normal CP and extended CP, respectively. The notation R_p is used to denote an RE used for RS transmission on antenna port p. In the case of two transmit antennas; RSs are inserted from each antenna where the RSs on the second antenna are offset in the frequency domain by three subcarriers [2]. To allow the UE to accurately estimate the channel coefficients, nothing is transmitted on the other antenna at the same time–frequency location as the RSs.

The reference symbols have complex values, which are determined according to the symbol position as well as that of the cell [2]. LTE specifications refer to this as a two-dimensional reference-signal sequence, which indicates the LTE cell identity. There are 510 RS sequences, corresponding to 510 different cell identities. The RSs are derived from the product of a two-dimensional pseudo-random sequence and a two-dimensional orthogonal sequence. There are 170 different pseudo-random sequences corresponding to 170 cell-identity groups, and three orthogonal sequences each corresponding to a specific cell identity within the cell identity group.

To summarize, there are five types of DL RS defined as follows:

- Cell-specific reference signals (CRSs),
- MBSFN (multimedia broadcast over a single frequency network) RSs,
- Positioning reference signals (PRSs),
- Channel-state information reference signals (CSI-RSs), and
- UE-specific RSs (DM-RS).

There is one RS transmitted per downlink antenna port. RE (k, l) used for transmission of CRSs on any of the antenna ports in a slot is not used for any transmission on any other antenna port in the same slot and is set to zero. MBSFN RSs are transmitted on antenna port 4. MBSFN RSs are defined for extended CP only.

6.3.3.2 Uplink Reference Signals

There are two types of RSs in the uplink of the LTE. The first are the demodulation reference signals (DM-RSs), which are used to enable coherent signal demodulation at the eNB (evolved Node B) side, similar to the UE-specific RSs (DM-RS) in the DL. These signals are time multiplexed with uplink data and are transmitted on the fourth or third SC-FDMA symbol of an uplink slot for normal or extended CP, respectively, using the same bandwidth as the data [2].

The second is the sounding reference signal (SRS) which is used to allow channel dependent (i.e., frequency selective) uplink scheduling as the DM-RSs cannot be used for this purpose

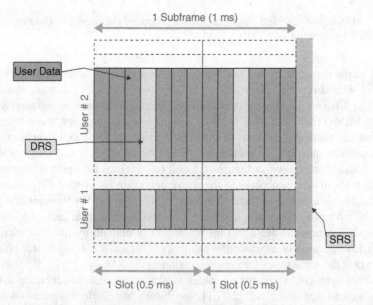

Figure 6.7 Uplink demodulation and sounding channel reference signals with normal cyclic prefix. (Source: [2] Telesystem. Reproduced with permission of Telesystem Innovations Inc.)

since they are assigned over the assigned bandwidth to a UE. The SRS is introduced as a wider band RS, typically transmitted in the last SCFDMA symbol of a 1 ms subframe, as shown in Figure 6.7 [2]. User data transmission is not allowed in this block, which results in about 7% reduction in UL capacity. The SRS is an optional feature and is highly configurable to control overhead – it can be turned off in a cell. Users with different transmission bandwidths share this sounding channel in the frequency domain.

To summarize, there are two types of UL RSs that are supported:

- DM-RS, associated with transmission of PUSCH (physical uplink shared channel) or PUCCH (physical uplink control channel)
- SRS, not associated with transmission of PUSCH or PUCCH.

The same set of base sequences is used for DM-RS and SRS.

6.3.3.3 Synchronization Signals (3GPP)

A UE wishing to access the LTE system follows a cell search procedure which includes a series of synchronization stages by which the UE determines time and frequency parameters that are necessary to demodulate DL signals, to transmit with correct timing and to acquire some critical system parameters [2]. There are three synchronization requirements in LTE: symbol timing acquisition by which the correct symbol start is determined; carrier frequency synchronization, which mitigates the effect of frequency errors resulting from Doppler shift and errors from electronics; and sampling clock synchronization. There are two cell search procedures in LTE: one for initial synchronization and another for detecting neighbor cells in preparation for handover. In both cases, the UE uses two special signals broadcast on each cell: primary synchronization sequence (PSS) and secondary synchronization sequence (SSS). The detection of these signals allows the UE to complete time and frequency synchronization and to acquire useful system parameters, such as cell identity, CP length, and access mode (FDD/TDD). At this stage, the UE can also decode the physical broadcast control channel (PBCH) and obtain important system information. Synchronization signals are transmitted

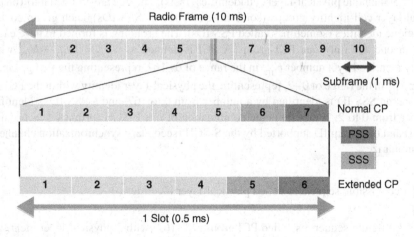

Figure 6.8 Synchronization signal frame/slot structure. (Source: [2] Telesystem. Reproduced with permission of Telesystem Innovations Inc.)

twice per 10 ms radio frame. The PSS is located in the last OFDM symbol of the first and eleventh slot of each radio frame, which allows the UE to acquire the slot boundary timing independent of the type of CP length. The PSS signal is the same for any given cell in every subframe in which it is transmitted (the PSS uses a sequence known as Zadoff–Chu (ZC)).

The location of the SSS immediately precedes the PSS – in the next to last symbol of the first and eleventh slot of each radio frame. The UE would be able to determine the CP length by checking the absolute position of the SSS. The UE would also be able to determine the position of the 10 ms frame boundary as the SSS signal alternates in a specific manner between two transmissions (the SSS uses a sequence known as M-sequences).

In the frequency domain, the PSS and SSS occupy the central six RBs, irrespective of the system channel bandwidth, which allows the UE to synchronize to the network without a priori knowledge of the allocated bandwidth. The synchronization sequences use 62 subcarriers in total, with 31 subcarriers mapped on each side of the DC subcarrier, which is not used. This leaves five subcarriers at each extremity of the six central RBs unused.

In a nutshell, there are two synchronization signals transmitted once every 5 ms as shown in Figure 6.8:

1. PSS
    – Subframe #0 and #5
    – Mapped on 72 subcarriers in the middle of the band
    – OFDM symbol #6
2. SSS
    – Subframe #0 and #5
    – Mapped on 72 subcarriers in the middle of the band
    – OFDM symbol #5.

6.4   PCI and TA Planning

6.4.1   PCI Planning Introduction

There are 504 unique physical-layer cell identities (PCIs). The PCI are grouped into 168 unique physical-layer cell-identity groups (sometimes referred as SSS IDs), each group containing three unique identities (sometimes called PSS IDs). The grouping is formed to have each PCI as part of one and only one PCI group. A PCI is defined as $N_{ID}^{cell} = 3N_{ID}^{(1)} + N_{ID}^{(2)}$ which is uniquely identified by a number $N_{ID}^{(1)}$ in the range of 0–167, representing the PCI group, and a number $N_{ID}^{(2)}$ in the range of 0–2, representing the physical-layer identity within the PCI group. Therefore, an SSS ID is identified by a number from 0 to 167, and a PSS ID is identified by a number from 0 to 2. A cell ID is the combination of one P-SCH (primary synchronization channel) and the group ID supported by the S-SCH (secondary synchronization channel).

To summarize,

1. PSS
    – Three possible sequences called physical-layer identities (0–2)
2. SSS
    – 168 different sequences called PCI groups (0–167) with 3 physical-layer identities per group
3. PCI
    – $168 \times 3 = 504$ PCI.

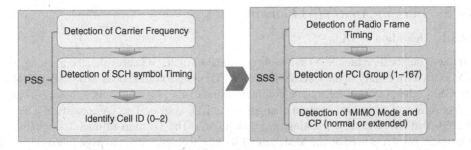

Figure 6.9 PSS and SSS planning procedures.

Figure 6.9 summarizes the PSS and SSS procedure.

6.4.2   PCI Planning Guidelines

1. *Frequency shift for reference signal* RSs are transmitted with one of six possible frequency shifts (modulo 6) to avoid time–frequency collisions between common RSs from up to six adjacent cells. The frequency shift is related to assigned PCI, as shown in Table 6.4.
2. *PCI assignment*
    (a) Avoid PCI collision and confusion by allocating different PCIs to neighbor cells and thus avoid problems in cell search and selection.
    (b) Use different PSS at neighbor cells.
    (c) Assign different PSS ID at neighbor cells.
    (d) The PCI planning can be conducted manually considering the above guidelines or automatically using the vendor planning tool.
3. *Summary of PCI planning recommendation*
    (a) The distance between cells with the same PCI should be maximized to prevent any UE from receiving the same PCI from two different cells.
    (b) Cells belonging to the same eNB should have PCI from the same group.
    (c) We should not consume all PCI resources from day one and we should have reserved buffer for future expansion to avoid full PCI re-planning.
    (d) If the same frequency is being used by two adjacent countries, then PCI coordination needs to be considered on the border, similar to overshooting coordination. If the same frequency is used in the neighbor countries, then PCI planning needs to be considered although it belongs to different PLMNs (public land mobile networks).

Table 6.4   Frequency shift versus PCI assignment

Frequency shift for RS	PCIs
0	0, 6, 12, 18, …
1	1, 7, 13, 19, …
⋮	⋮
5	5, 11, 17, 23, …

6.4.3 Tracking Areas (TA) Planning

6.4.3.1 Definition of TA

Similar to the location area (LA) and routing area (RA) in 2G/3G networks, the tracking Area (TA) is used for paging in the LTE. A TA in the LTE system corresponds to the RA in UMTS and GSM (global system for mobile communications). TA planning aims to reduce location update signaling caused by location changes in the LTE system. The TA consists of a cluster of eNBs that have the same tracking area code (TAC). The TA aims to track UE location in idle mode. The TA is used by the MME (mobility management entity) to page (notify) the idle UE of an incoming data session. The MME provides a list of TAs where the UE is allowed to register during the tracking area update (TAU) procedure, which occurs periodically or when entering a new TAC that is not part of the TA list.

The TA list can be used to limit the mobility of the UE in the case of using LTE for fixed broadband by limiting the UE access to certain TAC and, in this case, the list of TA will contain only one TAC where the UE is allowed to register. In this case the TAC needs to be overlapped to avoid issues on the TAC border or more than one TAC can be allowed.

A TAIs (tracking area identifiers) list is one that identifies the TAs that the UE can enter without performing a TAU procedure. The TAIs in a TAI list assigned by an MME to a UE pertain to the same MME area. Additionally, the TAIs in a TAI list assigned by an MME to a CS (circuit switched) fallback-capable UE pertain to the same LA. In this case, the defining of the relationship between the LTE TAs and the UMTS LAs is operator specific and depends on LTE/UMTS network topologies and vendors recommendation. In the LTE system, if a UE changes the TAs in the TAI list, TAU will not be triggered. The concepts of TAC and the TA list are shown in Figure 6.10.

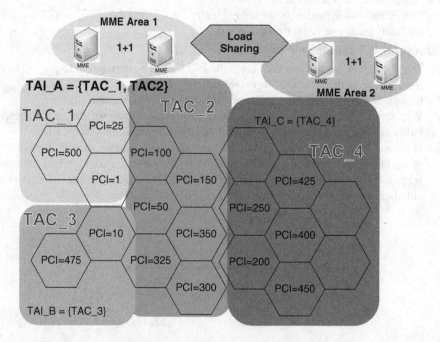

Figure 6.10 Tracking area code/list concept.

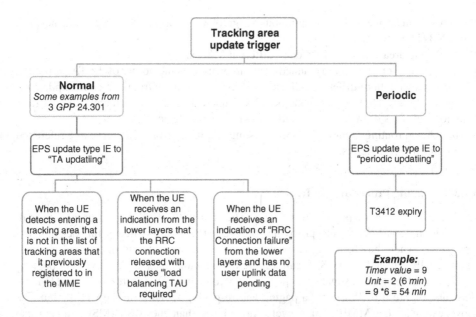

Figure 6.11   Tracking area update types.

In the example demonstrated in Figure 6.10, TAC_1 and TAC_2 belongs to same TA list while TAC_3 and TAC_4 belong to different TA Lists. In this scenario, if the UE performs EPS (evolved packet system) registration from TAC_1, the MMEs send TAC_2 in the TA list, implying that the UE can roam between the eNBs belong to these TA lists without having to re-register with the EPS network. The UE re-registers and performs the TAU procedure if the UE enters the coverage areas of eNB that are part of TAC_3 or TAC_4 which belong to different TA list.

Similar to the RA in UMTS/GSM, the smaller TA size may lead to frequent TAU procedures, thus increasing the MME load and reducing the UE battery back-up time. Moreover, the paging success rate will be degraded since the possibility of paging during the TAU procedure will be decreased when the UE cannot answer the paging during the TAU. On the other hand, a large TA list will lead to an increase in the paging load. As a result, the TA dimensioning is an important topic that identifies the TAs and TA lists to balance between the TAU rate and the paging load. The types of TAU are summarized in Figure 6.11 and refer to Chapter 2 for more details.

Why TA Planning?

The aim of TA planning is to prevent the ping-pong effect of TAU to achieve trade-off between the paging load, registration overhead, UE battery, and paging success rate.

The guidelines for TA dimensioning are summarized as follows:

- A TA should be medium size. The limitations by the MME and eNB must be considered. The minimum number of eNB values should be selected for TA. A relatively large TA is recommended as long as the paging load is not high.
- When the suburban area and urban area are covered discontinuously, a different TA is used for the suburban area.

- A TA should be planned for a continuous geographical area to prevent segmental networking of eNBs in each TA.
- The paging area cannot be located in different MMEs.
- The mountain or river in the planned area can be used as the border of the TA, to reduce the overlapping depth of different cells in two TAs. This will control the location update on the edge of the TA. On the other hand, the TA boundary should not run close to or parallel to major highways or railways to avoid high updates on the border.
- The LA and routing boundaries in the existing 2G/3G networks can serve as a reference for planning TA boundaries.

6.4.3.2 Paging Procedures in LTE

There are two types of paging in the LTE. The first is the EPS paging procedure and the second is the non-EPS paging procedure (i.e., CSFB (circuit switched fallback)). The first type occurs when the UE is paged for data session connectivity and the second type occurs when the UE is paged for a CSFB call. Figure 6.12 demonstrates the two types of paging.

It is important to investigate the paging and re-paging procedures in the LTE network elements. Figure 6.13 illustrates a paging and re-paging scenario in the LTE system. In the above example, the MME paging cycles are higher than the MSS (MSC Server) paging cycles, which can waste MME paging capacity.

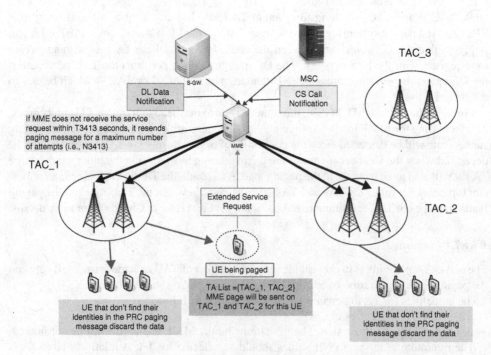

Figure 6.12 EPS paging and non-EPS paging (CSFB) procedure in LTE.

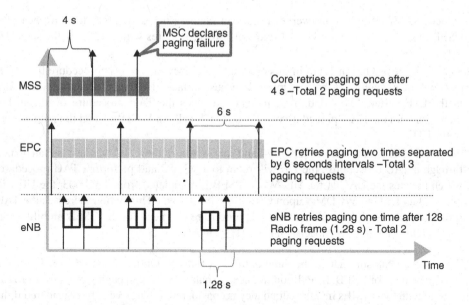

Figure 6.13   A paging and re-paging scenario in the LTE network.

Tracking Area (Option 1)	Tracking Area (Option 2)	Tracking Area (Option 3)	TAC	LAC
1		1	601	3501
1		2	602	3502
3	1	3	603	3511
3	1	4	604	3512
5		5	605	3521
5		6	606	3522

Figure 6.14   TAC-LAC planning options.

6.4.3.3   TAC–LAC Planning for CSFB

There are three possible options for TAC–LAC (location area code) planning. We will demonstrate these options via a practical example by considering an area with six UMTS LACs. We can plan the TAC and TAI list for this area via three options, as illustrated in Figure 6.14. In this exercise, the TAC–LAC planning is one to one. If this restriction does not imposed then another mapping scenarios would be possible which provide more flexibility in TAC-LAC planning.

Option 1: If UE is camped on TAC 604 and this is where the LAC was recently updated (i.e., through the TAU procedure), the UE will roam to TAC 603 without a TAU procedure. When CSFB is completed from TAC 603, the UE will need to do LAU (location area updating) in WCDMA (wideband code division multiple access) upon CS call establishment. This option reduces the TAU procedure overhead, but can increase the LAU procedure overhead for CSFB mobility users due to lack of frequent TAU procedures within TACs in the same TA lists

Option 2: Same as Option 1, but in this case, the frequency of the LAU procedure for CSFB calls will be much higher than in Option 1 because of lack of a TAU procedure for mobility in the LTE system. This option significantly reduces the TAU procedure overhead, but increases both the LAU procedure overhead for CS calls in WCDMA and the paging load in the LTE.

Option 3: If UE is camped on TAC 604 and this is where the LAC was last updated (i.e., through a TAU procedure), the UE will roam to TAC 603 and perform a TAU procedure, which updates the LAC in the UE. When CSFB is completed from TAC 603, the UE will not need an LAU in WCDMA upon CS call establishment. This option increases the TAU procedure overhead, but reduces the LAU procedure overhead for the CSFB mobility user and increases the paging capacity in the LTE system.

Therefore we can summarize the three options as follows: Option 1 should be sufficient at initial deployment of CSFB. In addition, we need to balance between paging capacity and TAU procedure with two MMEs if Geo-redundancy is considered. Option 3 can be considered if the CSFB performance tends to be more important than paging capacity in LTE. This option will save MME paging capacity (since eNB paging capacity may not be an issue initially). Option 2 is not recommended for CSFB enabled networks since the LAU procedure will increase in the network and the paging load will also increase.

The TAC to LAC planning depends on:

1. MME capability, such as one to one limitation.
2. LTE paging capacity for MME and eNB
3. TAU procedure overhead
4. LAU overhead impact on WCDMA call setup time and WCDMA signaling load.

The TA/LA dimensioning depends on:

1. LTE paging load and TAU overhead,
2. The percentage of mobility users, and the number of MMEs and users per eNB,
3. Paging intensity.

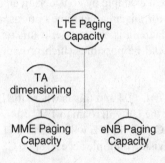

Figure 6.15   LTE paging capacity and TA dimensioning.

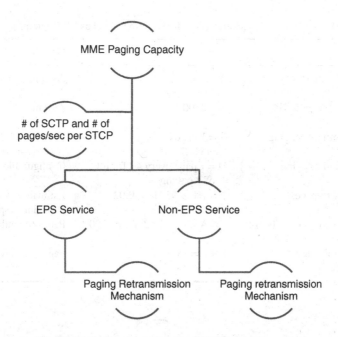

Figure 6.16 Factors impacting MME paging capacity.

6.4.3.4 LTE Paging Capacity

The paging capacities of the MME and eNB have an impact on TA dimensioning. Figure 6.15 summarizes the paging capacities. The MME paging capacity is summarized in Figure 6.16.

Table 6.5 MME paging capacity exercise 1 (MME paging does not match MSC paging)

Parameter	Value	Comments
Number of SCTP processing boards per MME	$A = 3$	Assumption
Number of pages/s per SCTP in MME	$B = 2500$	Assumption
EPS service paging/re-paging	$C = 1$ page every T3413 for N3413	N3413 = 4 (configurable)
Non-EPS paging/re-paging	$D = 1$ page every 6 s for total of 4 paging	Configurable
Total number of re-pages/s	$E = 1\% \times N3414 = 0.04$	Estimated at 1% MME no paging response
MME paging capacity (number of pages/s)	$F = A \times B = 3 \times 2500 = 7500$	Pages/s
MME paging blocking rate during busy hour	$F \times E = 300$	Pages/s

SCTP, stream control transmission protocol.

Table 6.6   MME paging capacity exercise 2 (MME paging matches MSC paging)

Parameter	Value	Comments
Number of SCTP boards per MME	$A = 3$	Assumption
Number of pages/s per SCTP in MME	$B = 2500$	Assumption
EPS service paging/re-paging	$C = 1$ page every T3413 for N3413	T3413 $= 6$ s, N3413 $= 4$ (configurable)
Non-EPS paging/re-paging	$D = 1$ page every 4 s for total of 4 paging	Configurable
Total number of re-pages/s	$E = 1\% \times \text{N3414} = 0.02$	Estimated at 1% MME no paging response
MME paging capacity (number of pages/s)	$F = A \times B = 3 \times 2500 = 7500$	Pages/second
MME paging blocking rate during busy hour	$F \times E = 150$	Pages/second

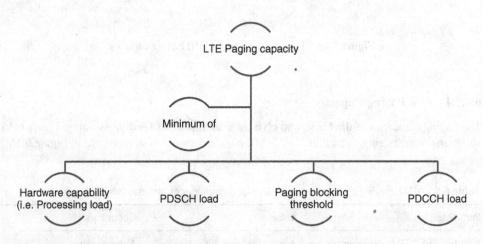

Figure 6.17   Factors impacting eNB paging capacity.

Table 6.5 illustrates the MME paging capacity exercise for the scenario demonstrated in Figure 6.13 (exercise 1).

As shown in Table 6.5, the MME paging of exercise 1 does not match the MSS paging which wastes MME paging resources. If we consider changing the MME paging in Table 6.5 to match the MSS paging (exercise 2) as shown in Table 6.6, the paging blocking rate of exercise 2 is reduced by 50% compared to scenario 1.

The eNB paging capacity depends on many factors, as demonstrated in Figure 6.17. The eNB paging capacity depends on the following factors:

1. The eNB CPU processing capability.

2. The PDSCH (physical downlink shared channel) and PDCCH (physical downlink control channel) load
3. The blocking rate target.

In order to calculate the eNB paging capacity, we need to calculate it for each of the above factors and select the minimum value as the overall paging capacity. We will not elaborate more on the eNB capacity as it is vendor specific and the reader can refer to the vendor's' related material for more details.

6.4.3.5 Tracking Area Dimension Exercise

The TA size is the minimum of MME paging capacity and eNB paging capacity. A TA dimensioning exercise for an LTE system with 20 MHz channel BW is conducted based on the parameters in Table 6.7.

Based on the above assumptions, the paging intensity can be estimated as follows:

1. Paging Intensity $(H) = (E*C + F*D)/3600 = 0.0280$ pages/s per subscriber
2. Paging intensity per 2500 subscriber per second $(I) = H*A = 70.081$ pages/s
3. Assume 5% of the users need four paging (5% of the users are at the edge of coverage)
4. Overall paging intensity $(J) = I + (0.05*4*I)/A = 0.03364$ pages/s/user.

The MME paging capacity can be estimated as follows:

Assuming that the number of attached users in the two MMEs are equally split between the two MMEs, then

1. The number of active users per MME $(K) = A/B = 1250$ users
2. The MME paging capacity is assumed to be 9000
3. Total number of eNBs per TA list per MME $(L) = 7500/(K*J) = 214$.

Table 6.7 Input to the TA dimensioning exercise

Item	Value	Name
Number of attached subscribers in one city	2500	A
Number of MMEs in LTE network	2	B
Number of dongle subscribers	30%	C
Number of smartphone/tablet subscribers	70%	D
Rate of dongle user paging requests	0.39 pages/h per subscriber	E
Rate of smartphone user paging requests	144 pages/h per subscriber[a]	F
Total number of attached users per eNB	1% on each eNB $= 0.01 \times A = 25$	G

[a]Total of 144 transaction, each of 14 s duration, with intervals of 70 s between transaction including CS pages, with 20 s inactivity timer to keep the users in the LTE idle mode most of the time. The 20 s inactivity timer is selected in line with the analysis in Chapter 4 for battery saving without DRX (discontinuous reception) feature.

The eNB paging capacity can be estimated as follows:

1. eNB paging capacity estimated (M) = 200 pages/s (for 5 MHz, and much higher for 20 MHz)
2. Total number of eNB per TA List (N) = M/(G*J) = 238.

Therefore, the bottleneck in this exercise is the MME paging capacity. As a result, the maximum number of eNBs to include per TA list = min (L, N) = 214 eNBs per one TA list. In this exercise, the MME paging capacity is the limiting factor in the TA dimensioning. In other scenarios with high capacity MME, the eNB may become the limiting factor. For example, if the MME paging capacity is 1000 then the maximum number of eNBs per TA for MME and eNB will be the same, that is, 238. If the MME paging capacity is above 10 K then the eNB paging capacity will be the limiting factor. For the sake of comparison with UMTS, the paging capacity is limited by the RNC (radio network controller) load and the PCCH (paging control channel) load (air interface). In most scenarios of UMTS, the limiting factor for paging capacity is the PCCH load, which determines the LA/URA size.

6.5   PRACH Planning

The physical random access channel (PRACH) always consumes six RBs in the UL. The PRACH overhead depends on the random access frequency. It is crucial to plan the sequences used for the RACH to meet the RACH success rate KPI, minimize collisions, and reduce the signaling load on the eNB. The UE is not guaranteed to be time aligned with the eNB when it transmits the random access preamble on the PRACH. The eNB may use the random access to calculate and send time alignment commands to the UE. Contention can also occur on the PRACH with multiple UEs transmitting preambles at the same time. Proper planning for the PRACH will enhance the user experience as follows:

1. Shorten the transition time from idle to connected mode,
2. Improve the handover success rate and therefore decrease the data latency,
3. Reduce the uplink interference and accordingly the UL interference margin,
4. Improve the call retainability. Retainability is the percentage of calls that were terminated by the customer (i.e., normal release) divided over the total number of calls that include the calls terminated by the network. Service providers target above 98% for this KPI.
5. Reduce the processing load of the eNB CPU.

PRACH is the UL channel that is used by the UE for initial cell and scheduling requests. Within one radio frame (10 ms) there is one or multiple PRACH resources, as controlled by the higher layers. A single PRACH resource consumes six RBs. The overhead introduced by the PRACH depends on the number of PRACH per radio frame. The number of PRACH per frame depends on the deployment scenario and vendor recommendation and capability. The random access preambles are generated from ZC sequences with zero correlation zone. There are 64 available preamble sequences in each cell. The 64 preamble sequences are first generated from a root ZC sequence using a cyclic shift. If less than 64 preamble sequences are generated, the remainder are generated from the root ZC sequence corresponding to the logical index. The previously mentioned root corresponds to the logical root sequence index, which is sent to the UE through the SIB2 (system information block).

6.5.1   Zadoff-Chu Sequence

The random access preambles are generated from ZC sequences with zero correlation zone, generated from one or several root ZC sequences. The network configures the set of preamble sequences the UE is allowed to use. There are 64 preambles available in each cell. The set of 64 preamble sequences in a cell is found by including first, in the order of increasing cyclic shift, all the available cyclic shifts of a root ZC sequence with the logical index, which is broadcast as part of the System Information SIB2. Additional preamble sequences, in case 64 preambles cannot be generated from a single root ZC sequence, are obtained from the root sequences with the consecutive logical indexes until all the 64 sequences are found. The logical root sequence order is cyclic: the logical index 0 is consecutive to 837 [1, 22].

The uth root ZC sequence is defined by

$$x_u(n) = e^{-j\frac{\pi u n(n+l)}{N_{ZC}}}, \quad 0 \le n \le N_{ZC} - 1 \tag{6.2}$$

where the length N_{ZC} of the ZC sequence is given by Table 6.8. From the uth root ZC sequence, random access preambles with zero correlation zones of length $N_{CS} - 1$ are defined by cyclic shifts according to

$$x_{u,v}(n) = x_u((n + C_v) \bmod N_{ZC}) \tag{6.3}$$

where the cyclic shift is given by

$$C_v = \begin{cases} vN_{CS} & v = 0, 1, \ldots, \lfloor N_{ZC}/N_{CS} \rfloor - 1, \ N_{CS} \ne 0 & \text{for unrestricted sets} \\ 0 & N_{CS} = 0 & \text{for unrestricted sets} \\ d_{start}\lfloor v/n_{shift}^{RA}\rfloor \\ \quad + (v \bmod n_{shift}^{RA})N_{CS} & v = 0, 1, \ldots, n_{shift}^{RA} \ n_{group}^{RA} + \bar{n}_{shift}^{RA} - 1 & \text{for restricted sets} \end{cases} \tag{6.4}$$

and N_{CS} is given by Table 6.8 for preamble formats 0–3 and 4, where the parameter zero-CorrelationZoneConfig (i.e., Ncs configuration) is provided by higher layers. The parameter *high-speed-flag* provided by higher layers determines if an unrestricted or restricted set should be used.

The variable d_u is the cyclic shift corresponding to a Doppler shift of magnitude $1/T_{SEQ}$ and is given by

$$d_u = \begin{cases} p & 0 \le p < N_{ZC}/2 \\ N_{ZC-p} & \text{otherwise} \end{cases} \tag{6.5}$$

where p is the smallest non-negative integer that fulfills $(pu)\bmod N_{ZC} = 1$. The parameters for restricted sets of cyclic shifts depend on d_u. For $N_{CS} \le d_u < N_{ZC}/3$, the parameters are given by

$$n_{shift}^{RA} = \lfloor d_u/N_{CS} \rfloor$$
$$d_{start} = 2d_u + n_{shift}^{RA}N_{CS}$$
$$n_{group}^{RA} = \lfloor N_{ZC}/d_{start} \rfloor$$
$$\bar{n}_{shift}^{RA} = \max(\lfloor (N_{ZC} - 2d_u - n_{group}^{RA}d_{start})/N_{CS}\rfloor 0) \tag{6.6}$$

Table 6.8 N_{CS} and N_{ZC} for preamble generation (preamble formats 0–3, 4) [1]

	N_{CS} value for formats 0–3		N_{CS} value for format 4
Random access preamble sequence length N_{ZC}	839		139
zeroCorrelationZoneConfig (N_{cs} configuration)	Unrestricted set (low speed cell)	Restricted set (high speed cell)	
0	0	15	2
1	13	18	4
2	15	22	6
3	18	26	8
4	22	32	10
5	26	38	12
6	32	46	15
7	38	55	N/A
8	46	68	N/A
9	59	82	N/A
10	76	100	N/A
11	93	128	N/A
12	119	158	N/A
13	167	202	N/A
14	279	237	N/A
15	419	–	N/A

For $N_{ZC}/3 \leq d_u \leq (N_{ZC} - N_{CS})/2$, the parameters are given by

$$n_{shift}^{RA} = \lfloor (N_{ZC} - 2d_u)/N_{CS} \rfloor$$

$$d_{start} = N_{ZC} - 2d_u + n_{shift}^{RA} N_{CS}$$

$$n_{group}^{RA} = \lfloor d_u/d_{start} \rfloor$$

$$\overline{n}_{shift}^{RA} = \min(\max(\lfloor (d_u - n_{group}^{RA} d_{start})/N_{CS} \rfloor, 0), n_{shift}^{RA}) \tag{6.7}$$

For all other values of d_u, there are no cyclic shifts in the restricted set.

6.5.2 PRACH Planning Procedures

In LTE, the RACH procedure is used to acquire UL synchronization and access the network for transmitting signaling and data. The RACH procedure can be initiated from idle or connected modes (e.g., handovers); failures or delay in executing RACH procedures can have a noticeable impact on the user experience and the network KPIs. Conducting proper PRACH planning can possibly improve the system KPIs and the user experience, as explained above.

The UE initiated the RACH procedure by sending preamble to the eNB (termed as MSG1) which is a complex value sequence based on a ZC sequence. ZC sequences maximize the orthogonality between users performing an RACH procedure at the same time due to the low cross correlation between different ZC sequences and zero autocorrelation between cyclic-shifted versions of the same sequence. This can be achieved by selecting roots generated from the same base sequence instead of the roots generated from different base sequences.

Contention-based and contention-free are two types of RACH procedures that can occur in the LTE system. In the contention-free RACH procedure, the network explicitly assigns the preamble to be used to the UE, as in handover scenarios. In this scenario the network is responsible for avoiding RACH collisions by maximizing orthogonality between users conducting RACH at same time. On the other hand, the UE selects a preamble from a pool of preambles that can be up to 64 in the contention-based RACH procedure.

The preambles are derived from ZC sequences. Orthogonality can be achieved either by having different ZC root sequences or by having a cyclic shift using the same root sequence. The number of root sequences needed by a UE to generate 64 preambles is decided by the amount of cyclic shift configured in the cell and the UE's speed, which are configured by the N_{cs} configuration parameter and the "highspeedflag" respectively, as shown in Table 6.8. The N_{cs} has 16 possible values where the index can take a value from 0 to 15 for an unrestricted set with low-speed cell and up to 14 with a restricted set with high-speed cell. For PRACH format 4, it can take a value up to 6. The amount of cyclic shift based on the N_{cs} configuration index is defined in Table 6.8. Use of a restricted set is not very common in field deployments, as it is intended for very high-speed users that could reach speeds around 150 km/h or above. Cells on a rural high way or serving a high-speed train route may be configured using the restricted set to serve users with very high speed.

6.5.3   Practical PRACH Planning Scenarios

The radio network planner needs to select the appropriate preamble format based upon the cell range. The typical preamble format is format 0, which allows a cell range of up to 15 km. Formats with longer guard times are suitable for larger cell ranges, which is the case for configurations 1–3. The preamble configuration used in a cell is signaled as part of the system information (SIB2). The preamble format is used with the link budget for PRACH. As shown in Table 6.9, PRACH formats 0 and 2 are 0.8 ms length and can extend cell ranges up to 15 and 29 km, respectively. PRACH formats 1 and 3 are 1.6 ms length and can extend cell ranges up to 78 and 100 km, respectively.

The different preamble format structures with the CP and the guard period for each format are shown in Figure 6.18 [23].

PRACH N_{cs} and Physical Root Planning Steps
- Identify the cell radius in kilometers.
- Select the preamble format (N_{cs}) based on cell radius and maximum delay spread using Equation 6.8 hereunder.
- Using PRACH format and cell radius tables find the corresponding N_{cs} value.
- Identify the number of physical roots based on the N_{cs} value.

$$N_{CS} > 1.04875 \times (20d/3 + T_{seq} + 2) \qquad (6.8)$$

N_{cs}: minimum required cyclic shift.

T_{seq}: ZC sequence length (μs) where delay spread is 5–16 μs

d: cell radius (km).

Table 6.9   PRACH formats and corresponding parameters

PRACH format	Cell radius (km)	CP length (ms)	Sequence length (ms)	Guard time (ms)	Total length (ms)	Maximum delay spread (µs)
Format 0	0–15	0.10	0.8	0.1	1	6.25
Format 1	0–29	0.68	0.8	0.52	2	16.76
Format 2	0–78	0.20	1.6	0.20	2	6.25
Format 3	0–100	0.68	1.6	0.72	3	16.76

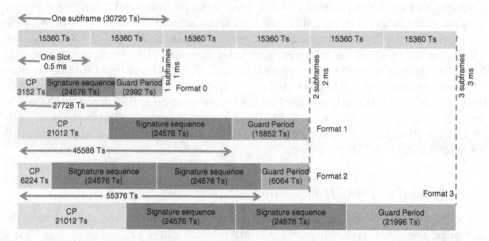

Figure 6.18   PRACH preamble formats (0–3). (Source: [22] Roke Manor Research 2012. Reproduced with permission of Roke Manor.)

Practical PRACH planning exercise for low speed cell with cell radius equal 10 km:

1. The N_{cs} value is determined by the cell radius from Equation 6.8. If the cell radius is 10 km and T_{seq} is 6 µs, then the N_{cs} value is 78.3. The N_{cs} is rounded to the nearest value from Table 6.9, which is 76.

2. As there are 838 root ZC sequences where the length of each root sequence is 839. The number of preambles generated per sequence is 839/76 which is rounded down to 11, that is, each index can generate 11 preamble sequences. The number of sequences per cell to generate 64 preambles is 64/preamble per sequence. In this case, six root sequence indexes are required to generate 64 preamble sequences.

3. The root index value = 838/number of sequences to generate 64 preambles. Therefore, the number of available root sequence indexes is 139 (0, 6, 12 … 828).

4. The available root sequence indexes are assigned to cells. The assignment principles are similar to those for PCIs.

Table 6.10 summarizes the above exercise and provides an additional two exercises for different cell radii.

The planning method of a high-speed cell is similar to that of a low-speed cell. The algorithm for determining available root sequence indexes, however, is more complex.

Table 6.10 PRACH planning exercises for different cell radius

R (km)	Ncs	Ncs configuration	No of preamble per sequence	No of sequence to generate 64 preambles	Available root index	Speed
10	77	76	11	6	139 (0, 6, 12, …)	Low speed cells (Unrestricted set)
50	357	419	2	32	26 (0, 32, 64, …)	
100	706	0	1	64	13 (0, 64, 128, …)	
30	217	237	1	64	To be obtained by planning tool	High speed cell (Restricted set)

6.6 Coverage Planning

6.6.1 RSSI, RSRP, RSRQ, and SINR

In this section, we analyze the main parameters for the LTE system coverage evaluation. The first parameter is the measured RSRP, which is comparable to the CPICH (common pilot indicator channel) RSCP (received signal code power) measurement in WCDMA. This measurement of the signal strength of an LTE cell helps to rank between the different cells as input for handover and cell reselection algorithms. The RSP (received signal power) is defined as the linear average over the power contributions (in watts) of the REs that carry CRSs within the considered measurement frequency bandwidth. Therefore it is only measured in the OFDM symbols that carry the reference symbols.

The second parameter is the RSSI which is the total received wideband power including all interference and thermal noise. The third parameter is the reference signal received quality (RSRQ) measurement. The RSRQ provides additional information when the RSRP is not sufficient to make a reliable handover or cell reselection decision. The RSRQ is the ratio between the RSRP and the RSSI. It depends on the measurement bandwidth, that is, the number of RBs. Since the RSRQ combines the signal strength and the interference level, then it provides additional input for mobility decisions algorithms. Therefore, the RSRP provides information about signal strength and the RSSI determines the interference and noise information. This is the reason why the RSRQ is estimated based on both the RSRP and the RSSI.

More important is the fourth parameter, which is the SINR. The SINR is the ratio of the average received modulated carrier power to the sum of the average co-channel interference power and the noise power from other sources. The UE is be able to identify new intra-frequency cells and perform the RSRP and the RSRQ measurements of identified intra-frequency cells without an explicit intra-frequency neighbor list containing PCIs [3]. The minimum RSRP and corresponding SINR depend on the frequency band, as illustrated in Table 6.11. As evident from Table 6.11, the minimum SINR is −4 for all bands while the RSRP depends on the frequency band. This is why the SINR is more important and it provides precious indication on coverage and expected throughput. Therefore, the coverage maps generated by the SINR is more accurate than RSRP or RSSI coverage maps. Also, the throughput maps is generated from mapping table with SINR.

We should not expect LTE coverage for RSRP → −121 dBm for band 3, that is, 180 MHz. However, if the interference level is well controlled and the RF parameters are optimized, then

Table 6.11   Minimum threshold for SINR and RSRP at different LTE bands

RSRP (dBm)	SINR (dB)	LTE bands
−124	−4	1, 4, 6, 10, 11, 18, 19, 21, 24, 33, 34, 35, 36, 37, 38, 39, 40, 41, 42, 43
−123	−4	9
−122	−4	2, 5, 7
−121	−4	3, 8, 12, 13, 14, 17, 20

the SINR usually has greater coverage than RSRP, for example, the area with SINR ≥ −4 dB is larger than the area with RSRP ≥ −121 dBm for the same number of sites, as highlighted in the case study at the end of the chapter. The target RSRP threshold is between −114 and −116 dBm, based on field test results. The RSRP has no direct relationship with the SINR, however, better RSRP brings better SINR at large inter-site distance like suburban and rural areas, but in dense urban areas where the inter-site distance is very small, the SINR is worse.

In this context and to further clarify the above parameters we consider a practical urban cluster with good LTE coverage. The CDF (cumulative distribution function) and PDF (probability density function) for RSRP, RSRQ, RSSI, and SINR are shown in Figures 6.19–6.22, respectively for this cluster. Indeed, the parameters represent an urban cluster with good LTE

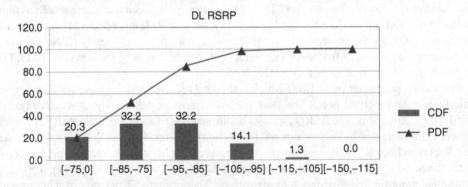

Figure 6.19   PDF and CDF of the measured LTE RSRP in the mobility route.

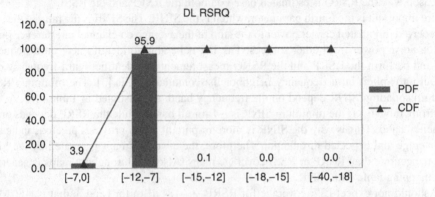

Figure 6.20   PDF and CDF of the measured LTE RSRQ in the mobility route.

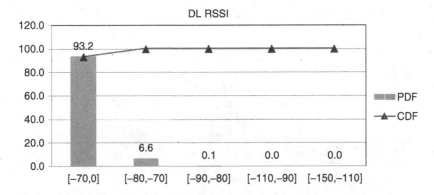

Figure 6.21 PDF and CDF of the measured LTE RSSI in the mobility route.

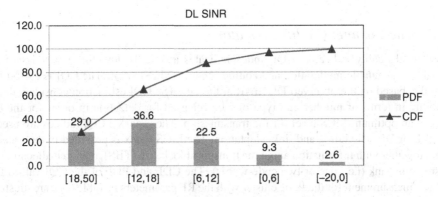

Figure 6.22 PDF and CDF of the measured LTE SINR in the mobility route.

coverage as evident from the values. The key statistical values (average, min, max, and standard deviation) for the four parameters (i.e., RSRP, RSRQ, RSSI, and SINR) for DL and UL are demonstrated in Table 6.12.

Table 6.12 Average values of the RF parameters (DL and UL)

Link	LTE RF measurements	Minimum	Maximum	Standard deviation	Average values
DL	Serving cell RSRP (dBm)	−112.28	−56.31	10.2	−83.84
	Serving cell RSRQ (dB)	−14.56	−2.88	0.89	−8.69
	RSSI (dBm)	−82.27	−27.54	10.22	−54.94
	SINR (dB)	−11.2	26.8	6.09	13.77
UL	Serving cell RSRP (dBm)	−114.5	−53.38	11.15	−84.04
	Serving cell RSRQ (dB)	−18.54	0	1.64	−4.8
	RSSI (dBm)	−84.52	−27.63	11.03	−58.93
	SINR (dB)	−17.3	30	7.66	15.24

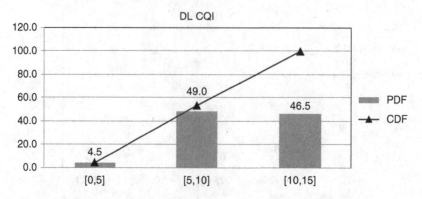

Figure 6.23   PDF and CDF for CQI values from field measurement.

6.6.2   The Channel Quality Indicator

The channel quality indicator (CQI) index indicates a suitable downlink transmission data rate, that is, a suitable modulation and coding scheme (MCS) value. The CQI is a 4-bit integer and is based on the observed DL SINR at the UE. The CQI estimation considers the UE capability in terms of number and type of receiver used for detection in order for the eNB to select an optimum MCS level for the transmission. The reported CQI indices are used by the eNB for DL scheduling and link adaptation. The MCS index is provided in the transport format together with information about the transport block size (TBS), resource allocation and transmission rank (i.e., one or two code-words). The CDF and PDF of the CQI values from field-test measurement for the same cluster with the RF parameters in Table 6.12 are illustrated in Figure 6.23. The average CQI index is 9.6 assuming normal distribution of the CQI index in each range of Figure 6.23. Although the CQI index is always an integer we consider the average for the sake of comparison. Therefore, the average CQI index of 9.6 corresponds to an SINR of ~13.77 dB, as explained above.

The DL CQI indices and their interpretations are given in Table 6.13. Based on an unrestricted observation interval in time and frequency, the UE derives for each CQI value reported in the UL subframe the highest CQI index between 1 and 15 in Table 6.13. The corresponding modulation, code rate, spectral efficiency, DL throughput per RB in kbps, and the estimated DL SINR in dB are all shown in Table 6.13. A similar look-up table for the UL is shown in Table 6.14. The spectral efficiencies and throughput per RB in kbps for DL and UL versus the estimated SINR is shown in Figures 6.24 and 6.25, respectively. These figures represent the minimum performance of PDSCH and PUSCH channels as per [4]. It is recommended to use the vendor-specific curves during link budget estimation. For accurate results, these curves can be obtained from field measurement for best mapping. Obtaining these curves from field measurement will allow the operator to evaluate the vendor performance versus 3GPP minimum performance. These curves are key criteria to benchmark the network vendor's performance. As evident from Table 6.13, the average CQI index is 10 (rounded up to the next integer) corresponding to an average spectral efficiency of 2.731 and an average SINR of 13.27 dB. Therefore, the cluster we have demonstrated is slightly conservative compared to 3GPP standard benchmarking where at an average SINR of 13.77 it reports an average CQI of 9.6.

Table 6.13 Four-bit DL CQI table

CQI index	Modulation	Code rate × 1024	Spectral efficiency (bps/Hz)	DL TP per RB (kbps)	Estimated DL SINR idB
1	QPSK	78	0.1523	19.1898	−7.28
2	QPSK	120	0.2344	29.5344	−4.78
3	QPSK	193	0.3770	47.502	−2.04
4	QPSK	308	0.6016	75.8016	0.66
5	QPSK	449	0.8770	110.502	2.84
6	QPSK	602	1.1758	148.1508	4.73
7	16QAM	378	1.4766	186.0516	6.38
8	16QAM	490	1.9141	241.1766	8.78
9	16QAM	616	2.4063	303.1938	11.49
10	64QAM	466	2.7305	344.043	13.27
11	64QAM	567	3.3223	418.6098	16.52
12	64QAM	666	3.9023	491.6898	19.71
13	64QAM	772	4.5234	569.9484	23.12
14	64QAM	873	5.1152	644.5152	26.37
15	64QAM	948	5.5547	699.8922	28.79

(Source: [5] 3GPP TS. Reproduced with permission of ETSI.)

Table 6.14 Look up table for efficiency, UL throughput, and SINR

Efficiency in bps/Hz	UL throughput	Estimated UL SNR in dB
	0	−4.96
0.1523	21.10878	−4.96
0.2344	32.48784	−2.50
0.3770	52.2522	0.21
0.6016	83.38176	2.87
0.8770	121.5522	5.37
1.1758	162.96588	8.00
1.4766	204.65676	10.64
1.9141	265.29426	14.48
2.4063	333.51318	18.81
2.7305	378.4473	21.65
3.3223	460.47078	26.85
3.9023	540.85878	31.94
4.5234	626.94324	37.40
5.1152	708.96672	42.60
5.5547	769.88142	46.46

(Source: [5] 3GPP TS. Reproduced with permission of ETSI.)

The spectral efficiency in Table 6.13 is defined as follows: Efficiency = the number of information bits/total number of symbols. The number of information bits + parity bits = total number of bits = total number of symbols × modulation order, so the efficiency = bitrate × modulation order. For example, with LTE, the CQI index 1, QPSK

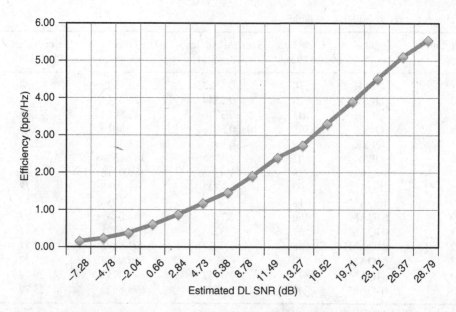

Figure 6.24   Spectral efficiency and corresponding throughput per RB versus DL SINR.

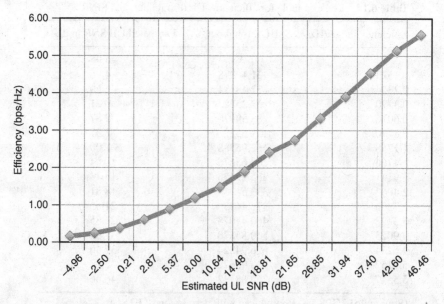

Figure 6.25   Spectral efficiency and corresponding throughput per RB versus UL SINR. (Source: [6] 3GPP TS. Reproduced with permission of ETSI.)

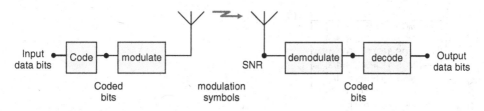

Figure 6.26 Coding and modulation for transmission of data over a radio link. (Source: [4] 3GPP. Reproduced with permission of ETSI.)

(quadrature phase shift keying), the modulation order of 2, code rate $= 78/1024 = 0.0762$, then the efficiency $= 0.1523 = 0.0762 \times 2$. More specifically, the $78/1024 = 0.076$ is the ratio of the information symbols (78) to the total number of symbols (1024). Then the efficiency equals 0.076×2 (QPSK modulation, one symbol occupies two information bits) $= 0.152$.

6.6.3 Modulation and Coding Scheme and Link Adaptation

In this section, we will analyze the MCS for LTE FDD. Figure 6.26 shows a typical radio transmitter and receiver with modulation and coding at Tx and Rx [4]. The throughput over a radio link is the number of data bits that can be successfully transmitted per modulation symbol. Coding (more specifically, forward error correction) adds redundant bits to the data bits, which can correct errors in the received bits. The degree of coding is determined by its *rate*, the proportion of data bits to coded bits. This typically varies from 1/8th to 4/5th. Coded bits are then converted into modulation symbols. The order of the modulation determines the number of coded bits that can be transmitted per modulation symbol. Typical examples are QPSK, 16QAM (quadrature amplitude modulation), and 64QAM which have 2, 4, 8 bits per modulation symbol, respectively.

The efficiency of a given MCS is the product of the rate and the number of bits per modulation symbol. Throughput has units of data bits per modulation symbol. This is commonly normalized to a channel of unity bandwidth, which carries one symbol per second. The units of efficiency then become bits per second, per hertz.

A given MCS requires a certain SNIR (measured at the receiver antenna) to operate with an acceptably low BER (bit error rate) in the output data. An MCS with a higher throughput needs a higher SNIR to operate. AMC (adaptive modulation and coding) works by measuring and feeding back the channel SNIR to the transmitter, which then chooses a suitable MCS from a "codeset" to maximize throughput (efficiency) at that SNIR and to maintain a target BER. A codeset contains many MCSs and is designed to cover a range of SNRs (signal to noise ratios). An example of a codeset is shown in Figure 6.27. Each MCS in the codeset has the highest throughput for a $1-2$ dB range of SNIR.

The Shannon bound represents the maximum theoretical throughput that can be achieved over an AWGN channel for a given SINR, as shown in Figure 6.27. The example of the AMC system achieves around $0.75\times$ the throughput of the Shannon bound, over the range of SNR which it operates. Figure 6.28 shows the baseline E-UTRA DL (evolved universal terrestrial radio access) and UL spectral efficiency (bps per Hz) versus SINR, based on the parameters in 3GPP together with the Shannon bound.

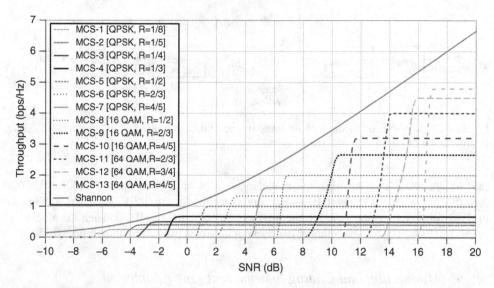

Figure 6.27 Throughput of a set of coding and modulation combinations, AWGN channels assumed. (Source: [4] 3GPP. Reproduced with permission of ETSI.)

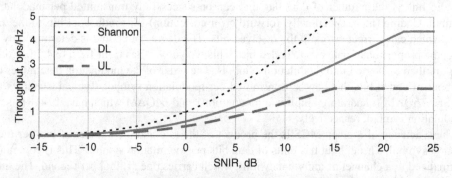

Figure 6.28 Spectral efficiency versus SNIR for baseline E-UTRA. (Source: [4] 3GPP TS. Reproduced with permission of ETSI.)

6.6.3.1 Link Adaptation

The link adaptation is mandatory to select the optimum MCS according to the channel conditions and UE capability. Link adaptation is mandatory to guarantee performance metrics such as data rate, packet error rate, and latency. The AMC is the most famous link adaption technique where various modulation schemes and channel coding rates can be adopted, based on the channel conditions. The users near to the eNB with high SINR values will be assigned a higher MCS and the assigned MCS will be decreased as the user moves away from the eNB as the SINR is degraded and the interference is increased. The same MCS should be applied to all groups of RBs belong to the same Layer 2 (L2) protocol data unit (PDU) scheduled to one user within one TTI by a single antenna stream.

By considering the same cluster with the RF parameters in Table 6.12, the CDF and PDF of the DL and UL MCS are shown in Figures 6.29 and 6.30, respectively. Table 6.15 provides the

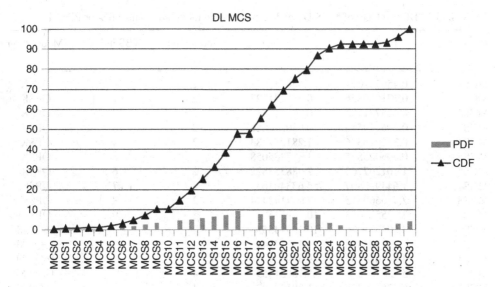

Figure 6.29   PDF and CDF for DL MCS.

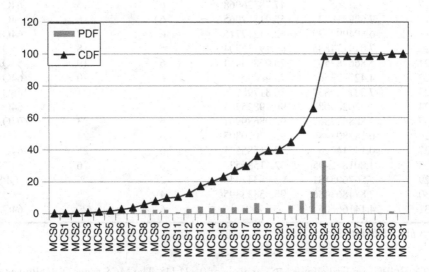

Figure 6.30   PDF and CDF for UL MCS.

values of the CDF and PDF for the DL MCS together with the modulation order and the TBS. The CDF and PDF of the modulation order in the DL and UL are illustrated in Figures 6.31 and 6.32, respectively. The MCS figures indicate the median average at 50% of MCS17 and MCS22 for the DL and UL, respectively. These are typical practical values for MCS for a cluster with very good RF converge and low interference. The MCS17 in the DL is in line with the average CQI index and the average SINR discussed earlier. With regard to higher order modulation, based on the reported CQI, the eNB scheduler selects the MCS assigned to the

Table 6.15   PDF and CDF of MCS DL and corresponding modulation and TBS for PDSCH

MCS index	PDF (%)	CDF (%)	Modulation order	TBS index	Modulation
MCS0	0.456781071	0.456781071	2	0	QPSK
MCS1	0.141560656	0.598341727	2	1	QPSK
MCS2	0.125771583	0.72411331	2	2	QPSK
MCS3	0.206129546	0.930242856	2	3	QPSK
MCS4	0.351016326	1.281259183	2	4	QPSK
MCS5	0.606025875	1.887285058	2	5	QPSK
MCS6	1.101267764	2.988552822	2	6	QPSK
MCS7	1.644928807	4.63348163	2	7	QPSK
MCS8	2.460660815	7.094142444	2	8	QPSK
MCS9	3.075573053	10.1697155	2	9	QPSK
MCS10	3.00553E-05	10.16974555	4	9	16QAM
MCS11	4.265403888	14.43514944	4	10	16QAM
MCS12	4.925679654	19.36082909	4	11	16QAM
MCS13	5.7483845	25.10921359	4	12	16QAM
MCS14	6.276176323	31.38538992	4	13	16QAM
MCS15	7.105823954	38.49121387	4	14	16QAM
MCS16	9.296026774	47.78724065	4	15	16QAM
MCS17	2E-05	47.78726068	6	15	64QAM
MCS18	7.62991897	55.41717965	6	16	64QAM
MCS19	6.694997499	62.11217715	6	17	64QAM
MCS20	7.079174885	69.19135204	6	18	64QAM
MCS21	5.866782507	75.05813454	6	19	64QAM
MCS22	4.422723561	79.4808581	6	20	64QAM
MCS23	7.337129857	86.81798796	6	21	64QAM
MCS24	3.38026408	90.19825204	6	22	64QAM
MCS25	1.888647567	92.08689961	6	23	64QAM
MCS26	0.183808446	92.27070805	6	24	64QAM
MCS27	0.14514726	92.41585531	6	25	64QAM
MCS28	1.00184E-05	92.41586533	6	26	64QAM
MCS29	0.567605131	92.98347046	2	[a]	QPSK
MCS30	2.874863486	95.85833395	4	[a]	16QAM
MCS31	4.14166605	100	6	[a]	64QAM

[a]Reserved for retransmission.

user. The 3GPP standard allows MCS indices of 0–31 [1]. The MCS range 0–9 allows QPSK modulation, MCS 10–16 allows 16QAM modulation, MCS 17–28 allows 64QAM modulation usage, and the range of MCS 29–31 allows special operation during retransmissions [1]. The UE derives the TBS in bits for each stream from the MCS.

The DL modulation order in Figure 6.31 indicates that 46% of the route is with 64QAM modulation. This demonstrates strong 64QAM utilization with the LTE system. For the sake of comparison with the HSPA+, Figure 6.33 indicates the CDF and PDF of the first and second carrier modulation with DC-HSDPA+ (high speed downlink packet access) for a similar cluster. Figure 6.33 indicates that HSPA+ offers only 7.4% of 64QAM modulation and around 40% of the route with QPSK modulation. This difference is due to many factors but the most important is the OFDM nature compared to WCDMA where the propagation

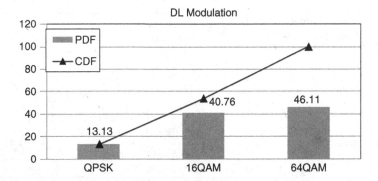

Figure 6.31   PDF and CDF of the LTE DL modulation order.

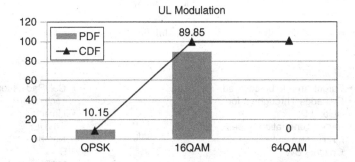

Figure 6.32   PDF and CDF of the LTE UL modulation order.

channel with OFDM is a frequency non-selective channel due to the orthogonality of the narrow band overlapped subcarriers, while the channel of the WCDMA system is frequency selective due to the spreading in the time domain and the carrier of the WCDMA is 5 MHz. As a result, the OFDM system has a less complicated receiver design, better detection capability, and higher spectral efficiency.

6.6.4   LTE Link Budget and Coverage Analysis

The aim of the link budget is to identify the maximum allowable path loss (MAPL) between the transmitter and receiver for the UL and DL. Therefore, the cell radius can be calculated for different terrain morphologies (i.e., dense urban, urban, suburban, and rural) based on the appropriate propagation model. The minimum SINR requirements in both the UL and DL are achieved with the MAPL and maximum transmit power. The LB considers many factors, such as building penetration loss, feeder loss, antenna gain, and the interference margin of radio links, to calculate all gains and losses that affect the final cell coverage. The cell radius of an eNB can be obtained according to the MAPL under a tuned propagation model. The cell radius can be used to estimate the total number of sites that needed to provide the RF coverage that meets the predefined coverage objectives. Figure 6.34 summarizes the link budget inputs and outputs.

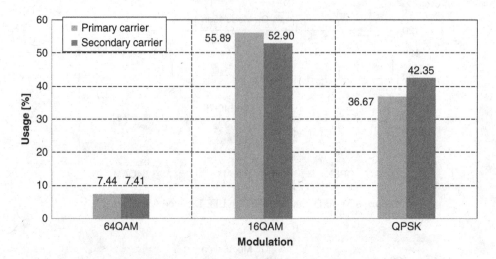

Figure 6.33 PDF of the DC-HSDPA+ modulation for the first and second carriers.

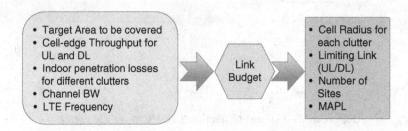

Figure 6.34 Inputs and outputs of the LB.

Since LTE Release 8 is a data centric technology, then the critical coverage constraint for the link budget is the data rate at the cell edge rather than the received signal level. The outcome of the link budget calculations enables the network designer to determine the expected coverage calculated in theory and compare it with the measured values in the field. Table 6.16 provides typical design targets for the LTE 1800 MHz link budget. Table 6.17 provides typical link budgets for an LTE system at 1800 MHz band (i.e., 3GPP band 3) with 20 MHz channel bandwidth. The formulas used in the link budget calculations are provided in Figure 6.35. The link budgets are calculated at different morphologies and for the UL and DL. A tuned version of the COST231-Hata model is used to estimate the path loss. The cell radius for each clutter is determined based on the smallest cell radius from the UL and DL. The link budgets for all morphologies demonstrate that the LTE system is uplink limited (i.e., MAPL of UL < MAPL of DL) and there is ~3–4 dB between the MAPLs for the UL and DL at typical CETs (i.e., DL at 512 kbps and UL at 128 kbps). To further illustrate the link budget concept, Figures 6.36 and 6.37 demonstrate the gain and losses in the UL and DL paths. The eNB outout power and the link budget demonstrated in Table 6.17 adopted the physical shared channels in the DL and UL, that is, PDSCH/PUSCH to determine the MAPL. Other major physical LTE channels may be included in the link budget estimation, such as RS, PDCCH, PBCH, and PDSCH in DL and

Table 6.16 Typical coverage design targets at LTE 1800 MHz

Criteria	Target
RSRP	≥ -116
Area coverage probability	95%
Cell-edge coverage probability	90%
SINR	≥ -3 dB
Cell edge throughput DL/UL (kbps)	1024/512

Item	Formulas
Data Channel Type	Physical UL/DL Shared Channel
Duplex Mode	FDD: Frequency Division Duplexing
User Environment	
System Bandwidth (MHz)	
Cell Edge Rate (kbps)	
Tx	
Max Total Tx Power (dBm)	A
Allocated RB	B
RB to Distribute Power	C
Subcarriers to Distribute Power	$D = 12{}^{*}C$
Subcarrier Power (dBm)	$E = A-10{}^{*}\text{Log}10(D)$
Tx Antenna Gain (dBi)	G
Tx Cable Loss (dB)	H
Tx Body loss (dB)	I
EIRP per Subcarrier (dBm)	$J = E+G-H-I$
Rx	
SINR (dB)	K
Rx Noise Figure (dB)	L
Receiver Sensitivity (dBm)	$M = K+L-174+10{}^{*}\text{Log}10(15000)$
Rx Antenna Gain (dBi)	N
Rx Cable Loss (dB)	O
Rx Body loss (dB)	P
Target Load	
Interference Margin (dB)	Q
Min. Signal Reception Strength (dBm)	$R = M-N+O+P+Q$
Path Loss & Cell Radius	
Indoor Penetration Loss (dB)	S
Std. Dev. of Shadow Fading (dB)	
Area Coverage Probability	
Shadow Fading Margin (dB)	T
Maximum Allowable Path Loss (dB)	$U = J-R-S-T$
eNodeB/UE Antenna Height (m)	
Cell Radius (km)	

Figure 6.35 LTE link budget formulas.

PUCCH, and PRACH in the UL. Link budgets for these channels will be analyzed later in this chapter. In this next section, we will analyze the link budget parameters and formulas.

6.6.4.1 eNB Output Power

This is one of the main factors that impact the link budget. In the link budget illustrated in Table 6.17, we considered 46 dBm output power per each branch of the MIMO 2×2 which

Table 6.17   UL and DL LTE 1800 link budgets for different clutters

	LTE link budget for best effort services							
Morphology	Dense urban		Urban		Suburban		**Rural**	
Data channel type	PUSCH	PDSCH	PUSCH	PDSCH	PUSCH	PDSCH	PUSCH	PDSCH
Duplex mode	FDD		FDD		FDD		FDD	
User environment	Indoor		Indoor		Indoor		Indoor	
System bandwidth (MHz)	20.0		20.0		20.0		20.0	
Cell edge rate (kbps)	128.00	512.00	128.00	512.00	128.00	512.00	128.00	512.00
Transmitter								
Maximum total Tx power (dBm)	23.00	46.00	23.00	46.00	23.00	46.00	23.00	46.00
Allocated RB	3	19	3	19	3	19	3	19
RB to distribute power	3	100	3	100	3	100	3	100
Subcarriers to distribute power	36	1200	36	1200	36	1200	36	1200
Subcarrier power (dBm)[a]	7.44	15.21	7.44	15.21	7.44	15.21	7.44	15.21
Tx antenna gain (dBi)	0.00	17.00	0.00	17.00	0.00	17.00	0.00	17.00
Tx cable loss (dB)	0.00	0.50	0.00	0.50	0.00	0.50	0.00	0.50
Tx body loss (dB)	0.00	0.00	0.00	0.00	0.00	0.00	0.00	0.00
EIRP per subcarrier (dBm)	7.44	31.71	7.44	31.71	7.44	31.71	7.44	31.71
Receiver								
SINR (dB)	−4.19	−5.37	−4.19	−5.37	−2.33	−4.94	−2.20	−4.43
Rx noise figure (dB)	2.30	7.00	2.30	7.00	2.30	7.00	2.30	7.00
Receiver sensitivity (dBm)	−134.13	−130.61	−134.13	−130.61	−132.26	−130.18	−132.14	−129.67
Rx antenna gain (dBi)	17.00	0.00	17.00	0.00	17.00	0.00	17.00	0.00
Rx cable loss (dB)	0.50	0.00	0.50	0.00	0.50	0.00	0.50	0.00
Rx body loss (dB)	0.00	0.00	0.00	0.00	0.00	0.00	0.00	0.00
Target load (%)	75.00	90.00	75.00	90.00	75.00	90.00	75.00	90.00
Interference margin (dB)	0.89	2.72	0.89	2.72	1.46	3.13	2.71	3.74
Minimum signal reception strength (dBm)	−149.74	−127.89	−149.74	−127.89	−147.31	−127.05	−145.93	−125.93
Path loss and cell radius								
Indoor penetration loss (dB)	19.00	19.00	15.00	15.00	11.00	11.00	8.00	8.00
Standard deviation of shadow fading (dB)	11.70	11.70	9.40	9.40	7.20	7.20	6.20	6.20
Area coverage probability (%)	95.00	95.00	95.00	95.00	95.00	95.00	90.00	90.00
Shadow fading margin (dB)	9.43	9.43	8.04	8.04	5.99	5.99	1.87	1.87
Maximum allowable path loss (dB)	128.74	131.16	134.13	136.56	137.76	141.77	143.50	147.77
eNodeB/UE antenna height (m)	25.00	1.50	30.00	1.50	40.00	1.50	50.00	1.50
Cell radius (km)	0.47	0.55	0.87	1.02	2.13	2.78	5.64	7.54

[a]Subcarrier power is estimated assuming that the Tx power is equally distrusted across the total bandwidth.

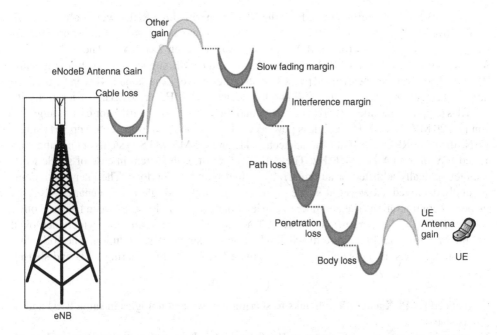

Figure 6.36 Downlink budget estimation.

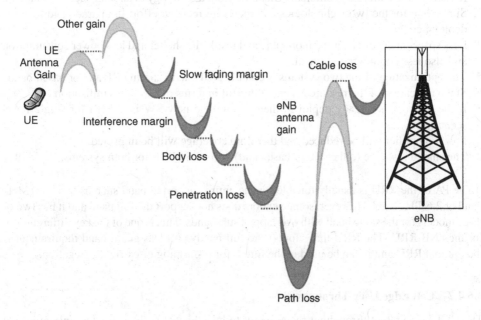

Figure 6.37 Uplink link budget estimation.

means 2×40 watts. Therefore, if we have the RRU scenario, three RRUs with 2×40 watt each are deployed to provide three sectors with MIMO 2×2 operation. Most of the suppliers currently supports 2×60 watt for MIMO 2×2 operation and even 2×80 watt. The higher power RRU can be used for either MIMO 4×4 evolution if the eNB support this evolution in same HW i.e., 2×60 watt can be evolved to 4×30 watt. Alternatively, the same RRU can be used for carrier aggregation in same band with another adjacent 10 MHz or 20 MHz carrier. Majority of RRUs support continues 40 MHz or higher and therefore it can be used for carrier aggregation i.e., 20 MHz + 20 MHz with less power i.e., 4×30 watt. Also, if the eNB supports single RAN, then the RRU can share the adjacent LTE and GSM/UMTS systems within the supported BW of the PA i.e., 40 MHz. The output power of each system in case of single RAN is either statically adjusted or automatically shared based on the load. The later is not commercially deployed. However, it is one of the key features in single RAN scenario. Also, the supported configurations (i.e., number of carriers for each technology) between LTE and other systems (GSM/UMTS) are vendor specific. The single RAN scenario is not widely deployed and the operator should be careful when adopting the single RAN solution. The following points summarizes the pros and cons of the single RAN including sharing RRU and sharing based band:

1. Reduced CAPEX and OPEX thanks to sharing the two technologies in same RRU and or base band.
2. Dynamic allocation of RF power and/or the base band resources offer efficient utilization for network resources and efficient offloading from technology to another such as 3G to 4G.
3. Single SW for the two technologies and therefore reduced effort in upgrade and features deployment.
4. Less space on the tower or the monopole and inside the shelter and less power consumption and also less cooling requirement.
5. The optimization of the two systems is complicated due to sharing RF and/or based band.
6. Static allocation of the resources is not efficient in terms of resources utilization.
7. If the network is already deployed then a complete new SW is needed for single RAN scenario.
8. Power amplifier will be reduced and therefore coverage will be impacted.
9. The shared resource (either RF or base band) will be a SPOF for both systems.

The PA of the RRU is usually tunable for the entire supported band such as the 1800 MHz band or 2.6 GHz band. However, some vendors does not support the full band and it has two or three models for the same band with overlapped sub-bands. This is one of the key differentiator for the eNB RRU. The RRU that support the full bandwidth less guard band requirement is the optimal RRU and it can be used in the future for re-farming or carrier aggregation.

6.6.4.2   Cell Edge User Throughput

The CET is the target throughput requirement to be achieved at the cell edge; minimum net single UE throughput requirement in DL and UL. The CET determines the service that can be provided at the cell edge. Accordingly it can limit the minimum MCS to be used. The CET is usually provided by the network operator according to the required services at cell edge. It depends on the operator strategy and the business requirement. A typical DL CET for

HSPA+ network is 128 kbps, while for the LTE network it can be 512 kbps to 1 Mbps, and possibly up to 2 Mbps. A typical DL CET for HSPA+ network is 256 kbps and for LTE it can be 1 Mbps up to 4 Mbps. The CET is a key input to the link budget and it directly impacts the cell radius and accordingly the number of required sites to cover a certain area. The CET used in the link budgets in Table 6.17 is 512 in the DL and 128 in the UL for the sake of maximum cell radius estimation.

The CET adopted in the link budget exercise is for illustration purposes. In a practical exercise, a typical CET would be 512 kbps for the UL and 1 Mbps for the DL. Alternatively, an aggressive requirement may consider 1 Mbps for the UL and 2 Mbps for the DL.

Table 6.18 Link budget for VoLTE services

	VoLTE link budget with RoHC			
Morphology	Dense urban		Urban	
Data channel type	PUSCH	PDSCH	PUSCH	PDSCH
Duplex mode	FDD		FDD	
User environment	Indoor		Indoor	
System bandwidth (MHz)	20.0		20.0	
Cell edge rate (kbps)	56.00	336.00	56.00	336.00
Transmitter				
Maximum total Tx power (dBm)	23.00	46.00	23.00	46.00
Allocated RB	1	14	1	14
RB to distribute power	1	100	1	100
Subcarriers to distribute power	12	1200	12	1200
Subcarrier power (dBm)	12.21	15.21	12.21	15.21
Tx antenna gain (dBi)	0.00	18.00	0.00	18.00
Tx cable loss (dB)	0.00	0.50	0.00	0.50
Tx body loss (dB)	3.00	0.00	3.00	0.00
EIRP per subcarrier (dBm)	12.21	32.71	12.21	32.71
Receiver				
SINR (dB)	−1.60	−5.16	−1.60	−5.16
Rx noise figure (dB)	2.30	7.00	2.30	7.00
Receiver sensitivity (dBm)	−131.54	−130.40	−131.54	−130.40
Rx antenna gain (dBi)	18.00	0.00	18.00	0.00
Rx cable loss (dB)	0.50	0.00	0.50	0.00
Rx body loss (dB)	0.00	3.00	0.00	3.00
Target load	50.00%	75.00%	50.00%	75.00%
Interference margin (dB)	1.11	2.27	1.11	2.27
Minimum signal reception strength (dBm)	−147.93	−128.13	−147.93	−128.13
Path loss and cell radius				
Indoor penetration loss (dB)	19.00	19.00	15.00	15.00
Standard deviation of shadow fading (dB)	11.70	11.70	9.40	9.40
Area coverage probability (%)	95.00	95.00	95.00	95.00
Shadow fading margin (dB)	9.43	9.43	8.04	8.04
Maximum allowable path loss (dB)	131.71	132.41	137.10	137.80
eNodeB/UE antenna height (m)	25.00	1.50	30.00	1.50
Cell radius (km)	0.57	0.60	1.06	1.11

It is important now to conduct a link budget for VoLTE services. If we consider RoHC (robust overhead compression) and assume one RB in the UL then we can use 56 kbps in the UL for VoLTE and 336 kbps in the DL, which needs 14 RB in the DL. If we consider the lower number of the RBs in the DL, then we have to use high MCS, which leads to cell radius shrink. Table 6.18 demonstrates the link budget for this exercise for the UL/DL and for dense urban and suburban clutters. As shown from the link budget, it is also UL limited, similar to data only scenario, and the cell radius of the VoLTE is larger than the data cell radius demonstrated in Table 6.17.

6.6.4.3 Channel Models

The channel model is characterized by delay profiles that are selected to be representative of low, medium, and high delay spread environments. The delay profiles of the channel models are defined in Table 6.19 and the tapped delay line models are defined in Table 6.20 [6]. The most typical UE speed and multipath profiles are considered according to the type of environment (e.g., dense urban, urban, rural, etc.). In the link budget example in Table 7.17, the ETU (extended typical urban) at 3 km/h is used for dense urban and urban morphologies. The channel model ETU at 120 km/h and EVA (extended vehicular A) at 120 km/h are used for suburban and rural, respectively. Selecting a multipath channel model

Table 6.19 Delay profiles for E-UTRA channel models

Model	Number of channel taps	Delay spread (r.m.s.) (ns)	Maximum excess tap delay (span) (ns)
Extended pedestrian A (EPA)	7	45	410
Extended vehicular A model (EVA)	9	357	2510
Extended typical urban model (ETU)	9	991	5000

Table 6.20 Typical propagation channel Models used for LTE

Channel model		Extended pedestrian A model (EPA)		Extended vehicular A model (EVA)		Extended typical urban model	
Number of Taps		7		9		9	
Excess tap delay (ns)	Relative power (dB)	0	0	0	0	0	−1
		30	−1	30	−1.5	50	−1
		70	−2	150	−1.4	120	−1
		90	−3	310	−3.6	200	0
		110	−8	370	−0.6	230	0
		190	−17.2	710	−9.1	500	0
		410	−20.8	1090	−7	1600	−3
				1730	−12	2300	−5
				2510	−16.9	5000	−7

(Source: [9] 3GPP TS. Reproduced with permission of ETSI.)

Table 6.21 Channel model parameters [6]

Model	Maximum Doppler frequency (Hz)
EPA 5 Hz	5
EVA 5 Hz	5
EVA 70 Hz	70
ETU 30 Hz	30
ETU 70 Hz	70
ETU 300 Hz	300

Table 6.22 Antenna heights, coverage probability, and shadowing standard deviation for different morphology

Morphology	DU	MU	SU	RU
Antenna height (m)	25	30	40	50
Coverage probability (%)	95	95	95	90
Shadowing standard deviation (dB)	11.7	9.4	7.2	6.2
Channel model	ETU 3 km/h	ETU 3 km/h	ETU 120 km/h	EVA 120 km/h

for a certain morphology is adopted for simulation purpose only. In practice, various multipath channel conditions occur concurrently within the same cell environment. Therefore, a mix of multipath channel models represents the real channel characteristics. However, for link budget purposes, the worst-case channel model is considered and, therefore, the worst-case cell radius is estimated for each clutter.

Table 6.21 shows propagation conditions that are used for the performance measurements in a multipath fading environment for low, medium, and high Doppler frequencies. The coverage

Table 6.23 Specifications of LTE 1800 MHz four port antenna with two cross-polarized pairs

Frequency Band (MHz)	1710–1880	1850–1990	1920–2170	2300–2500	2500–2690
Gain (dBi)	17.7	17.7	17.7	17.7	18.2
Beamwidth, horizontal (°)	65	65	65	65	61
Beamwidth, vertical (°)	7.5	7	6.5	5.5	5.1
Beam tilt (°)	0–10	0–10	0–10	0–10	0–10
Front-to-back ratio at 180° (dB)	28	30	28	27	27
Isolation (dB)	30	30	30	30	30
VSWR|return loss (dB)	1.5|14.0	1.5|14.0	1.5|14.0	1.5|14.0	1.5|14.0
PIM, third order, 2 × 20 W (dBc)	−150	−150	−150	−150	−150
Input power per port, maximum (W)	350	350	350	300	300
Polarization	±45°	±45°	±45°	±45°	±45°
Impedance (Ω)	50	50	50	50	50

PIM, passive intermodulation.
(Source: [7] Commscope. Reproduced with permission of Commscope.)

probability, the associated shadowing standard deviation, and the channel model for different morphologies for the link budgets presented in Table 6.17 are summarized in Table 6.22.

6.6.4.4   eNB Antenna Gain

The antenna gain is proportional to the antenna size, LTE band, and beamwidth of the antenna patterns (horizontal and vertical). A large antenna with narrow beamwidth provides a high gain while a short antenna with wider beamwidth provides less gain. The selection of antenna gain and beamwidth depends on the clutter type and coverage requirement. The low gain antenna (15–17 dBi) can be used in dense urban and urban clutters while a high gain antenna (18–20 dBi) can be used in rural areas and highways to extend the RF converge. An urban clutter with multi-story buildings (10 floors) needs a wide vertical beamwidth antenna to cover the floors and a suburban clutter with short houses/villas needs a narrow vertical BW to avoid unnecessary overshooting and, therefore, increase the interference margin due to inter-cell interference. A typical LTE 1800 MHz antenna with two dual polarized antennas (four antenna ports) that can accommodate LTE 1800 MHz and GSM 1800 or LTE 1800 MHz with 4RX diversity or MIMO 4×4 is summarized in Table 6.23 and its pattern is demonstrated in Figure 6.38. This antenna can also accommodate 3G signals in the 1710/2100 MHz band, and 4G signals in the 2600 MHz band.

As shown in Table 6.23, the gain increases with increase in the frequency band while the beamwidth of the horizontal and vertical is reduced.

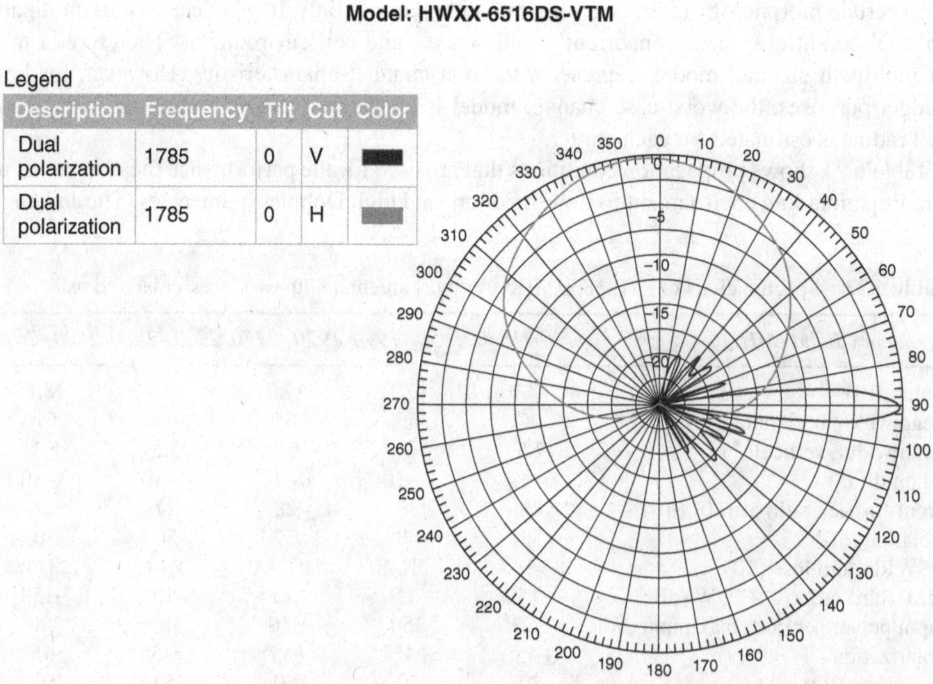

Model: HWXX-6516DS-VTM

Legend

Description	Frequency	Tilt	Cut	Color
Dual polarization	1785	0	V	■■■
Dual polarization	1785	0	H	■■■

Figure 6.38   Vertical and horizontal beam patterns for the antenna specified in Table 6.23. (Source: [7] Commscope. Reproduced with permission of Commscope.)

This antenna comes with a beam tilt range from $0°$ to$10°$. The beam tilt can be adjusted separately for the two technologies that are sharing the same antenna. If the antenna is used for MIMO 4×4 or 4RX diversity then the same tilt should be applied. The beam tilt depends on the converge requirement and interference limitation. It can be adjusted manually or automatically by adding an RET (remote electrical tilt) system. The RET system is an important element in a SON (self-optimizing network) to optimize the network for less interference and best converge. The majority of mobile operators did not adopt RET with 2G/3G systems and with early deployment of the LTE system. However, with the evolution of the SON and the hetNet (heterogeneous network), it will be an important factor to optimize the LTE network and control the interference.

6.6.4.5   MCS Selection

The selection of minimum MCS for the link budget depends on the required CET that is usually provided by the operator. A robust MCS should be selected to guarantee the required CET under the worst RF channel condition. With the LTE-FDD system and due to change in the UL and DL channels and change in power allocation, the selection of the MCS for UL and DL is different at the cell edge in the link budget and also in the practical scenario. Each MCS is mapped to a certain modulation group (QPSK, 16QAM, 64QAM), and each MCS index is assigned a TBS index. The TBS reflects the amount of user data bits sent during one TTI (1 ms) and the TBS depends on the number of scheduled RBs.

6.6.4.6   Equivalent Isotropic Radiated Power (EIRP)

The EIRP (equivalent isotropic radiated power) indicates the power that would be radiated by the theoretical isotropic antenna to achieve the peak power density observed in the direction of maximum antenna gain. The power radiated by a directional antenna is transposed into the radiated power of an isotropic antenna by consideration of antenna gain and power at the antenna input. For the LTE system, the EIRP per subcarrier in the DL and UL are calculated, respectively, as follows:

$$\text{EIRP}_{DL}^{SC} = P_{eNB(sc)} + AG_{eNB} - FL + MG \tag{6.9}$$

where $P_{eNB(sc)}$ is the power per subcarrier in the DL, AG_{eNB} is the antenna gain of the eNB, FL is the feeder loss, and MG is the MIMO gain.

$$\text{EIRP}_{UL}^{SC} = P_{U(sc)} + AG_{UL} - BL \tag{6.10}$$

where $P_{U(sc)}$ is the power per subcarrier in the UL, AG_{UL} is the antenna gain of the UE, and BL is the body loss.

As a result the EIRP in the DL and UL is calculated, respectively, as follows:

$$\text{EIRP}_{DL}^{SC} = 15.2 + 17 - 0.5 + 0 = 31.7 \text{ dBm} \tag{6.11}$$

$$\text{EIRP}_{UL}^{SC} = 7.44 + 0 - 0 = 7.44 \text{ dBm} \tag{6.12}$$

Similarly, the EIRP could be calculated per RB as follows:

$$\text{EIRP}_{DL}^{RB} = 46 - 10\log_{10}(100) + 17 - 0.5 + 0 = 42.5 \text{ dBm} \tag{6.13}$$

$$EIRP_{UL}^{RB} = 23 - 10\log_{10}(3) + 0 - 0 = 18.23\,dBm \tag{6.14}$$

The EIRP in the DL is calculated based on the total number of RB due to the OFDMA (orthogonal frequency division multiple cccess) while in the UL, the allocated RBs (i.e., three RBs) are only used due to the SC-FDMA.

6.6.4.7 Receiver Sensitivity

The eNode-B receiver sensitivity is the signal level/threshold at which the RF signal can be detected with a certain quality. This threshold refers to the antenna connector and should take into account the further demodulation and the required output signal quality. The receiver sensitivity depends on the following factors:

- Data rate targeted at cell edge,
- Target quality/HARQ (hybrid automatic repeat request) (i.e., block error rate (BLER), maximum number of retransmissions),
- Radio environment conditions (multipath channel, mobile speed), and
- Noise figure (NF) of the eNB receiver.

The receiver sensitivity per subcarrier is calculated as follows:

$$RxSen_{subcarrier} = SINR + NF + NP + 10\log_{10}(15\,000) \tag{6.15}$$

where SINR is the threshold of the receiver that can demodulate the signal and it is related to the MCS for the UL and DL, respectively, the BLER target, MIMO gain, and HARQ setting. The SINR is obtained from the system simulation results. The SINR values are vendor specific and depend on the receiver design. The NP is the density of the thermal white noise power, which is $-174\,dBm/Hz$ and estimated as follows:

$$NP = \log_{10}(290 \times 1.38 \times 10^{-23} \times 103) \tag{6.16}$$

In the LTE system, a single subcarrier is $15\,kHz$. Therefore, the thermal noise power per subcarrier for the DL and UL are calculated as below:

$$NP_{subcarrier} = NP + 10\log_{10}(15 \times 1000) = -132.24\,dBm \tag{6.17}$$

As a result, the DL and UL receiver sensitivity are estimated based on the number of sub-carriers allocated in the DL and UL as follows:

$$RxSen = SINR + NF + NP_{subcarrier} + N_{subcarrier} \tag{6.18}$$

where $N_{subcarrier}$ is $10\log10$ (the number of allocated subcarriers). The number of allocated subcarriers equals the total number of received subcarriers in the available system bandwidth of $20\,MHz$ in the case of the DL as all the RB are used in the DL, as shown in the LB due to the OFDMA transmission. In the UL with SC-FDMA transmission, the total number of allocated subcarriers equals the number of subcarriers allocated to the user to achieve the target CET of $128\,kbps$ at cell edge which is 36 subcarriers (3 RB × 12 subcarriers per RB) as shown in the link budget in Table 6.17. Therefore, $N_{\text{subcarrier}}$ in the DL and UL are estimated respectively as follows:

$$N_{\text{subcarrier}}^{\text{DL}} = 10\log_{10}(100 \times 12) = 30.8\,\text{dB} \tag{6.19}$$

$$N_{\text{subcarrier}}^{\text{UL}} = 10\log_{10}(3 \times 12) = 15.56\,\text{dB} \tag{6.20}$$

The adopted link budget in this chapter is based on subcarrier receiver sensitivity and accordingly the EIRP used is based on the output power per subcarrier. In other approaches, the receiver sensitivity on the DL and UL is used and in this case total power for all allocated subcarriers should be used in the EIRP.

6.6.4.8 Noise Figure

The NF in dB is the ratio of the input SINR at the input end to the output SINR at the output end of the receiver. It is a key factor to measure the receiver performance. Therefore, the receiver sensitivity together with the NF should be considered to benchmark the eNB receiver performance. The NF depends on the bandwidth and the eNB capability. For LTE terminals, the NF is between 6 and 8 dB and it is considered as 7 dB in the link budget provided in Table 6.17.

Therefore, the final receiver sensitivities in the DL and UL are estimated respectively as follows:

$$\text{RxSen}_{\text{DL}} = -132.24 - 5.37 + 7 + 30.8 = -99.81\,\text{dB} \tag{6.21}$$

$$\text{RxSen}_{\text{UL}} = -132.24 - 4.19 + 2.3 + 15.56 = -118.57 \tag{6.22}$$

UE Characteristics

The transmitter characteristics are specified at the antenna connector of the UE with a single or multiple transmit antenna(s). For the UE with an integral antenna only, a reference antenna with a gain of 0 dBi is assumed [6]. The maximum transmit power of an LTE UE depends on the power class of the UE. Currently, only one power class is defined in 3GPP TS 36.101 [6], which is 23 dBm output power with 0 dBi antenna gain. In the link budget exercise, we have considered a USB dongle with 1.5 m height and 23 dBm output power. It is recommended to consider a safety factor for UE output power in the link budget, which depends on the frequency band as per Table 6.24.

6.6.4.9 SINR Performances

The SINR figures are derived from link level simulations or, better, from equipment measurements (lab or field measurements). The SINR depends on the eNB performance, radio

Table 6.24 UE antenna gain for different LTE bands

LTE frequency band (MHz)	UE antenna gain (dBi)
700/800	−5
1800	−3
2600	0

Table 6.25 Penetration loss range for different clutters

Clutter type	Penetration loss range (dB)	Typical values in the LB (dB)
Dense urban	19–25	19
Urban	15–18	15
Suburban	10–14	11
Rural	5–8	8

conditions (multipath fading profile, mobile speed), receive diversity configuration (two branch by default and optionally four branch with MIMO 4×4 or MIMO 4×2), targeted data rate, and the required QoS. In the LTE system the required SINR for the PDSCH replaces the required Eb/No of the UMTS Rel. 99 DCH (dedicated channel) link budget. This is mainly because the Eb/No is not accurate with fast link adaptation. We demonstrated in Section 6.5.2 the SINR values versus the throughput per RB and spectral efficiency for uplink and downlink in Figures 6.24 and 6.25 respectively.

6.6.4.10 Penetration Loss

The penetration loss indicates the fading of radio signals from an indoor terminal to a base station due to obstruction by a building and vice versa. The penetration loss depends on the type of the clutter and the nature of the buildings in the target coverage area. Table 6.25 summarizes typical penetration losses range for different clutters.

6.6.4.11 Body Loss

Body loss is the loss generated due to signal blocking and absorption when a terminal antenna is close to the body of the user. This affects handsets in particular. The body loss depends on the position of the terminal and the user. For terminals such as a USB dongle, a mobile WiFi device, and an LTE fixed router the position of such terminals is far from the user's body and, therefore, the body loss can be ignored. This is the case with the link budget in Table 6.17 as we have considered a data dongle in this data only scenario. For mobile terminals such as smartphones, the body loss must be considered in the link budget, as shown in the VoLTE link budget in Table 6.18 and it is about 3 dB as the phone will be adjacent to the user body. Therefore, the body loss will be considered only in the case of a smartphone or voice handset and will not be considered with other devices as well as the eNB, since the antennas of the eNB are installed away from the user's body.

6.6.4.12 Feeder Loss

Feeder loss is the losses due to RF feeders, RF jumpers, and connecters in the path between the antenna and the eNB. The feeder loss is calculated according to the feeder type, feeder material, length, and diameter. If the eNB is installed inside the shelter and a feeder system is used between the antenna and the eNB, then all losses from the feeders, jumpers, combiners (if any), splitters (if any), and so on, are considered in the link budget. On the other hand, if the

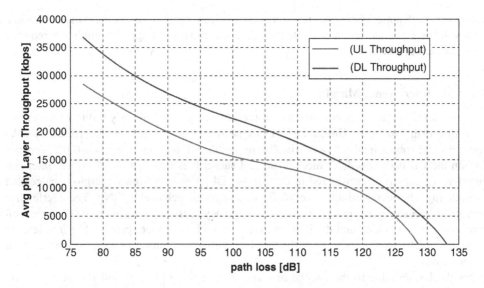

Figure 6.39   Average layer DL/UL throughputs for LTE system versus path loss.

distributed eNB with baseband unit and remote radio units (RRUs) is considered, then only the loss of the jumper between the RRU and antenna is considered and it is about 0.5 dB loss in the link budget exercise in Table 6.17. With a full feeder system the losses may be 3 dB or more according to the aforementioned characteristics.

6.6.4.13   Link Budget Validation

In order to validate the LB in Table 6.17, a field measurement is conducted. Figure 6.39 illustrates the average UL and DL throughputs versus the path loss from the field test results for a commercial LTE system in dense urban terrain. It is evident that the LTE system is UL link limited and the difference in path losses is almost similar to the theoretical one (i.e., 3–4 dB difference between the MAPLs of the DL and UL). Table 6.26 provides the link budgets results for the four clutters. By comparing the path loss from Figure 6.39 with the MAPL for dense urban clutter from Table 6.26, we can observe that the UL throughput converges to zero at approximately 128 dB path loss and the MAPL at 128 kbps is around 128.74 dB from Table 6.26. We

Table 6.26   Link budget results

Environment	UL cell edge rate (kbps)	MAPL (dB)	Cell range (km)
DU	**128**	**128.74**	**0.47**
MU	128	134.13	0.87
SU	128	137.76	2.13
RU	128	143.50	5.64

can conclude that the link budget is accurate and indeed it represents the field measurements. This validation is important to validate the cell radius and accordingly the number of required sites for each clutter.

6.6.4.14 Interference Margin

The interference margin is encountered in the link budget due the possibility of noise rise according to the load level. Unlike the UMTS system, LTE has no intra-cell interference thanks to the OFDM subcarriers' orthogonality. Therefore, the UL RB load can reach 100%, depending on the UL eNB scheduler mechanism and subscribers distribution. However, inter-cell interference should be considered in the UL and DL. The interference margin is calculated considering other-cell loading, target SINR, and minimum achievable SINR. The interference margin is affected by many factors, such as frequency reuse scheme, inter-site distance, cell load, system bandwidth, and the ICIC algorithm, which is vendor specific. The interference margin accounts for the increase in the terminal noise level caused by the interference from other users.

For the UL and due to the change in scenario on a per TTI basis and non-deterministic distribution of users, it is recommended to estimate the interference margin using an actual dynamic simulation. While for the DL, the interference margin can be computed analytically as the relation between signals received with and without interference and can be estimated as follows:

$$IM_{DL} = -10\log_{10}(1 - Load_{DL}I_N 10^{0.1SINR_{PDSCH}}) \tag{6.23}$$

where $Load_{DL}$ is the DL load, I_N is the adjacent cells interference factor, and $SINR_{PDSCH}$ is the required SINR for PDSCH detection.

In practice, the interference margin depends on the planned capacity and coverage, so there is a trade-off between capacity and coverage similar to other cellular technologies. Specifically,

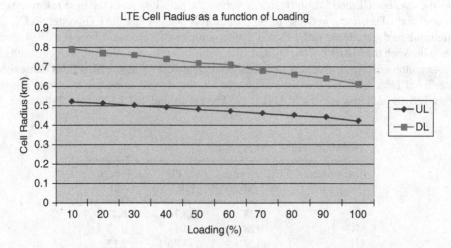

Figure 6.40 Impact of system loading on system radius for the LTE system at 1800 MHz band with urban indoor scenario at 128 kbps/512 kbps cell edge throughputs for UL/DL, respectively.

when the system capacity increases, the interference margin also increases and accordingly the cell radius will be decreased. The orthogonal nature of the LTE allows a smaller provisioning of cell breathing, as shown in Figure 6.40 and the interference margin when compared with WCDMA/HSPA+. Figure 6.40 demonstrates the UL and DL cell radius versus system loading for the LTE system at 1800 MHz band for urban indoor scenario. The cell radius is estimated using a link budget tool similar to Table 6.17 by changing the target load. The interference margin indicates the degradation of system receive performance caused by internal interference in the system due to the system load. In fact, due to the frequency division nature of LTE, there is also a close correlation between actual traffic load and interference margin experienced by the network. Various techniques (e.g., femtocell, small cell (i.e., micro and pico), UL and DL coordinated multi-point, and ICIC) are being adopted to reduce the level of interference over thermal increase, which will further improve the capacity offered by the LTE system. Enhanced interference cancelation techniques, such as dynamic/semi-static ICIC, will also have significant impact on the interference margin reduction. 3GPP introduced also LTE Release 9 that offers improvements to the throughput by introducing the HetNet concept. LTE-Advanced also supports advanced interference management schemes, such as enhanced inter-cell interference coordination (eICIC) and coordinated multipoint transmission (CoMP). LTE-Advanced is designed to improve cell capacity and CET when compared to Release 8 LTE [8].

6.6.5 Comparative Analysis with HSPA+

6.6.5.1 Loading Impact

One of the important factors that may impact the cell radius is the system loading. Unlike the legacy UMTS system, it is not expected that the LTE system will be severely degraded with increase in the system loading thanks to the OFDM technique. Moreover, the LTE system has no intra-cell interference thanks to subcarriers orthogonality. Since the LB has been validated, we can estimate the cell radius based on different loading scenarios to analyze the impact of the loading on the cell radius. Figure 6.40, provides the theoretical DL/UL cell radii versus loading for an LTE system in an urban terrain with the same parameters of the LB in Table 6.17. The figure reconfirms that the LTE is a UL-limited system. The UL cell radius reduction is about 10% at 100% UL load and about 5% at 50% UL load (i.e., a typical practical UL load). This is a very graceful degradation compared to the HSPA+ system illustrated hereunder.

For the sake of comparative analysis, the same exercise is conducted with the HSPA+ system at the 2.1 GHz band (i.e., the most popular band for the UMTS system). Figure 6.41 illustrates the loading impact on the cell radius for the UL and DL of the HSPA+ system. The figure indicates that the DL is the limiting link in HSPA+ as long as the UL loading is less than 90%. The DL cell radius has been decreased by 45% at 100% load while the UL cell radius goes close to 0 (from the CDMA (code division multiple access) theory) at 100% UL load. This illustrates the well-known cell breathing drawback of the UMTS system. To further analyze the cell breathing in the UMTS system, the cell radius is illustrated versus the allocated power per user to guarantee 512 kbps DL throughput at the cell edge. Figure 6.42 demonstrates the cell radius and the number of users with 512 kbps DL throughput at the cell edge versus allocated RF power percentage per user. The figure indicates that the cell radius shrinks as a function of the number of users at the cell edge with 512 kbps throughput and only a maximum of six users can achieve 512 kbps throughput at the cell edge while the cell radius is stretched to 200 m.

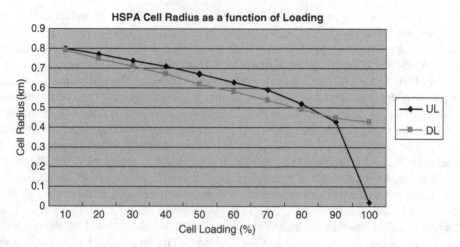

Figure 6.41 .Impact of cell loading on cell radius for HSPA+ at 2.1 GHz in urban indoor scenario at 128 kbps/512 kbps cell edge throughputs for UL/DL, respectively.

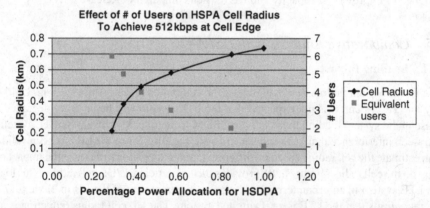

Figure 6.42 Cell radius and number of users with 512 kbps DL throughput at cell edge versus power allocation per user.

Therefore, one native HSDPA carrier (i.e., only HSDPA data) with 40 W power amplifier can serve only six users with 512 kbps at the cell edge with a cell radius of 200 m. This limitation is a major differentiator for the LTE system thanks to the OFDM technique.

The following summarizes the loading impact on the LTE and the UMTS systems:

- Loading has significant impact on UMTS coverage
- About 13% reduction in UMTS cell radius when the loading increases from 50 to 70%
- LTE is more robust against loading and cell radius reduction is only around 2.5% when the loading increases from 50 to 70%.

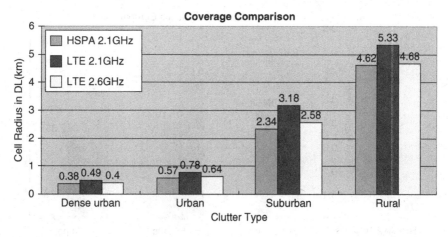

Figure 6.43 DL cell radius comparison at cell edge throughput of 512 kbps, indoor coverage and 90% cell loading.

6.6.5.2 Coverage Comparison between LTE and HSPA+

Cell Radius Comparison

In this section we conduct a comparative analysis between LTE coverage and HSPA+ coverage. Figure 6.43 compares the DL cell radius at different clutters for HSPA+ at 2.1 GHz and LTE at 2.1 and 2.6 GHz. As shown from the figure, the LTE system at 2.1 GHz provides 15–35% better DL coverage based on the clutter type compared to the HSPA+ at the same frequency. Moreover, the LTE at 2.6 GHz provides DL coverage similar to the HSPA at 2.1 GHz. However, HSPA+ at 2.1 GHz has better coverage in the UL compared to LTE 2.6 GHz because of the lower band gain, as shown in Figure 6.44. As a result, LTE at 2.6 GHz can reuse existing 3G sites; no additional LTE standalone sites are required at the beginning of the LTE deployment. However, other constraints on LTE deployment and future expansion may be considered. Therefore, the 2.6 GHz band cannot be considered as the only frequency band for nationwide coverage but rather it can be used as hotspots coverage. Also, some handset vendors opt not to implement 2.6 GHz initially. The LTE deployment at 1800 MHz is the optimum spectrum for LTE nationwide rollout and can be combined with LTE 2.6 GHz for hotspots converge as a capacity layer, or with 800 MHz, in which case the 1800 MHz will be used for converge in dense urban and urban clutters while the 800 MHz will be used mainly for the suburban and rural converge. We will provide coverage prediction scenarios at the end of the chapter at different frequency bands.

Cell-Edge Throughput Comparison

Figure 6.45 demonstrates the DL CET at different cell radiuses for HSPA+ at 2.1 GHz with a 5 MHz carrier and for LTE at 2.6 GHz using 5 and 20 MHz carriers. As shown from the

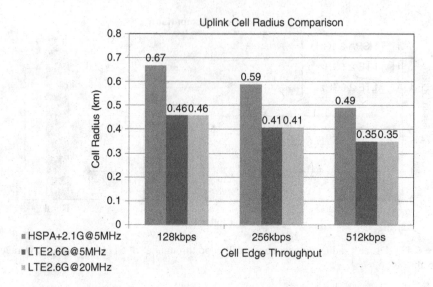

Figure 6.44   UL cell radius at different cell edge throughput for HSPA+ and LTE.

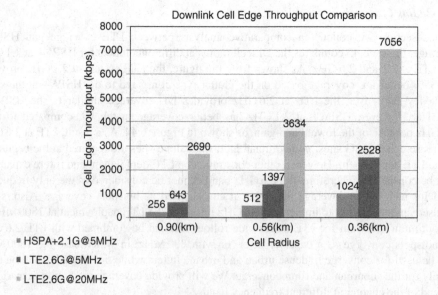

Figure 6.45   DL cell edge throughput at different cell radius for HSPA+ and LTE.

Figure 6.44, the LTE system provides better CET compared to HSPA+. More specifically, the LTE system at 2.6 GHz provides around 50% CET gain compared to HSPA+ at 2.1 GHz at the same channel BW, that is, 5 MHz. Moreover, the LTE at 2.6 GHz with 20 MHz channel BW provides around seven times the HSPA+ CET.

6.6.6 Link Budget for LTE Channels

In the previous sections of link budget, we focused on the shared traffic channel in the DL and UL (i.e., PDSCH, PUSCH). In this section we further analyze the link budget by comparing the link budgets for different LTE channels and identifying the limiting channel in the UL and DL. The PUSCH is included in this analysis as it was the limiting link as highlighted before. Table 6.27 provides the LTE channels that are considered in this analysis. Table 6.28 summarizes the common parameters used in the link budget analysis and comparison.

Table 6.29 provides the characteristics of the different clutters, including indoor penetration loss at different frequency bands, log normal fading (LNF) standard deviation and cell edge coverage reliability.

The eNB antenna gain and UE antenna gain at different frequency bands are provided in Table 6.30.

Figures 6.46 and 6.47 illustrate the maximum allowable path loss and the corresponding cell radius, respectively, for the LTE channels with different clutter types.

Table 6.27   Major LTE channels for link budget

LTE channels for LB	
Downlink	PDCCH
	RS
	PBCH
Uplink	PUSCH
	PRACH

Table 6.28   LTE parameters for the link budget comparison

Parameter	Value	Units
LTE frequency band	1800	MHz
Cell edge throughput	500	kbps
LTE channel bandwidth	20	MHz
LTE downlink power per cell	47.78	dBm
UE transmission power	23	dBm
Interference margin	2	dB
PRACH format	Format 0	
Number of Rx antennas	2	
Body loss	3	dB
eNB jumper loss	0.5	dB
UE jumper loss	0	dB
eNB antenna height	30	m
UE height	1.5	m
eNB noise figure	2.5	dB
UE noise figure	7	dB

Table 6.29 Indoor penetration loss, converge reliability, and LNF of different clutters

Indoor penetration loss (dB)				
Band/clutter (MHz)	DU	U	SU	R
700	15	15	12	6
1800	18	18	14	7
2600	24	24	19	11
Clutter characteristics				
Item/clutter	DU	U	SU	R
Cell edge reliability (%)	95	95	90	90
Log normal fading standard deviation	7.7	7.7	6.7	6.7

Table 6.30 Antenna gain of eNB and UE at different frequency bands

Band/antenna gain (MHz)	eNB antenna gain	UE antenna gain
700	15	−5
1800	17	−3
2600	21	0

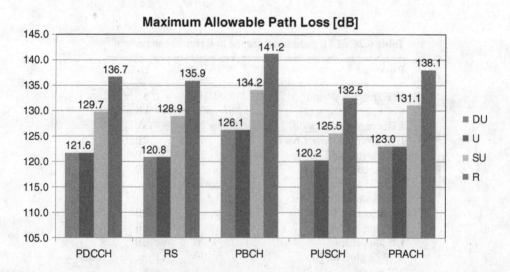

Figure 6.46 Maximum allowable path loss for LTE channels at different clutter type.

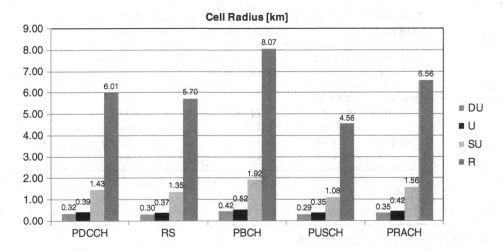

Figure 6.47 Maximum allowable path loss for LTE channels at different clutter type.

Table 6.31 Link budget results for LTE channels at DU clutter with the same parameters as Table 6.27 except the cell edge throughput is 256 kbps

LTE channels for LB		Maximum allowable path loss (dB)	Cell radius (km)
Downlink	PDCCH	121.64	0.39
	RS	**120.82**	**0.37**
	PBCH	126.14	0.52
Uplink	PUSCH	124.25	0.46
	PRACH	122.98	0.42

The figures demonstrate that the PUSCH is the limiting link for this scenario as it yields the smallest MAPL in all scenarios and, therefore, the shortest cell radius. This is in line with the same results proved earlier in this chapter. Nevertheless, if we consider 256 kbps CET, Table 6.31 summarizes the results of the link budgets for LTE channels in Table 6.27. Table 6.31 demonstrates that the limiting link is the DL and the limiting channel is the RS channel. This is because the RS channel link budget considers the entire 20 MHz in the DL while the PDCCH considers the effective BW corresponding to 100 RB, which is 18 MHz. Also, in the UL the PRACH became the limiting channel in the UL rather than the PUSCH. Therefore, it is important to consider all LTE channels during the link budget for accurate coverage prediction.

To further analyze this topic, Table 6.32 provides the link budgets for LTE channels with the same parameters as in Table 6.28 except for the frequency band of 700 MHz. Surprisingly, the RS is the limiting link in this scenario as well, even if the CET is increased up to 1 Mbps, as seen from the table.

It is interesting now to analyze the limiting link and the limiting channel for the LTE system at different frequency bands, at different clutters, and for different CET. Table 6.33 provides

Table 6.32 Link budget results for LTE channels at DU clutter with same parameters of Table 6.28 except the frequency is 700 MHz

LTE channels for LB		Maximum allowable path loss (dB)	Cell radius (km)
Downlink	PDCCH	120.64	0.68
	RS	**116.82**	**0.53**
	PBCH	125.14	0.91
Uplink	PUSCH	125.25	0.92 (0.58[a])
	PRACH	123.98	0.84

[a](0.58) is the cell radius of PUSCH at 1 Mbps UL throughput.

Table 6.33 LTE LB at different bands, different cell edge throughput, and different clutters

LTE LB at UL cell edge TP of 256 kbps				
Band/clutter (MHz)	DU	U	SU	R
700	0.53 (RS)	0.64 (RS)	1.88 (RS)	6.58 (RS)
1800	0.3 (RS)	0.37 (RS)	1.35 (RS)	5.7 (RS)
2600	0.21 (PRACH)	0.26 (PRACH)	1.01 (PUSCH)	4.81 (PUSCH)
LTE LB at UL cell edge TP of 512 kbps				
Band/clutter (MHz)	DU	U	SU	R
700	0.53 (RS)	0.64 (RS)	1.88 (RS)	6.58 (RS)
1800	0.29 (PUSCH)	0.35 (PUSCH)	1.08 (PUSCH)	4.56 (PUSCH)
2600	0.18 (PUSCH)	0.22 (PUSCH)	0.77 (PUSCH)	3.59 (PUSCH)
LTE LB at UL Cell Edge TP of 1024 kbps				
Band/clutter (MHz)	DU	U	SU	R
700	0.53 (RS)	0.64 (RS)	1.71 (PUSCH)	6 (PUSCH)
1800	0.24 (PUSCH)	0.29 (PUSCH)	0.89 (PUSCH)	3.75 (PUSCH)
2600	0.15 (PUSCH)	0.18 (PUSCH)	0.63 (PUSCH)	3.03 (PUSCH)

the link budget for LTE at different frequency bands (700 MHz, 1800 MHz, and 2.6 GHz) with four different clutters (DU, U, SU, and R). The parameters in Tables 6.28–6.30 are used with this exercise. Also, three CET targets are considered, that is 256 kbps, 512 kbps, and 1024 kbps.

The outcome from this exercise is outlined as follows:

1. Neither the limiting link nor the limiting channel is consistent in LTE.
2. At the lower frequency band, the LTE system tends to be DL limited and the RS is the limiting channel. The LB for 800 MHz band may be slightly deviated where the DL and UL frequency order is reversed. Specifically, the lower frequency is used for DL while the higher frequency for UL.
3. At the higher frequency band, the LTE system tends to be UL limited and the PUSCH is the limiting link.
4. The target CET is a key parameter to determine the limiting link and the limiting channel.
5. At aggressive UL CET the LTE system tends to be UL limited and PUSCH is the limiting channel.
6. At relaxed UL CET, the LTE system tends to be DL limited and RS is the limiting channel.

Table 6.34 PRACH link budget for different formats and different clutters

PRACH format	DU	U	SU	R[a]
Format 0	0.35	0.42	1.56	9.79
Format 1	0.38	0.42	1.54	9.65
Format 2	0.39	0.48	1.78	11.24
Format 3	0.4	0.48	1.79	11.32

[a]eNB height is 60 m for rural clutter while with other clutters the eNB height is 30 m.

With regard to the PRACH, and as highlighted before, there are four formats of the PRACH and these formats impact the link budget of the PRACH. Table 6.34 summarizes the cell radius for PRACH with the different formats and at different clusters with the same parameters of Table 6.28. As shown in Table 6.34, the higher format extends the LTE converge; however it is not up to the maximum cell radius range provided in Table 6.9. Therefore, an extended cell radius with higher PRACH format 1–3 should consider the link budget for PRACH as well as other LTE channels.

6.6.7 RF Propagation Models and Model Tuning

For LTE RF coverage prediction using a planning tool, it is important to select the appropriate propagation model. Also, we have to tune the model for a specific region and specific clutter and for an LTE frequency band if the selected model does not cover the LTE band. There are many propagation models and different models have been developed to meet the needs of realizing the propagation behavior in different conditions. Each propagation model is valid in a specific scenario and specific frequency. If the model is not chosen correctly, the model will either overestimate or underestimate the path loss and accordingly the predicted coverage. If the model overestimate the path loss, that means the user may need lower signal energy to reach services. Underestimating the path loss means the user may need more power, also inter-cell interference will not be predicted correctly. So path loss prediction should be as accurate as possible. Radio propagation models are classified as outdoor and indoor propagation models. In this section, we will analyze the outdoor models. The propagation models can also be divided into empirical models and deterministic models as follows [9]:

1. **Deterministic Models (Theoretical Models)** The deterministic model is based on theoretical analysis, so it can be applied in different scenarios without affecting its accuracy. However, the realization of a theoretical model is based on a large database of scenario features and an accurate 3D digital map, which sometimes is impractical or even impossible [3].
2. **Empirical or Semi-empirical Models (Statistical Model)** Empirical models based on extensive measurement data. In empirical models, all the environment factors are implicitly considered.

We will summarize the main RF propagation models that are currently used with LTE path loss prediction. All the propagation models mentioned below are empirical in nature, model coefficients are developed based on field measurements. Measurements are typically performed in the field to measure path loss, delay spread, or other channel characteristics. For accurate prediction, it is necessary to calibrate/tune these coefficients based on actual field measurements for a specific area, clutter, and frequency.

6.6.7.1   Okumura–Hata Related Models

1. **Okumura Model:** The Okumura method is semi-empirical and based on extensive measurements performed in the Tokyo area [10]. It is a pure experience statistical model, so its statistics are represented by curves without a specific formula.
2. **Okumura–Hata:** The Okumura model was intended for manual use. Hata, in 1980 [11], derived semi-empirical formulas from Okumura's curves for computational use. The Okumura–Hata model applies well for large cells. In the configuration of large cells, the antenna of the base station is usually higher than the surrounding buildings or obstacles. The main propagation loss for the Okumura–Hata model is the diffraction and scattering over rooftops near the mobile station. This model can be applied to the following scenario:
   (a) Frequency range 400–1500 MHz
   (b) Terrain morphologies (DU, U, SU, and RU)
   (c) Mobile antenna height from 1 to 10 m
   (d) BTS antenna height from 30 to 200 m
   (e) Cell radius 1–20 km (macro sites).
3. **The COST231 Hata Model:** COST is a European Union forum for cooperative scientific research. The COST231 group extended the studies of Okumura–Hata. The Okumura–Hata propagation model works for frequencies below 1500 MHz and thus does not work, for example, for the 2100 MHz band. Okumura's propagation curves have been analyzed in the upper frequency band to find a suitable expression for 2100 MHz [12]. Therefore COST231–Hata, which is an empirical model, extends the frequency range of the Hata model from 1500 to 2000 MHz. The path loss of the COST231–Hata model is estimated as follows:

$$PL = -46.3 - 33.9 \log(f) - 13.82 \log(h_{BS}) - a(h_{UE})$$
$$- [44.9 - 6.55 \log(h_{BS})] \log(d) - C_0 \qquad (6.24)$$

where f is the carrier frequency in MHz, h_{BS} is the base station antenna height in meters, h_{UE} is the mobile station height in meters, and d is the cell radius in km. The terms $a(h_{UE})$ and C_0 are used to account for different trains. The term $[44.9 - 6.55 \log(h_{BS})]$ is the slope in dB/decade. The slope is a factor indicating how severe the loss becomes as a function of distance from the base station. Therefore, the path loss can be defined in a general form as follows:

$$PL = -PL_0 - s \cdot \log(d) \qquad (6.25)$$

where PL_0 is the intercept and s is the slope. Typical values for the slope and intercept at different morphologies and at the main LTE bands are shown in Table 6.35. These values are obtained by considering the eNB height equal to 35 m and the UE height to 1.5 m. Moreover, the Okumura–Hata model path loss versus distance from the eNB (d) is shown for low bands (700 MHz, 800 MHz) in Figure 6.48 and the COST231–Hata path loss versus the distance from eNB for higher bands (1800 MHz, 2.1 GHz, and 2.6 GHz) is shown in Figure 6.49. The cell radius in Figures 6.48 and 6.49 is chosen in two different ranges (100 km : 1 km) with 100 m steps and (1 km : 3 km) with 1 km steps to highlight the dense urban, urban, and suburban scenarios. In this context, typical formulas for the slope and intercept of Hata and COST231–Hata models at different clutters are summarized in Table 6.36 [13–15].

4. **Ericsson 9999 model:** There have been several suggestions to improve the Okumura–Hata and COST231–Hata models to take more account of the propagation environment. One of the improved Okumura–Hata models is the Ericsson 9999 model [9] that adds clutter adjustments for other morphology types. Even if the validity of the model here is stated to be up to 2 GHz, it is possible to adapt Ericsson 9999 to higher frequency bands by tuning the model against measurements. This model can be applied to the following scenario:

   (a) Frequency range 150 MHz to 2 GHz and it can be extended beyond 2 GHz via model tuning
   (b) Terrain morphologies (DU, U, SU, and RU)
   (c) Mobile antenna height from 1 to 5 m
   (d) BTS antenna height from 20 to 200 m
   (e) Cell radius from 0.2 to 100 km (macro sites).

   This model was developed by Ericsson and it has considered all terrain types and covers a wide range of all other factors. The PL of this model can be estimated by the following equation:

$$PL = A_0 + A_1 \cdot \log(d) + A_2 \cdot \log(h_{eNB}) + A_3 \cdot \log(h_{eNB}) \log(d)$$
$$- 3.2(\log(11.75.h_{UE}))^2 + g(f)$$

where $g(f) = 44.49.\log(f) - 4.78(\log(f))^2$, h_{eNB} and h_{UE} are the effective height of the eNB antenna and the UE, respectively, and f is the frequency in MHz. The parameters A_0, A_1, A_2, and A_3 are constants, which can be tuned for better fitting of specific propagation conditions. Default values are: $A_0 = 36.2$, $A_1 = 30.2$, $A_2 = -12.0$, and $A_3 = 0.1$. An example of a tuned version of the Ericsson 9999 model at 900 and 1800 MHz is shown in Table 6.37.

   Moreover, it has been shown in [16], that Ericsson the 9999 model with default parameters provides the optimal prediction results for WiMAX (worldwide interoperability for microwave access) at 3.5 GHz in a LOS (line-of-sight) scenario compared to other models such as COST231–HATA. Therefore, it is possible to extend this model to a higher frequency band such as 3.5 GHz, which is a candidate spectrum for LTE-TDD for fixed wireless broadband access and also as an offloading layer with hetNet structure.

5. **Standard Propagation Model:** The standard propagation model (SPM) is based on empirical formulas and a set of parameters that are set to their default values. SPM is a model (deduced from the Hata formula) particularly suitable for prediction in the 150–3500 MHz band over a long distance (1 km $< d < 20$ km) and can be used with different technologies, such as GSM900/1800, UMTS, CDMA2000, WiMAX, and LTE. This model uses the terrain profile, diffraction mechanisms (calculated in several ways), and can take into account clutter classes and effective antenna heights in order to calculate the path loss. This model offers several additional features to improve its flexibility and accuracy, such as the

Table 6.35 Slope and intercept at main LTE bands and for different clutters

Propagation model	Frequency band (MHz)	Clutter	Slope (dB)/ decade	Intercept
Okumura–Hata	700	Dense urban	34.8	125.64
Okumura–Hata	700	Urban	34.8	122.63
Okumura–Hata	700	Suburban	34.8	113.32
Okumura–Hata	700	Rural	34.8	100.15
COST231–Hata	1800	Dense urban	34.8	138.32
COST231–Hata	1800	Urban	34.8	135.27
COST231–Hata	1800	Suburban	34.8	123.33
COST231–Hata	1800	Rural	34.8	108.35
COST231–Hata	2600	Dense urban	34.8	143.73
COST231–Hata	2600	Urban	34.8	140.67
COST231–Hata	2600	Suburban	34.8	127.53
COST231–Hata	2600	Rural	34.8	111.58

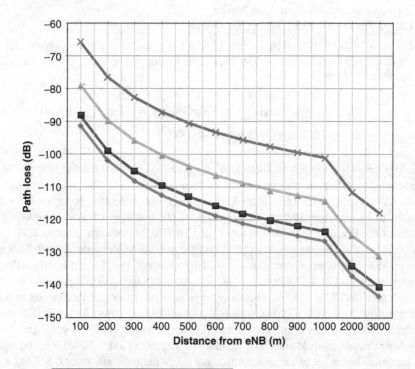

Figure 6.48 The Okumura–Hata model path loss versus distance from the eNB for low bands.

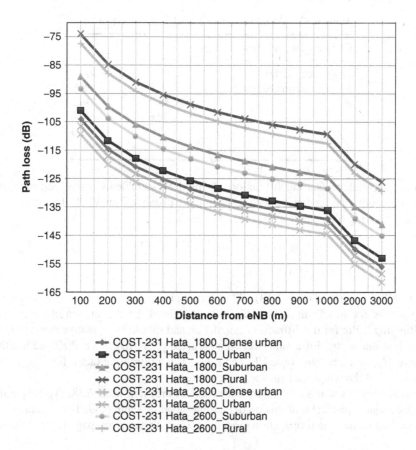

Figure 6.49 COST231–Hata path loss versus the distance from eNB for higher bands.

Table 6.36 Slope and intercept of Hata and COST231–Hata models

Path loss model	Slope	Intercept
Free space	20	$20\log(f) + 32.44$
Hata (rural)	$44.9 - 6.55\log(h_b)$	$33.6 + 7.8\log(f) - 13.8\log(h_{eNB}) - 4.8[\log(f)]^2$ $- 3.2[\log(11.8 h_{UE})]^2$
Hata (suburban)	$44.9 - 6.55\log(h_b)$	$59.2 + 26.2\log(f) - 13.8\log(h_{eNB}) - 3.2[\log(11.8\,h_{UE})]^2$ $- 2[\log(f/28)]^2$
Hata (urban)	$44.9 - 6.55\log(h_b)$	$68.8 + 26.2\log(f) - 13.8\log(h_{eNB}) - [1.1\log(f) - 0.7]$ $h_{UE} + 1.6\log(f)$
Hata (dense urban)	$44.9 - 6.55\log(h_b)$	$74.5 + 26.2\log(f) - 13.8\log(h_{eNB}) - 3.2[\log(11.8\,h_{UE})]^2$
Cost231 (rural)	$44.9 - 6.55\log(h_b)$	$10.3 + 15.6\log(f) - 13.8\log(h_{eNB}) - 3.2[\log(11.8\,h_{UE})]^2$ $- 4.8[\log(f)]^2$
Cost231 (suburban)	$44.9 - 6.55\log(h_b)$	$45.9 + 33.9\log(f) - 13.8\log(h_{eNB}) - 3.2[\log(11.8\,h_{UE})]^2$ $- 2[\log(f/28)]^2$
Cost231 (urban)	$44.9 - 6.55\log(h_b)$	$45.5 + 33.9\log(f) - 13.8\log(h_{eNB}) - [1.1\log(f) - 0.7]h_{UE}$ $+ 1.6\log(f)$
Cost231 (dense urban)	$44.9 - 6.55\log(h_b)$	$54.3 + 33.9\log(f) - 13.8\log(h_{eNB}) - 3.2[\log(11.8\,h_{UE})]^2$

Table 6.37 Tuned version of model 9999 for urban city

Calibrated model for urban city		
Frequency (MHz)	1800	900
Mobile antenna height (m)	1.5	1.5
A_0	36.2	36.2
A_1	30.6	30.6
A_2	−12.0	−12.0
A_3	0.1	0.1

inclusion of clutter offset and diffraction. The model is suitable for macro-cell environments and can incorporate dual-slope with respect to the distance from eNB, if needed.

The following formula describes the SPM model

$$L_{SPM} = K1 + K2\log(d) + K3\log(H_{Tx}) + K4\text{DiffLoss} + K5\log(d)\log(H_{Rx})$$
$$+ K6\log(H_{Rx}) + K_{clutter}F_{clutter} \tag{6.26}$$

where $K1$ is a constant offset (dB), $K2$ is a multiplying factor for log(d), d is the distance between the Rx and Tx in meters, H_{Tx} is the height of the Tx antenna in meters, $K4$ is a multiplying factor for the diffraction calculation and should be a positive number, DiffLoss is the loss due to the diffraction over an obstructed path (dB), $K5$ and $K6$ are multiplying factors, H_{Rx} is the height of the UE, and $K_{clutter}$ is a multiplying factor for $F_{clutter}$ which is an average of the weighted loss due to the clutter.

The $K1$ factor depends on the frequency band, as shown in Table 6.38. Typical parameters and the ranges for $K2:K6$ of this model are provided in Table 6.39. These parameters can be adjusted to tune the propagation model according to actual propagation conditions.

Table 6.38 $K1$ factor of the standard propagation model

Frequency (MHz)	900 (GSM900)	1800 (GSM/LTE)	1900	2100 (UMTS)
$K1$	12.5	22	23	23.8

Table 6.39 Standard propagation model parameters range

Constant	Minimum	Typical	Maximum
$K2$	20	44.9	70
$K3$	−20	5.83	20
$K4$	0	0.5	0.8
$K5$	−10	−6.55	0
$K6$	−1	0	0

6.6.7.2 B-Walfisch–Ikegami Related Models

1. *COST231–Walfisch-Ikegami Model (COST231–WIM)*: This model is adopted for urban areas with a large population and densely located buildings where the cell radius is usually smaller than 1 km due to the capacity requirement. The error in using the Hata model in such micro cells is large. To enable the Hata model to apply for areas with densely located high buildings, Cost231 proposes the COST231–WIM based on numerous onsite tests and model analysis [17]. COST231–WIM is a combination between the two models described in the Ikegami model and the Walfisch–Bertoni model, including a LOS component. In urban areas with a large population and densely located buildings, the eNB antenna is usually higher than the average height of the surrounding buildings but lower than the tallest building. The COST231–WIM, based on theoretical Walfisch–Bertoni model [9, 17] calculates the multiple screen forward diffraction loss of antenna of a high base station. It uses the test-based data for antenna of a low base station. Therefore, this model can be applied to the following scenario:

   (a) Frequency range 800 –2000 MHz
   (b) Terrain morphologies (DU and U)
   (c) Mobile antenna height from 1 to 3 m
   (d) BTS antenna height from 4 to 50 m
   (e) Cell radius 100 m to 5 km (mainly for micro cells).

2. *Advanced WIM*: There have been several suggestions for improving the WIM to take more propagation environment into account. This model works in the same physical conditions as the COST231–WIM described before. Like COST231–WIM, the advanced WIM works best for uniform building separation and building heights and flat ground. It requires height information for the calculation of effective antenna height and knife-edge diffraction [9].

In order to highlight the impact of the model type on the RF prediction, we have conducted RF prediction (using DL RSRP) for a certain LTE 1800 cluster with four different models. Figures 6.50–6.53 demonstrate the RF prediction using the tuned model, Okumura urban model, COST231 urban/suburban model, and SPM model with the parameters in Table 6.39. The tuned model provides better coverage compared to the non-tuned models. The non-tuned models illustrate shrinking in the coverage compared to the tuned model and, therefore, if such models are adopted in the RF prediction, they will lead to a greater number of sites, which is not needed. This clearly indicates the importance of the model tuning for coverage prediction and accordingly for the whole RF coverage planning exercise during the initial planning phase.

As demonstrated above, the majority of RF propagation models are empirically based on field measurements under certain geographic conditions, such as clutter type and terrain elevation. Therefore, these default models engender major deviations from the actual models for new clutters and terrains. This leads to prediction errors in the path loss calculations and accordingly major deviation in the RF prediction and the number of required sites. Moreover, such prediction errors will lead to wrong assumptions in the initial planning phase and accordingly the number of sites that are needed to cover the target area and hence increase inter-cell interference or gaps in coverage. The non-tuned model may overestimate or underestimate the path loss and hence the RF coverage.

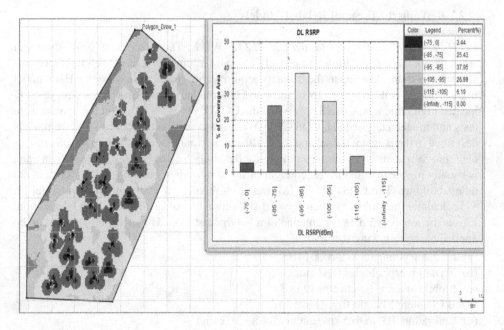

Figure 6.50   DL RSRP prediction using tuned model.

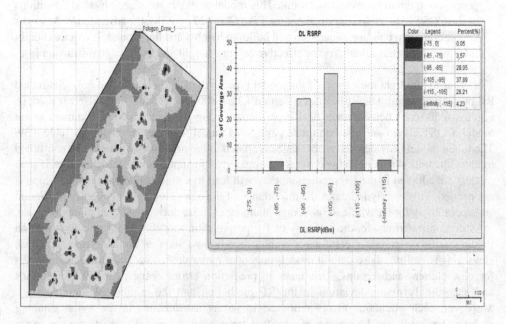

Figure 6.51   DL RSRP prediction using Okumura urban model.

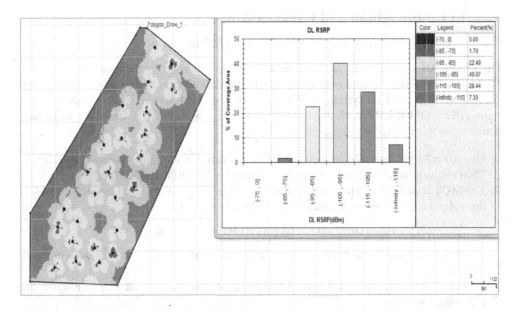

Figure 6.52   DL RSRP prediction using COST231 urban/suburban model.

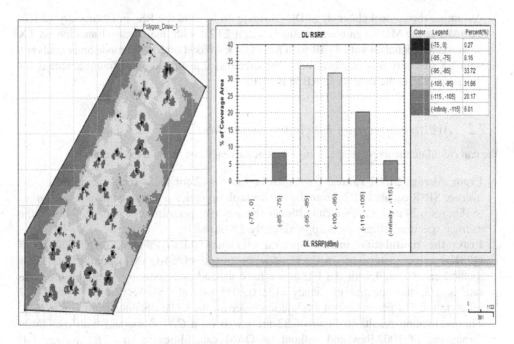

Figure 6.53   DL RSRP prediction using SPM model with the parameters in Table 6.39.

6.7 LTE Throughput and Capacity Analysis

6.7.1 Served Physical Layer Throughput Calculation

The DL and UL capacities (achievable throughput) are impacted by the total BW (1.4, 3, 5, 10, 15, and 20 MHz), the total overheads, and the spectral efficiency, which is determined by the DL SINR. In the LTE system, the UL capacity is divided between control channels and signals (SRS, DM-RS, PRACH) and the traffic channel, that is, PUSCH. Based on the RE/RB definition in this chapter, we can define the following:

1. One RB = 84 RE, one subframe contains 168 RE, and one subframe contains two RB.
2. Then one subframe = 12 (subcarriers) × 7 symbols × 2 (Slot) = 168 RE (normal CP)
3. The MCS is defined as MCS = CodeRate × CodeBits, then the served physical layer UL throughput can be estimated based on the following formula

$$\text{TP}_{\text{UL}}(\text{kbps}) = (1 - \text{BLER}) \times (168 - 24) \times \text{MCS} \times N_{\text{UL}}^{\text{RB}} \tag{6.27}$$

assuming the control channels/signals in the UL consumes 24 RE and $N_{\text{UL}}^{\text{RB}}$ is the number of RB in the UL.
4. Similarly, the served physical layer DL throughput can be calculated according to the following formulas:

$$\text{TP}_{\text{DL}}(\text{kbps}) = (1 - \text{BLER}) \times (168 - 36 - 12) \times \text{MCS} \times N_{\text{DL}}^{\text{RB}} \times \text{MG} \tag{6.28}$$

Assuming the control channels in DL consume 36 RE and the RSs in DL occupy 12 RE. The MG is the MIMO gain and it equals = 2, if 2T2R with dual stream transition on TX antennas and equals 4 with 4T4R with MIMO 4 × 4 based on MIMO mode and equals = 1 with TX diversity modes where only one stream is transmitted on different antennas for diversity. $N_{\text{DL}}^{\text{RB}}$ is the number of RB in the DL.

6.7.2 Average Spectrum Efficiency Estimation

We can calculate the average spectrum efficiency using three methods as follows:

1. **From average SINR values:** As shown in Table 6.12 for the demonstrated cluster, the average SINR equals 13.77. From the vendor look-up-tables or curves similar to the curves in Figures 6.24 and 6.25, we can estimate the average spectrum efficiency as follows: the average spectrum efficiency is approximately 2.8 bits/s/Hz.
2. **From the modulation order:** Spectral efficiency (bits/symbol) = [2 × percentage of QPSK + 4 × percentage of 16-QAM + 6 × percentage of 64-QAM] × coding rate. This method can be used with the UL as we have a fixed code rate, by assuming the code rate is 1/3, the spectral efficiency = [2 × 0.1015 + 4 × 89.15 + 6 × 0] × 100 × 1/3 = 1.27 bits/symbol. The percentage of modulation order for the UL is obtained from Figure 6.32. The modulation results in Figure 6.32 are based on a CAT 3 modem with maximum throughput of 100 Mbps and without 64 QAM capabilities in UL. The average UL spectral efficacy is above the target UL spectral efficiency of 1.2, which is the baseline for LTE-Advanced.
3. **From MCS values:** We can estimate the average DL efficiency from the MCS PDF by taking the MCS percentile from Table 6.15 and the spectral efficiency from Table 6.13, the average spectral efficiency can be estimated as follows:

(a) From Table 6.13, the average spectral efficiency of QPSK modulation = 0.5697, the average spectral efficiency of 16QAM modulation = 1.9323, the average spectral efficiency of 64QAM modulation = 4.1914,

(b) Then, the average DL spectral efficiency based on the actual percentile of each MCS = $0.5697 \times (\text{sum} (\text{MCS0:MCS9}) + \text{MCS29}) + 1.9323 \times (\text{sum} (\text{MCS10:MCS16}) + \text{MCS30}) + 4.1914 \times (\text{sum} (\text{MC17:MCS28}) + \text{MCS31}) = 2.8878$.

The achieved average DL spectral efficiency (2.8) based on the mobility test results for the illustrated practical cluster with the RF parameters in Table 6.12, is better than the baseline spectral efficiency estimated in [18] and substantially exceeds the target 1.53 bit/s/Hz per sector in DL set by 3GPP at the start of the LTE design [8], and it is even above the target average user spectral efficiency for LTE-Advanced of 2.4. It is worth noting that this cluster is lightly loaded and therefore the spectral efficiency is expected to be degraded with a loaded scenario.

6.7.3 Average Sector Capacity

The average sector throughput is an important factor in the LTE dimensioning capacity. The available channel bandwidth for PDSCH is calculated as follows:

1. Available RB for PDSCH = total RB − fixed overhead − control channel overhead − paging channel overhead.
2. Available symbols for PDSCH = available RB for PDSCH × 12 × number of symbols per subcarrier.
3. Total PDSCH throughput capacity per sector-carrier is calculated as: the available symbols for the PDSCH × the median spectral efficiency.

The number of subcarriers depends on the system BW (i.e., $1.4 \to 72$, $3 \to 180$, $5 \to 300$, $10 \to 600$, $15 \to 900$, $20 \to 1200$). The number of DL symbols is 7 or 6 according to the normal or extended CP, respectively. The total number of RE is the number of subcarriers according to the BW multiplied by twice the number of symbols (7 or 6).

Therefore, the average sector throughput at a certain DL loading can be calculated as follows:

Average sector throughput = available PDSCH RB% × DL loading% × average spectral efficiency × number of subcarriers per symbol × number of symbols × number of total RB/1000 = xx Mbps.

If we consider a sector with 20 MHz channel bandwidth, then the average sector throughput according to the previous formula equals $0.8 \times 0.80 \times 2.8 \times 12 \times 14 \times 100/1000 = 30$ Mbps. Although, the average user throughput is higher than this value but it is better to use 30 Mbps as average sector throughput in dimensioning to count for loading, interference margin, and other users on cell edge which will decrease the average sector throughput.

6.7.4 Capacity Dimensioning Process

The LTE capacity dimensioning process consists of the following steps:

1. Traffic profile to determine the target capacity. The traffic profile consists of the following inputs:

(a) Number of LTE subscribers for data and voice

(b) Type of services

(c) Traffic usage per subscriber for voice (mErlang) and data (kbps)

(d) Activity factor for voice and data.

2. Estimate the average sector throughput. This can be obtained via simulation or based on field measurement.

3. Calculate the number of eNB/sectors needed to cater for the total traffic demand.

An alternative way is for the startup operator to provide the total number of eNB (outdoor and indoor) needed to meet the coverage targets and the designer will then provide the maximum number of subscribers that the system can support based on the provided traffic profile. Table 6.40 outlines a typical traffic profile for a commercial LTE network dimensioning.

The end-to-end network dimension process is summarized in the form of a flowchart in Figure 6.54. The flowchart illustrates the capacity and converge planning. It also indicates

Table 6.40   Traffic profile for LTE network dimensioning

#	Item	Value	Unit	Comments
1	Number of outdoor sites with 1/1/1 configuration	2000	Site	Number of macro sites
2	Equivalent number of indoor sites with 1/1/1 configuration	500	Site	This will be estimated by dividing all indoor site sector over 3
3	Data subscriber active factor including overbooking	0.2		Only 20% active and other 80% only online
4	Active factor for VOIP subscriber including overbooking	0.1		Equal to VOIP traffic per sub, that is, 100 mErlang
5	DL throughput per sub at busy hour	150	kbps	DL throughput of data subscriber at busy hour
6	UL throughput per sub at busy hour	50	kbps	UL throughput of data subscriber at busy hour
7	VOIP traffic per subscriber	60	kbps	VOIP subscriber traffic
8	Average DL throughput per site	90	Mbps	Simulation results for FDD LTE 1800 with 20 MHz BW and 1/1/1 configuration at urban clutter
9	Average DL throughput per sector	30	Mbps	Simulation results for FDD LTE 1800 with 20 MHz BW and 1/1/1 configuration at urban clutter
10	Average UL throughput per site	48	Mbps	Simulation results for FDD LTE 1800 with 20 MHz BW and 1/1/1 configuration at urban clutter
11	Average UL throughput per sector	16	Mbps	Simulation results for FDD LTE 1800 with 20 MHz BW and 1/1/1 configuration at urban clutter
12	S1 with overhead	1.3		To consider the overhead at S1 interface
13	S1 overbooking	20		To be considered in dimensioning the SGW capacity

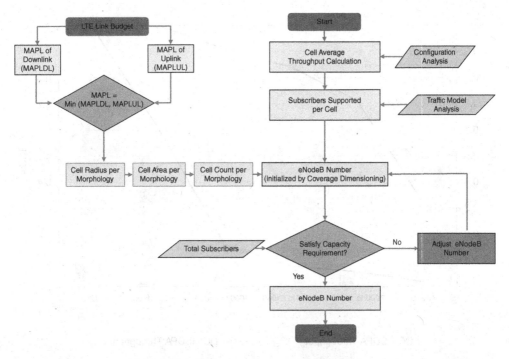

Figure 6.54 LTE network dimensioning flowchart.

how we can estimate the cell radius, and the number of eNB to meet the coverage and capacity requirement.

As highlighted before, the most important factor during the LTE capacity dimensioning is the average sector throughput in the UL and DL. In this exercise we have estimated the average throughput for the LTE system based on a commercially deployed LTE network with CAT 3 UE. Figure 6.55 shows the CDF and PDF of DL throughput performance of allocated LTE/DC-HSDPA systems from commercially deployed networks. This cluster is a mix between dense urban, urban, and suburban clusters. The average throughput is estimated to be around 33 Mbps over the entire route, with mobility at 80 km/h. Without losing the generality, we assumed the average sector throughput equals the average user throughput (i.e., 33 Mbps). To be more conservative, we will consider 30 Mbps as the average sector throughput, which is in line with earlier recommendation. It is worth mentioning that, the average mobility throughput of a lightly loaded and optimized LTE network in good RF conditions can hit 50 Mbps. Figure 6.56 provides the CDF and PDF of the mobility DL throughput for the urban clutter with the RF conditions shown in Table 6.12. The average DL throughput in this clutter is around 50 Mbps. Figure 6.57 provides the CDF and PDF for the UL throughput in the same clutter. The average UL throughput in this clutter is around 35 Mbps. Nevertheless, we assumed the average UL throughput of 16 Mbps in the traffic profile provided in Table 6.40. This is to account for a loaded scenario with inter-cell interference. The achieved 35 Mbps mobility throughput came as a result of a single user in the cluster with very high output power, which is not the typical case with a multi-user scenario in a loaded system. Typical average sector throughputs of an LTE system at different clutters and different BWs are shown in Table 6.41.

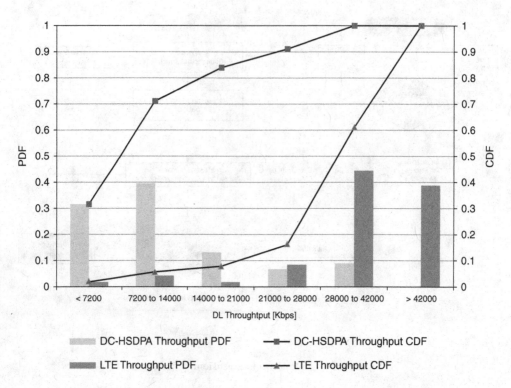

Figure 6.55   LTE/DC-HSDPA DL throughputs in DU, U, and SU clusters.

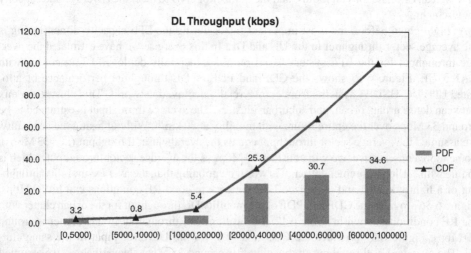

Figure 6.56   PDF and CDF of DL mobility throughput for SU cluster with RF parameters in Table 6.12.

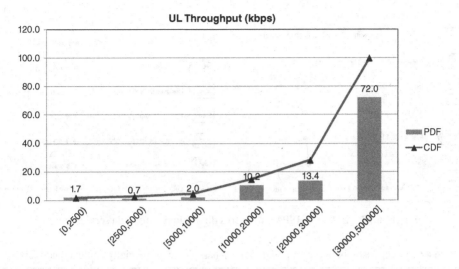

Figure 6.57 PDF and CDF of UL mobility throughput for SU cluster with RF parameters in Table 6.12.

Table 6.41 Average DL/UL throughputs for LTE at urban and suburban clutter

Bandwidth (MHz)	Scenario	Cell average throughput	
		DL (Mbps)	UL (Mbps)
5	Urban	8	5
	Suburban	6	3
10	Urban	17	10
	Suburban	13	7
15	Urban	25	15
	Suburban	20	10
20	Urban	33	20
	Suburban	26	14

6.7.5   *Capacity Dimensioning Exercises*

6.7.5.1   **LTE Dimensioning Based on Data Service Only**

In this section, we will initially conduct a simple dimensioning exercise for a data only LTE network without voice. Figure 6.58 shows the LTE dimensioning exercise and a comparison with a DC-HSPA+ system. The key input to the dimensioning exercise is the average sector throughput and the outcome is the number of subscribers based on the provided traffic profile. The average sector throughput of LTE is 33 Mbps, as estimated in the previous section. The average sector throughput of the HSPA+ system is estimated to be 12.3 Mbps based on collocated LTE/HSPA+ sites as depicted in Figure 6.55. A complete comparative comparison between the collocated LTE and HSPA+ systems with DL throughputs demonstrated in

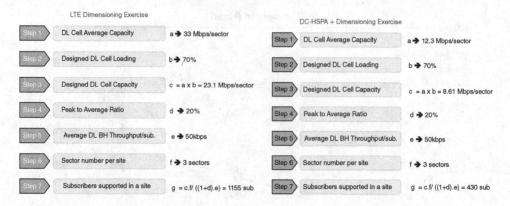

Figure 6.58   LTE and HSPA+ networks dimensioning for data services only.

Table 6.42   Comparison between LTE and HSPA+ based on commercially deployed networks

Criteria	DC-HSPA+ (2.1 GHz)	LTE (1800 MHz)
Mobility average throughput	9 Mbps with DC (2×5 MHz)	33 Mbps with 20 MHz channel BW
Average scheduling rate (%)	73[a]	100
Normalized mobility average throughput	12.3 Mbps with DC (2×5 MHz)	Same as above
Mobility spectrum efficiency	1.23 (i.e., 12.3 Mbps/10 MHz)	1.65 (i.e., 33 Mbps/20 MHz)
Throughput (%)	2.1% of the route > 21 Mbps	50% of the route > 28 Mbps
Number of serving cells	100	67
Estimated cell radius (m)	390	500 (28% improvement)
64QAM utilization (%)	8% of the route	40% of the route
MIMO usage (%)	MIMO + DC is not available yet	62% of the route

[a]Estimated from the number of successful high speed-shared control channels (HS-SCCH) decoded by the UE.

Figure 6.55 is summarized in Table 6.42 [8]. A typical DL loading of 70% is used in the dimensioning exercise in Figure 6.58 and the traffic profile of the user is assumed to be 50 kbps at busy hour (i.e., typical user throughput if all users accessed the system at the same time). For the sake of a comparative analysis, the HSPA+ system is considered with two native HSPA+ carriers (i.e., no R99 traffic) and using the DC-HSPA+ feature where two carriers are aggregated to provide an instantaneous peak throughput of 42 Mbps. A peak-to-average margin of 20% is considered to accommodate burst traffic. Figure 6.58 indicates that the LTE system offers a capacity improvement of 34% compared to the HSPA+, which is directly reflected by the average sector throughput gain depicted in Table 6.42 and thus as spectrum efficiency gain [8]. In this exercise we considered the DL throughput requirement only. In a practical exercise, the DL and UL should be considered and the number of supported subscribers for the DL and UL estimated and then the minimum selected. We shall consider this in a comprehensive exercise later in this chapter.

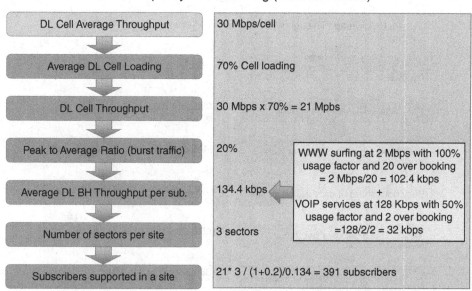

Figure 6.59   LTE dimensioning exercise based on voice and data traffic.

6.7.5.2   LTE Data and Voice Dimensioning Exercise

Figure 6.59 illustrates an LTE diminishing exercise based on voice and data. Assuming user data rate for Internet access of 2 Mbps and assuming 100% usage factor and 20 overbooking. In addition, assuming VOIP (voice over Internet protocol) traffic of 128 kbps with 50% usage factor and overbooking of 2. The busy hour user throughput from voice and data is 134.4 kbps, as illustrated in Figure 6.59. Similar steps are adopted as in Figure 6.58, the total number of users per cell will be 391 users with voice and data traffic profile as provided earlier.

6.7.5.3   LTE Dimensioning Based on Monthly Bundle

In this exercise, we assume the user bundle (the average actual consumption) per month is 10 GB. We conduct a reverse engineering dimensioning technique to estimate the maximum number of subscribers per cell, as shown in Figure 6.60. The average daily usage per user is $10/30 = 0.333$ GB per day. The daily traffic can be estimated as a percentage of the busy hour traffic. In this dimension exercise we assume the busy hour traffic is 20% of the daily traffic. The busy hour traffic is estimated based on 70% load of the average sector throughput of 30 Mbps. Therefore, we convert the 21 Mbps to a volume in GB per hour as follows: $[21 \text{ Mbps}/(8 \times 1024)] \times 60 \times 60 = 9.23$ GB. As a result, the daily traffic from the site is $3 \times 9.23/20\% = 138.43$ GB. The maximum number of subscribers per site according to the 10 GB average usage per subscriber is $(138.43/0.333) = 415$ subscribers. The total number of the supported subscribers per network is 415 multiplied by the total number of sites.

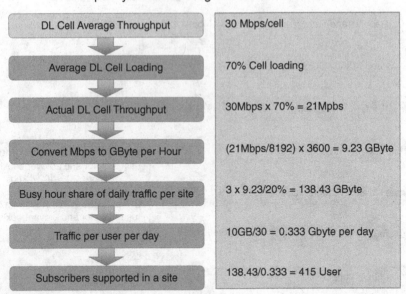

Figure 6.60   LTE dimensioning exercise based on traffic volume per month.

6.7.5.4   LTE Network Comprehensive Dimensioning Exercise

In this exercise we conducted a complete network dimensioning for a nationwide LTE network. Based on the traffic profile and the inputs in Table 6.40, the complete network dimensioning is demonstrated in Table 6.43. As demonstrated in this exercise, the maximum number of subscribers per cell is calculated based on the limiting link (i.e., DL or UL) in terms of capacity. In this scenario, the DL is the limiting capacity link with a maximum number of subscribers equal to $30/(0.2 \times 150/1024 + 0.1 \times 60/1024) = 853$ (data + VOIP). An overbooking of 30 is considered to calculate the SGW required capacity. Table 6.44 summarizes the outcome of this dimensioning exercise.

6.7.6   Calculation of VoIP Capacity in LTE

LTE supports VoIP to provide voice services. A simplified analysis is conducted to approximately estimate the achievable VoIP capacity per cell. Refer to [19] for a comprehensive simulation-based analysis of VoIP capacity. Assuming a full-rate 12.2 kbps adaptive multi-rate (AMR) speech codec is used, every 20 ms, AMR speech codec generates (12.2 kbps × 20 ms = 244 bits) during the voice active interval. These bits are placed in an RTP/UDP/IP (real-time transport protocol/user datagram protocol/Internet protocol) packet with about 3 bytes (= 24 bits) of overhead. The IP RoHC is assumed to be active. The VoIP packet includes about (244 speech bits + 24 IP-related header bits = 268) bits before the air interface protocol stack. The VoIP packets pass through three layers PDCP (packet

Table 6.43 Complete LTE network dimensioning

Item	Value	Unit	Formulas	Remarks and clarifications
Number of macro sites	2 000	Site	M	Number of macro outdoor sites
Number of IBS sites	500	Site	I	Number of IBS indoor sites
Number of active sub per cell	255	Sub	AS = 3*SUM(AD:AV)	Active data subscribers + active VOIP subscribers, every S111 site
Data subscriber active factor with overbooking	0.2	%	D	Only 20% active, other 80% only online
Active factor for VOIP subscribers with overbooking	0.1	%	V	Equal to VOIP traffic per sub, that is, 100 mErlang
Active data subscribers per cell	170	Sub	AD = ROUNDDOWN(HL*D,0)	Total number of data subscriber*data subscriber active factor
Active VOIP subscribers per cell	85	Sub	AV = ROUNDDOWN(HL*V,0)	HLR VOIP subscriber*active factor for VOIP subscriber
Maximum number of subscribers in DL per cell	853	Sub	MDL = (DLSS/(D*DS/1024 + V*VS/1024))	Maximum number of supported subscribers in DL based on traffic Mix
Maximum number of subscribers in UL per cell	1024	Sub	MUL = D25/((D*US/1024) + (V*VS/1024))	Maximum number of supported subscribers in UL based on traffic Mix
Maximum number of HSS subscribers per cell	853	Sub	HL = ROUNDDOWN(Min(MDL,MUL))	Select the limiting link in terms of capacity
Total number of active/connected data subscribers at BH based on the below traffic profile	1 275 000	Sub	TAD = M*3*AD + I*3*AD	This is the maximum number of simultaneously active/connected VOIP subscribers according to maximum system capacity
Total number of active/connected VOIP subscribers at BH based on the below traffic profile (on top of the above data sub)	637 500	Sub	TAV = M*3*AV + I*3*AV	

(continued overleaf)

Table 6.43 *(continued)*

Item	Value	Unit	Formulas	Remarks and clarifications
Total number of supported HSS subscribers	5 118 000	Sub	$TH = HL*3*M$	This the number of SIM card that can be defined in the HSS according to the traffic profile
Maximum number of active/connected sub at BH	3 000	Sub	MA	Based on the maximum hardware capacity
Maximum number of scheduled sub at BH	1 000	Sub	MS	
Total number of supported active subscribers	1 912 500	Sub	$TAS = TAD + TAV$	Total number of active and connected subscribers out of the total defined Sub in the HSS
DL throughput per sub at BH	150	kbps	DS	DL TP of data subscriber
UL throughput per sub at BH	50	kbps	US	UL TP of data subscriber
VOIP per sub	60	kbps	VS	VOIP subscriber traffic
Average DL throughput per site	90	Mbps	DLS	Simulation results for FDD LTE 1800 with 20 MHz BW and 1/1/1 configuration at urban clutter
Average DL throughput per sector	30	Mbps	DLSS	Simulation results for FDD LTE 1800 with 20 MHz BW and 1/1/1 configuration at urban clutter
Average UL throughput per site	48	Mbps	ULS	Simulation results for FDD LTE 1800 with 20 MHz BW and 1/1/1 configuration at urban clutter
Average UL throughput per sector	16	Mbps	ULSS	Simulation results for FDD LTE 1800 with 20 MHz BW and 1/1/1 configuration at urban clutter

	Value	Unit	Formula	Comment
Aggregated S1-U throughputs without oversubscription	481	Gbps	$ASO = AST$	Total throughput of all sites
Active data subscribers per site	510	Sub	$ADS = AD*3$	Total number of active Sub per site
DL throughput of data subscribers per site	77	Mbps	$DLD = ADS*DS/1000$	Average traffic/peak traffic $(38/45) = 85\%$
UL throughput of data subscribers per site	26	Mbps	$ULD = ADS*US/1000$	Actual UL throughput coming from data sub
Active VOIP subscribers per site	255	Sub	$AVS = 3*AV$	Total number of voice subscriber per site
VOIP throughput per site	45.90	Mbps	$VTS = AVS*3*VTSu/1000$	Total throughput from VOIP Sub per site
Total throughput per site	147.90	Mbps	$TTS = SUM(DLD, ULD, VTS)$	Total throughput per site from data + VOIP
Total traffic throughput	369.75	Gbps	$TTT = TTS*SUM(M:1)/1000$	Total data + VOIP throughput from the network
S1 with overhead	1.3		S	To count for overhead
Aggregated S1-U (DL + UL) throughput	480.68	Gbps	$AST = TTT*S$	Total traffic throughput*S1 with overhead
S-GW throughput capacity with oversubscription	16.02	Gbps	$SGW = AST*0.1$	Overbooking of 30

IBS, indoor building system; HSS, home subscriber service; SIM, subscriber identity module.

Table 6.44 Summary of LTE dimensioning exercise

Summary of the Dim exercise	Value
Total number of active Sub (voice + data) Sub	1 912 500
Total number of HSS subscribers	5 118 000
Total number of sites	2 500
S1 net TP (Mbps)	147.90
S1 actual capacity (Mbps)	240.34
EPC capacity (Gbps)	16.0

Table 6.45 Fixed reference channel parameters for performance requirements

Modulation/code rate	QPSK 1/3	16QAM 3/4	64QAM 5/6
Allocated resource blocks	1	1	1
DFT-OFDM symbols per subframe	12	12	12
Modulation	QPSK	16QAM	64QAM
Code rate	1/3	3/4	5/6
Payload size (bits)	**104**	**408**	**712**
Transport block CRC (bits)	24	24	24
Code block CRC size (bits)	0	0	0
Number of code blocks – C	1	1	1
Coded block size including 12 bits trellis termination (bits)	396	1308	2220
Total number of bits per subframe	288	576	864
Total symbols per subframe	144	144	144

CRC, cyclic redundancy check.

data convergence protocol), RLC (radio link control), and MAC before reaching the PHY layer. Adding RoHC 4 bytes (= 32 bits) to account for headers added by these three layers. Therefore, the total VoIP packet payload becomes (268 + 32 = 300) bits before the PHY layer.

We need to calculate how many RBs are needed to carry the target payload of 300 bits. As per Table 6.45 of [20], 1 RB can carry the payload of 104 bits when the modulation scheme is QPSK and the coding rate is (1/3). This payload is from the MAC layer to the PHY layer. Therefore, three RBs can carry (104 bits per PRB × 3 PRBs = 312 bits), which would be sufficient for the VoIP payload of 300 bits. On the other hand, if the channel conditions allow the modulation scheme of 16-QAM and the coding rate of (3/4), 1 RB can carry 408 bits, as shown in Table 6.45 [20], which can carry the VoIP payload. The users are distributed across the cell, some would have good channel conditions and can support 16-QAM, coding rate = $^3/_4$; others may have bad channel conditions and would require a more robust QPSK, coding rate = 1/3. According to Figure 6.32 we have 90% of users able to use 16-QAM, coding rate = $^3/_4$, and 10% of users need QPSK, coding rate = 1/3, the average number of RBs consumed by a typical VoIP user in a cell would be (0.10 × 3 RBs + 0.90 × 1 PRB) = 1.2 RBs. In 1 ms subframe, there are 100 RBs and assuming 80 net RBs after removing the signaling and control overheads and assuming 80% load, allowing (80 × 0.80 RBs/1.2 PRBs per user) = 53 users. Since the AMR speech codec generates a new speech frame every 20 ms, during a span of 20 ms, we can have 20 subframes carrying VoIP packets for (20 subframes × 53 users per subframe) = 1060 users.

These calculations assume that every single packet with a specific modulation scheme and certain amount of coding is received without any errors all the time. However, in practice, some packets would be lost, requiring HARQ retransmission. If we need one (additional) retransmission on average, RBs would need to be allocated to a given VoIP user twice per 20 ms interval instead of just once per 20 ms interval. Since a VoIP user is now consuming twice as many 1.2 RBs during the 20 ms interval, the number of VoIP users would be reduced by half (i.e., 1060/2 = 530). In summary, for the assumptions made here, the VoIP capacity in LTE is 530 in the case of a 20 MHz channel bandwidth assuming only 1 HARQ retransmission.

Several factors would increase the VoIP capacity estimated above. If many users can work with a reduced degree of channel coding, then the capacity would be higher. We did not use 64-QAM in the analysis above, because only UE Category 5 can support such ascheme high modulation scheme in the UL, and we may not see such UEs for quite some time and also the percentage of the users with very good RF conditions that achieve 64QAM will be very low and may be less than 10%. Adopting advanced antenna techniques would also increase the VoIP capacity. Consideration of the voice activity factor would also increase the capacity because there is no need to send hundreds of unnecessary bits during the silence intervals. Some factors would decrease the VoIP capacity estimated above. If many users in the cell need more redundancy than that provided by (1/3) coding, we would need more RBs per user, reducing the capacity. If semi-persistent scheduling is not used, a higher control channel overhead would decrease the achievable VoIP capacity. In summary, a 20 MHz channel bandwidth could support about 500 VoIP users in LTE. From a practical point of view, the 500 VOIP users per site is the maximum capability of an eNB. A safe assumption is to assume 20 users per 1 Mbps/cell. Therefore, the average VOIP throughput assuming 500 users per 20 MHz cell is 30 Mbps/500, which equals 60 kbps.

6.7.7 LTE Channels Planning

In this section, we will provide the key dimensioning guidelines for LTE channels in the DL and UL. The focus in this section is mainly on three main channels: PDCCH, PDSCH, and PUSCH. The PDCCH is the DL control channel and PxSCH is the physical DL/UL shared channels that are used for user payload. The aim of this study is to validate the dimensioning exercise that we have conducted in the previous section and to make sure that the total number of sectors can cater for the traffic and signalling load based on the provided traffic profile.

6.7.7.1 DL (PDCCH/PDSCH) Dimensioning

In OFDM time-frequency grid, up to four symbols of the frame are used for the control channels. These channels are the PCFICH (physical control format indicator channel), the PHICH (physical hybrid automatic repeat request indicator channel), and the PDCCH. The standard allows dynamic or static symbol assignment. The PDCCH capacity depends on the channel BW (number of REs) and the required REs for the scheduled UE. The PDCCH consists of control channel elements (CCEs), each CCE is 36 RE. The CCEs form the control region per subframe and can occupy up to four symbols. The number of supported users will depend on the number of CCE. The number of PDDCH symbols depends on the control format index (CFI), for example, assume a 20 MHz channel BW, then the number of symbols used by the PDCCH can be calculated as shown in Table 6.46. The table provides a complete exercise for PDCCH total overheads with different CFI and Ng values.

Table 6.46   Number of overhead symbols used by PDCCH

Inputs

Channel BW	10MHz	20MHz	Formulas
Number of PCFICH symbols in frame	160		PCFICH occupies four REs per RB and is distributed across four RBs
Number of RS symbols in frame	2000	4000	
Ng	Number of PHICH groups in 10MHz	Number of PHICH groups in 20MHz	CEILING(Ng×(number of RB/8,1)
1/6	2	3	
1/2	4	7	
1	7	13	
2	13	25	
Number of PHICH symbols in frame	10MHz	20MHz	Each PHICH groupe consists of three REG (i.e. 3×4 REs) in one frame
1/6	240	360	
1/2	480	840	12×10× Number of PHICH groups
1	840	1560	
2	1560	3000	
Total number of symbols for normal CP	7000	14000	Number of RB×10×number of symbols per frame (i.e., 14)

Results

CFI	10MHZ				20MHZ				Formulas
	NG=1/6	NG=1/2	NG=1	NG=2	NG=1/6	NG=1/2	NG=1	NG=2	
1	4600	4360	4000	3280	9480	9000	8280	6840	Total symbols for normal CP – number of PHICH symbols in frame – number of PCFICH symbols in frame – # of PHICH groups
2	11600	11360	11000	10280	23480	23000	22280	20840	
3	18600	18360	18000	17280	37480	37000	36280	34840	
Total number of symbols per frame	84000				168000				Total number of RBs × number of REs/RB × total number of symbols in Normal CP × 10ms TTI

PDCCH overhead %	10MHZ				20MHZ				Formulas
CFI	NG=1/6	NG=1/2	NG=1	NG=2	NG=1/6	NG=1/2	NG=1	NG=2	
1	5.48%	5.19%	4.76%	3.90%	5.64%	5.36%	4.93%	4.07%	Total number of PDCCH overhead symbols/total number of symbols per frame
2	13.81%	13.52%	13.10%	12.24%	13.98%	13.69%	13.26%	12.40%	
3	22.14%	21.86%	21.43%	20.57%	22.31%	22.02%	21.60%	20.74%	

The PDCCH can consume up to 22% of the channels BW, as shown in Table 6.46 and, therefore, it seriously impacts the available capacity for the PDSCH. The selection of the CFI depends on the number of subscribers and other relevant activities, as highlighted hereunder. The PDCCH resource diminishing determines the number of required sectors/carriers. The PDCCH resources depend on the DL and UL traffic demand, paging, and SIB scheduling. The RE availability for the PDCCH is determined by CFI 1, 2, or 3. The SINR values (i.e., RF condition and MCS) of the user determine the CCE requirement and thus the number of users that can be served by the PDCCH. A user at the cell edge with low SINR may need eight CCEs while a user near the cell center may only need one CCE. The CCEs are assigned in 1, 2, 4, 8 steps.

The UL dimensioning process encompasses: (i) determining the number of sectors required for PDCCH demand, and (ii) determining the number of sectors required for the PDSCH demand and selecting the maximum number of sectors.

The number of sectors for PDCCH is estimated by dividing the total required CCEs per subframe by the CCE capacity per sector per subframe, which depends on the average SINR, paging, antenna configuration, channel BW, and CP. The PDCCH demand consists of traffic demand and signaling demand.

Considering the dimensioning exercise in Table 6.43 where we have 1 275 000 data subscribers with 150/50 kbps in DL/UL and 637 500 VOIP subscribers. Assuming 20 mErlang traffic (data and voice) at BH for each subscriber and BHCA (busy hour call attempts) of 2 for data and voice. The data and voice activity factor are provided in Table 6.43, then:

The PDCCH Erlang demand for data and voice = DL Erlang demand for data and voice + UL Erlang demand for data and voice.

DL Erlang demand from data traffic = DL Erlang per user at BH × BHCA_data × DL activity factor for data × total number of data subscribers = $0.02 \times 2 \times 0.2 \times 6\,397\,500 = 51\,180$ Erlang

DL Mbps demand from voice traffic = DL Erlang per user at BH × BHCA_data × DL activity factor for data × total number of data subscribers = $0.02 \times 2 \times 0.1 \times 6\,397\,500 = 25\,590$ Erlang

Total DL demand = $51\,180 + 25\,590 = 76\,770$ Erlang

UL Mbps demand from data traffic = UL Erlang per user at BH × BHCA_data × UL activity factor for data × total number of data subscribers = $0.02 \times 2 \times 0.2 \times 6\,397\,500 = 51\,180$ Erlang

UL Mbps demand from voice traffic = UL Erlang per user at BH × BHCA_data × UL activity factor for data × total number of data subscribers = $0.02 \times 2 \times 0.1 \times 6\,397\,500 = 25\,590$ Erlang

Total UL Erlang demand = $51\,180 + 25\,590 = 76\,770$ Erlang

Therefore the total DL + UL demand = $153\,540$ Erlang

Assuming the signaling demand is 3% of the total traffic (voice + data) demand. Then, the total demand (traffic + signaling) = $1.03 \times 153\,540 = 158\,146.2$ Erlang

Assuming an average 3 CCE per user, therefore, the total number of required CCE = $3 \times 158\,146.2 = 474\,438.6$ Erlang.

For 2×2 MIMO with CFI = 3, we assume that the maximum CCE capacity for PDCCH = 65 per sector.

Table 6.47 PDCCH dimensioning based on the exercise in Table 6.43

#	Item	Value	Unit
1	BHCA for data subscribers	2.0	A
2	BHCA for voice subscribers	2.0	B
3	DL activity factor for data subscribers	0.2	C
4	UL activity factor for data subscribers	0.2	D
5	VOIP subscriber activity factor	0.1	E
5	Total number of data subscribers	6 397 500.0	F
6	Erlang traffic of data subscribers (mErlang)	20.0	G
7	Erlang traffic of voice subscribers (mErlang)	20.0	H
8	DL Erlang demand from data traffic	51 180.0	$I = (G/1\,000)*C*F*A$
9	DL Erlang demand from voice traffic	25 590.0	$J = (H/1\,000)*E*F*B$
10	UL Erlang demand from data traffic	25 590.0	$K = (G/1\,000)*E*F*A$
11	UL Erlang demand from voice traffic	51 180.0	$L = (H/1\,000)*D*F*B$
12	Total Erlang traffic demand	153 540.0	$M = I + J + K + L$
13	Total Erlang demand from traffic and signaling (+3%)	158 146.2	$N = M \times 1.03$
14	Average CCE per subscriber	3.0	O
15	Total number of available CCE	474 438.6	$P = N \times O$
16	Total number of CCE per sector (CFI 3)	26.0	Q
17	Total number of required sectors	18 248	$W = P/Q$
18	Total number of required sites	6 083	$Z = W/3$

Therefore, the total number of required sectors is 18 248 sectors and, therefore, the total number of required sites is 6083. Table 6.47 outlines this exercise.

As shown in Table 6.47, the total number of sites is very high compared to the actual number of sites. Therefore, the PDCCH is a limiting channel for this exercise and although the 2500 sites can meet the user demand in terms of traffic, they will not meet the user demand in terms of PDCCH requirement. An iterative approach is adopted to modify the traffic profile to meet the user demand without changing the number of sites. Table 6.48 provides the outcome of this process, which yields a modified activity factor of 10% for DL data and 5% for UL data and 5% for voice.

As an alternative approach to meet the traffic profile of Table 6.43, we can use CFI 3 where approximately 65 CCE resources are available for the PDCCH. Table 6.49 summarizes this exercise.

6.7.7.2 PDSCH Dimensioning

The PDSCH capacity varies from 65 to 85% of the available RB based on the channel BW and the CFI value. The PDCCH consumes up to 20% of the available RB based on the CFI value and channel BW, therefore impacting the PDSCH capacity. Assuming an average sector throughput of 30 Mbps and 70% load. Table 6.50 summarizes the PDSCH dimensioning. The same input to the PDCCH exercise is used. The target user throughput is 1.5 Mbps and the data activity factor is 20% and the data traffic is 20 mErlang. The voice subscriber activity factor is

Table 6.48 PDCCH dimensioning based on the exercise in Table 6.43 with modified activity factors

#	Item	Value	Unit
1	BHCA for data subscribers	2.0	A
2	BHCA for voice subscribers	2.0	B
3	DL activity factor for data subscribers	0.10	C
4	UL activity factor for data subscribers	0.05	D
5	VOIP subscriber activity factor	0.05	E
5	Total number of data subscribers	6 397 500.0	F
6	Erlang traffic of data subscribers (mErlang)	20.0	G
7	Erlang traffic of voice subscribers (mEranlg)	20.0	H
8	DL Erlang demand from data traffic	25 590.0	I = (G/1 000)*C*F*A
9	DL Erlang demand from voice traffic	12 795.0	J = (H/1 000)*E*F*B
10	UL Erlang demand from data traffic	12 795.0	K = (G/1 000)*E*F*A
11	UL Erlang demand from voice traffic	12 795.0	L = (H/1 000)*D*F*B
12	Total Erlang traffic demand	63 975.0	M = I + J + K + L
13	Total Erlang demand from traffic and signaling (+3%)	65 894.3	N = M × 1.03
14	Average CCE per subscriber	3.0	O
15	Total number of available CCE	197 682.8	P = N × O
16	Total number of CCE per sector	26.0	Q
17	Total number of required sectors	7 603	W = P/Q
18	Total number of required sites	2 534	Z = W/3

Table 6.49 PDCCH dimensioning based on the exercise in Table 6.43 with CFI = 3

#	Item	Value	Formula
1	BHCA for data subscribers	2.0	A
2	BHCA for voice subscribers	3.0	B
3	DL activity factor for data subscribers	0.20	C
4	UL activity factor for data subscribers	0.10	D
5	VOIP subscriber activity factor	0.10	E
5	Total number of data subscribers	6 397 500.0	F
6	Erlang traffic of data subscriber (mErlang)	20.0	G
7	Erlang traffic of voice subscriber (mErlang)	20.0	H
8	DL Erlang demand from data traffic	51 180.0	I = (G/1 000)*C*F*A
9	DL Erlang demand from voice traffic	38 385.0	J = (H/1 000)*E*F*B
10	UL Erlang demand from data traffic	25 590.0	K = (G/1 000)*E*F*A
11	UL Erlang demand from voice traffic	38 385.0	L = (H/1 000)*D*F*B
12	Total Erlang traffic demand	153 540.0	M = I + J + K + L
13	Total Erlang demand from traffic and signaling (+3%)	158 146.2	N = M × 1.03
14	Average CCE per subscriber	3.0	O
15	Total number of available CCE	474 438.6	P = N × O
16	Total number of CCE per sector	65.0	Q
17	Total number of required sectors	7 299	W = P/Q
18	Total number of required sites	2 433	Z = W/3

Table 6.50 PDSCH dimensioning

Item	Value	Formula
BHCA for data subscribers	2.0	A
BHCA for voice subscribers	2.0	B
DL activity factor for data subscribers	0.20	C
Target DL user throughput (Mbps)	1.50	D
VOIP subscriber activity factor	0.10	E
Total number of data subscribers	6 397 500.0	F
Erlang traffic of data subscriber (mErlang)	20.0	G
Average throughput of voice subscriber (kbps)	60.0	H
DL Mbps demand from data traffic (Mbps)	76 770.0	$I = (G/1\,000)*C*F*A*D$
DL Mbps demand from voice traffic (Mbps)	76 770.0	$J = (H/1\,024)*E*F*B$
Total Mbps traffic demand (Mbps)	153 540.0	$M = I + J$
Average throughput per sector at 80% load	21.0	$N = 30 \times 0.7$
Total number of required sectors	7 311	$W = M/N$
Total number of required sites	2 437	$Z = W/3$

Table 6.51 PUSCH dimensioning exercise

#	Item	Value	Formula
1	BHCA for data subscribers	2.0	A
2	BHCA for voice subscribers	2.0	B
3	UL activity factor for data subscribers	0.10	C
4	Target UL user throughput (Mbps)	1.00	D
5	VOIP subscriber activity factor	0.10	E
6	VOIP subscriber average throughput (Mbps)	0.06	K
7	Total number of data subscribers	6 397 500.0	F
8	Average UL data call duration (s)	90.0	G
9	Average UL voice call duration (s)	40.0	H
10	UL demand from data traffic (Mbps)	31 987.5	$I = G*C*F*A*D/3\,600$
11	UL demand from voice traffic (Mbps)	853.0	$J = H*E*F*B*K/3\,600$
12	Total UL Mbps traffic demand (Mbps)	32 840.5	$M = I + J$
13	Average throughput per sector at 60% load	9.6	$N = 16 \times 0.7$
14	Total number of required sectors	3 421	$W = M/N$
15	Total number of required sites	1 140	$Z = W/3$

10% and the average throughput of a voice call is 60 kbps. As shown in Table 6.50, the total number of sites is 2437 sites, which means the total number of sites meets the user requirement based on the assumption in Table 6.43. This exercise indicates that the PDSCH is the limiting channel in this dimensioning exercise if we consider CFI = 3 where the PDCCH demand is less than the PDSCH.

6.7.7.3 PUSCH Dimensioning

The uplink shared channel (UL-SCH) is carried over the PUSCH. Unlike the UMTS system, most of the user and control plane channels are carried over the UL-SCH. Therefore, the PUSCH is the main channel in the UL that needs to be properly dimensioned. Other channels are not a bottleneck in terms of capacity dimensioning. The PUCCH consumes up to four RBs, based on the channel BW (two with 5 MHz and four with other channel BWs). The SRS bandwidth is spread over four RBs with 2 ms periodicity. The RB availability for PUSCH varies from 65 to 80% based on the fixed channels requirement (SRS and RS) and the control channel requirement that depends on the traffic. The total PUSCH demand throughput can be estimated for voice and data as follows:

Total PUSCH demand = number of Sub × (BHCA per data sub × average call duration of data sub (s) × activity factor of data sub × target user throughput (Mbps) + BHCA per voice sub × average call duration of voice sub (s) × activity factor of voice sub × average VOIP user throughput (Mbps))/3600.

Table 6.51 demonstrates a dimensioning exercise based on the input in Table 6.43. As shown in Table 6.51, the required number of sites to cater for PUSCH demand is less than the actual number of sites. Therefore, the PUSCH is not a limiting factor in this exercise.

6.8   Case Study: LTE FDD versus LTE TDD

In this case study we will compare FDD-LTE 1800 MHz with 10 and 20 MHz channel BW, with TDD-LTE 2.3 GHz with 20 MHz. This exercise is conducted for a nationwide converge with 1000 sites. The following summarizes the input to this exercise:

1. Indoor coverage
2. Penetration loss: 19 dB (dense urban)/15 dB (urban)/11 dB (suburban)/8 dB (rural)
3. Coverage probability: 95% (dense urban/urban), 90% (suburban/rural)
4. 20 MHz bandwidth, MIMO 2 × 2 with 2 × 60 W and RS subcarrier power 20.01 dBm
5. 10 MHz bandwidth, MIMO 2 × 2 with 2 × 60 W, RS subcarrier power 23.02 dBm.

Table 6.52 summarizes the outcome of the link budgets (i.e., cell radius) for different clutters and for DL and UL FDD-LTE 1800 MHz with 10 and 20 MHz and TDD-LTE 2300 MHz with 20 MHz. With regard to the FDD scenario and since it is UL limited based on the PUSCH then 10 and 20 MHz bandwidth have the same cell radius. The FDD-LTE 1800 MHz offers around 25% cell radius gain compared to TDD-LTE 2300. As shown from the table the FDD-LTE and TDD-LTE are UL limited in this scenario.

Three performance indicators are adopted to analyze this exercise: RSRP, SINR, and DL MAC throughput. Figures 6.61 and 6.62 illustrate the RSRP and SINR, respectively, for the FDD-LTE 1800 MHz and TDD-LTE 2300. Figure 6.63 illustrates the MAC layer through-

Table 6.52 LET link budget outcome for FDD-LTE 1800 with 10 and 20 MHz channel BW and TDD-LTE 2300

Frequency band, channel BW, and access modes	UL/DL load	Cell edge rate	Dense urban cell radius (km)		Urban cell radius (km)		Suburb cell radius (km)		Rural cell radius (km)	
			UL	DL	UL	DL	UL	DL	UL	DL
FDD-1800 MHz with 10 MHz BW	UL: 50% load DL: 100% load	UL at 512 kbps DL at 1 Mbps	0.33	0.69	0.49	1.15	1.58	4.04	3.90	8.59
FDD-1800 MHz with 20 MHz BW	UL: 50% load DL: 100% load	UL at 512 kbps DL at 1 Mbps	0.33	0.57	0.49	0.95	1.58	3.31	3.90	7.02
TDD-2300 MHz with 20 MHz TDD	UL: 50% load DL: 100% load	UL at 512 kbps DL at 1 Mbps	0.23	0.43	0.39	0.76	1.26	2.65	3.00	5.91

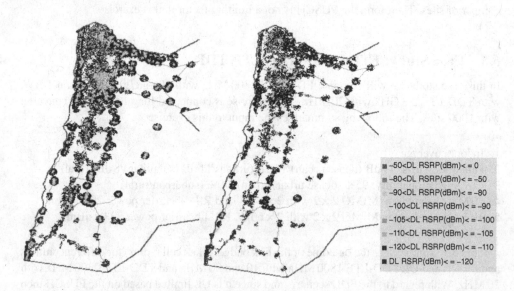

■ −50<DL RSRP(dBm)< = 0
■ −80<DL RSRP(dBm)< = −50
  −90<DL RSRP(dBm)< = −80
  −100<DL RSRP(dBm)< = −90
■ −105<DL RSRP(dBm)< = −100
  −110<DL RSRP(dBm)< = −105
■ −120<DL RSRP(dBm)< = −110
■ DL RSRP(dBm)< = −120

Figure 6.61 DL RSRP for FDD-LTE 1800 MHz and TDD-LTE 2300 MHz for same number of sites.

put for the FDD-LTE 1800 MHz with 20 and 10 MHz channel BWs and TDD-LTE 2300. Table 6.53 lists the RSRP, SINR, and DL MAC throughput for LTE 1800 MHz with 10 and 20 MHz. Table 6.54 lists the RSRP, SINR, DL MAC throughput for TDD-LTE 2300 MHz with 20 MHz. To further analyze the throughputs we have provided PDF and CDF for MAC DL The throughput for FDD-LTE 1800 MHz with 10 and 20 MHz and TDD-LTE 2300 with 20 MHz is shown in Figure 6.64.

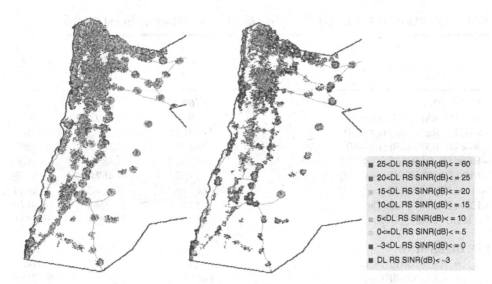

Figure 6.62 DL SINR for FDD-LTE 1800 MHz and TDD-LTE 2300 for same number of sites.

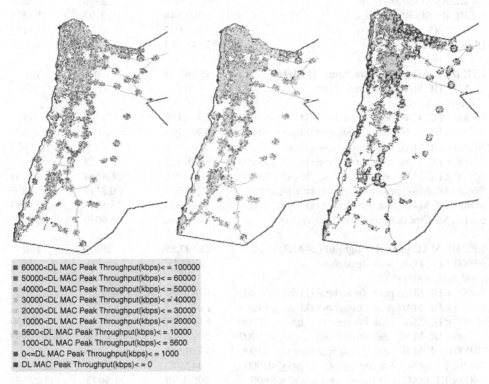

Figure 6.63 DL MAC throughput for FDD-LTE 1800 MHz with 20 and 10 MHz channels and TDD-LTE 2300 MHz with 20 MHz.

Table 6.53   RSRP, SINR, DL MAC throughput for LTE 1800 MHz with 10 and 20 MHz

	Coverage area (km²)	Coverage area (%)	Cumulate coverage area (%)
LTE DL RSRP	**15 769.13**	**100**	**100↑**
−50 < DL RSRP(dBm) ≤ 0	0	0.0002	100
−80 < DL RSRP(dBm) ≤ −50	189.1045	1.1991	100
−90 < DL RSRP(dBm) ≤ −80	533.5986	3.3838	98.8007
−100 < DL RSRP(dBm) ≤ −90	1 337.375	8.4811	95.4169
−105 < DL RSRP(dBm) ≤ −100	1 177.525	7.4674	86.9358
−110 < DL RSRP(dBm) ≤ −105	1 664.683	10.5565	79.4684
−120 < DL RSRP(dBm) ≤ −110	4 613.873	29.2588	68.9119
DL RSRP(dBm) ≤ −120	6 252.976	39.6531	39.6531
LTE DL RS SINR	**15 765.86**	**100**	**100**
25 < DL RS SINR(dB) ≤ 60	1 463.641	9.2838	100
20 < DL RS SINR(dB) ≤ 25	1 085.826	6.8871	90.7163
15 < DL RS SINR(dB) ≤ 20	1 688.428	10.7094	83.8292
10 < DL RS SINR(dB) ≤ 15	2 291.621	14.5354	73.1198
5 < DL RS SINR(dB) ≤ 10	2 719.09	17.2466	58.5844
0 < DL RS SINR(dB) ≤ 5	2 509.848	15.9197	41.3378
−3 ≤ DL RS SINR(dB) ≤ 0	1 362.037	8.6391	25.4181
DL RS SINR(dB) < −3	2 645.364	16.779	16.779
LTE DL MAC peak throughput (10 MHz)	**15 769.27**	**100**	**100↑**
60 000 < DL MAC peak throughput (kbps) ≤ 100 000	0	0	100
50 000 < DL MAC peak throughput (kbps) ≤ 60 000	206.6178	1.3101	100
40 000 < DL MAC peak throughput (kbps) ≤ 50 000	489.9793	3.107	98.6899
30 000 < DL MAC peak throughput (kbps) ≤ 40 000	786.9841	4.9905	95.5829
20 000 < DL MAC peak throughput (kbps) ≤ 30 000	1 708.827	10.8363	90.5924
10 000 < DL MAC peak throughput (kbps) ≤ 20 000	4 488.918	28.4664	79.7561
5 600 < DL MAC peak throughput (kbps) ≤ 10 000	3 281.812	20.8114	51.2897
1 000 < DL MAC peak throughput (kbps) ≤ 5 600	2 187.853	13.8744	30.4783
0 < DL MAC peak throughput (kbps) ≤ 1 000	2 618.2788	16.6039	16.6039
LTE DL MAC peak throughput (20 MHz)	**15 547.89**	**100**	**100↑**
60 000 < DL MAC peak throughput (kbps) ≤ 100 000	1 290.41	8.2994	100
50 000 < DL MAC peak throughput (kbps) ≤ 60 000	678.4934	4.3639	91.7006
40 000 < DL MAC peak throughput (kbps) ≤ 50 000	1 043.715	6.7128	87.3367
30 000 < DL MAC peak throughput (kbps) ≤ 40 000	1 504.177	9.6746	80.6239
20 000 < DL MAC peak throughput (kbps) ≤ 30 000	2 993.596	19.254	70.9493
10 000 < DL MAC peak throughput (kbps) ≤ 20 000	3 710.658	23.8659	51.6953
5 600 < DL MAC peak throughput (kbps) ≤ 10 000	1 442.782	9.2797	27.8294
1 000 < DL MAC peak throughput (kbps) ≤ 5 600	305.8599	1.9672	18.5497
0 < DL MAC peak throughput (kbps) ≤ 1 000	2 578.1987	16.5825	16.5825

Table 6.54   RSRP, SINR, DL MAC throughput for TDD-LTE 2300 MHz with 20 MHz

	Coverage area (km^2)	Coverage area (%)	Cumulate coverage area (%)
LTE DL RSRP	**10 382.97**	**100**	**100↑**
$-50 <$ DL RSRP(dBm) ≤ 0	0	0	100
$-80 <$ DL RSRP(dBm) ≤ -50	22.9575	0.2214	100
$-90 <$ DL RSRP(dBm) ≤ -80	126.7256	1.2206	99.7787
$-100 <$ DL RSRP(dBm) ≤ -90	352.6277	3.396	98.5581
$-105 <$ DL RSRP(dBm) ≤ -100	300.0878	2.8899	95.1621
$-110 <$ DL RSRP(dBm) ≤ -105	431.9952	4.1604	92.2722
$-120 <$ DL RSRP(dBm) ≤ -110	1 621.195	15.6142	88.1118
DL RSRP(dBm) ≤ -120	7 527.381	72.4976	72.4976
LTE DL RS SINR	**10 382.91**	**100**	**100**
$25 <$ DL RS SINR(dB) ≤ 60	331.7036	3.1945	100
$20 <$ DL RS SINR(dB) ≤ 25	286.7068	2.7612	96.8056
$15 <$ DL RS SINR(dB) ≤ 20	478.9597	4.6128	94.0444
$10 <$ DL RS SINR(dB) ≤ 15	791.379	7.6221	89.4316
$5 <$ DL RS SINR(dB) ≤ 10	1 172.408	11.2916	81.8095
$0 <$ DL RS SINR(dB) ≤ 5	1 565.114	15.0742	70.5179
$-3 \leq$ DL RS SINR(dB) ≤ 0	2 063.948	19.8784	55.4437
DL RS SINR(dB) < -3	3 692.687	35.5653	35.5653
LTE DL MAC peak throughput	**10 382.97**	**100**	**100↑**
$60\,000 <$ DL MAC peak throughput (kbps) $\leq 100\,000$	0	0	100
$50\,000 <$ DL MAC peak throughput (kbps) $\leq 60\,000$	15.152	0.146	100
$40\,000 <$ DL MAC peak throughput (kbps) $\leq 50\,000$	66.708	0.6423	99.8538
$30\,000 <$ DL MAC peak throughput (kbps) $\leq 40\,000$	194.6799	1.875	99.2115
$20\,000 <$ DL MAC peak throughput (kbps) $\leq 30\,000$	529.2039	5.097	97.3365
$10\,000 <$ DL MAC peak throughput (kbps) $\leq 20\,000$	1 654.91	15.9384	92.2395
$5\,600 <$ DL MAC peak throughput (kbps) $\leq 10\,000$	1 623.819	15.639	76.3011
$1\,000 <$ DL MAC peak throughput (kbps) $\leq 5\,600$	1 453.54	13.9993	60.6621
$0 <$ DL MAC peak throughput (kbps) $\leq 1\,000$	4 844.9572	46.663	46.6628

The following summarizes the outcome of this case study:

- 2×10 MHz and 2×20 MHz FDD 1800 MHz will have the same coverage and interference level if the RS power is set to the same value, for example, with 2×60 W, 50.8 dBm, the RS power is 20.01 dBm. Therefore, we did not provide the RSRP and SINR for the FDD-LTE 1800 MHz with 10 MHz as they are the same in Figures 6.61 and 6.62, respectively.
- If the RS power is increased, the RSRP will increase and the SINR will get worse in dense urban, but is good for a rural area. The RS power is a compromise between coverage and interference.

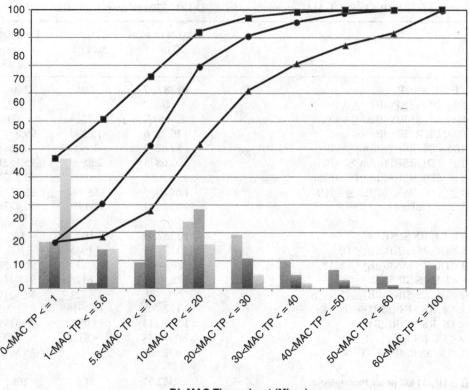

DL MAC Throughput (Mbps)

  ▬ FDD-LTE1800MHz with 20MHz (PDF)   ▬▲▬ FDD-LTE1800MHz with 20MHz (CDF)

  ▬ FDD-LTE1800MHz with 10MHz (PDF)   ▬●▬ FDD-LTE1800MHz with 20MHz (CDF)

  ▬ TDD-LTE2300MHz with 20MHz (PDF)   ▬■▬ FDD-LTE1800MHz with 20MHz (CDF)

Figure 6.64   PDF and CDF for MAC DL throughput for FDD-LTE 1800 MHz with 10 and 20 MHz and TDD-LTE2300.

- FDD-LTE 1800 MHz with 20 MHz bandwidth has obvious DL MAC throughput gain compared to FDD-LTE 1800 MHz with10 MHz bandwidth and TDD-LTE 2300 MHz with 20 MHz, as evident from Figure 6.63.
- FDD-LTE 1800 MHz with 10 MHz bandwidth has obvious throughput and coverage gain compared to TDD-LTE 2300 MHz with 20 MHz.
- If we consider SINR = −3 dB is the cell-edge SINR, then the 1000 sites of FDD-LTE 1800 MHz provide converge for ~13 000 km² while the 1000 sites of TDD-LTE 2300 MHz provides coverage for 5538 km² which is 42.6% of the coverage provided by the FDD-LTE 1800.
- The average DL throughputs for FDD-LTE 1800 with 20 and 10 MHz are 24 and 12.3 Mbps, respectively, while for TDD-LTE 1800 with 20 MHz, the average DL throughput is 6 Mbps. We assumed that the clutter size for FDD-LTE is 15 769 km² while for TDD-LTE it is

10 382.97 km^2. We assumed the points within each DL throughout range are normally distributed and, therefore, the center of the range is used for average throughput estimation by multiplying the center of each range by the corresponding area covered by this range and then adding all results and dividing the results by the total area.

- We should not expect reliable LTE converge for RSRP < -121 dBm. However, if the interference is well controlled and the RF parameters are fully optimized, the SINR coverage will be better than the RSRP. This clarifies the difference in coverage between SINR and RSRP.
- There is no direct relation between the RSRP and the SINR; in suburban and rural clutter better RSRP brings better SINR due to high inter-site distance but in dense urban and urban clutters with small inter-site distance, better RSRP brings worse SINR due to the interference.

References

[1] 3GPP (2013) Physical Channels and Modulations. TS 36.211, V10.7.0.
[2] Telesystem Innovations. (2010) LTE in a Nutshell, White Paper.
[3] 3GPP (2011) Requirements for Support of Radio Resource Management. TS 36.133 version 10.1.0.
[4] 3GPP (2009) Radio Frequency (RF) System Scenarios. TS 36.942 V 8.2.0.
[5] 3GPP (2011) Physical Layer Procedures. TS 36.213 V 10.1.0.
[6] 3GPP (2011) User Equipment (UE) Radio Transmission and Reception. 36.101, V8.15.0.
[7] COMMSCOPE http://www.commscope.com/catalog/andrew/product_details.aspx?id=15653&tab=2 (accessed 23 September 2013).
[8] Elnashar, A. and El-Saidny, M.A. (2013) A Practical Performance Analysis of The Long Term Evolution (LTE) System Based on Field Test Results. IEEE Vehicular Technology Magazine (Sept. 2013).
[9] Seker, S.S., Yelen, S., and Kunter, F.C. (2010) Comparison of propagation loss prediction models of UMTS for an URBAN AREAS. 18th Telecommunications forum TELFOR 2010, pp. 902–905.
[10] Okumura, Y., Ohmori, E., Kawano, T. and Fukuda, K. (1968) Field strength and its variability in VHF and UHF land-mobile radio service. *Review of the Electrical Communications Laboratory*, **16**(9-0), 825–873.
[11] Hata, M. (1980) Empirical formula for propagation loss in land mobile radio services. *IEEE Transactions on Vehicular Technology*, **VT-29**, 317–325.
[12] COST Action 231 (1999) Digital Mobile Radio Towards Future Generation Systems, Final Report. Technical Report EUR 18957. European Communities.
[13] Wong, T. (1999) Some technique for CDMA capacity enhancement. PhD dissertation, The University of Texas at Arlington.
[14] Rappaport, T.S. (1996) *Wireless Communications, Principles and Practice*, Prentice Hall PTR, Upper Saddle River, NJ ISBN: 0-13-375536-3.
[15] Parsons, D. (1994) *The Mobile Radio Propagation Channel*, Halsted Press. ISBN: 0-470-21824, TK 6570.M6P38.
[16] Milanovic, J., Rimac-Drlje, S., and Bejuk, K. (2007) Comparison of propagation models accuracy for WiMAX on 3.5 GHz. Electronics, Circuits and Systems, 2007. ICECS 2007. 14th IEEE International Conference on Electronics, Circuits and Systems.
[17] Walfisch, J. and Bertoni, H.L. (1988) A theoretical model of UHF propagation in urban environments. *IEEE Transactions on Antennas and Propagation*, **36**(12), 1788–1796.

[18] Irmer, R., Mayer, H.-P. and Weber, A. (2009) Multisite field trial for LTE and advanced concepts. *IEEE Communications Magazine*, **47**, 92–98.

[19] 3GPP Performance Evaluation Checkpoint: VoIP Summary. R1-072570.

[20] 3GPP (2009) Base Station (BS) Radio Transmission and Reception. TS 36.104 V8.7.0.

[21] 3GPP (2009) Evolved Universal Terrestrial Radio Access (E-UTRA) and Evolved Universal Terrestrial Radio Access Network (E-UTRAN); Overall Description. TS 36.300 V8.5.0. 23.

[22] Roke Manor Research (2012) LTE Uplink Physical Layer Behavioural Model. White Paper.

7

Voice Evolution in 4G Networks

Mahmoud R. Sherif

Although LTE (long term evolution) is a packet-only technology, the standard was defined to efficiently handle voice and multimedia services, providing support for more than mobile Internet access. This chapter describes some of the main practical aspects on how to support telephony services over LTE/EPC (evolved packet core). An end-to-end approach is followed in this chapter outlining the requirements in the LTE radio access, EPC, the IMS (IP multimedia subsystem) core, and the voice application server. The chapter includes a description of the radio features for voice, how QoS (quality of service) is provisioned to support conversational speech, IMS features for voice telephony, media handling, roaming architecture, and how to take advantage of any existing circuit-switched infrastructure.

A general view of the network transformation toward VoLTE (voice over long term evolution) and all the added-on services is shown in Figure 7.1.

7.1 Voice over IP Basics

To understand how LTE can support voice services in the packet-switch domain, this section explains the basics of the VoIP (voice over Internet protocol).

7.1.1 VoIP Protocol Stack

To understand the "basics" of VoIP, one must be familiar with the IP (Internet protocol) stack and the difference between different protocols and the function of each.

Like most networking software, TCP (transmission control protocol)/IP is modeled in layers [1]. This layered representation leads to the term protocol stack, which is synonymous with protocol suite.

As shown in Figure 7.2, the VoIP is modeled in three layers. The following sections will describe the function of each of the layers.

Design, Deployment and Performance of 4G-LTE Networks: A Practical Approach, First Edition. Ayman Elnashar, Mohamed A. El-saidny and Mahmoud R. Sherif.

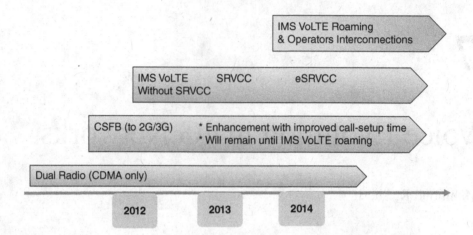

Figure 7.1   Evolution path to support VoLTE.

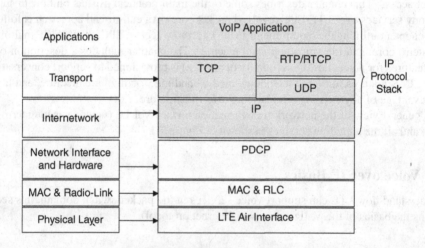

Figure 7.2   VoIP protocol stack.

7.1.1.1   Internetwork Layer

The internetwork layer, also called the Internet layer or the network layer, provides the "virtual network" image of an Internet (that is, this layer shields the higher levels from the physical network architecture below it). IP is the most important protocol in this layer. It is a connectionless protocol that does not assume reliability from the lower layers [1]. IP does not provide reliability, flow control, or error recovery; these functions are provided at a higher level.

Part of communicating messages between computers is a routing function that ensures that messages will be correctly delivered to their destination. IP provides this routing function. A message unit in an IP network is called an IP datagram. This is the basic unit of information transmitted across TCP/IP networks.

7.1.1.2 Transport Layer

The transport layer provides the end-to-end data transfer. Multiple applications can be supported simultaneously. The transport layer is responsible for providing a reliable exchange of information. The main transport layer protocol is TCP.

Another transport layer protocol is UDP (user datagram protocol), which provides a connectionless service, in contrast to TCP, which provides a connection-oriented service. That means that applications using UDP as the transport protocol have to provide their own end-to-end flow control and their own error detection/recovery. Usually, UDP is used by applications that need a fast transport mechanism.

Transmission Control Protocol (TCP)

TCP is a standard protocol with STD number 7 and is described by RFC 793-TCP [2]. TCP provides considerably more facilities for applications than UDP, notably, error recovery, flow control, and reliability. TCP is a connection-oriented protocol, unlike UDP which is connectionless. Most of the user application protocols, such as Web browsing, e-mail, and FTP (file transfer protocol), use TCP. TCP can be characterized by the following facilities it provides for the applications using it:

- **Data transfer**: From the application's viewpoint, TCP transfers a contiguous stream of bytes through the network. The application does not have to bother with chopping the data into basic blocks or datagrams. TCP does this by grouping the bytes in TCP segments, which are passed to the IP for transmission to the destination. Also, the TCP itself decides how to segment the data and it can forward the data at its own convenience.
- **Reliability**: TCP assigns a sequence number to each packet transmitted and expects a positive acknowledgment (ACK) from the receiving TCP. If the ACK is not received within a timeout interval, the data is retransmitted. Since the data is transmitted in blocks (TCP segments) only the sequence number of the first data byte in the segment is sent to the destination host. The receiving TCP uses the sequence numbers to rearrange the segments when they arrive out of order, and to eliminate duplicate segments.
- **Flow control**: The receiving TCP, when sending an ACK back to the sender, also indicates to the sender the number of bytes it can receive beyond the last received TCP segment, without causing overrun and overflow in its internal buffers. This is sent in the ACK in the form of the highest sequence number it can receive without problems. This mechanism is also referred to as a window-mechanism.
- **Multiplexing**: This section introduces the concepts of port and socket, which are necessary to exactly determine which local process at a given host actually communicates with which process at which remote host using which protocol. Each process that wants to communicate with another process identifies itself to the TCP/IP protocol suite by one or more ports. A port is a 16-bit number, used by the host-to-host protocol to identify to which higher level protocol or application program (process) it must deliver incoming messages. On the other hand, a socket is a special type of file handle which is used by a process to request network services from the operating system. Each side of a TCP connection has a socket that can be identified by the triple <TCP, IP address, port number>. Therefore, multiplexing is achieved through the use of ports, just as with UDP.
- **Logical connections**: The reliability and flow control mechanisms described above require that TCP initializes and maintains certain status information for each data stream. The

combination of this status, including sockets, sequence numbers, and window sizes, is called a logical connection. Each connection is uniquely identified by the pair of sockets used by the sending and receiving processes.

- **Full duplex**: TCP provides concurrent data streams in both directions.

User Datagram Protocol (UDP)

UDP is a standard protocol with STD number 6 and is described by RFC 768 – UDP [3]. UDP is basically an application interface to IP. It adds no reliability, flow-control, or error recovery to IP. It simply serves as a multiplexer/demultiplexer for sending and receiving datagrams. All VoIP applications rely on UDP as the transport layer for the bearer traffic (speech packets).

Real-Time Transport Protocol (RTP)

RTP (real-time transport protocol) is defined by IETF (http://www.ietf.org) in RFC 3550 and 3551 [4, 5]. RTP provides generic transport capabilities for real-time multimedia applications. It supports both conversational and streaming applications such as:

- Internet radio
- Internet telephony
- Music-on-demand
- Videoconferencing
- Video-on-demand and
- Applications may include multiple media streams.

RTP is critical in providing the following major functions:

- Identifies encoding scheme,
- Facilitates playout at appropriate times,
- Synchronizes multiple media streams,
- Indicates packet loss,
- Provides performance feedback, and
- Indicates frame boundary.

RTP normally runs over UDP and usually runs with the companion protocol RTCP (real-time transport control protocol) on consecutive ports. For the sake of clarity, the RTCP handles feedback, synchronization, and user interface. One of the best descriptions in [1] of the RTP and RTCP protocols is that they are "transport protocols implemented in the application layer."

7.1.1.3 VoIP Application Layer

A typical VoIP application consists of two main parts; (i) control and (ii) bearer. The control part is responsible for the signaling portion of the VoIP application, necessary for establishing and releasing the call, as well as many other functions, such as adding another party (conference call), call waiting, call forwarding, and many other capabilities. On the other hand, the bearer part is responsible for establishing a virtual circuit in the form of an IP connection between the calling party and the called party.

In summary, a typical VoIP application would consist of the following two components:

- Signaling which mainly uses the protocol named session initiation protocol (SIP) [6] which uses TCP as the transport layer. This is shown in Figure 7.3.

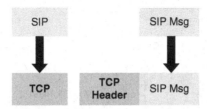

Figure 7.3   Session initiation protocol (SIP) over TCP.

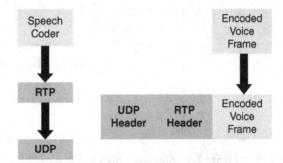

Figure 7.4   Encoded speech over RTP over UDP.

- Bearer traffic (the actual packetized voice) encoded speech over RTP over UDP. This is shown in Figure 7.4.

7.1.2   VoIP Signaling (Call Setup)

VoIP signaling is handled using the SIP [6]. A simple depiction of a typical call setup is shown in Figure 7.5.

A more typical SIP call setup scenario is shown in Figure 7.6.

It is important to note that the SIP protocol is a "chatty" protocol which requires some special features when applied on any radio access network (RAN):

- A faster "access" (uplink signaling channel) and "control/paging" (downlink signaling channel) is definitely needed. This can be in the form of shorter access cycle duration as well as a shorter/faster paging cycle (handset wake-up cycle much more often – but can affect battery life).
- Signaling compression (SIGCOMP) [7, 8]:
  - Typical SIP messages are 1000–2500 bytes.
  - SIGCOMP compresses SIP messages by up to 90% [9]. This leads to a significant reduction in the transit time and the associated signaling.

7.1.3   VoIP Bearer Traffic (Encoded Speech)

VoIP encoded speech traffic is also carried/encapsulated within an IP/UDP packet. The following section provides the background on the VoIP traffic encapsulation and the related headers for each protocol layer.

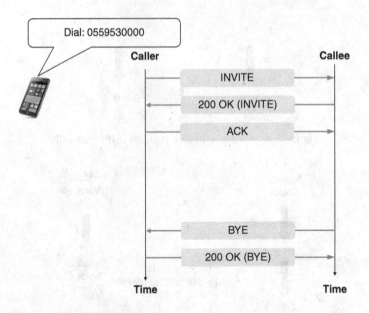

Figure 7.5 Simplified SIP call setup.

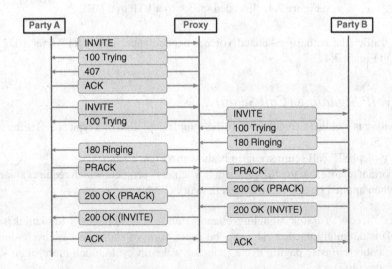

Figure 7.6 Typical SIP call setup.

As seen in Figure 7.7, the overhead due to the RTP, UDP, and IP headers is huge when compared to the actual VoIP payload. It is therefore absolutely necessary to deploy a technique to compress those headers.

Robust header compression (RoHC) [10–12] is used for VoLTE to compress the RTP/UDP/IP headers from 60 bytes (IPV6) or 40 bytes (IPV4) into a few bytes (~4 bytes) in order to improve voice efficiency (payload size to header size ratio). A detailed description of the RoHC is described later in this chapter.

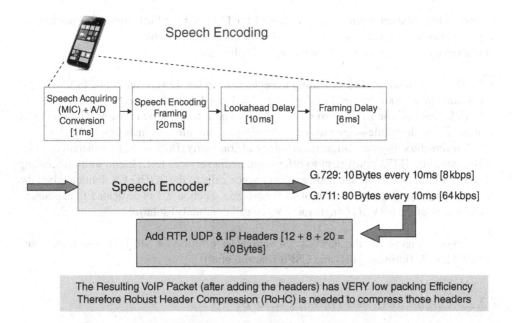

Figure 7.7   Speech encoding and UDP/IP encapsulation.

7.2   Voice Options for LTE

The road to providing voice services over an LTE network has a couple of options. For some operators, the target architecture for VoLTE is simply going directly with a VoLTE solution based on IMS.

The roadmap operators follow to IMS will be based on their individual business, economic and regional competitive requirements, which of course vary widely. So, while some operators will target initial deployment of VoLTE services with IMS, there will be other operators looking to leverage the LTE network for voice prior to investing in IMS. For these operators, there is a very viable option for utilizing their existing 2G or 3G circuit-switched voice network to cater for the voice services for their LTE subscribers. This is known as circuit-switched fallback (CSFB) [13].

7.2.1   SRVCC and CSFB

It is worth noting here that the two options mentioned above are also referred to as the "single-radio" solution. This is because, in general, there are two approaches that have been taken to enable providing voice services while offering 4G packet switched (PS) data services: dual radio solutions and single radio solutions. Dual radio solutions use two always-on radios (and supporting chipsets), one for PS LTE data and one for circuit switched (CS) telephony, and as a data fallback where LTE is not available.

Dual radio solutions have emerged for the LTE-CDMA2000 network interworking, driven by time-to-market pressures. Single radio solutions use one radio to handle both types of traffic, and use network signaling to determine when to switch from the PS network to the CS network. This solution is universally accepted for LTE 3GPP (third generation partnership project) network interworking solutions and is the focus of this book.

In summary, there are two 3GPP standardized solutions:

1. CSFB, as detailed in [14], providing a mechanism to hand over voice calls to the 2G/3G CS domain without the need to deploy any dedicated infrastructures to support voice in LTE. In that case the LTE network can be viewed as a data-centric network. This handover applies to both mobile-originated calls (MOCs) and mobile-terminated calls (MTCs).
2. IP multimedia subsystem single radio voice call continuity (IMS SRVCC) as detailed in [15, 16], based on IMS deployment to offer multimedia services to LTE end users (including voice) within LTE coverage and hand over voice calls to the 2G/3G CS domain when the user equipment (UE) moves out of LTE coverage. Section 7.3 is dedicated to explaining the details of the SRVCC (single radio voice call continuity) solution.

As shown in Figure 7.8, the 3GPP-based options for offering the VoLTE network are based on IMS SRVCC (ultimate goal) and CSFB (interim goal).

7.2.2   Circuit Switched Fallback (CSFB)

In simple terms, the CSFB enables the device that is operating in LTE mode to hand over to either the 3G or 2G circuit-switched networks to either receive or place a voice call [14].

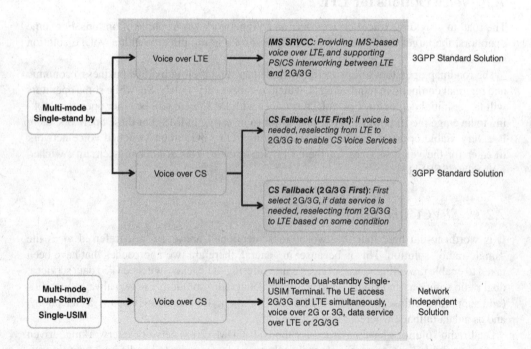

Figure 7.8   Offering voice service over LTE.

To clarify, let us take the example of an incoming call while the device is already on LTE with an established data connection. The LTE network will page the device. The device will respond with a service request message to the network, accordingly the network will signal the device to move (fall back) to 2G/3G to accept the incoming call. Similarly for outgoing calls, the same service request is used to move the device to 2G/3G to place the outgoing call.

While CSFB enables single radio voice solutions for LTE handsets, the switching requirements impose some technical challenges for which solutions, described in detail below, have been developed. CSFB addresses the requirements of the first phase of the evolution of mobile voice services, which began on a commercial scale in 2011. CSFB is the solution to the reality of mixed networks today and throughout the transition to ubiquitous all-LTE networks in the future phases of LTE voice evolution.

Table 7.1 summarizes the pros and cons of the CSFB.

As clearly detailed in Chapter 4, the CSFB and SMS (short message service) over SGs (serving grants) in the EPS (evolved packet system) function are realized by using the SG's interface mechanism between the MSC (mobile services switching center) server and the MME (mobility management entity). The SG's interface functionality is based on the mechanisms specified for the Gs interface in [17, 18].

Table 7.1 Pros and cons of CSFB

Pros	Cons
1. Re-use 2G/3G CS domain for voice calls (limited investment for supporting voice in LTE network) 2. 3GPP SGs interface mechanism is used 3. All SS and IN services are fully supported since Day 1 due to 2G/3G CS domain re-use. End users do not see differences to compare to today's 2G/3G voice service 4. Emergency call is supported from Day 1 5. Legal interception is supported from Day 1 6. Keep unchanged charging (to compare to 2G/3G)	**Equipment impact** 1. Both the MME and MSC-S/VLR are impacted. Especially, existing MSC and MSC-server should support SGs interface 2. SGW enhancement is needed if the ongoing PS sessions need to be maintained while the UE falls back to 2G/3G CS domain to initiate or respond to a voice call. There is a risk that it will be needed to suspend or interrupt on-going data sessions when making a voice call **Features and signaling impact** 3. Need to perform (two times) handover to 2G/3G network for voice call (in-coming and out-going) 4. Lot of additional signaling is generated to set up a voice call **Performance impact** 5. A mapping mechanism between the TAI and LAI is needed. The success rate of an MTC call may be impacted because of inaccurate mapping 6. Delay to set up a voice call (compared to 2G/3G) will be increased

TAI, tracking area identity.

7.2.2.1 CSFB Deployment

In this section, we want to highlight a practical aspect regarding deploying CSFB for existing operators. A typical operator would have multiple MSCs. To deploy CSFB, there are two options that can be used, as depicted in Figure 7.9:

1. Deploy the SGs interface per the MSC-server pool.
2. Upgrade all existing MSCs to support the SGs interface.

Table 7.2 summarizes a high-level comparison between the two options.

In summary, for solution "A," the MSC server and MME need to support the SG's interface. Furthermore, it is better to deploy a pair of high capacity MSC servers to provide CSFB service. As for solution "B," it is mandatory to upgrade all 2G/3G MSC units in the LTE coverage area.

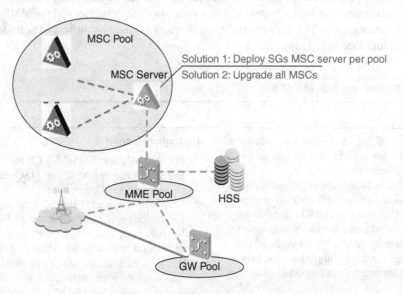

Figure 7.9   Options for deployment of CSFB.

Table 7.2   Deployment options for CSFB

Solution A	Pros	There is little impact on the existing network, because only MSC server needs to be upgraded or added
	Cons	Need to support MSC pool
Solution B	Pros	No need to support MSC pool
	Cons	Need to upgrade all MSC units in the LTE coverage area

7.2.2.2   Co-existence between CSFB and IMS/SRVCC

Both CSFB and IMS SRVCC can co-exist within the same network:

- CSFB is useful during the migration phase when operators are in the process of migrating the voice customers from CSFB to the IMS SRVCC solution. This is fully expected, especially as the CSFB-enabled handsets are already readily available while the IMS SRVCC-enabled handsets are still in the pipeline.
- As both CSFB and IMS SRVCC are standardized solutions, it is definitely the natural migration path to support both CSFB and IMS SRVCC in parallel for roaming purposes.
- According to [14], IMS SRVCC always has the highest priority over CSFB (MOC and MTC). In other words, an LTE UE will first check the IMS capabilities to setup the voice call.

7.3   IMS Single Radio Voice Call Continuity (SRVCC)

This section explains the SRVCC, which is based on the availability of IMS. First, we start by going through the IMS architecture and the standard-defined interfaces, then we go through the SRVCC and how it interacts with the different IMS components, showing the typical call flow and how the call continuity is being maintained.

The main concept of voice call continuity (VCC) is depicted in Figure 7.10. Wherever the LTE coverage is not ubiquitous and there is an ongoing VoLTE call, if the subscriber moves to a location that does not have LTE coverage, the VoLTE call has to be seamlessly and with minimal interruption to the ongoing voice call (VoLTE) moved to the existing 3G/2G CS network.

As depicted in Figure 7.11, the legacy networks provide different services via CS and PS domains. Namely, the CS domain provides voice and SMS services while the PS domain provides the data services.

After the LTE and IMS are introduced, the VoIP service and high-speed data service are simultaneously offered over LTE. Due to the lack of a call control function, the EPC needs the IMS system for call control and service provision. The IMS system is crucial as it provides

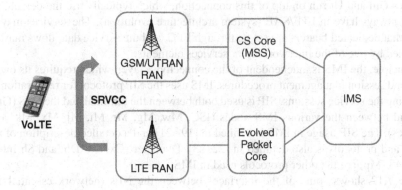

Figure 7.10   Voice call continuity – VoLTE to 3G/2G.

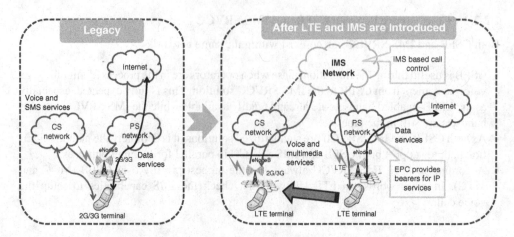

Figure 7.11   VoLTE – after LTE and IMS are introduced.

voice continuity from the LTE to the CS by anchoring the voice call. In the meantime, the LTE terminal uses the same MSISDN in the LTE and CS networks.

This section will present the capabilities described in Table 7.3 which are absolutely necessary for any operator to successfully deploy VoLTE.

7.3.1   IMS Overview

The IMS can be used to provide services over fixed and wireless accesses alike. The IMS architecture and functionality are defined in [19, 20]. In this book, we focus on the functional architecture of IMS as presented in Figure 7.12, and a short description of the main functions follows below.

IMS is an overlay service layer on top of the IP connectivity layer that the EPS provides. Figure 7.12 shows a line from the UE to the P-GW (PDN gateway) that represents the UE's IP connectivity to IMS and other external networks through the RAN and EPC. The signaling interfaces Gm and Ut run on top of this connection, which typically use the default bearer a UE will always have in LTE/SAE (system architecture evolution). The services may further require that dedicated bearers are set up through EPC, and the service data flows may need to be handled by one of the interworking or services elements.

In principle, the IMS is independent of the connectivity layer, which requires its own registration and session management procedures. IMS uses the SIP protocol for registration and for controlling the service sessions. SIP is used both between the terminal and the IMS (Gm interface) and between the various IMS nodes (ISC, Mw, Mg, Mr, Mi, Mj, Mx, Mk, and Mm Interfaces). The SIP usage in IMS is defined in [19–21] and a detailed description of the procedures and protocols is also detailed in [22, 23]. Diameter (Cx, Dx, Dh, and Sh Interfaces) and H.248 (Mp) are the other protocols used in IMS.

Figure 7.12 shows some of the interfaces between the NEs (network essentials) in the IMS-based VoLTE solution.

Table 7.3   Crucial capabilities required to successfully deploy VoLTE

Network capability	Main function	Why needed
IMS	Call control and anchoring function	Due to the lack of call control function, the EPC needs the IMS system for call control and service provision
Voice call continuity (VCC)	Voice call continuity from E-UTRAN to 3G/2G CS and PS domain	Absolutely necessary in the case of non-continuous LTE coverage (as an overlay on top of an existing 3G/2G network)
Enhanced SRVCC (eSRVCC)	Improved handover from LTE to 3G/2G by introducing the ATCF and ATGW	Less probability of interruption and call drops of VoLTE calls when handing over to 3G/2G
End-to-end QoS	RAN support for special bearer (QCI = 1), transport network (DSCP), EPC, and IMS	Without a guaranteed end-to-end QoS, voice quality can be compromised and will fail to meet user experience expectations
Semi-persistent scheduler (SPS)	Special dynamic scheduling for VoIP traffic	Lack of SPS can drastically affect the capacity (number of VoIP users supported per eNB)
TTI bundling	Ability to bundle four TTIs in the UL (especially for VoIP)	Significant improvement in VoIP link budget (improve indoor coverage)
CDRX	Connected mode short DRX allowing the UE to sleep between frames	Significant battery savings and talk-time
ROHC	RTP/UDP/IP header compression for VoIP	Capacity and efficiency
Vocoder (NB and WB) and associated de-jitter buffer	Narrowband AMR and wideband AMR (HD) plus a capable de-jitter buffer to cope with the variation in delay associated with any voice over PS service	Regular NB-AMR quality, HD voice (WB-AMR), and smart de-jitter buffer

The typical IMS infrastructure consists of the following areas:

- IMS core and applications, consisting of:
  - Media resource and access layer, containing the elements that support the handling of media and interaction with access networks such as:
    - Media gateway (IM-MGW)
    - Border gateway function (BGF)
    - Media resource function processor (MRFP).

- Control layer, containing the elements that support session and signaling control such as:
  - Call session control function (CSCF)
  - Media gateway control (MGC)
  - Session border control (SBC)
  - ENUM translation and DNS server.
- Application layer, containing the application servers that enable services in the network such as:
  - Multimedia telephony application server (MTAS)
  - Presence and group management (PGM)
  - IMS messaging gateway.
- Subscriber management, consisting of:
  - Home subscriber server (HSS) which handles the profiles of the IMS users.
  - Provisioning manager. Optionally used to hide network complexity, hence speeding up the integration process.
- User interface, consisting of IMS clients.

For the purpose of this book, the main focus is not to describe the entire IMS system components, functions, and interfaces. Therefore, we will describe only the main IMS components involved in VoLTE calls:

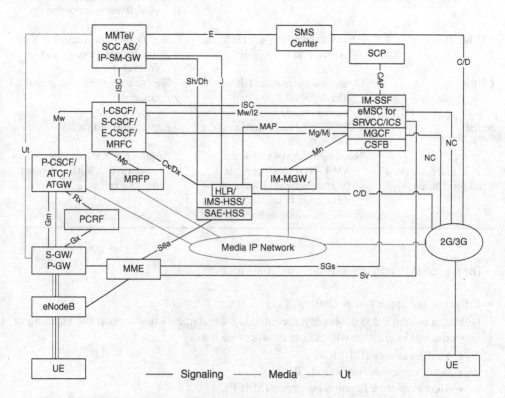

Figure 7.12   Interfaces between NEs in the IMS-based VoLTE solution.

1. **Call session control function (CSCF)**

   The CSCF handles session establishment, modification, and release of IP multimedia sessions using the SIP/SDP (session description protocol) protocol suite. CSCF is an essential node in IMS for processing signaling. It provides support for IPs on a scalable and high-performance platform. The CSCF has the following different roles in the network:

   (a) Proxy call session control function (P-CSCF).
   (b) Interrogating call session control function (I-CSCF).
   (c) Serving call session control function (S-CSCF).
   (d) Emergency call session control function (E-CSCF).
   (e) Break-out gateway control function (BGCF).
   (f) Break-in control function (BCF).

   These logical entities are designed to execute traffic in either collocated or standalone mode, based on network configurations that can be used in different network scenarios.

2. **Session border gateway (SBG)**

   The SBG is a high-performance carrier-class session controller product providing the ability to correlate all signaling and media streams (such as audio and video) that pass the network borders, hence providing a comprehensive suite of functions required to access and interconnect IP core domains (for example, IMS) and other IP networks with preserved security, privacy, and QoS.

   The SBG ensures network security, bandwidth fraud protection, topology hiding, QoS, service level agreements, hosted network address translation traversal, address, and port translation (NAPT), and other critical functions for real-time IP streams.

   Furthermore, the SBG enables lawful intercept, charging, IP-PBX business trunking, handling of the geographical location of users, access network connection admission control, and functionality for handling geographical redundancy of other nodes, as well as admission control and bearer authentication.

3. **Media gateway controller (MGC)**

   The MGC is a flexible system that can be integrated to different solutions and adapted to the various needs of the operator and the particular end users involved. The use of open standards and protocols, such as, SIP, ISUP, H.323, RTSP, and H.248, makes it possible for it to be integrated to different types of solutions.

   The MGC is a node that provides the signaling level interworking function between the CS telephone network and the PS multimedia networks. The MGC adapts the call and session level signaling between the two networks and controlling MGW nodes, as needed for setting media level connections between the CS circuits and the PS media streams. The MGW is responsible for handling the media payload received from/to the CS network. All CS payload circuits terminate on an MGW and the job is to adapt the payload carried on these circuits into RTP packets that are suitable for transport over an IP network.

   The MGC implements the media gateway controller function (MGCF) for the PSTN (packet switched telephone network) gateway subsystem for interworking with the Ericsson IMS core system.

   The CS interworking function provides the signaling conversion between the signaling protocols used in the CS networks and the session control protocols used in the IMS networks, as well as the conversion between the CS TDM (time-division multiplexing) bearer circuits and the PS VoIP media streams used in IMS networks.

4. **IM – Media gateway**

The MGW function is located at the boundary between CS networks and PS networks, providing adaptation of the media path between the networks. It is located at the edge of an IP network. It receives requests from the controller (MGC), and performs actions according to the requests received. It establishes connections using RTP/IP bearers and performs media conversion between the CS and the IP network. It also notifies the controller about events occurring in the MGW or the networks.

In addition, the MGW provides echo cancelation, law conversion (mu-law to A-law and vice versa), tone generation, and dual-tone multi-frequency (DTMF) tone detection and generation capabilities, including DTMF relay (RFC4733). It also provides transmission of fax.

The IM-MGW establishes connections between an IP network using RTP/IP bearers and the CS network and provides media conversion. The MGW also supports IP to IP interconnection, including media conversion (between different codecs).

5. **Multimedia resource function (MRF)**

The MRF function provides network-based media services in IMS, such as announcement and conference services. The MRF is split into a multimedia resource function controller (MRFC) and a multimedia resource function processor (MRFP).

6. **Telephony application server (TAS)**

The TAS is an IMS application server for first-line voice and multimedia communication services. It is truly a convergent application server with which operators are able to run a first-line communication service for both fixed and mobile networks.

TAS delivers high quality voice with support for the PSTN equivalent regulatory and supplementary services over IP. TAS does not store user subscription data. Instead, the HSS transparent data repository is used as a centralized storage of all TAS data (subscriber service profiles, supplementary services data).

Table 7.4 describes only the "essential" interfaces that are utilized during a typical VoLTE call. For a more comprehensive list of all the interfaces and their description, see Ref. in [20].

7.3.2 VoLTE Call Flow and Interaction with IMS

A typical call flow for a VoLTE call establishment starts with the UE power ON procedure including the UE attachment to the EPC which was explained in depth in Chapter 4. The control procedure specific to VoLTE attachment is shown in Figure 7.13. The steps are shown as follows:

1. UE is powered ON.
2. UE sends an attach request message to the MME.
3. MME executes a location registration with the HSS. In addition, it fetches the user profile for that VoLTE subscriber including the VoLTE APN (access point name). Furthermore, MME determines the PGW based on the VoLTE APN data fetched from the HSS.
4. MME sends a request to the PGW to set up the VoLTE bearer.
5. PGW assigns an IP address for the UE and also determines the P-CSCF to be assigned for this call.

Table 7.4 IMS interfaces

Interface	Location	Function	Protocol compliance
Gm	Located between an IMS UE and the P-CSCF	The Gm interface is used for registration and session control for IMS subscribers	SIP
Mw	Located between two CSCFs	The Mw interface is used for message exchange between CSCFs during IMS registration and session processes and also functions as a proxy	SIP
Cx	Located between the CSCF and the HSS	The CSCF and the HSS use the Cx interface to perform the following operations: • the I-CSCF queries the HSS for information required for selecting an S-CSCF • The CSCF queries the HSS for routing information • The CSCF queries the HSS for roaming authorization information • The CSCF downloads from the HSS security parameters required for IMS subscriber authentication • The HSS sends initial filter criteria (iFC) data associated with the IMS session to the CSCF	Diameter
Dx	Located between the CSCF and the SLF (subscriber locator function)	If multiple HSSs are deployed on an IMS network, the CSCF uses the Dx interface to query the SLF for the HSS that stores subscription data of a subscriber. The Dx interface is not required when only one HSS is deployed	Diameter
Mg	Located between the CSCF and the MGCF	The CSCF uses the Mg interface to interwork with other non-IMS networks, such as a PSTN, wireless CS network, IP-based 3G R4 network, and fixed next generation network (NGN)	SIP

(*continued overleaf*)

Table 7.4   (*continued*)

Interface	Location	Function	Protocol compliance
Mi	Located between the S-CSCF and the BGCF	The S-CSCF uses the Mi interface to forward session control signaling to the BGCF so that the BGCF can select an appropriate MGCF. This interface implements interworking between an IMS network and a PSTN or between an IMS network and a 3G or 2G CS network	SIP
Mj	Located between the BGCF and MGCF	The BGCF and MGCF use the Mj interface to exchange session control signaling to implement interworking between an IMS network and a PSTN or CS domain	SIP
Mk	Located between two BGCFs	The Mk interface is used to forward session control signaling from the BGCF that resides in the same network as the originating S-CSCF to the BGCF that resides in the same network as the MGCF	SIP
Mm	Located between the CSCF and a multimedia IP network	The CSCF uses the Mm interface to interwork with a multimedia IP network, including an IMS network	SIP
Mr	Located between the CSCF and the MRFC	By using the Mr interface, the S-CSCF instructs the MRFC to provide services such as announcement playback, digit collection, conference bridges, and video streams	SIP
Mx	Located between the I-BCF and the CSCF or BGCF	The Mx interface is used for interworking between the IMS network and a PS network	SIP
ISC	Located between the S-CSCF and the AS	The S-CSCF determines whether to trigger a service based on: • iFC template data obtained from the HSS • SIP service request sent from an IMS UE	SIP

Table 7.4 (*continued*)

Interface	Location	Function	Protocol compliance
		If the service needs to be triggered, the S-CSCF routes the session request over the ISC interface to a specific AS for final processing	
Rf	Located between the CCF (charging collection function)and other entities such as the CSCF, MRFC, BGCF, and AS	The Rf interface is used for session-related offline charging	Diameter
Ro	Located between the OCS and other entities such as the AS and open charging gateway (OCG)	The Ro interface is used for session-related online charging	Diameter
Rx	Located between the PCRF and the application function (AF)	The Rx interface is used for session information exchange at the application layer. Based on the session information, the PCRF determines the policies and charging control schemes to be used	Diameter
Gx	Located between the PCRF and the policy and charging enforcement function (PCEF)	The PCRF uses the Gx interface to deliver charging policies to the PCEF, and the PCEF uses the Gx interface to report bearer events to the PCRF	Diameter
Sh	Located between the HSS and the SIP AS or open service access-service capability server (OSA-SCS)	The AS uses the Sh interface to query the HSS for value-added service data and to synchronize data to the HSS	Diameter
Dh	Located between the SLF and the AS or OSA-SCS	The AS uses the Dh interface to identify the HSS that stores subscriber data based on the subscriber identity and home domain information	Diameter
Mn	Located between the MGCF and the IM-MGW	The MGCF uses the Mn interface to control media streams transmitted on the IM-MGW and allocation of special resources	H.248

(*continued overleaf*)

Table 7.4 (*continued*)

Interface	Location	Function	Protocol compliance
Mp	Located between the MRFC and the MRFP	The MRFC uses the Mp interface to control the MRFP for announcement playback, conferences, and dual tone multiple frequency (DTMF) digit collection	H.248
Mb	Located between an IMS network and an IP network	The Mb interface allows IMS user- and control-plane data to be transmitted over the IP network	IP
Ut	Located between a UE and the AS	A UE uses the Ut interface to manage and customize services on the MMTel AS	XML configuration access protocol (XCAP)
Sv	Located between the SRVCC-IWF and the MME	The Sv interface is used to control and implement SRVCC services	GTPv2
Mw/I2	Located between the ATCF and the SRVCC-IWF or ICS enhanced MSC server	The Mw/I2 interface is used to control eSRVCC handovers or to allow ICS subscribers to access an IMS network	SIP
E	Located between the IP-SM-GW and the SMSC	The E interface is used for SMS interworking between an IMS network and a CS domain	Mobile application part (MAP)
J	Located between the IP-SM-GW and the HLR	The J interface is used to obtain SMS routing information	MAP
C/D	Located between the HLR and the VMSC, GMSC, SMSC, or eMSC (SRVCC and ICS)	The C/D interface is used to transfer the location and routing information about mobile subscribers	MAP
Nc	Located between MSCs	The Nc interface is used to control inter-MSC calls	BICC/ISUP/TUP
Mc	Located between the MSC and MGW or the eMSC (SRVCC IWF (interworking function) or ICS) and the MGW	The Mc interface is used to control bearer resources	H.248
CAP	Located between the IM-SSF and the SCP	The CAP interface allows IMS subscribers to use traditional IN services	C application part (CAP)

OCS, online charging system.

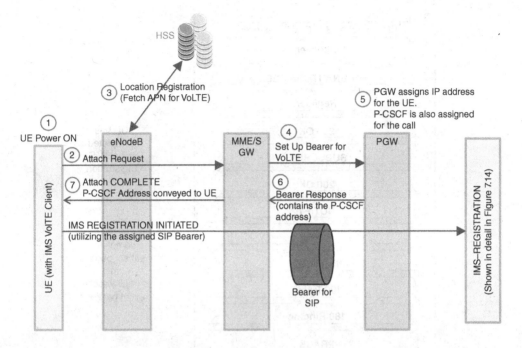

Figure 7.13 VoLTE UE attach procedure.

6. The PGW sets the P-CSCF address in the protocol configuration option and communicates it to the SGW (serving gateway) in the bearer response. SGW passes the same address to the MME.
7. MME sends the attach complete to the UE containing the P-CSCF address. This represents the completion of the attach procedure and the start of the registration procedure.

Next, the UE IMS client will start the IMS call flow which consists of three parts. The first part starts after the UE is attached to the ePC and represents the registration phase. This is where the UE establishes and completes the initial IMS registration. The second part is where the UE completes the subscription to the registration event package. The third and last part is where a VoLTE call is established between the UE and the IMS which starts with the invite, followed by the ringing until the call is established and the VoIP RTP traffic is established between the two endpoints.

The IMS registration procedure between a VoLTE UE and the IMS is shown in Figure 7.14.

In Figure 7.15, a more detailed call flow is shown in steps as follows: before sending any SIP requests, the UE must perform "P-CSCF discovery" where the UE identifies the P-CSCF that will handle the VoLTE call. The next part of the procedural flow includes IMS registration, event subscription, and call connection.

After authentication, security and UE capability requests, the network accepts the attach request and activates the EPS bearer context.[1] Accordingly a PDP (packet data protocol)

---

[1] The detailed call flow is outlined in Chapter 4.

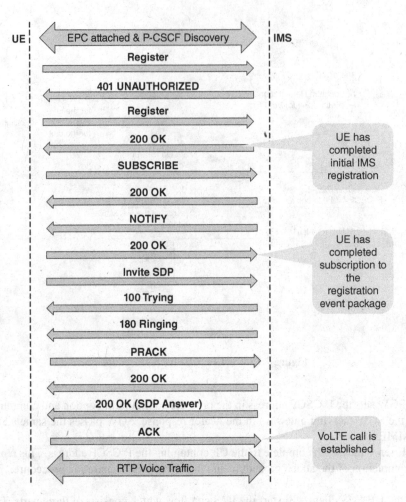

Figure 7.14   VoLTE registration and call setup.

context is now established for this UE. The typical IMS SIP client registration call flow is as follows:

1. The IMS client attempts to register by sending a register request to the P-CSCF.
2. The P-CSCF forwards the register request to the I-CSCF.
3. The I-CSCF queries the HSS to decide which S-CSCF should handle the register request and decides which S-CSCF to forward the register to.
4. The I-CSCF forwards the register request to the chosen S-CSCF.
5. The S-CSCF then sends the P-CSCF a 401 (unauthorized) response as well as a challenge string in the form of a "number used once" or "nonce."
6. The P-CSCF forwards the 401–unauthorized response to the UE. Both the UE and the network have stored some shared secret data (SSD) which is stored on the UE's ISIM or USIM (universal subscriber identity module) and correspondingly stored on the HSS.

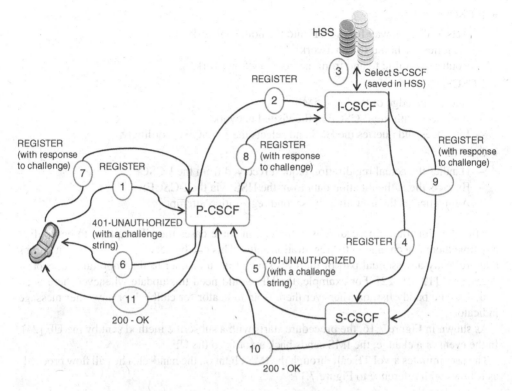

Figure 7.15 Detailed call flow for VoLTE registration.

7. The UE sends a register request to the P-CSCF. This time the request includes the nonce and SSD.
8. The P-CSCF forwards the new register request to the I-CSCF.
9. The I-CSCF forwards the new register request to the S-CSCF.
10. The S-CSCF polls the HSS (via the I-CSCF) for the SSD, hashes it against the nonce and determines whether the UE should be allowed to register. If the values match, the S-CSCF sends 200–OK response to the P-CSCF. At this point an IPSec (Internet protocol security) association is established by the P-CSCF
11. The P-CSCF forwards the 200–OK response to the UE.

For the purpose of this book, the intent is not to explain the function of every single IMS network element, however, it is very important for the reader to understand the functions of the following network elements and their essential role during the IMS registration:

• UE with the IMS Client:
  – Attaches to the LTE Network and activates the PDP context
  – Discovers the P-CSCF to use, makes a deliberate unauthenticated registration attempt, then waits for the expected 401 response
  – Extracts the nonce from the response and hashes it with the SSD and includes all in the second registration request.

- P-CSCF:
  - Acts as the gateway for the UE into the home network
  - Identifies the home IMS network
  - Routes the traffic to and from the home IMS network.
- I-CSCF:
  - Acts as the edge of the home IMS
  - Interfaces with the P-CSCF in the visited network
  - Interfaces and queries the HSS and selects the S-CSCF accordingly.
- S-CSCF:
  - Handles the actual regidtration request received from the I-CSCF
  - Extracts the authentication data from the HSS (via the I-CSCF)
  - Authenticates the user after the second registration attempt.

After the registration procedure is complete, in some cases the UE requests to be notified any time there is a change in the registrationstatus. This can be necessary in certain situations that are really an essential part of the IMS since it is an enabler to the important concept of "presence" [19, 21, 22]. For example, the UE would need this update whenever there is an update to the buddy list, or whenever there is an indicator for callback or any other message indicator.

As shown in Figure 7.16, the procedure starts with a subscribe method sent by the UE [24]. In the event of a change, the IMS sends back a notify to the UE.

The user initiates a VoLTE call through the IMS client on the handset. The call flow proceeds as follows with reference to Figure 7.17:

1. This triggers the call origination.
2. An invite message is sent to the terminating terminal via the P-CSCF, the S-CSCF, and the TAS. That invite message has several pieces of information including the originating and terminating terminals IDs and the supported codecs of the originating terminal.
3. The terminating terminal responds with a "183 session progress" message. This message contains the information of the supported codecs on the terminating terminal.
4. When the above message is received by the P-CSCF, it signals to the PCRF (policy and charging rules function) to set up the dedicated bearer for the voice media.

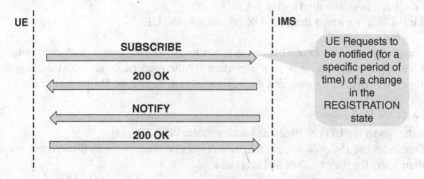

Figure 7.16   UE request to be notified of a change in registration.

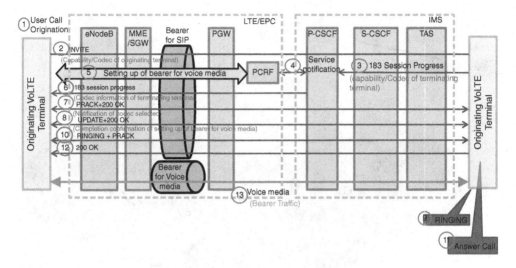

Figure 7.17 VoLTE call origination.

5. The dedicated bearer for the voice media is set up by the PGW, SGW, eNodeB (eNB), and originating terminal. The appropriate QCI (QoS class identifier) (QCI = 1; conversational speech) is assigned to such a dedicated bearer.

6. In parallel, the P-CSCF forwards the "183 session progress" message to the originating terminal. Accordingly, the originating terminal compares the codec capabilities of the terminating terminal to its own and decides the codec to be used.

7. The originating terminal replies back to the terminating terminal using PRACK notifying it of the codec to be used, accordingly the terminating terminal responds with "200 OK."

8. Both terminals confirm the setup of the voice bearer through the "update" message from the originating terminal and the corresponding response of "200 OK" from the terminating terminal.

9. Following the above confirmation, the ringtone starts playing at the terminating terminal.

10. The terminating terminal sends back a "PRACK" confirming the ringing.

11. The ser at the terminating terminal answers the call.

12. The terminating terminal sends a "200 OK" to the originating terminal informing it that the call has been answered and that the session is now established and the actual voice traffic can now start through the already established voice dedicated bearer utilizing the assigned QCI (QCI = 1; conversational speech).

7.3.3   Voice Call Continuity Overview

As previously mentioned, the VoLTE evolution uses the SRVCC that can seamlessly maintain the Voice Calls as the VoLTE mobile users move from LTE to non-LTE coverage areas as shown in Figure 7.18. This transfer of the VoLTE call in progress to the legacy 2G/3G CS voice network needs to be done without compromising the voice quality, the latency, and the overall QoS levels available currently in legacy networks [15].

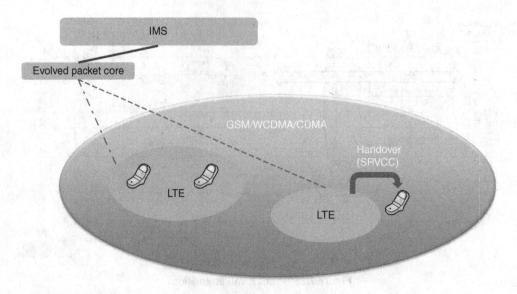

Figure 7.18   Voice call continuity.

Since its introduction in Release 8, SRVCC has evolved with each new release, a brief summary of SRVCC capability and enhancements is given below.

1. **3GPP Release 8**

   It introduces SRVCC for voice calls that are anchored in the IMS core:

   (a) From E-UTRAN (evolved universal terrestrial radio access network) access to 3GPP2 1xCS.

   (b) From E-UTRAN/UTRAN (HSPA, high speed packet access) access to 3GPP UTRAN/GERAN (GSM/EDGE radio access network) CS accesses.

   To support this functionality 3GPP introduced a new protocol interface and procedures between the MME and the MSC (Sv interface) for SRVCC from E-UTRAN to 3GPP UTRAN/GERAN, between the SGSN (serving GPRS support node) and the MSC (Sv interface) for SRVCC from UTRAN (HSPA) to 3GPP UTRAN/GERAN and between the MME and the 1xCS IWS (S102 Interface) for SRVCC from EUTRAN to CDMA 2000 1xRTT.

2. **3GPP Release 9**

   It introduces the SRVCC support for emergency calls that are anchored in the IMS core. IMS emergency calls, placed via LTE access, need to continue when SRVCC handover occurs from the LTE network to GSM/UMTS/CDMA (global system for mobile communications/universal mobile telecommunications system/code division multiple access) networks. This evolution resolves a key regulatory exception. This enhancement supports IMS emergency call continuity:

   (a) From E-UTRAN to CDMA2000 1xCS.

   (b) From E-UTRAN/UTRAN (HSPA) to 3GPP UTRAN/GERAN CS.

   Functional and interface evolution of EPS entities are needed to support the IMS emergency call with SRVCC.

3. **3GPP Release 10**

It introduces procedures of enhanced SRVCC:

(a) Support of mid-call feature during SRVCC handover (eSRVCC).

(b) Support of SRVCC PS-CS transfer of a call in the alerting phase (aSRVCC).

The MSC server-assisted mid-call feature enables PS–CS access transfer for UEs not using the IMS centralized service (ICS) capabilities, while preserving the provision of mid-call services (inactive sessions or sessions using the conference service).

SRVCC in the alerting phase feature adds the ability to perform access transfer of media of an IM session in the PS to CS direction in the alerting phase for access transfers.

4. **3GPP Release 11**

It introduces two new capabilities:

(a) SRVCC for 3G-CS (vSRVCC).

(b) SRVCC from UTRAN/GERAN to E-UTRAN/HSPA (rSRVCC).

The vSRVCC feature provides support of video call handover from E-UTRAN to UTRAN-CS for service continuity when a video call is anchored in IMS and the UE is capable of transmitting/receiving on only one of those access networks at a given time.

Service continuity from UTRAN/GERAN CS access to E-UTRAN/HSPA was not specified in 3GPP Releases 8, 9, and 10. To overcome it, Release 11 provides support of VCC from UTRAN/GERAN to E-UTRAN/HSPA.

To enable video call transfer from E-UTRAN to UTRAN-CS, IMS/EPC is evolved to pass relevant information to the EPC side and S5/S11/Sv/Gx/Gxx interfaces are enhanced for video bearer related information transfer.

To support the SRVCC from GERAN to E-UTRAN/HSPA, GERAN specifications are evolved to enable the MS (mobile station) and BSS (business support system) to support seamless service continuity when MS handovers from GERAN CS access to E-UTRAN/HSPA for voice call.

To support SRVCC from UTRAN to E-UTRAN/HSPA, UTRAN specifications are evolved to enable the RNC (radio network controller) to perform rSRVCC HO and provide relative UE capability information to the RNC.

7.3.4 SRVCC from VoLTE to 3G/2G

The SRVCC handover mechanism is fully controlled by the network. Accordingly, the call remains under the control of the IMS core network, which maintains access to subscribed services implemented in the IMS service engine before, during, and after the handover, as shown in Figure 7.19. To achieve this goal, there are two essential phases that need to be successfully completed:

1. IRAT (inter-radio access technology) handover: handover the VoLTE terminal from LTE radio access to WCDMA/GSM (wideband code division multiple access) radio access.
2. A mechanism to move access control and voice media anchoring from the EPC to the MSC-S.

The combination of the above two steps has raised concerns in the telecom industry regarding the ability of the technology to meet the required performance levels. Experiments illustrated in [25] using commercial infrastructure and test phones with commercial chipsets show that

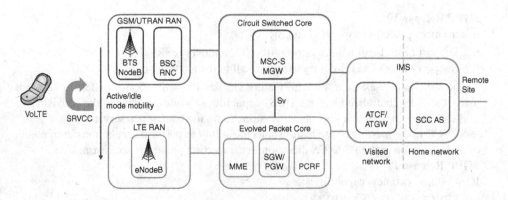

Figure 7.19   Centralized control and voice media anchoring in SRVCC.

the SRVCC mechanism can match the requirements for performance standards – with voice interruption of less than 300 ms and a handover success rate greater than 99%.

Since its initial specification in 3GPP Release 8, SRVCC has evolved continuously. To ensure interoperability of the various implementations, the GSMA has provided a set of guidelines for SRVCC [26], detailing the requirements for networks and terminals.

The GSMA guidelines recommend 3GPP Release 10 architecture for SRVCC as shown in Figure 7.19 as it supports the reduction of both voice-interruption delay during handover and the dropped-call rate compared with earlier configurations.

The Release 10 configuration includes all components needed to manage time-critical signaling between the terminal and the network, and between network elements in the serving network (visited network). Signaling, therefore, follows the shortest possible path and is as robust as possible. As a result, voice-interruption time caused by switching from the packet core to the circuit core is minimized and consistent, irrespective of whether the terminal is located in its home network or roaming.

3GPP has specified in [15, 20] the principles for centralization and continuity of services in the IMS in order to provide consistent services to the user regardless of the attached access type. In order to support this principle, originated and terminated sessions via the CS or PS domains need to be anchored in the service centralization and continuity application server (SCC AS) in the IMS. In other words, the anchoring point controlling the call becomes the SCC AS.

The other important interface is the Sv interface, which is introduced between the MSC-S and the MME, and through which the call control can be handed over from the MME to the MSC-S.

The next section explains the typical procedure of the SRVCC handover from the LTE to the 3G/2G [15]. The Sv interface requires an upgrade to the legacy MSC Server (to become MSC-S SRVCC ready).

For facilitating session transfer (SRVCC) of the voice component to the CS domain, the IMS multimedia telephony sessions needs to be anchored in the IMS.

The high level concept of the SRVCC from E-UTRAN to UTRAN/GERAN is shown in Figure 7.20. For SRVCC from E-UTRAN to UTRAN/GERAN, MME first receives the handover request from E UTRAN with the indication that this is for SRVCC handling, and

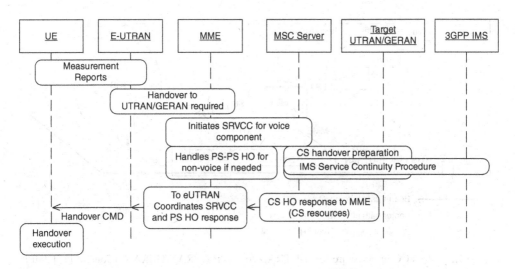

Figure 7.20 High level concepts for SRVCC from E-UTRAN to UTRAN/GERAN. (Source: [15] 3GPP. Reproduced with permission of ETSI.)

then triggers the SRVCC procedure with the MSC server enhanced with SRVCC via the Sv reference point if MME has the SRVCC session transfer number for SRVCC (STN-SR) information for this UE. The MSC server, enhanced for SRVCC, then initiates the session transfer procedure to IMS and coordinates it with the CS handover procedure to the target cell. The MSC server, enhanced for SRVCC, then sends the PS-CS handover response to MME, which includes the necessary CS HO command information for the UE to access the UTRAN/GERAN.

Handling of any non-voice PS bearer is done by the PS bearer splitting function in the MME. The MME starts the handover of the non-voice PS bearer during the SRVCC procedure, based on the information received from E UTRAN. The handover of non-voice PS bearer(s), if performed, is done according to the IRAT handover procedure, as defined in [18]. The MME is responsible for coordinating the forward relocation response from the PS-PS handover procedure and the SRVCC PS to the CS response.

The SRVCC architecture and the additional function provided by the MSC server (MSC server enhanced for SRVCC) are shown in Figure 7.21. This architecture also applies to the roaming scenario (i.e., S8, S6a are not impacted due to SRVCC). The Sv interface is the main reference point to interface the MME with the MSC server.

An MSC server which has been enhanced for SRVCC provides the following functions as needed for support of SRVCC:

- Handling the relocation preparation procedure requested for the voice component from MME/SGSN via the Sv reference point;
- Behaving as the MSC Server enhanced for ICS as defined in TS 23.292 [19] if supported and when the ICS flag received via the Sv reference point is set to true and, optionally, if the MSC server is configured to know that the VPLMN (visited public land mobile network) has a suitable roaming agreement with the HPLMN (home public land mobile network) of the UE;

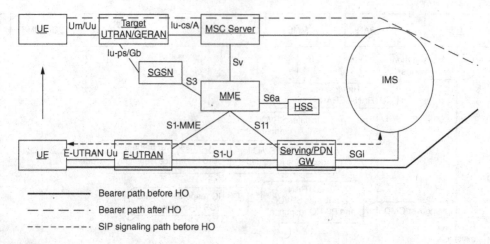

Figure 7.21   SRVCC architecture for E-UTRAN to 3GPP UTRAN/GERAN. (Source: [15] 3GPP. Reproduced with permission of ETSI.)

- Invoking the session transfer procedure or emergency session transfer procedure from IMS to CS according to TS 23.237 [20];
  - Coordinating the CS handover and session transfer procedures;
- Handling the MAP update location procedure without it being triggered from the UE.

The MME follows the rules and procedures described in TS 23.401 [18] with the following additions and clarifications:

- Performing the PS bearer splitting function by separating the voice PS bearer from the non-voice PS bearers.
- Handling the non-voice PS bearers handover with the target cell according to the IRAT handover procedure as defined in TS 23.401 [18].
- Initiating the SRVCC handover procedure for handover of the voice component to the target cell via the Sv interface and including an emergency indication if this is an emergency session. If there are multiple voice bearers and one of those is for IMS emergency session then MME only executes the SRVCC for emergency. It is worth noting that the UE may have two voice PS bearers if both emergency and normal IMS voice sessions are ongoing. Only the PS voice bearer associated with the IMS emergency session will be executed for SRVCC.
- Coordinating PS handover and SRVCC handover procedures when both procedures are performed.
- Sending the equipment identifier to the MSC server during the handover procedure for the case of UEs operating in limited service mode.

The SRVCC UE indicates to the network that the UE is SRVCC-capable when being configured for using the IMS speech service supported by the home operator, for example, the IMS Multimedia Telephony Service for bi-directional speech, as described in TS 22.173 [21], and the operator policy on the SRVCC UE, as specified in TS 23.237 [20] does not restrict the session transfer.

The SRVCC STN-SR, C MSISDN, and optional ICS flag per VPLMN for the subscriber are downloaded to MME from HSS during the E-UTRAN attach procedure. HSS also informs the MME when the STN-SR is modified or removed from the subscription.

The PCRF enforces the architecture principle to use QCI = 1 (and traffic-class conversational with source statistics descriptor = "speech") for voice bearer with IMS sessions anchored in the SCC AS, based on the service the session relates to. This may be achieved by deploying the S9 reference point, or configuration and roaming agreements.

The E-UTRAN attach procedure for the 3GPP SRVCC UE is performed as defined in TS 23.401 [18] with the following additions:

- SRVCC UE includes the SRVCC capability indication as part of the "MS network capability" in the attach request message and in tracking area updates. MME stores this information for SRVCC operation. The procedures are as specified in TS 23.401 [18]. If the service configuration on the UE is changed (e.g., the user changes between an IMS speech service supported by the home operator and a PS speech service incompatible with SRVCC), the UE can change its SRVCC capability indication as part of the "MS network capability" in a tracking area update message.
- SRVCC UE includes the GERAN MS Classmark 3 (if GERAN access is supported), MS Classmark 2 (if GERAN or UTRAN access or both are supported) and supported codecs IE (if GERAN or UTRAN access or both are supported) in the attach request message and in the non-periodic tracking area update messages. MS Classmark 2, MS Classmark 3, STN-SR, C-MSISDN, ICS indicator, and the supported codec IE are not sent from the source MME to the target MME/SGSN at inter CN-node idle mode mobility.
- HSS includes SRVCC STN-SR and C MSISDN as part of the subscription data sent to the MME. If the SRVCC STN-SR is present, it indicates the UE is SRVCC subscribed. If a roaming subscriber is determined by the HPLMN not allowed to have SRVCC in the VPLMN then HSS does not include SRVCC STN-SR and C MSISDN as part of the subscription data sent to the MME.
- MME includes a "SRVCC operation possible" indication in the S1 AP initial context setup request, meaning that both the UE and the MME are SRVCC-capable.

Intra-E-UTRAN S1-based handover and E-UTRAN to UTRAN (HSPA) Iu mode IRAT handover procedures for 3GPP SRVCC UE are performed, as defined in TS 23.401 [18] and as detailed in Chapter 4, with the following additions:

- MS Classmark 2, MS Classmark 3, STN-SR, C-MSISDN, ICS indicator, and the supported codec IE are sent from the source MME to the target MME/SGSN if available.
- The target MME includes a "SRVCC operation possible" indication in the S1-AP handover request message, meaning that both the UE and the target MME are SRVCC-capable.
- The target SGSN includes a "SRVCC operation possible" indication in the RANAP (radio access network application part) common ID message, meaning that both the UE and the target SGSN are SRVCC-capable.
- For X2-based handover, the source eNB includes a "SRVCC operation possible" indication in the X2-AP handover request message to the target eNB, as specified in TS 36.423 [27].

A detailed call flow for SRVCC from E-UTRAN to GERAN without DTM support is described in [15]. For the purpose of practicality, this book will only discuss the E-UTRAN to UTRAN with PS HO or GERAN with DTM HO support.

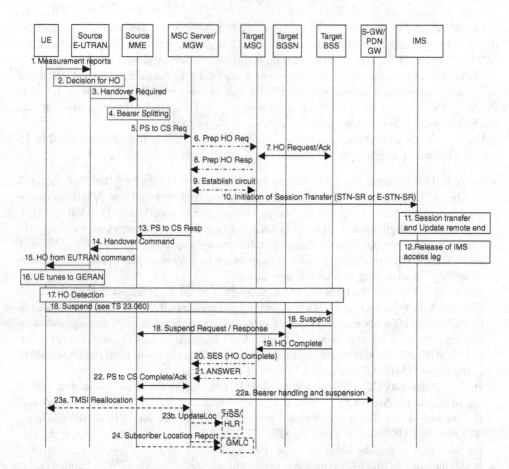

Figure 7.22 SRVCC call flow for E-UTRAN to 3GPP UTRAN/GERAN. (Source: [15] 3GPP. Reproduced with permission of ETSI.)

Figure 7.22 shows a call flow for SRVCC from E-UTRAN to UTRAN or GERAN with DTM HO support, including the handling of the non-voice component.

The flow requires that the eNB can determine that either the target is UTRAN with PS HO or the target is GERAN with DTM support and the UE is supporting DTM. The SRVCC procedure is as follows:

1. The UE sends measurement reports to E-UTRAN.
2. Based on the UE measurement reports the source E UTRAN decides to trigger an SRVCC handover to UTRAN/GERAN.
3. If the target is UTRAN, the source E UTRAN sends a handover required (target ID, generic source to target transparent container, SRVCC HO indication) message to the source MME. SRVCC HO indication indicates to the MME that this is for CS + PS HO. When the source E-UTRAN indicates using SRVCC HO indication that the target is both CS- and PS-capable and this is a CS + PS HO request, the source MME sends the single received transparent container to both the target CS domain and the target PS domain. If the target is GERAN, the source E UTRAN sends a handover required (target ID, generic

source to target transparent container, additional source to target transparent container, SRVCC HO indication) message to the source MME. The E UTRAN places the "old BSS to new BSS information IE" for the CS domain in the additional source to target transparent container. In this case, the MME identifies from the SRVCC HO indication that this is a request for a CS + PS handover.

4. Based on the QCI associated with the voice bearer (QCI 1) and the SRVCC HO indication, the source MME splits the voice bearer from all other PS bearers and initiates their relocation toward the MSC Server and SGSN, respectively.

5. This step consists of the following:

(a) Source MME initiates the PS-CS handover procedure for the voice bearer by sending a SRVCC PS to CS request (IMSI (international mobile subscriber identity), target ID, STN-SR, C MSISDN, source to target transparent container, MM context, emergency indication) message to the MSC server. The emergency indication and the equipment identifier are included if the ongoing session is an emergency session. Authenticated IMSI and C MSISDN is also included if available. The message includes information relevant to the CS domain only. The MME received STN-SR and C MSISDN from the HSS as part of the subscription profile downloaded during the E UTRAN attach procedure. MM context contains security related information.

(b) The MSC Server interworks the PS-CS handover request with a CS inter MSC handover request by sending a prepare handover request message to the target MSC. If the target system is GERAN, the MSC server assigns a default SAI (serving area interface) as the source ID on the interface to the target BSS, and uses BSSMAP encapsulated for the prepare handover request. If the target system is UTRAN, the MSC server uses RANAP encapsulated for the prepare handover request.

(c) The target MSC requests resource allocation for the CS relocation by sending the relocation request/handover request message to the target RNS (radio network subsystem)/BSS. If the target RAT (radio access technology) is UTRAN, the relocation request/handover request message contains the generic source to target transparent container. If the target RAT is GERAN, the relocation request/handover request message contains the additional source to target transparent container.

6. In parallel to the previous step, the source MME initiates relocation of the PS bearers. The following steps are performed:

(a) The source MME sends a forward relocation request (generic source to target transparent container, MM context, PDN (packet data network) connections IE) message to the target SGSN. If the target SGSN uses S4-based interaction with the S-GW and P-GW, the PDN connections IE includes bearer information for all bearers except the voice bearer.

(b) The target SGSN requests resource allocation for the PS relocation by sending the relocation request/handover request (source to target transparent container) message to the target RNS/BSS.

7. After the target RNS/BSS receives both the CS relocation/handover request with the PS relocation/handover request, it assigns the appropriate CS and PS resources. The following steps are performed:

(a) The arget RNS/BSS acknowledges the prepared PS relocation/handover by sending the relocation request acknowledge/handover request acknowledge (target to source transparent container) message to the target SGSN.

(b) The target SGSN sends a forward relocation response (target to source transparent container) message to the source MME.

8. In parallel to the previous step the following steps are performed:

    (a) The target RNS/BSS acknowledges the prepared CS relocation/handover by sending the relocation request acknowledge/handover request acknowledge (target to source transparent container) message to the target MSC.

    (b) The target MSC sends a prepare handover response (target to source transparent container) message to the MSC server.

    (c) Establishment of circuit connection between the target MSC and the MGW associated with the MSC server.

9. For a non-emergency session, the MSC server initiates the session transfer by using the STN-SR, for example, by sending an ISUP IAM (STN-SR) message toward the IMS. For an emergency session, the MSC server initiates the session transfer by using the locally configured E-STN-SR and by including the equipment identifier. Standard IMS service continuity or emergency IMS service continuity procedures are applied for execution of the session transfer.

10. During the execution of the session transfer procedure the remote end is updated with the SDP of the CS access leg, according to TS 23.237 [20]. The downlink flow of VoIP packets is switched toward the CS access leg at this point.

11. The source IMS access leg is released according to TS 23.237 [20].

12. The MSC server sends a SRVCC PS to CS response (target to source transparent container) message to the source MME.

13. Source MME synchronizes the two prepared relocations and sends a handover command (target to source transparent container) message to the source E UTRAN.

14. E UTRAN sends a handover from the E UTRAN command message to the UE.

15. The UE tunes to the target UTRAN/GERAN cell.

16. Handover detection at the target RNS/BSS occurs. The UE sends a handover complete message via the target RNS/BSS to the target MSC. If the target MSC is not the MSC server, then the target MSC sends an SES (handover complete) message to the MSC server. At this stage, the UE re-establishes the connection with the network and can send/receive voice data.

17. The CS relocation/handover is complete. The following steps are performed:

    (a) The target RNS/BSS sends relocation complete/handover complete message to the target MSC.

    (b) The target MSC sends an SES (handover complete) message to the MSC server. The speech circuit is through connected in the MSC server/MGW according to TS 23.009 [28].

    (c) Completion of the establishment procedure with the ISUP answer message to the MSC server according to TS 23.009 [28].

    (d) The MSC Server sends a SRVCC PS to CS complete notification message to the source MME. The source MME acknowledges the information by sending a SRVCC PS to CS complete acknowledge message to the MSC server.

    (e) The source MME deactivates the voice bearer toward S-GW/P-GW and sets the PS-to-CS handover indicator to delete bearer command message. This triggers the MME-initiated dedicated bearer deactivation procedure.

    (f) If the HLR (home location register) is to be updated, that is, if the IMSI is authenticated but unknown in the VLR (visitor location register), the MSC server performs a TMSI (temporary mobile subscriber identity) reallocation toward the UE using its own non-broadcast LAI (location area identification) and, if the MSC server and other MSC/VLRs serve the same (target) LAI, with its own network resource identifier (NRI). The TMSI reallocation is performed by the MSC server toward the UE via target MSC.

(g) If the MSC server performed a TMSI reallocation in step 17f, and if this TMSI real-location was completed successfully, the MSC server performs a MAP (mobile appli-cation part) update location to the HSS/HLR.

18. In parallel to the previous step, the PS relocation/handover is completed. The following steps are performed:

(a) The target RNS/BSS sends relocation complete/handover complete message to the target SGSN.

(b) The target SGSN sends a forward relocation complete message to the source MME. After having completed step 17e, the source MME acknowledges the information by sending a forward relocation complete acknowledge message to the target SGSN.

(c) The target SGSN updates the bearer with S GW/P GW/GGSN (gateway GPRS support node) as specified in TS 23.401 [18].

(d) The MME sends a delete session request to the SGW.

(e) The source MME sends a release resources message to the source eNB. The source eNB releases its resources related to the UE.

19. For an emergency services session after handover is complete, the source MME or the MSC server may send a subscriber location report carrying the identity of the MSC Server to a GMLC (gateway mobile location center) associated with the source or target side, respectively, to support location continuity.

From a user-experience point of view, SRVCC needs to meet two KPIs (key performance indicators): (i) voice-interruption time and (ii) call-retention probability. A detailed perfor-mance measurement was illustrated in [25] that revealed the following:

1. **Voice-interruption time:**

Both the IRAT handover in the RAN and the session transfer in the core network con-tribute to voice-interruption time as they break and remake the connection. As shown in Figure 7.23, the SRVCC minimizes overall interruption by initiating these procedures in parallel.

Measurement reveals that the session-transfer process is fast, on the order of 0.01 s. Given that the time to execute the media-redirection step (moving the media anchor to the MSS (MSC server) domain) is relatively short, voice interruption is dominated by the IRAT han-dover delay.

However, voice interruption is not equal to the IRAT handover delay. The IRAT handover delay is the time it takes for the terminal to receive the handover command from the network (as shown in Figure 7.23 as IRAT handover execution) to the time it is synchronized on the new radio access and an ACK message has been received. The voice interruption is dominated by IRAT handover delay but is measured from when the last voice packet is sent over LTE until voice media is sent over CS access.

Another interesting performance metric is the time it takes for the network to prepare the SRVCC handover. This is the time between the moment the terminal measures and reports bad LTE reception and the moment the terminal receives and executes the handover com-mand. Handover preparation delay does not directly impact the voice stream experienced by

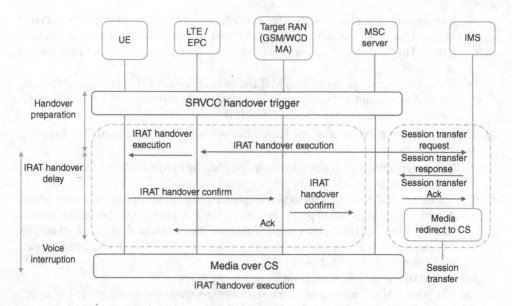

Figure 7.23   SRVCC – IRAT and session transfer parallel procedure.

users, but it must be short otherwise the handover may fail due to poor LTE reception when handover execution occurs. Long preparation delay may lower call-retention probability.

As indicated in [25], the handover preparation time had an average of 520 ms, handover procedure delay of 180 ms and average voice interruption time of 190 ms. All figures were very close to the expected typical CS IRAT handovers (between WCDMA and GSM) and inter-frequency handovers within a RAT (occurs when the RAN orders the terminal to retune the radio to a new frequency).

2. **Call-retention probability:**
Based on data collected from existing commercial deployments [25], the percentage of successful handovers for IFHO in well-planned networks is in the 98–99% range. Most of the calls for which handover fails return successfully to the original cell and continue, so the actual retention-failure rate is typically significantly less than 0.5%. Similar statistics are probable for SRVCC. Should call retention become a problem in certain areas, retainability can be improved potentially by tuning the handover parameter settings.

7.3.5   Enhanced SRVCC (eSRVCC)

In the SRVCC, the IMS call control is performed in the home network [29]. This is illustrated in Figure 7.24 where the voice media anchoring is done at the terminating terminal.

As shown, the voice bearer traffic path starts from the originating UE1, passes through the eNB, SGW, PGW, optional SBC (if media anchoring option is enabled), and terminates at the UE2. As for the SIP signaling path, it starts at the originating terminal UE1, passes through the eNB, SGW, PGW, optional SBC, CSCF, SCC AS, CSCF, and terminates at the UE2. When the LTE radio coverage starts to diminish, the eNB decides to switch the radio system from LTE to 3G. Therefore, the eNB commands the initiation of the SRVCC processing by

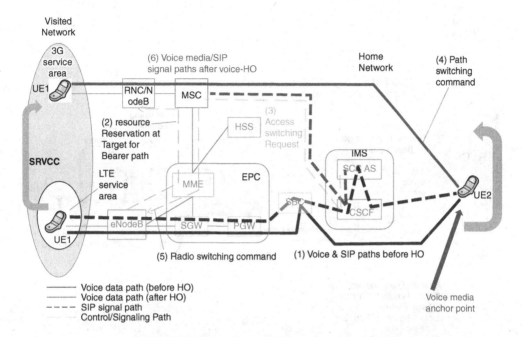

Figure 7.24   SRVCC – media anchoring can cause processing delays.

sending a switching request to the MME. Accordingly, the MME requests the MSC to allocate resources on the CS side to handle the ongoing call. The MSC then proceeds to reserve the bearer path resources together with the RNC to control the call after the handover is complete. The MSC then sends a switching request to the SCC-AS to transfer the call from LTE to 3G. Accordingly, the SCC-AS requests the terminating terminal UE2 to change the voice bearer traffic path from the PGW to the MSC. In parallel, the MSC confirms the resource allocation at the 3G side by sending a response to the MME, which relays the same confirmation to the eNB. The eNB then instructs the originating terminal UE1 to switch to the 3G radio. The voice bearer traffic path and the SIP signaling path after the handover is complete are both shown in Figure 7.24. Accordingly, an improvement, eSRVCC, in the call flow was introduced to control the voice handover in the visited network. Hence, the visited network anchors the voice data and SIP signaling paths at the time of calling and receiving of the VoLTE calls [29].

As illustrated above, in SRVCC there is a potential increase in processing delays due to the switching of the communication path being controlled by the home network (this is especially clear in the roaming scenarios). To improve this situation and decrease this extra latency, eSRVCC introduces a new functional entity, namely the access transfer control function (ATCF) and the access transfer gateway (ATGW), as shown in Figure 7.25.

The function of the ATCF and the ATGW is mainly to realize the anchoring of the SIP signaling and the voice bearer path within the visited network without the need to do this within the home network, thus eliminating significant processing delays during the handover. More specifically, the ATCF is responsible for relaying the SIP signaling and, during call setup, the ATCF allocates an ATGW depending on the media information. The ATGW, in turn, is responsible for relaying the voice traffic bearer and the voice media anchoring.

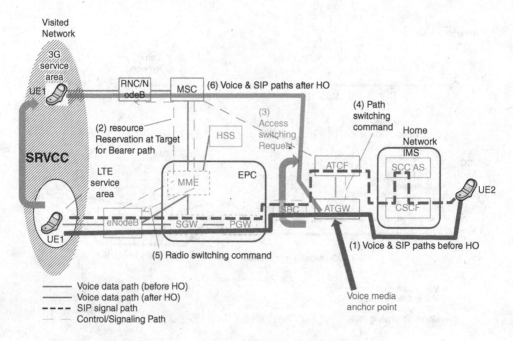

Figure 7.25 eSRVCC – improving the HO delay through the ATCF and ATGW.

When the LTE radio coverage starts to diminish, the eNB decides to switch the radio system from LTE to 3G. Therefore, the eNB commands the initiation of the SRVCC processing by sending a switching request to the MME. Accordingly, the MME requests the MSC to allocate resources on the CS side to handle the ongoing call. The MSC then proceeds to reserve the bearer path resources together with the RNC to control the call after the handover is complete. In parallel, the MSC sends a switching request to the ATCF to transfer the call from LTE to 3G. Accordingly, the ATCF sends a request to change the voice and SIP signaling path from the PGW to the MSC, to the CSCF/SCC-AS. Accordingly, the same request is relayed to the terminating terminal UE2. In parallel, the MSC confirms the resource allocation at the 3G side by sending a response to the MME which relays the same confirmation to the eNB. The eNB then instructs the originating terminal UE1 to switch to the 3G radio immediately upon receiving the response. In this way, the eSRVCC improves the handover delay compared to the SRVCC since the ATGW becomes the anchor for the media.

7.4 Key VoLTE Features

7.4.1 End-to-End QoS Support

Without a guaranteed end-to-end QoS, voice quality can be compromised and will fail to meet user experience expectations, especially since VoLTE is practically a VoIP call over LTE access, which means that the voice payload is carried within IP packets that can easily experience the common problems of the IP-based transmission networks, including packet delay, jitter, and packet loss.

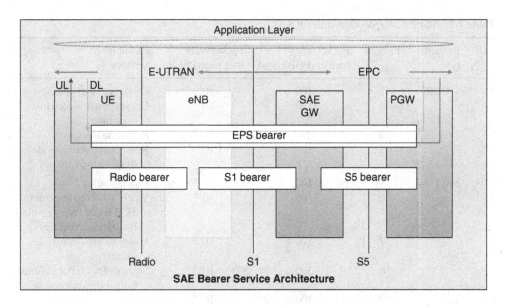

Figure 7.26   End-to-end SAE bearer service architecture.

The QoS indicates the expected service class in terms of packet delay tolerance, acceptable packet loss rates, and required minimum bit rates during network communication. QoS ensures that the request and response of a user (i.e., inter-user QoS) or application (i.e., intra-user QoS) correspond to a certain predictable service class. The QoS is a general term that is used for various conditions with service supplies and demands to assess the capability of meeting customer service requirements. The assessment is not based on accurate scoring, but on analysis of service quality in different conditions. Specific measures can be taken to improve service quality. The most common problems of the IP-based transmission networks are packet delay, jitter, and packet loss. Additionally, different applications require different bandwidths. Therefore, the problem becomes more severe and a robust QoS mechanism is mandatory. The QoS provides a comprehensive solution in such a situation. Figure 7.26 illustrates the end-to-end bearer service architecture.

An EPS bearer uniquely identifies packet flows that receive a common QoS treatment between the UE and the PGW (i.e., same scheduling, queue, management/rate, shaping, and policy). As shown in Figure 7.26, the EPS bearer consists of the radio bearer between the UE and the eNB, the S1-bearer between the eNB and the SGW, and the S5/S8 bearer between the SGW and PGW. An EPS bearer can be a guaranteed bit rate (GBR) or non-guaranteed bit rate (non-GBR) [2].

Table 7.5 provides the 3GPP QCIs for different applications with corresponding QoS requirements. The QCI is further used within the LTE access network to define the control packet-forwarding treatment from an end-to-end perspective [18]. It also ensures a minimum standard level of QoS to ease the interworking between the LTE networks, mainly in roaming cases and in a multi-vendor environment. The packet delay budget (PDB) defines an upper bound delay that a packet is allowed to experience between the UE and the PGW. As highlighted in Table 7.5, the VoLTE is mapped to QCI = 1 which is the conversational coice service.

Table 7.5   3GPP LTE QoS class index (QCI)

QCI	Resource type	Priority	Packet delay budget (PDB) (ms)	Packet error loss rate (PELR)	Examples of services
1	GBR	2	100	10^{-2}	Conversational voice
2		4	150	10^{-3}	Conversional video (live streaming)
3		3	50	10^{-3}	Real-time gaming
4		5	300	10^{-6}	Non-conversational video (buffered streaming)
5	non-GBR	1	100	10^{-6}	IMS signaling
6		6	300	10^{-6}	Video (buffered streaming), TCP-based (www, e-mail, ftp, p2p file sharing)
7		7	100	10^{-3}	Voice, video, interactive gaming
8		8	300	10^{-6}	Same as QCI 6 but used for further differentiation
9		9	300	10^{-6}	

The key QoS parameters [30] attached to a bearer are outlined as follows:

1. QCI: (for inter/intra-user QoS) is used to control packet-forwarding treatment (e.g., scheduling weights, admission thresholds, queue management thresholds, link layer protocol configuration, etc.), and is typically pre-configured by the operator.
2. Allocation and retention priority (ARP) (for inter-user QoS): the ARP is stored in the subscriber profile in HSS on a per APN basis (at least one APN must be defined per subscriber) and it can take a value between 1 and 15, based on the user priority (i.e., gold, silver, and bronze). The primary purpose of the ARP is to decide if a bearer establishment/modification request can be accepted or rejected (i.e., admission control) in case of resource limitation.
3. GBR and maximum bit rate (MBR) – this parameter is defined for the GBR bearer only.
4. Aggregate maximum bit rate (AMBR) sums all non-GBR bearers per terminal/APN.

To explain the extent of complexity that typical operators may face when attempting to deploy full end-to-end QoS, Figure 5.3 illustrates a typical high-level network architecture diagram.

The main challenges in deploying QoS can be briefly summarized as follows:

• Many network elements are "shared" with other "existing" networks (e.g., 2G, 3G, and HSPA+):
  – HLR/HSS (specifically when a single database solution is adopted for LTE/3G/2G)
  – CS core (for any voice services) – CSFB
  – PS core (e.g., PCRF)
  – Transport network (backhaul and backbone): the transport network is the common backbone for many services. Those services can be the fixed, IPTV, broadband, and

mobile services, all being served on this common backbone network. Any change in the configuration in this backbone needs to be carefully managed so as not to affect any other legacy service utilizing this same backbone.

- IPSec Gateway represents another layer that needs to be carefully designed and deployed. The tunnel between the eNB and the ePC needs to be configured in such a way as to allow certain fields in the inner headers of the IP payload to be copied to the outer (VPN, virtual private network) header of the tunnel.
- Redundancy is a requirement, thus practically doubling the amount of reconfiguration of network elements to support QoS.
- Multiple vendors are a fact of life nowadays for most operators. Interoperability between different network elements from different vendors must be thoroughly tested before any commercial deployment.

The eNB guarantees the downlink GBR associated with a GBR bearer and enforces the downlink AMBR associated with a group of non-GBR bearers. In order to maintain the same priority of the QCI over end-to-end implementation, the QCI should be mapped to the IP transport network. A key QoS parameter used for this purpose is the diffserv code point (DSCP) that is encapsulated within each IP packet. Figure 7.27 illustrates the QoS mapping on the transport network layer.

A typical mapping between QCI and DSCP is also shown in Table 7.6. The signaling is mapped to the highest priority (i.e., DSCP = 46) while other applications with different QCIs are mapped accordingly. Furthermore, the QCI/DSCP need to be mapped to the transmission network, such as the microwave (MW) links air interface. The mapping to the MW air interface is based on a queuing technique and the number of queues depends on the MW manufacturer. Assuming a MW with eight queues, then the mapping is illustrated also in Table 7.6.

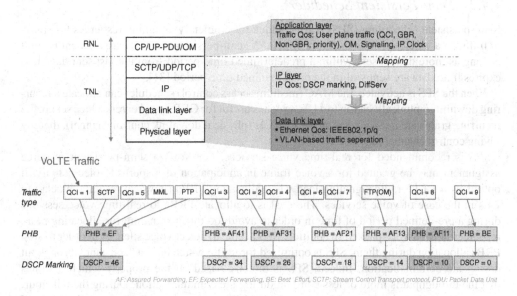

Figure 7.27 QoS mapping on transport network layer.

Table 7.6 QCI mapping to DSCP and MW queuing

GBR/non-GBR	QCI	Priority	Packet delay budget (ms)	Packet loss rate	DSCP	MW queuing	Service sample
GBR	1	2	100	10^{-2}	46	7	Conversational voice
	2	4	150	10^{-3}	26	4	Conversational video (live streaming)
	3	3	50	10^{-3}	34	5	Real-time gaming
GBR non-GBR	4	5	300	10^{-6}	26	4	Non-conversational video (buffer streaming)
	5	1	100	10^{-6}	46	7	IMS signaling
	6	6	300	10^{-6}	18	2	Video (buffer streaming) TCP based (www, e-mail, chat, ftp)
	7	7	100	10^{-3}	18	2	Voice, video (live streaming) interactive streaming
non-GBR	8, 9	8, 9	300	10^{-6}	0	0	Video (buffer streaming) TCP-based (www, e-mail, chat, ftp)

7.4.2 Semi-Persistent Scheduler

Semi-persistent scheduling (SPS) is a technique for efficiently assigning resources for spurts of traffic in a wireless communication system. A semi-persistent resource assignment is valid as long as more data is sent within a predetermined time period from the last sent data, and expires if no data are sent within the predetermined time period [31].

When the SPS is activated, the MAC (medium access control) scheduler can allocate a recurring downlink/uplink allocation to a UE, which is useful for GBR-type bearers. Once set up, the recurring grants are not signaled on the PDCCH (physical downlink control channel), thereby saving control channel resources.

SPS is recommended for real-time voice services. For VoIP, a semi-persistent resource assignment may be granted for a voice frame in anticipation of a spurt of voice activity. It optimizes the LTE radio signaling load when allocating PRBs (physical resource blocks) to UEs in the case of voice services. The goal is to allocate PRBs to real-time voice sessions during a pre-defined period of time in order to avoid too much signaling for allocating radio resources to each VoIP packet. In addition, SPS is able to detect voice silences in order to stop PRBs allocation during them. SPS is optimized for traffic which have a "constant" throughput during the communication. Indeed, SPS is directly based on the property which assumes that the throughput is more or less constant during the communication. During the talk spurt (assuming AMR (adaptive multi-rate) codec), VoIP packets normally arrive at intervals of 20 ms; during the silent period, silence indicator (SID) packets arrive at an interval of 160 ms.

SPS is useful for VoIP services and provides the following three benefits:

- Guarantees a better QoE (quality of experience) for VoIP services.
- Reduces the control signaling overhead for VoIP transmission.
- Maximizes the resource utilization by dynamically activating/deactivating resource allocation according to the transition between a silent period and talk spurt.

The SPS algorithm allocates a certain amount of resources (such as resource blocks) for the voice call during the call setup period through RRC (radio resource control) signaling. As shown in Figure 7.28, the allocation is semi-persistent and does not need to be requested again through UL/DL control signaling until the call ends and the resources are released.

To allow the maximum resource utilization during the silent period, the resource allocation will be deactivated by means of explicit signaling exchanged over the PDCCH. When the VoIP call transits from the silent period to the talk spurt, similar PDCCH signaling is used to activate the semi-persistent resource allocation. The SPS significantly reduces the PDCCH overhead and ensures the QoS for VoIP services by reserving the resources in a semi-persistent fashion. It also improves the resource utilization by dynamically activating or deactivating resource allocation activities between talk spurt and silent periods.

When compared to dynamic scheduling, the SPS can handle up to 35% more VoLTE users (compared to dynamic scheduling utilizing the same number of control channels) [32].

As shown in Figure 7.28, we can summarize the main aspects regarding the SPS as follows:

- For voice packets, persistent allocation is used for new transmissions (a sequence of chunks located every 20 ms are allocated for one user "persistently" until this user switches into DTX (discontinuous transmission) periods) and dynamic allocation for retransmissions.
- All SID (silence insertion descriptor) packets (initial transmission and retransmission) should use dynamic scheduling.

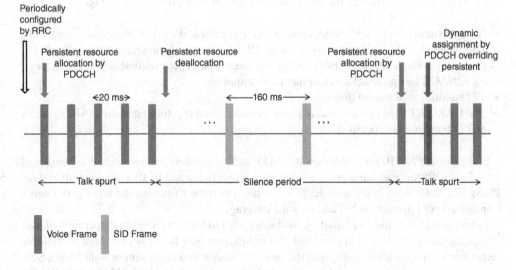

Figure 7.28   Semi-persistent resource allocation for VoLTE.

- Scheduling priority: new voice frames transmission should have the highest priority, followed by the voice Retransmission, then by the SID packets.
- Talk-spurt-based: eNB reallocates or releases resources when UE switches from active periods into DTX periods.

Therefore there are two options to deal with real-time voice services scheduling:

1. If real-time voice services are allocated to a dedicated EPS bearer then SPS can be activated.
2. If real-time voice services are allocated to an EPS bearer (e.g., default EPS bearer) where both data and voice services are mixed together then DS is used. Indeed SPS is not suitable for data services.

Table 7.7 provides some details for SPS.

7.4.3 TTI Bundling

Uplink TTI (time transmission interval) bundling has been introduced in LTE to improve the uplink coverage for real-time voice services. By definition, LTE UEs at the cell edge with weak uplink SINR (signal-to-interference-noise ratio) can retransmit the same data block in continuous subframes by means of TTI bundling. UEs do not always have to activate TTI bundling and it is limited to scenarios with weak uplink radio coverage. The activation and deactivation of TTI bundling transmission is controlled by an RRC signaling message. As shown in Figure 7.29, TTI bundling is especially useful in extending the cell coverage for the indoor VoLTE users by adding an additional gain in the UL link budget in addition to the choice of MCS (modulation and coding scheme) for UL transmissions for users at the cell edge. The decision on which users should use TTI bundling is based on path loss measurements and on availability of resources.

TTI bundling is specified as follows in Release 8/9/10 to improve UL coverage:

- A single transport block is channel coded and transmitted in a set of four consecutive TTIs.
- The bundled TTIs are treated as a single UL resource assignment, where a single UL grant and a single PHICH (physical hybrid automatic repeat request indicator channel) ACK/NACK (negative acknowledgment) are required.
- TTI bundling is activated through the RRC.
- The HARQ RTT (hybrid automatic repeat request round trip time) and the HARQ process of TTI bundling are specified.

In the Release 8/9/10 specifications, the TTI bundling mechanism is restricted to bundles of four TTIs, QPSK (quadrature phase shift keying) modulation and to allocations of up to three PRBs. For VoIP, these constraints leave some room to further improve the amount of energy transmitted per information bit, and thus the coverage.

To understand how the TTI bundling works, we need to explain the normal operation of how an uplink transport block is transmitted. An uplink transport block is converted to multiple redundancy versions after coding and the first redundancy version is sent in a subframe. Subsequent uplink transmissions of the transport block are dependent on the HARQ ACK/NACK which is sent for four subframes durations after the first transmission.

Table 7.7 SPS analysis

Analysis item	Semi-persistent scheduling
Increase of VoIP capacity	To save the PDCCH signaling overhead To allow more VoIP users to be supported SPS is able to minimize the amount of signaling messages to be exchanged between eNodeBs and UEs for narrowband real-time voice services where throughput is more or less constant during the voice communication Increase the capacity of VoIP Note: RoHC may have an impact on SPS performance. Indeed, SPS needs to detect/determine the compressor states of RoHC and guarantees a margin when allocating PRB resources in order to reduce the influence of RoHC compressor states changes
Voice quality	L2 link adaptation ability is not so efficient because the MCS is unchangeable during the semi-persistent scheduling. The closed-loop power control for PUSCH channel and slow AMC for PDSCH channel can solve this problem The voice quality is slightly degraded during the intra-LTE handover procedure. It can be solved by using the following mechanisms: • adaptive switch strategy between dynamic scheduling and semi-persistent scheduling • closed-loop power control for PUSCH channel • slow AMC for PDSCH channel
Coverage	The coverage is slightly degraded. The adaptive switch strategy between dynamic scheduling and semi-persistent scheduling can solve this problem
Mixed services	The maximum MCS of semi-persistent scheduling is set to MCS15 (because the resources of SPS are configured at the beginning. When UE moves, the eNodeB has no idea to recalculate SPS based on UE's CQI and SNIR). That means more PRB should be used. Also if there are VoIP sessions and other services in parallel within the cell, then the overall cell throughput maybe decreased
UE dependence	UE must support semi-persistent scheduling

PUSCH, physical uplink shared channel; AMC, adaptive modulation and coding; PDSCH, physical downlink shared channel; and CQI, channel quality indicator.

In TTI bundling, as shown in Figure 7.30, the different redundancy versions can all be sent in consecutive subframes without waiting for the HARQ ACK/NACK feedback and a combined ACK/NACK can be sent after processing all the uplink transmissions of a transport block. Indeed, within a TTI bundle, HARQ retransmissions are non-adaptive and are performed without waiting for feedbacks (e.g., NACK or ACK) related to previous transmissions according to the TTI "fixed bundle size" of four subframes.

The TTI bundling process is typically triggered by the UE informing the eNB about its power limitations in the present state. This could, for example, happen at the edge of a cell when the LTE terminal has to send high power but is limited by the power capability of the terminal. This triggers the eNB to transmit the various redundancy versions of the same transport block

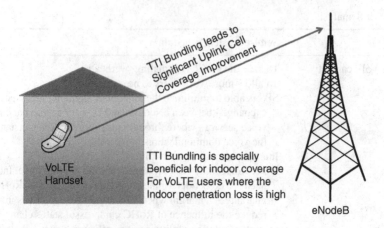

Figure 7.29   Benefits of TTI bundling.

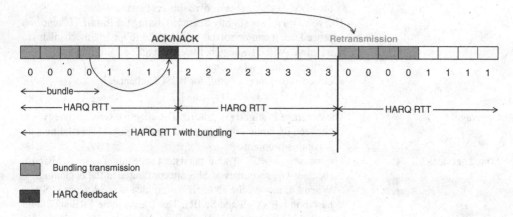

Figure 7.30   TTI bundling.

in consecutive subframes or TTIs, giving rise to the name TTI bundling. A single PDCCH allocation is sufficient for the multiple transmissions, thus saving control overhead as compared to the RLC (radio link control) segmentation approach. A single HARQ ACK/NACK for the combined transmissions is generated after processing the TTI bundle, which can reduce the error rate of the transport block as compared with processing a single redundancy version. This approach can also reduce the delay in the HARQ process compared to transmissions of the redundancy versions separated in time using the normal approach.

As shown in Figure 7.31, uplink TTI bundling enables up to four redundancy versions of the same transport block to be sent in four consecutive subframes [33].

A single RLC PDU (protocol data unit) is transmitted as multiple redundancy versions in consecutive subframes using a single common allocation. The channel coding used in LTE enables easy generation of the multiple redundancy versions from which the transmissions in the TTI bundle are generated. A common RLC header is shared across the TTI bundle and the same HARQ process identity is used for multiple transmissions in the same TTI bundle.

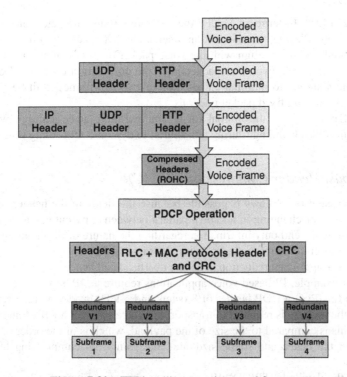

Figure 7.31 TTI bundling mapping to subframes.

Combined processing of the redundant transmissions over multiple subframes leads to a better probability of detection of the transport block. Thus, with limited power, the UE has a better chance of a successful transmission with lower latency using the TTI bundling method.

The low transmission power of some handsets, the short TTI length, and the long RTT of the HARQ transmissions ($\sim 8-10$ ms) makes TTI bundling extremely important for any operator to deploy before commercializing the VoLTE services. TTI bundling is expected to improve the UL coverage of applications like narrowband real-time VoLTE when low power handsets are likely to be involved. Simulation results indicate an expected gain in the range of $1-5$ dB due to TTI bundling on the uplink [31, 34]. TTI bundling also helps to achieve an efficient latency performance for VoIP sessions, even at the edge of the LTE cell. Furthermore, it is important to note that TTI bundling is UE dependent.

7.4.4 Connected Mode DRX

A constantly-on voice session can quickly reduce battery life. Since VoLTE traffic is predictable (e.g., 20 ms codec packets), a UE receiver does not have to constantly monitor the physical-layer control channel, and the receiver can essentially be turned off between receptions. This must be carefully configured, though, since missing ACKs or HARQ messages can add unacceptable latency.[2]

---

[2] Refer to Chapter 4 for a detailed overview and analysis of the DRX capabilities in LTE.

As outlined in [35], In terms of VoLTE the most important parameters are the DRX (discontinuous reception) cycle, the on-duratiount imer, and DRX-inactivity timer parameters. The DRX cycle should be set to align with the voice packet rate or its multiple (e.g., 20 ms, or preferably 40 ms bundling two voice packets) and the on-duration timer and DRX-inactivity timer parameters are set to minimal values (one or two sub-frame) to fit best to the traffic pattern of VoLTE having fixed packet intervals.

From the UE point of view, the optimal VoLTE configuration can be achieved by dynamically switching from SPS scheduling (during talk spurts) to dynamic scheduling (during SID).

7.4.5   Robust Header Compression (ROHC)

Header compression of IP flows is possible because the fields in the headers of IP packets are either constant or changing in a known pattern between consecutive packets in the same flow. It is possible to send only information regarding the nature of the changing fields of the headers with respect to the reference packet in the same IP flow.

The benefit is a significant reduction in header overhead and hence an increase in bandwidth efficiency. For example, IP-based voice applications require an IP header: 20 octets for IPv4 and 40 octets for IPv6, a UDP header of 8 octets and a RTP header of 12 octets. A total of 40 octets for the headers is required to transport the voice payload for IPv4 and 60 octets for IPv6. When this is compared to the size of the payload, which is of the order of 15–32 bytes (depending on the codec and frame size/rate), the gain from compressing the headers is quite apparent.

Looking at the detailed RTP/UDP/IP headers, as shown in Figure 7.32, you can note the classification of each field within the RTP/UDP/IP headers in terms of the field being static

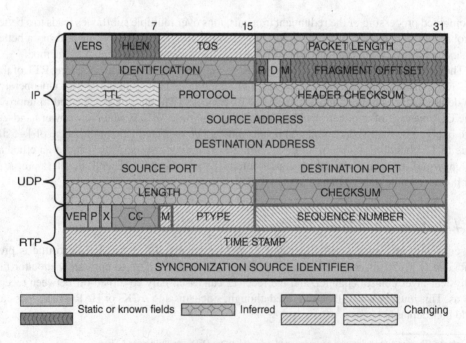

Figure 7.32   IP/UDP/RTP header fields' classification.

(and thus can be easily compressed/eliminated altogether from the compressed header), changing, or can be inferred from the remaining fields.

The main changing header fields (voice) are:

- RTP sequence number (SN) (16 bits) – incremented by one for each VoIP packet sent.
- RTP timestamp (TS) (32 bits) – incremented by a constant value even during the silence period.
- RTP marker (M-bit) (1 bit) – for voice the marker bit should be set only in the first packet of a talkspurt.
- UDP checksum (16 bits) – optional for IPv4. If disabled, its value is zero. If enabled, its value depends on the payload, changing with every packet.
- IPv4 identification (IP-ID) (16 bits) – identifies the fragments of a datagram when reassembling fragmented datagrams.

As shown in Figure 7.33, in order for header compression to work there must be a compressor and a de-compressor. A context is basically a snapshot of the complete (uncompressed) headers of an IP flow. This context is always exchanged from the compressor to de-compressor at initialization of the header compression scheme. After that the context is updated according to some criterion that is dependent on the header compression scheme.

During normal operation, the compressor will always try to send compressed headers instead of full headers. The compressed header represents the relative changes to a reference packet in the same IP flow and, therefore, the changes are relatively small.

Some header compression schemes may employ feedback from the de-compressor to the compressor to indicate the current context state in the de-compressor. A result of this could

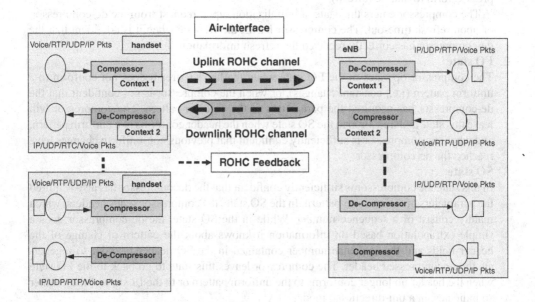

Figure 7.33   ROHC compressor/de-compressor system.

be to send sufficient information to update the context in the de-compressor. With the basic tools of header compression it is possible to define a protocol that will work on any link layer technology.

The detailed operation of the, "ROHC" protocol is specified in IETF RFC 3095 [10] and the related radio bearer enhancements are detailed in [36].

The compressor starts in the lowest compression state and gradually transitions to higher compression states. The general principle is that the compressor will always operate in the highest possible compression state, under the constraint that the compressor has sufficient confidence that the de-compressor has the information necessary to decompress a compressed header.

In the reliable mode, that confidence comes from receipt of ACKs from the de-compressor. Otherwise, it comes from sending the information a certain number of times, utilizing a CRC (cyclic redundancy check) calculated over the uncompressed RTP/UDP/IP header, and from not receiving NACKs. The compressor may also transition back to a lower compression state when necessary.

For IP/UDP/RTP, IP/UDP, and ESP/IP (encapsulated security protocol) compression pro-files, the three compressor states are:

- Initialization/refresh (IR),
- First order (FO), and
- Second order (SO).

1. **IR state**
   The purpose of this state is to set up or refresh the context between the compressor and de-compressor. The information that is sent from the compressor may contain static and non-static fields in uncompressed form (full refresh), or just non-static fields in uncom-pressed form (dynamic refresh).
   The compressor enters this state at initialization, upon request from the de-compressor, or upon refresh time-out. The compressor leaves the IR state when it is confident that the de-compressor has correctly received the refresh information.

2. **FO state**
   The compressor operates in the FO state when the header stream does not conform to a uniform pattern (i.e., constant changes), or when the compressor is not confident that the de-compressor has acquired the parameters of the uniform pattern. The compressor will leave this state and transition to the SO state when the header conforms to a uniform pattern, and when the compressor is sufficiently confident that previous non-uniform changes have reached the de-compressor.

3. **SO state**
   In this state the compressor is sufficiently confident that the decompressor has also acquired the parameters of the uniform pattern. In the SO state, the compressor sends headers, which mainly consist of a sequence number. While in the SO state, the de-compressor does a simple extrapolation based on information it knows about the pattern of change of the header fields and the sequence number contained in the SO header in order to regener-ate the uncompressed header. The compressor leaves this state to go back to the FO state when the header no longer conforms to the uniform pattern or to the IR state, if the counter so indicates, in a uni-directional mode.

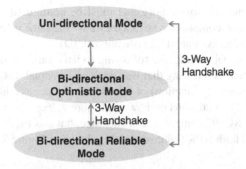

Figure 7.34 ROHC modes.

There are three modes of operation, each with the three states as described earlier:

- Uni-directional,
- Bi-directional optimistic, and
- Bi-directional reliable.

The possible transitions are shown in Figure 7.34. Compression always starts in the unidirectional mode and transits to any of the bi-directional modes depending on feedback from the de-compressor. The unidirectional mode implies that there is no feedback from the de-compressor to the compressor, while for the bi-directional optimistic there is irregular feedback, and periodic feedback for the bi-directional reliable mode.

A brief description of the modes is given below:

- U-mode is used when a feedback channel is not present or undesirable to use, and it should be known that the robustness and efficiency can never be as good as with feedback (if the channel is not completely error-free).
- O-mode is aiming for the highest compression efficiency while providing reasonable robustness.
- R-mode is almost completely robust but has slightly higher overhead and more feedback messages.

The allowed state transitions are shown in Figure 7.35 and the rules and packets formats that are required are described in more detail in [10].

ROHC has two kinds of parameters; configuration parameters that are mandatory , and implementation parameters that are optional and, when used, mandate how a ROHC implementation operates.

Figure 7.35 ROHC state transitions. (Source: [38] 3GPP. Reproduced with permission of ETSI.)

Configuration parameters are mandatory and must be configured (signaled by the RRC) between compressor and de-compressor, so that they have the same values at each. An example of a configuration parameter is context identification (CID).

Based on some field and lab testing, the following ROHC statistics shown in Tables 7.8 and 7.9 were noted from a near-cell testing that was done on an actual VoLTE call.

The CDF (cumulative density function) of the ROHC compressed header size for a typical VoLTE call under ideal RF conditions is shown in Figure 7.36.

A similar CDF of the ROHC compressed header size, but this time for a call under a much worse RF condition and high BER (bit error rate) is shown in Figure 7.37.

Table 7.8 VoLTE RTP/UDP/IP compressed header statistics

Mean	3.44
Standard error	1.51
Median	3
Minimum	3
Maximum	41
Count	10 490

Table 7.9 ROHC packet type statistics

Feedback	230
IR	8
Type-0/normal	7 816
Type-1	2 385
Type-2	51
Total	10 490

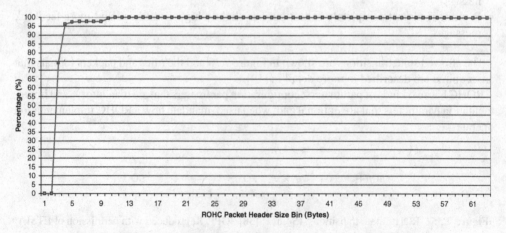

Figure 7.36 CDF for ROHC compressed header size under good RF.

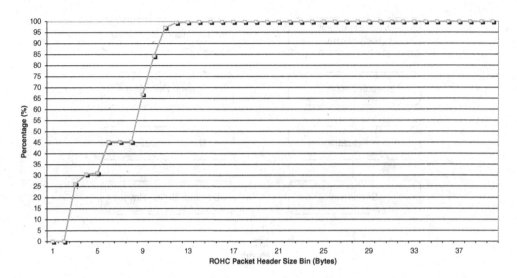

Figure 7.37   CDF for ROHC compressed header size under bad RF.

Since the ROHC function resides in the eNB, it is mandatory that ROHC context re-establishment is executed for every inter-eNB handover as follows:

- The UE has to transmit extra packets to re-establish context. This usually requires two or three IR packets which contain information such as IP address, UDP ports, and RTP SN and are 40 bytes for IPv4 and 60 bytes for IPv6.
- This additional overhead needs to be sent after each inter-eNB handover and consists of up to 188 bytes (assuming IPv6).

The other option that can be implemented by infrastructure vendors is to do a ROHC context transfer which reduces the overhead for handover events and improves the capacity and link budget. In the case of context transfer, the extra overhead reduces to 80 bytes (assuming IPv6). At the time of writing, the 3GPP does not allow context transfer of ROHC.

7.4.6   VoLTE Vocoders and De-Jitter Buffer

As explained, VoLTE is, in reality, VoIP over the LTE access network. As shown in Figure 7.38, the original voice signal is sampled and encoded to a specific bit rate digital stream at the end of the sending process. The compressed digital stream data are then encapsulated into IP/RTP/UDP packets to be sent to the receiving side, mainly through the IP network. Every packet contains the encoded speech (compressed voice data) together with information on the packet's origin, projected destination address, and must have the packet stream reconstructed in the correct order with the help of TS (in the RTP header).

Speech coders convert the analog signal to a compressed representation of the original voice signal in the form of a digital bitstream (this is called speech encoding) and, at the receiving end, another codec converts the digital bitstream back into an analog signal (this is called speech decoding). This results in a significant saving in network bandwidth and also supports

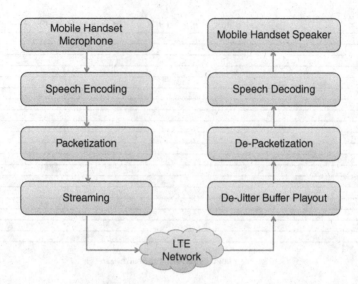

Figure 7.38   Speech coding and de-jitter buffer in VoLTE.

silence containment, where silence is not encoded or transmitted. The size of the resulting encoded data stream, the speed of the encoding/decoding operations and the quality and fidelity of the sound signal are the three most important factors to be optimized by codecs.

A typical example of the components of an IMS VoLTE client [37] is shown in Figure 7.39. A more detailed overview of the speech receiver, including the de-jitter buffer, is shown in Figure 7.40. Jitter buffer management (JBM) denotes the actual buffer as well as any control, adaptation, and media processing algorithm (excluding speech decoder) used in the management of the jitter induced in the transport channel.

The blocks "network analyzer" and "adaptation control logic", together with the information on buffer status from the actual buffer, control functionality, whereas the "speech decoder" and "adaptation unit" provide the media processing functionality.

The gray dashed lines in Figure 7.40 indicate the measurement points for the jitter buffer delay, that is, the difference between the decoder consumption time and the arrival time of the speech frame at the JBM.

The functional processing blocks are as follows:

- **Buffer:** The jitter buffer unpacks the incoming RTP payloads and stores the received speech frames. The buffer status may be used as input to the adaptation decision logic. Thehe buffer is also linked to the speech decoder to provide frames for decoding when they are requested for decoding.
- **Network analyzer:** The network analysis functionality is used to monitor the incoming packet stream and to collect reception statistics (e.g., jitter, packet loss) that are needed for jitter buffer adaptation. Note that this block can also include, for example, the functionality needed to maintain statistics required by the RTCP if it is being used.
- **Adaptation control logic:** The control logic, adjusting playback delay and operating the adaptation functionality, makes decisions on the buffering delay adjustments and required

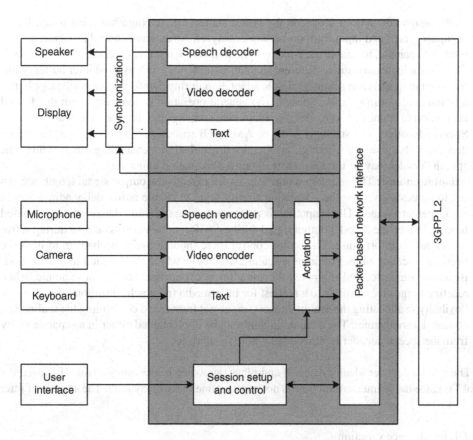

Figure 7.39 Components of an IMS VoLTE client. (Source: [38] 3GPP. Reproduced with permission of ETSI.)

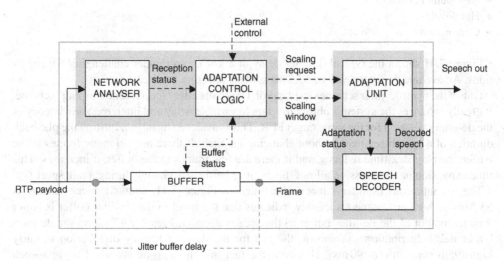

Figure 7.40 Typical speech receiver including the de-jitter buffer. (Source: [38] 3GPP. Reproduced with permission of ETSI.)

media adaptation actions based on the buffer status (e.g., average buffering delay, buffer occupancy, etc.) and input from the network analyzer. An external control input can also be used, for example, to enable inter-media synchronization or other external scaling requests. The control logic may utilize different adaptation strategies, such as fixed jitter buffer (without adaptation and time scaling), simple adaptation during comfort noise periods, or buffer adaptation also during active speech. The general operation is controlled with the desired proportion of frames arriving late, the adaptation strategy, and the adaptation rate.

- **Speech decoder:** The standard AMR or AMR-WB speech decoder. Note that the speech decoder is also assumed to include error concealment/bad frame handling functionality. The speech decoder may be used with or without the adaptation unit.
- **Adaptation unit:** The adaptation unit shortens or extends the output signal length according to requests given by the adaptation control logic to enable buffer delay adjustment in a transparent manner. The adaptation is performed using the frame-based or sample-based time scaling on the decoder output signal during comfort noise periods only, or during active speech and comfort noise. The buffer control logic should have a mechanism to limit the maximum scaling ratio. Providing a scaling window in which the targeted timescale modifications are performed improves the situation in certain scenarios – for example, when reacting to the clock drift or to a request for inter-media (re)synchronization – by allowing flexibility in allocating the scaling request on several frames and performing the scaling in a content-aware manner. The adaptation unit may be implemented either in a separate entity from the speech decoder or embedded within the decoder.

The speech decoder ideally receives and plays one voice frame every 20 ms. However, the VoLTE voice frame inter-arrival time is not constant due to the delay jitter. The sources of jitter include:

- UL interference variation
- DL scheduler
- RF channel conditions
- Handovers
- Core network.

Figure 7.41 shows the typical de-jitter delay of a VoLTE call under challenging RF conditions. An average de-jitter delay of 53.5 ms is observed from the measurements.

One of the main de-jitter schemes used in VoLTE is called the "speech time warping" scheme. It greatly enhances the system's ability to adapt to variable delay and jitter, reducing latency of the de-jitter operation for the same target PER. This is achieved through adjusting the playback duration of a speech segment without changing its pitch. If there are too many frames in the buffer, the playback time reduces, and if there are few frames in the buffer, it increases. This time expansion or compression allows the de-jitter buffer size change during a talk spurt.

The instantaneous de-jitter latency, illustrated in Figure 7.41, shows a mean latency is 53.5 ms. A high instantaneous latency indicates that the input to the de-jitter buffer is larger than the output of the de-jitter buffer to the speech decoder. Figure 7.42 shows the de-jitter buffer delay distribution. As shown, the tail for the de-jitter latency distribution is fairly high (95th percentile at 90 ms). However, the fact that this can be absorbed using speech

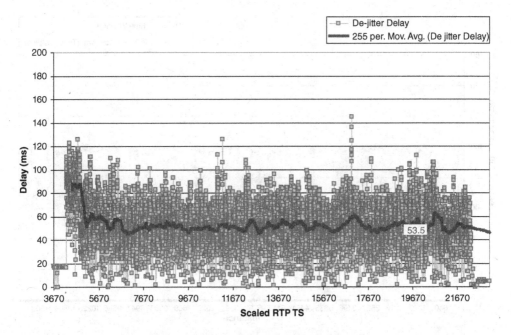

Figure 7.41 Typical de-jitter delay.

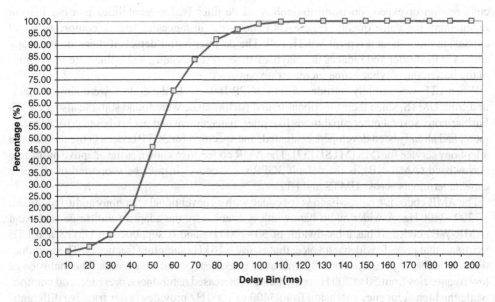

Figure 7.42 De-jitter delay distribution.

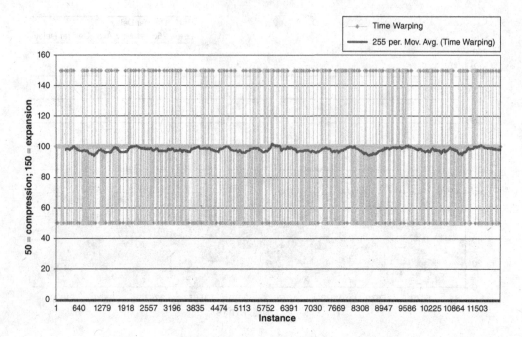

Figure 7.43   Speech time warping (compression/expansion).

compression or expansion using the enhanced de-jitter buffer capabilities is a big help in maintaining the voice quality. Figure 7.43 shows the instances of speech compression and expansion throughout a typical VoLTE call. The de-jitter buffer detects if there are too many frames in the buffer and reduces the playback time (compression), and if there are few frames in the buffer, the playback time increases (expansion).

When LTE was initially introduced into 3GPP Release 8, the codecs were simply inherited from UMTS, since their functionality and performance were found suitable also for LTE. Furthermore, several other capabilities like jitter management and packet-based error concealment, had already been developed, evaluated, and specified for 3GPP codecs in the multimedia telephony service for IMS (MTSI) [37]. The AMR codec, also known as the adaptive multi-rate narrowband (AMR-NB), is the default 3GPP voice codec. It must be supported in all voice capable terminals across UMTS and LTE.

The AMR codec is a narrowband voice codec with a conventional telephony audio bandwidth of 200–3400 Hz. A wider audio bandwidth is enabled by the adaptive multi-rate wideband (AMR-WB) codec. It has a bandwidth of 50–7000 Hz and is supported in UMTS and LTE terminals that provide wideband voice, that is, use a 16 kHz sampling frequency. The introduction of wideband gives substantially improved voice quality and naturalness. The inclusion of low frequencies from 50 to 200 Hz contributes to increased naturalness, presence, and comfort, while the high-frequency extension from 3400 to 7000 Hz provides better fricative differentiation and, therefore, higher intelligibility.

The wider audio bandwidth of AMR-WB adds a feeling of transparent communication and eases speaker recognition. Figure 7.44 shows a comparison between the narrowband AMR and

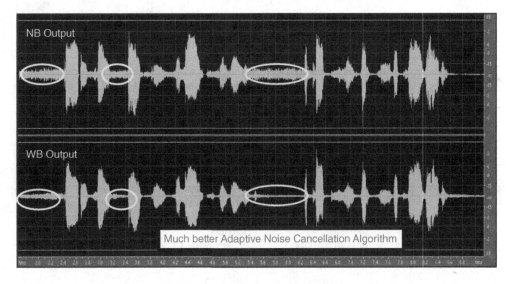

Figure 7.44 AMR-WB versus AMR-NB.

the wideband AMR. The inclusion of the higher frequencies in the wideband AMR has a major positive impact on the voice quality. Furthermore, testing has revealed that the AMR-WB has an even significantly better adaptive noise cancelation algorithm, where the background noise can be canceled in a much more efficient way.

7.5 Deployment Considerations for VoLTE

It is important to note that LTE operators with a smaller spectrum (5 or 10 MHz) who plan to deploy VoLTE will soon need another dual-band LTE to be able to successfully offer the VoLTE service. In this section we consider such a case where there is a need to have dual-band LTE and the options for deploying VoLTE.

The two options would be either to use both carriers as a "shared-carrier configuration" or dedicate one of the two carriers to VoLTE and the other to data, which we would call "assigned-carrier configuration." Table 7.10 summarizes the different aspects for each option.

It is worth noting that both configurations provide enough capacity to serve all types of traffic and both configurations deliver similar voice and data performance. However, the shared-carrier option has a simpler network configuration and shows more robustness in the case of a sudden increase in VoLTE and/or data traffic and can potentially show better metrics performance, more specifically in throughput.

In the case of the shared-carrier configuration, in idle mode, both carriers would have equal priority. Load balancing is maintained through cell reselection based on the RSRQ (reference signal received quality). Furthermore, the system information for both carriers is configured similarly. In connected mode, there would be no need for redirection from the carrier to another after connection establishment, as the call will resume on the current serving carrier. For inter-frequency handover between the carriers, it will be allowed, based on the RSRQ, (as well

Table 7.10 VoLTE deployment options

Shared-carrier	Assigned-carrier
VoLTE calls are evenly distributed between the two carriers	VoLTE calls will be divided between carriers as ~80% on the dedicated VoLTE carrier and ~20% on the second carrier
BE data volume is almost the same on both carriers	BE data call attempts while on VoLTE carrier will be ~20%
Same total/user average throughput on both carriers	BE data volume will be divided between carriers as ~80% on the dedicated data carrier and ~20% on the VoLTE carrier
BE data calls are almost the same on both carriers	Similar average throughput per user is expected on both carriers
BE data volume is almost the same on both carriers	BE data volume per user almost the same on both carriers
CQI reporting is the same for both carriers	Almost similar CQI reporting on both carriers

Table 7.11 Advantages and disadvantages of shared-carrier

Advantages	Disadvantages
No forced redirection to the other carrier which leads to less signaling overhead and faster call establishment	More complex network parameter configuration to meet the VoLTE QoS requirements which requires delicate and proper radio bearer priority configuration at the UE and the corresponding QoS configuration at the network which should be implemented properly to ensure availability of resources for VoLTE users
More efficient in resource allocation	More complex configuration of the measurement report to ensure proper trigger for inter-frequency HO
Same set of SIB configuration on both carrier	
Less susceptible to human error during eNB provisioning	

as the RSRP (reference signal received power)) and will be triggered either to maintain the load balancing or to maintain the integrity of the service (e.g., throughput). Table 7.11 summarizes the main advantages and disadvantages of the shared-carrier configuration.

In the case of assigned-carrier configuration, in idle mode, all VoLTE UEs would camp on the same carrier (Carrier1) and no inter-frequency cell reselection would be allowed to the second carrier. This is possible by proper configuration of the SIBs (system information blocks). In connected mode, if a voice call is requested, it will stay on the same carrier (Carrier1). If a data call is requested while the Voice call is on-going on the first carrier (Carrier1), it will also be established on Carrier1. No inter-frequency HO would be allowed unless insufficient resources are available on Carrier1, in which case the Carrier 2 resources would be shared for VoLTE calls. If a data call is being requested, the call will be redirected to the second carrier (Carrier 2). If a voice call is requested while the data call is on-going on the second carrier (Carrier 2), it will also be established on Carrier 2. Similarly, no inter-frequency HO is allowed unless insufficient resources are available on Carrier 2, in which case the Carrier 1 resources would be shared for data calls.

With the assigned-carrier configuration it is mandatory to configure the basic QoS and priority configuration, however, it is certainly easier to meet the QoS requirement for VoLTE calls since it is assigned to a specific carrier. In addition, there is less signaling overhead since less inter-frequency HO is expected. However, more complex configuration for the system information is required since there is a different set of system information configuration for each carrier. Furthermore, it is important to ensure proper inter-frequency HO will take place when no resources are available for VoLTE users on the assigned carrier. Similarly, it is important to redirect the BE user to the second carrier (Carrier 2) after establishing a call on the first carrier.

References

[1] Tanenbaum, A. and Wetherall, D. (2011) *Computer Networks*, Prentice Hall.
[2] Postel, J. (1981) Transmission Control Protocol, STD 7, RFC 793, September 1981.
[3] Postel, J. (1980) User Datagram Protocol, STD 6, RFC 768, August 1980.
[4] Schulzrinne, H., Casner, S., Frederick, R., and Jacobson, V. (2003) RTP: A Transport Protocol for Real-Time Applications. RFC 3550, July 2003.
[5] Schulzrinne, H. and Casner, S. (2003) RTP Profile for Audio and Video Conferences with Minimal Control. RFC 3551, July 2003.
[6] Rosenberg, J., Schulzrinne, H., Camarillo, G. *et al.* (2002) SIP: Session Initiation Protocol. RFC 3261, June 2002.
[7] Price, R., Borman, C., Christoffersson, J. *et al.* (2003) Signaling Compression (SigComp). RFC 3320, January 2003.
[8] Surtees, A., West, M., and Roach, A.B. (2007) Signaling Compression (SigComp) Corrections and Clarifications. RFC 4896, June 2007.
[9] Mäenpää, J. (2005) Performance of signalling compression in the third generation mobile networks. MS Thesis. Helsinki University of Technology, June 2005.
[10] Bormann, C., Burmeister, C., Degermark, M., *et al.* (2001) RObust Header Compression (ROHC): Framework and Four Profiles: RTP, UDP, ESP, and Uncompressed. RFC 3095, July 2001.
[11] Pelletier, G. and Sandlund, K. (2008) RObust Header Compression Version 2 (ROHCv2): Profiles for RTP, UDP, IP, ESP and UDP-Lite. RFC 5225, April 2008.
[12] Sandlund, K., Pelletier, G., and Jonsson, L.-E. (2010) The RObust Header Compression (ROHC) Framework. RFC 5795, March 2010.
[13] Qualcomm (2007) Circuit-switched Fallback. The first Phase of Voice Evolution for Mobile LTE Devices, White Paper, May 2007.
[14] 3GPP (2013) Circuit Switched Fallback in Evolved Packet System. TS 23.272.
[15] 3GPP (2013) Single Radio Voice Call Continuity (SRVCC). TS 23.216.
[16] 3GPP (2012) Evolved Packet System (EPS); 3GPP Sv interface (MME to MSC, and SGSN to MSC) for SRVCC. TS 23.205.
[17] 3GPP (2013) General Packet Radio Service (GPRS); Service Description. TS 23.060.
[18] 3GPP (2013) General Packet Radio Service (GPRS) enhancements for Evolved Universal Terrestrial Radio Access Network (E-UTRAN) Access. TS 23.401.

[19] 3GPP (2013) IP Multimedia Subsystem (IMS) Centralized Services. TS 23.292.

[20] 3GPP (2013) IP Multimedia Subsystem (IMS) Service Continuity. TS 23.237.

[21] 3GPP (2012) (IP Multimedia Core Network Subsystem (IMS) Multimedia Telephony Services and Supplementary Services. TS 22.173.

[22] Spirent (2012) IMS Procedures and Protocols: The LTE User Equipment Perspective.

[23] Tanaka, I. and Koshimizu, T. (2012) Overview of GSMA VoLTE profile. *NTT Docomo Technical Journal*, **13**(4), 45.

[24] Roach, A.B. (2012) Session Initiation Protocol (SIP)-Specific Event Notification. RFC 3265, June 2012.

[25] Ericsson (2012) Voice Handover in LTE Networks, White Paper, October 2012.

[26] GSMA (2013) IMS Service Centralization and Continuity Guidelines V6.0. IR.64, Febuary 2013.

[27] 3GPP (2013) Evolved Universal Terrestrial Radio Access Network (E-UTRAN); X2 Application Protocol (X2AP). TS 36.423.

[28] 3GPP (2012) Handover Procedure. TS 23.009.

[29] Koshimizu, T., Tanaka, I. and Nishida, K. (2012) Inter-domain Handover technologies in LTE for Voice (VoLTE) and TV phone. *NTT Docomo Technical Journal*, **13**, 4.

[30] Elnashar, A. Practical Aspects of LTE Network Design and Deployment. Springer Wireless Networks Journals (provisionally accepted).

[31] 4G Americas (2012) 4G Mobile Broadband Evolution: Release 10, Release 11 and Beyond – HSPA, SAE/LTE and LTE-Advanced, October 2012.

[32] Jiang, D., Wang, H., Malkamaki, E., and Tuomaala, E. (2007) Principle and performance of semi-persistent scheduling for VoIP in LTE system. Proceedings of the International Conference on Wireless Communications, Networking and Mobile Computing (WiCOM '07), September 2007, pp. 2861–2864.

[33] 3GPP (2013) Medium Access Control (MAC) Protocol Specification. TS 36.321.

[34] 3GPP (2012) LTE Coverage Enhancements. TR 36.824.

[35] Renesas Mobile Whitepaper (2013) VoLTE Power Consumption, Febuary 2013.

[36] 3GPP (2002) Radio Access Bearer Support Enhancements. TR 25.844.

[37] 3GPP (2013) IP Multimedia Subsystem (IMS); Multimedia Telephony; Media Handling and Interaction. TS 26.114.

8

4G Advanced Features and Roadmap Evolutions from LTE to LTE-A

Ayman Elnashar and Mohamed A. El-saidny

The standardization in 3GPP Release 8 defines the first specifications of LTE (long term evolution). The evolved packet system (EPS) is defined, mandating the key features and components of both the radio access network (E-UTRAN, evolved universal terrestrial radio access network) and the core network (evolved packet core, EPC). Orthogonal frequency division multiplexing (OFDM) is defined as the air interface with the ability to support multi-layer data streams using multiple input–multiple output (MIMO) antenna systems to increase spectral efficiency. LTE is defined as an all-IP network topology differentiated over the legacy circuit switch (CS) domain. However, Release 8 specification makes use of the CS domain to maintain compatibility with 2G and 3G systems by utilizing the voice calls circuit switch fallback (CSFB) technique for any of those systems. Other significant aspects defined in this initial 3GPP release are the self-organizing networks (SONs) and home base stations (home eNodeB, eNB), aiming at revolutionized heterogeneous networks (HetNets). Moreover, Release 8 provides techniques for the smartphone's battery saving, known as connected mode discontinuous reception (C-DRX).

LTE Release 9 provides improvements to Release 8 standards, most notably enabling improved network throughput by refining SON and improving eNB mobility [1]. Additional MIMO flexibility is introduced with multi-layer beam-forming. Furthermore, CSFB improvements have been introduced to reduce the voice call setup time delays.

The International Telecommunication Union (ITU) has created the term IMT-Advanced (International Mobile Telecommunications-Advanced) to identify mobile systems whose capabilities go beyond those of IMT-2000. In order to meet this new challenge, 3GPP's partners have agreed to expand the specification scope to include the development of systems

Design, Deployment and Performance of 4G-LTE Networks: A Practical Approach, First Edition. Ayman Elnashar, Mohamed A. El-saidny and Mahmoud R. Sherif.
© 2014 John Wiley & Sons, Ltd. Published 2014 by John Wiley & Sons, Ltd.

beyond 3G's capabilities. Some of the key features of IMT-Advanced are: Worldwide functionality and roaming, compatibility of services, interworking with other radio access systems and enhanced peak data rates to support advanced services and applications with a nominal speed of 100 Mbps for high mobility and 1 Gbps for low mobility users.

Release 10 defines LTE-A as the first standard release that meets the ITU's requirements of fourth generation, 4G [2]. The increased data rates up to 1 Gbps in the downlink and 500 Mbps in the uplink are enabled through the use of scalable and flexible bandwidth allocations up to 100 MHz, known as carrier aggregation (CA). Additionally, improved MIMO operations have been introduced to provide higher spectral efficiency. The support of HetNets and relays added to this 3GPP release also improve capacity and coverage. Lastly, a seamless interoperation of LTE and WLAN (wireless local area network) networks is defined to support traffic offload concepts.

Release 11 continues the evolutions toward the LTE-A requirements. Enhanced interference cancellation and CoMP (coordinated multi-point transmission) are means for further improving the capacity in the 4G networks.

The overall 3GPP feature evolution and timeline of the LTE system are represented in Figure 8.1. The operator's decision to evolve into the advanced LTE releases largely depends on factors like device capabilities, network vendor roadmaps, and the modeled business costs.

Figure 8.2 shows an illustration of IMT-Advanced requirements and how they are achieved from Release 8 (LTE) to Release 10 (LTE-A).

This chapter describes several LTE and LTE-A features in the evolution from Release 8 to Release 10. Some of LTE-A feature deployments have dependences on other LTE features, as in the case of HetNet with ICIC (inter-cell interference coordination) and SON. On the other hand, some of the LTE features have been introduced at the same time, and in some cases in the same product, as other LTE-A features. This is the case for category 4 capabilities being primarily introduced in the same devices having LTE-A CA features. Conceptually, the

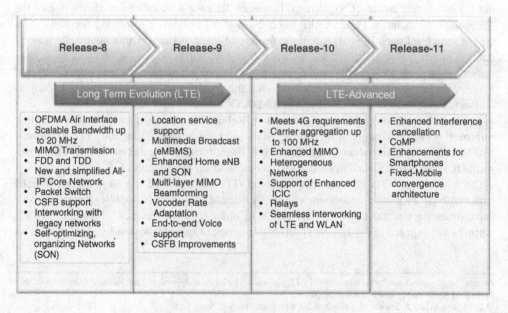

Figure 8.1 LTE evolution in the standard.

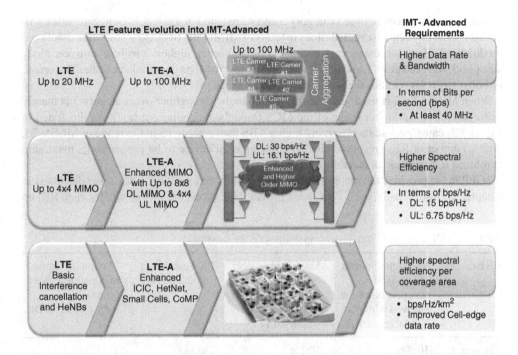

Figure 8.2   LTE-A main features.

advanced features in both LTE and LTE-A roadmap are not deployed as stand-alone features, or historically tied to an enhancement over previous releases (e.g., ICIC or some SON functions). Hence, this chapter covers the realistic LTE to LTE-A evolution and the features that have strong dependences, and have been recently introduced in commercial deployments, or are yet to come in the near future.

The chapter begins with a discussion on the evolution of LTE UE (user equipment) capabilities from category 3 to category 4 devices. It then discusses the main LTE-A features shown in Figure 8.2. It also provides an overview of several features initially introduced in Release 8 and then enhanced in Release 10 and beyond, such as SON. Finally, the concept of location-based services in the LTE network, applicable to Release 9 but seeing an increasing interest in the market, is described.

8.1   Performance Comparison between LTE's UE Category 3 and 4

The evolution in LTE networks requires progressive enhancements in the device capability in order to support the cutting-edge features. LTE networks have been initially deployed with category 3 devices. Then, as soon as the LTE system has established its performance solidarity, the device makers in the mean time have evolved to higher speed categories. One of the additions to LTE device capabilities is category 4 chipsets, supporting higher than 100 Mbps speed in the 20 MHz channel. Though category 4 is still part of 3GPP Release 8, it is discussed in the context of LTE evolution as it is being introduced in parallel with the progressing evolution into Release 10 networks (i.e., CA with category 4 devices).

The main differences between category 3 and 4 are listed in Table 8.1. Both categories provide the same physical layer capabilities in terms of MIMO and 64QAM (quadrature amplitude modulation). However, category 4 is capable of supporting higher downlink transport block size (TBS) than category 3, and hence reaching up to 150 Mbps downlink data rate. The uplink of both categories remains the same with up to 51 Mbps [3].

When MIMO 2×2 is used, the category 4 gain in the throughput over category 3 is mainly distinguished at the higher range of the MCS (modulation/coding scheme). As illustrated in Figure 8.3, category 4 starts observing higher TBS than category 3 in the range of MCS > 23, with a number of RBs (resource blocks) allowing each device to be capped to its maximum combined transport block capabilities [4].

3GPP in [4] provides the mapping between the MCS and the TBS index (also discussed in Chapter 3). Additionally, each TBS index provides the actual number of bits for each RB, up

Table 8.1 UE categories and device capabilities on the downlink

UE category	Maximum number of DL-SCH transport block bits received within a TTI (across all TBs)	Maximum number of bits of a DL-SCH transport block received within a TTI	Total number of soft channel bits	Maximum number of supported layers for spatial multiplexing in DL
Category 1	10 296	10 296	250 368	1
Category 2	51 024	51 024	1 237 248	2
Category 3	102 048	75 376	1 237 248	2
Category 4	150 752	75 376	1 827 072	2
Category 5	299 552	149 776	3 667 200	4

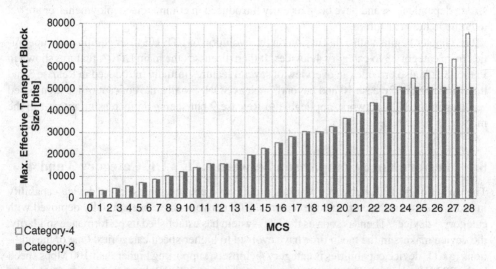

Figure 8.3 Transport block size and modulation/coding scheme (MCS) comparison between category 3 and 4.

to 100 RBs in 20 MHz. The eNB scheduler is therefore responsible for assigning the UEs with a combination of MCS and RBs supportable within their category capabilities. The eNB can schedule category 4 UEs with 100 RBs and set the MCS to 28 for each codeword, providing no more than 150 752 bits in a TTI (time transmission interval) with 75 376 bits for each codeword. With this scheduling, the UE can reach the maximum theoretical throughput of 150 Mbps with an effective coding rate of 0.876.

However, for category 3 UEs, the TBS is capped to no more than 102 048 bits in a TTI. Hence the eNB can schedule the UEs with different MCS/RBs combinations in order to achieve the maximum theoretical throughput. Let us take the following examples[1]:

- The eNB can schedule 100 RBs and set the MCS to 23 for each codeword. This combination provides a TBS of 51 024 bits in each codeword. However, this combination has an effective coding rate of 0.593.
- Alternatively, the eNB can schedule category 3 UEs with 80 RBs and set the MCS to 27 for each codeword. This combination also provides a TBS of 51 024 bits in each codeword. However, this combination has an effective coding rate of 0.742.
- Or, schedule 68 RBs and set the MCS to 28 for each codeword. This combination also provides a TBS of 51 024 bits in each codeword. This combination has an effective coding rate of 0.872, which is the optimal scheduled combination, closely matching the Category 4 effective coding rate.

Referring to the mapping of the CQI (channel quality indicator) to spectral efficiency in [4], we can produce the DL (downlink) throughput expected to be scheduled for each CQI and for each of the two device categories. This is shown in Figure 8.4. The throughput for each device category is calculated from the best matching combination of MCS/RBs providing an effective coding rate suitable for the spectral efficiencies for each CQI in the table mentioned in [4].

The calculations in Figure 8.4 assume 20 MHz bandwidth, two antenna ports, and a CFI control format indicator (PCFICH (physical control format indicator channel) CFI) suitable for the channel conditions, and normal cyclic prefix (i.e., 7 OFDM symbols/slot). It is assumed that for near cell conditions (i.e., higher CQI values), CFI = 1 or 2 will be scheduled more than CFI = 3 where fewer of PDCCH (physical downlink control channel) symbols are allocated to avail resources for the PDSCH (physical downlink shared channel). The opposite is assumed for far cell conditions (i.e., lower range of CQI). The effective coding rate is deduced from the following general equations:

$$\text{code rate} = \frac{\text{code block size plus}}{(\text{total number of resource elements}) \times (\text{modulation order})}$$

$$\text{code block size plus} = \text{TBS} + 24 + \frac{\text{TBS}}{6144 - 24} \times 24$$

$$\text{total number of RE} = f(\text{assigned number of RBs, CFI, cyclic prefix,}$$

$$\text{number of symbols for other control channels})$$

---

[1] Refer to Figure 3.28 in Chapter 3 for an overview of the eNB scheduler inputs, mechanisms, and outputs.

$$\text{modulation order} = \begin{cases} 2, & \text{QPSK} \\ 4, & \text{16QAM} \\ 6, & \text{64QAM} \end{cases}$$

$$\text{spectral efficiency} = \text{code rate} \times \text{modulation order}$$

It is shown from Figure 8.4 that a category 4 device's downlink throughput is expected to outperform that of a category 3 in the range of CQI > 12, where both devices are scheduled with MCS/TBS providing the same effective coding rate and hence similar spectral efficiency.

With the differences between the two device categories explained so far, and given the current status of LTE deployments in most markets, categories 3 and 4 will definitely coexist in the same network. In order to understand the benefits of category 4 over 3, it is thus important to assess the performance comparison of the two categories in different RF (radio frequency) conditions. This section discusses the performance aspects and the potential network optimization, in order to maintain a differentiated speed for devices with category 4 capabilities.

8.1.1   Trial Overview

The comparison test in this section was performed in a network with 1800 MHz frequency band and 20 MHz bandwidth. The network supports LTE data rate capabilities of up to 150 Mbps on the air interface with a suitable backhaul bandwidth. The eNBs are all equipped with 2×2 MIMO with transmission mode 3 configuration. The test was conducted during a time with low-load network.

During the trial, two different devices were used, one supporting category 4 and the other category 3. They were cabled into one RF connector and run in parallel. Hence, the test was set up to ensure the two devices were observing the same RF. The devices were evaluated simultaneously in three different cases: near cell stationary, far cell stationary, and mobility test.

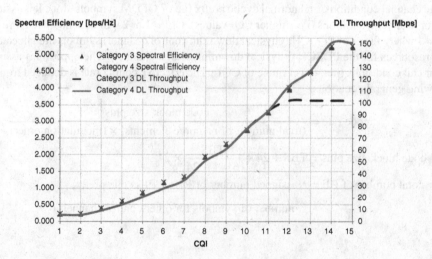

Figure 8.4   Downlink throughput versus CQI comparison between category 3 and 4 in the same spectral efficiency.

8.1.2   Downlink Performance Comparison in Near and Far Cell Stationary Conditions

In the near cell test, both devices were observing RSRP (reference signal received power) of -57 dBm. The average reported CQI from both of them was 14, reflecting the best channel quality. The two devices were reporting Rank-2 channel at all times, indicating a preference for two codewords scheduling.

From the discussions in previous sections regarding the MCS/RBS combination, CQI, and spectral efficiency, we will evaluate the actual scheduling behavior of the two devices compared to the theoretical discussion. As indicated before, in very good channel quality, it is expected that the eNB will use the MCS/RBs combination that maximizes the effective coding rate providing the intended spectral efficiency for each category. Hence, it is expected that the eNB will assign an MCS/RBs combination of [28, 68] and [28, 100] for categories 3 and 4, respectively. With this, let us evaluate the trial results and how the scheduler coped with these coding rates for different devices to keep the spectrum efficiency within the same range at near cell stationary and, at the same time, giving the category 4 the expected throughput gain over category 3.

As the two devices are running simultaneously with full-buffer download, the network scheduler shares the cell capacity (150 Mbps) between the two of them. Figure 8.5 illustrates the downlink physical layer throughput for both devices.

As observed in this test run, category 4 throughput is in the range of 50% higher than that of category 3. This is in line with the maximum theoretical throughput gain of category 4 over 3. The eNB keeps the throughput distinction between the two devices in such good RF conditions while scheduling both devices with an MCS of 28, as shown in the MCS distribution in Figure 8.6. Moreover, category 3 was assigned with an average of 68 RBs whilst category 4 was assigned with 100 RBs. These MCS/RBs combinations maintain the coding rate of 0.87

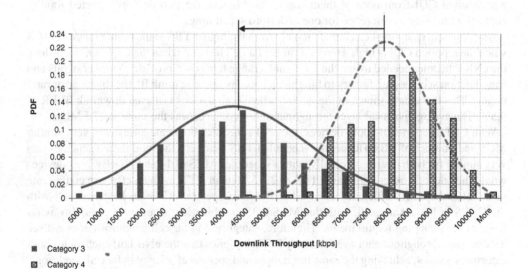

Figure 8.5   Downlink physical layer throughput distribution for categories 3 and 4 in near cell conditions.

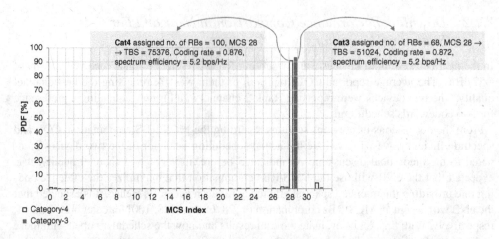

Figure 8.6   Downlink MCS distribution for categories 3 and 4 in near cell conditions.

and spectral efficiency of 5.2 bps/Hz for both devices at the same exact CQI of 14 reported by both of them. This is in line with the theoretical calculations shown before. However, category 4 achieved higher throughput by scheduling the UE more TTIs in the time domain.

In conclusion, for the near cell condition, the gain of category 4 throughput came purely from scheduling higher TBS resulting from the combination of MCS/RBs scheduled by the eNB. In the mean time, the eNB maintained the same spectral efficiency for the reported CQI from each device. This in turn indicates that the eNB was coping well with the device capability. In addition, the presence of category 4 in the cell increased the cell throughput. The average measured cell throughput (aggregated over the two users) in this test reached 130 Mbps.

On the other hand, in far cell testing, both devices observed RSRP of −97.8 dBm. The average reported CQI from both of them was 8. In this test, the two devices reported Rank-1 channel, indicating a preference for one codeword scheduling.

As discussed, category 4 can benefit mostly from the higher TBS at the higher range of MCS when the reported CQI is high. However, in the far cell test and when the CQI reported is low, the eNB scheduler needed to cap the MCS into a range that could provide a number of bits and a modulation scheme satisfactory to the channel quality and downlink BLER (block error rate) targets. This is the case shown in Figure 8.7, where both devices achieved downlink BLER of 9.5%. The average physical layer throughput of both devices was the same at 17.5 Mbps.

With CQI 8 reported by both devices, the eNB schedules both categories with very similar MCS and number of RBs. Hence, the scheduled TBS (and also the modulation scheme) was very similar for both devices. Actually, with a scheduled MCS of 14 and 86 RBs, both devices would then decode exactly the same TBS of 22152 bits in a TTI with an effective coding rate of 0.489 and spectral efficiency of 1.96 bps/Hz. This spectral efficiency matches well with the CQI 8 in [4]. As a result, with this MCS/RBS combination, category 3 capability is far from its TBS capping requirements. Therefore, category 4 in far cell conditions does not see a substantial throughput gain over category 3. This implies that the eNB fairly schedules both categories 3 and 4, achieving the same throughput and spectral efficiency in far cell conditions.

The results in both of these stationary conditions show that network planning to target higher SINR (signal-to-interference-noise atio) and thus a better reported CQI will definitely benefit category 4 devices in achieving a better throughput within the same spectral efficiency. It is

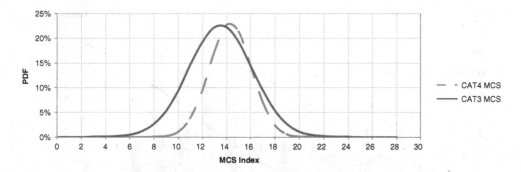

Figure 8.7 Downlink MCS distribution for categories 3 and 4 in far cell conditions.

then expected that an extra RF planning and cell overshooting cleanup is required when more category 4 devices become available in networks supporting 150 Mbps cell throughput. From the near cell test in particular, it is clear that, with category 4 devices, the cell throughput increases, a trend that is appealing to network operators for increasing the LTE capacity.

8.1.3 Downlink Performance Comparison in Mobility Conditions

From the results of the near and far cell tests, it is clear that category 4 can mainly achieve its throughput gains when the channel quality is good. It is then good to evaluate the gain, if any, during mobility conditions.

Figure 8.8 shows the downlink physical layer throughput distribution for both categories over the entire mobility route. The mobility test was conducted in an urban LTE deployment.

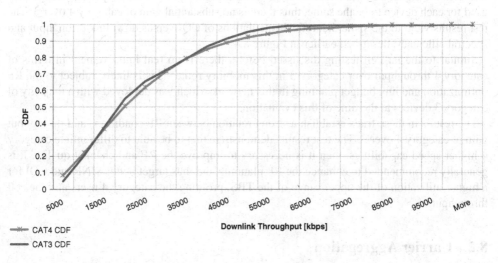

Figure 8.8 Downlink physical layer throughput distribution for categories 3 and 4 in mobility conditions.

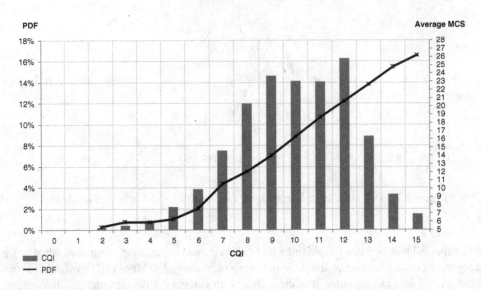

Figure 8.9   Category 4 CQI distribution and MCS vs. CQI in mobility route.

Both categories running simultaneously observed very similar throughput over the entire mobility route. Looking a step farther, the CQI distribution of the category 4 is illustrated in Figure 8.9. Additionally, the average MCS versus each reported CQI is plotted in the same figure.

It is clear from Figure 8.9 that MCS > 23 start to occur strongly at CQI > 12. In this route, CQI > 12 applies for only 13%. The average reported wideband CQI in this route is 10, providing an average scheduled MCS of 16. As the higher range of CQI is rarely reported by the UE, then the higher side of scheduled MCS has also limited usage. This caps the average TBS used for each device to be the same, thus there is no substantial gain of category 4 over 3. The test results yield to the same theoretical calculations of CQI versus downlink throughput and spectral efficiency, described earlier in Figure 8.4.

Similar to the far cell testing discussed before, there is no real improvement in terms of category 4 throughput over category 3 in this mobility route. This is likely subject to the RF optimization and link budget planning in this network which was deployed with a majority of category 3 devices at the time of this evaluation.

The results in these trials establish a strong relation between RF conditions and throughput gain of category 4 over 3. The test results demonstrate that to be able to differentiate category 4 higher speed capabilities, then it is necessary to improve the RF and channel quality. It is generally recommended to enhance the RF planning and thus target better SINR and CQI for a higher utilization of the upper range of the TBS, profiting the category 4 users and the cell throughput.

8.2   Carrier Aggregation

CA in 3GPP is not a new topic, and has been introduced for HSPA+ in Release 8. In addition to improvements in the data rate and latencies by combining multiple frequencies (carriers) during downlink (or later uplink) scheduling, there have been other motivations. In particular, greater trunking efficiency with joint scheduling of radio resources over multiple frequency carriers can greatly improve the cell capacity. If the number of users stays the same, aggregating of carriers can ultimately provide load balancing across the carriers by reducing the unused resources on each. This is mainly the case for users with higher burst data rate for bursty applications, such as HTTP Web browsing, gaming, and so on.

In practice, bursty traffic is more common than full-buffer traffic. In this case, all the UEs may not be actively downloading data continuously. Instead, the UEs download data in "bursts," and the data burst size is more like a truncated log normal distribution.

In HSPA+, multiple CA options are available in 3GPP, where Release 8 has introduced dual-cell high speed downlink packet access (DC-HSDPA). DC-HSDPA combines two 5 MHz *adjacent* carriers, hence the name dual-cell. The majority of HSPA+ networks worldwide have deployed the feature. Then, from there, 3GPP has been continuously evolving CA with different combinations of bandwidth and bands. Aggregating more carriers (up to 8×5 MHz) increases the data rate while expanding the carriers, being aggregated, across different bands would encourage operators with a different spectrum options to deploy. Hence, DC-HSDPA across different bands has been referred to as dual-band dual-cell high speed downlink packet access (DB-DC-HSDPA) and multi-carrier HSDPA is referred to MC-HSDPA. The combination of different bands supported has been limited to those most common in market deployments. For example, Release 9 supports a combination of UMTS 2100 and 900 MHz.

On the other hand, LTE has utilized the same concept in ways to meet the requirements of IMT-Advanced. CA in LTE-A allows the radio interface to combine a number of carriers, up to five, with several combinations of bandwidth and bands. LTE has already given the flexibility for the bandwidth usage up to 20 MHz. LTE-A hence adds the flexibility to aggregate the different LTE bandwidths within different bands [5]. Figure 8.10 shows the CA evolution in HSPA+ and LTE. Note that similar concepts have also been applied to the uplink of both LTE and HSPA+, for the same purpose.

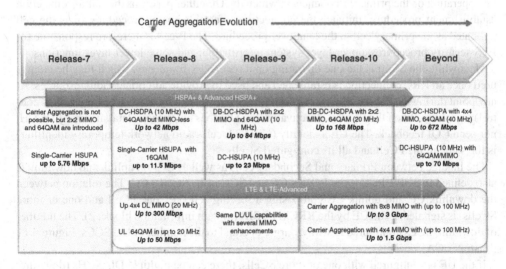

Figure 8.10 Carrier aggregation evolution in HSPA+ and LTE.

For future releases (beyond Release 11), it is expected that aggregation of carriers among both HSPA and LTE will merge where simultaneous reception of HSPA+ and LTE can be allowed. This will be an inter-RAT (radio access technology) CA that potentially provides gains of load balance between different RATs as well as increasing the user's data rate.

8.2.1   Basic Definitions of LTE Carrier Aggregation

The LTE downlink and uplink CA can be configured independently. However, 3GPP allows the number of uplink carriers being aggregated to be equal or less than the downlink.

Additionally, CA facilitates efficient use of a fragmented spectrum for contiguous and non-contiguous spectrum aggregation. The bandwidth of carriers being aggregated can vary from 1.4, 3, 5, 10, and 15–20 MHz [5]. This is particularly helpful for operators who do not have a contiguous 20 MHz spectrum but are looking for higher data rates. Therefore, 3GPP defines the term component carriers (CCs) which can have the bandwidth of 1.4, 3, 5, 10, 15, or 20 MHz, confining to the Release 8 numerology.

Given that LTE-A supports up to 100 MHz (i.e., 5×20 MHz), therefore, the maximum number of CCs allowed is 5. The UE may simultaneously receive or transmit one or multiple CCs depending on its capabilities. Each CC supports backward compatible operation, ensuring interoperability between CA and non-CA deployments. Only CCs that belong to the same eNB may be aggregated, and they are assumed to be synchronized (i.e., single timing advance command to control the UE's uplink transmission timing for all UL (uplink) CCs).

Similar to DC-HSDPA, LTE CA defines the primary cell (PCell) and secondary cell (SCell) concept. At least one of the CCs must have a control channel (PDCCH) where the RRC (radio resource control) informs which CC the UE will camp on for control messages. Additionally, CA does not apply for a UE in RRC-idle mode. Hence, idle mode mobility procedures of LTE Release 8 and 9 apply in a network deploying CA.

A UE that is configured for CA connects to one PCell and up to four SCells. The PCell is the one operating on the primary frequency, in which the UE either performs the initial connection establishment procedure, initiates the connection re-establishment procedure, or is the cell indicated as the primary cell in the handover procedure. The P-cell is therefore responsible for EPS security procedures, upper layer system information, and some lower layer functions.

Meanwhile the SCell is the one operating on a secondary frequency, which may be configured once an RRC connection is established and used to provide additional radio resources. At any point during the call, SCells can be activated or deactivated by MAC (medium access control) signaling. When the SCell is deactivated, the UE does not receive data, monitor PDCCH, nor send CQI feedback. The UE's identity (C-RNTI, cell-radio network temporary identifier) is the same in the PCell and all its configured SCells.

The linkage between Primary and Secondary cells as well as CCs on uplink and downlink is also defined in 3GPP. PCC is the PCell's CC and SCC is the SCell's CC. The relation between the downlink CC and uplink CC composing a serving cell (i.e., one PCell and one or more SCells) is signaled to the UE by the RRC (i.e., SIB (system information block) 2). The number of downlink SCCs configured is always larger or equal to the number of UL SCCs. Figure 8.11 illustrates CA concepts.

If the UE is configured with one or more SCells, there can be multiple DL-SCHs (downlink shared channels) and there may be multiple UL-SCHs (uplink shared channels) per UE. Hence,

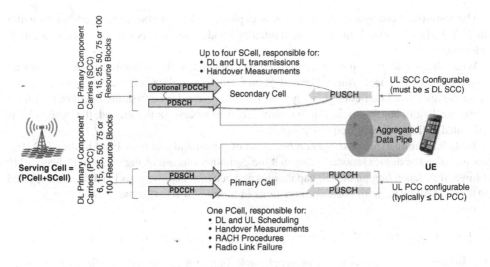

Figure 8.11 Carrier aggregation illustration.

there may be one DL-SCH and one UL-SCH on the PCell, one DL-SCH and zero or one UL-SCH for each SCell.

8.2.2 Band Types of LTE Carrier Aggregation

With CA's flexibility of deployment, several types are defined in 3GPP. Figure 8.12 illustrates the types of CA bands.

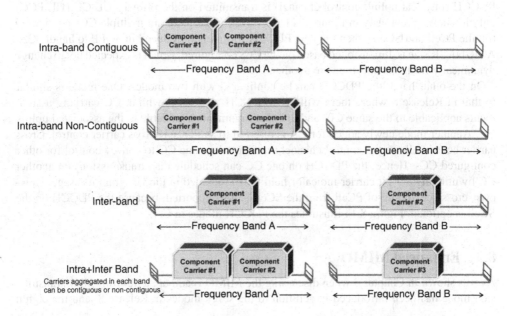

Figure 8.12 Band types of carrier aggregation.

The intra-band contiguous type of CA is deployed where a contiguous bandwidth wider than 20 MHz is available in the same frequency band. This type is rare in today's spectrum allocation.

When the operators acquire a fragmented LTE spectrum, then inter-band CA becomes an important choice of deployment. Depending on the spectrum allocated, the operator can choose from the intra-band non-contiguous type or the inter-band. In the first, the operator may have some bandwidth but fragmented in the same frequency band. In the second, the bandwidth is allocated within different frequency bands.

Table 8.2 summarizes possible combinations of intra-band and inter-band in the FDD (frequency division duplex) mode.[2] Note that the combinations are changing from Release 10 to 11 in 3GPP, depending on the UE capabilities and network deployments [5]. It is also expected that there will be more of these combinations in future 3GPP revisions.

8.2.3 *Impact of LTE Carrier Aggregation on Protocol Layers*

CA impacts several layers in the protocol stack. Figure 8.13 summarizes the impact on the NAS (non-access stratum), RRC, PDCP (packet data convergence protocol), RLC (radio link control), and MAC.

The major changes in the user-plane are basically in the MAC and PHY layers. Aggregation of CCs is performed on the MAC and PHY layers, and there is a single RLC and PDCP layer. MAC supports multiple transport changes (i.e., DL-SCH) when CA is enabled. Each CC will have an independent HARQ (hybrid automatic repeat request) entity for new transmissions and retransmissions, with eight HARQ processes per CC.

In the PHY layer, uplink and downlink channels are modified to handle CA operations. The ACK/NACK (positive acknowledgment/negative acknowledgment) feedback sent on the PUCCH (physical uplink control channel) is transmitted on the primary UL CC (UL PCC) only as shown previously in Figure 8.11. However, since there are multiple CCs configured (on the PCell and SCell), then the new PUCCH format is introduced in 3GPP to handle CA. ACK/NACK size is dimensioned based on the CCs configured, which is expected not to change dynamically from one sub-frame to another.

On the other hand, the PDCCH can be configured with two modes. One mode is similar to that in Release 8 where there will be a PDCCH on each downlink CC carrying assignments applicable to the same CC, and uplink assignments applicable to the associated uplink CC. Another mode newly added for CA is referred to as PDCCH cross carrier control. Cross carrier control allows the control channel transmitted on one CC to convey control for other configured CCs. Hence, the PDCCH on one CC can schedule data transmissions on another CC by utilizing a 3-bit carrier indicator field (CIF) included in the UL grant message. However, cross-scheduling of PCell from the SCell is not supported, because the PDCCH on the SCell is optional. Figure 8.14 illustrates the PDCCH modes in CA.

8.3 Enhanced MIMO

As was shown in Chapter 4 when discussing the MIMO roadmap, a new downlink transmission mode has been introduced in addition to the seven modes in Release 8, and the eighth

---

[2] Refer to Table 1.2 in Chapter 1 for LTE FDD Frequency Bands.

Table 8.2 CA configurations in 3GPP Releases 10 and 11

CA type	E-UTRA CA band	UE CA bandwidth class	Implementation
Intra-band contiguous	CA_1	{1C}	"C" means up to two CC with the following combination
	Band 1		15 + 15 MHz
			20 + 20 MHz
	CA_7	{1C}	"C" means up to two CC with the following combination
	Band 1		15 + 15 MHz
			20 + 20 MHz
Inter-band	CA_1-5	{1A, 5A}	"A" means up to one CC in each band
	Bands 1 and 5		Band 1: 10 MHz
			Band 5: 10 MHz
	CA_1-18	{1A, 18A}	"A" means up to One CC in each band
	Bands 1 and 18		Band 1: 5, 10, 15, or 20 MHz
			Band 18: 5, 10, or 15 MHz
	CA_1-19	{1A, 19A}	"A" means up to one CC in each band
	Bands 1 and 19		Band 1: 5, 10, 15, or 20 MHz
			Band 19: 5, 10, or 15 MHz
	CA_1-21	{1A, 21A}	"A" means up to one CC in each band
	Bands 1 and 21		Band 1: 5, 10, 15, or 20 MHz
			Band 21: 5, 10, or 15 MHz
	CA_2-17	{2A, 17A}	"A" means up to one CC in each band
	Bands 2 and 17		Band 2: 5 or 10 MHz
			Band 17: 5 or 10 MHz
	CA_3-5	{3A, 5A}	"A" means up to one CC in each band
	Bands 3 and 5		Band 3: 10, 15, or 20 MHz
			Band 5: 5 or 10 MHz
			"A" means up to one CC in each band
			Band 3: 10 MHz
			Band 5: 5 or 10 MHz
	CA_3-7	{3A, 7A}	"A" means up to one CC in each band
	Bands 3 and 7		Band 3: 5, 10, 15, or 20 MHz
			Band 7: 10, 15, or 20 MHz
	CA_3-20	{3A, 20A}	"A" means up to One CC in each band
	Bands 3 and 20		Band 3: 5, 10, 15, or 20 MHz
			Band 20: 5 or 10 MHz
	CA_4-12	{4A, 12A}	"A" means up to one CC in each band
			Band 4: 1.4, 3, 5, or 10 MHz
	Bands 4 and 12		Band 12: 5 or 10 MHz
	CA_4-13	{4A, 13A}	"A" means up to one CC in each band
	Bands 4 and 13		Band 4: 5, 10, 15, or 20 MHz
			Band 13: 10 MHz
	CA_4-17	{4A, 17A}	"A" means up to One CC in each band
	Bands 4 and 17		Band 4: 5 or 10 MHz
			Band 17: 5 or 10 MHz
	CA_7-20	{7A, 20A}	"A" means up to one CC in each band
	Bands 7 and 20		Band 7: 10, 15, or 20 MHz
			Band 20: 5 or 10 MHz

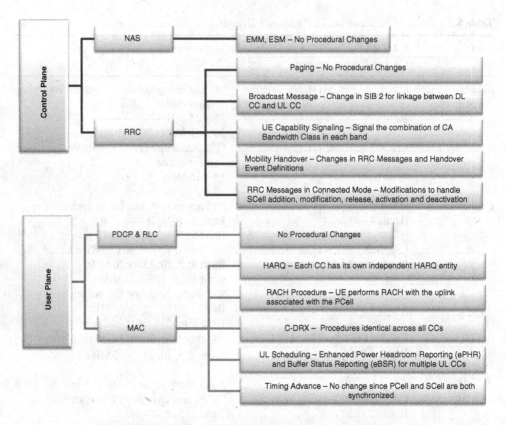

Figure 8.13 Impact of carrier aggregation on protocol layers.

one in Release 9. Release 10 has introduced transmission mode 9. Moreover, Release 10 has
introduced the support of downlink 8×8 MIMO and uplink SU-MIMO (single user multiple
input–multiple output) with up to four layers.

8.3.1   Enhanced Downlink MIMO

LTE-A aims at reducing the overhead for 8×8 MIMO configurations. Prior to Release 10,
cell-specific downlink reference signals (DL-RSs) were mapped into antennas based on the
configured number of antenna ports. The overhead of DL-RS in terms of resource elements
per RB can cause suboptimal performance when more antenna ports are added. Hence, LTE-A
introduces thechannel state information (CSI) concept. In principle, decoupling the RS for
channel-state information and RS for demodulation, generates a new DL-RS known as the
CSI-RS channel-state information reference signal.

The CSI-RS is transmitted on each physical antenna port with less overhead. It is used for
measurement purposes. CSI-RS is used with Release-10 Transmission Mode 9. Mode 9 supports both SU-MIMO and MU-MIMO (multiple user multiple input–multiple output) with
seamless switching between them. It supports up to eight layers with up to two codewords

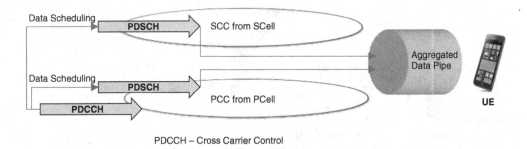

Figure 8.14 PDCCH transmission in carrier aggregation.

(similar to legacy Release 8/9 codewords). In addition to the CQI, RI (rank indicator), and PMI (precoding matrix indicator) feedback, the UE is required to feedback CSI as well. CSI feedback is also carried on the PUCCH or PUSCH (physical uplink shared channel).

8.3.2 Uplink MIMO

Uplink MIMO has been introduced in Release 10 to improve the uplink capacity as well as the user's data rate. It supports up to four layers (i.e., 4×4) with two PUSCH transmission modes.

Transmission Mode 1 supports a single antenna port and mode 2 is a MIMO mode. Hence, Mode 1 is similar to Release 8/9 operation. Mode 2 can utilize two or four layers with precoding. This implies changes in the PHY layer as well as the MAC. HARQ operation in the MAC is modified to support two transport blocks transmission.

8.4 Heterogeneous Network (HetNet) and Small Cells

HetNet is a multilayer, multi-technology, and possibly a multi-band network. More specifically, the HetNet is the next evolution of network structure that combines macro and small cell layers, including micro eNB, pico eNB, femto cell, WiFi hotspot, and a distributed antenna system for in-building solutions. It also incorporates multi-technology including 2G, 3G, HSPA+, LTE, LTE-Advanced, and WiFi. Furthermore, the HetNet structure may include multi-band

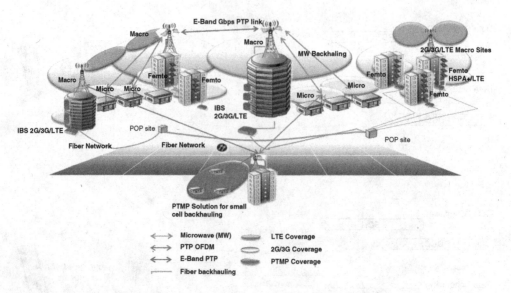

Figure 8.15   HetNet structure and small cell backhauling.

deployment, such as GSM 900 and GSM 1800, LTE 2.6 GHz and LTE 800, or LTE 1800, and WiFi at 2.4 GHz and 5 GHz.

The small cells are the central part of the HetNet structure, deployed to offer coverage or/and capacity enhancement. The main goal of the small-cell is to offload traffic from the macro sites. Hence, the macro layer provides coverage and the small cell layer provides capacity/throughput improvement. Such HetNet topology, compared to the macro sites alone, requires a non-traditional approach for small cell backhauling and robust interference management techniques. Figure 8.15 demonstrates a typical HetNet network deployment. The efficient management of this evolved complicated multilayer network widely allows a competitive position for the mobile operators.

The HetNet deployment in Figure 8.15 shows several backhauling options. The wireless backhauling is adopted to complement the fiber network that cannot be extended to every site or small cell location. The next section explains the backhauling options, taking into account the small cell deployments.

8.4.1   Wireless Backhauling Applicable to HetNet Deployment

There are typically two categories of wireless backhaulings:

- Point-to-point (PTP) consisting of the following two types of backhauling:
  - Traditional microwave (MW) PTP backhauling with either hybrid IP/TDM or native IP MW with up to 1 Gbps (depends on the frequency band and the supplier) throughput in MW bands from 6 GHz up to 38 GHz for macro sites. The net throughput depends on the channel bandwidth, antenna size, frequency band, and link budget. Additionally, these links can cover up to 30 km, depending on the frequency band and antenna size. These links deploy a channel bandwidth from 7 MHz up to 54 MHz and adopt higher

modulation techniques of up to 256QAM in good RF conditions and based on the link budget.
- Wireless GigE (known as E-band) links with up to 2.4 Gbps full duplex (PTP only with max. 5 km). These links use the E-band of 70/80 GHz (frequency bands 71–76 GHz and 81–86 GHz as duplex pairs) and a channel bandwidth of up to 1 GHz and modulation (of up to 16QAM). These links can serve short hubs of 3–5 km, depending on the rain density.
- Point-to-multipoint (PTMP) consisting of the following three types of backhauling:
  - PTP and PTMP using preparatory OFDM techniques that can provide a speed of 300 Mbps in 40 MHz with the TDD (time division duplex) mode. It can be used for long and mid-range backhauling, depending on the frequency bands. This technology adopts unlicensed and licensed bands, such as 5 GHz and 3.5 GHz, respectively. They typically work in line-of-sight (LoS) and near LoS scenarios.
  - PTMP MW backhauling in 26 GHz and 28 GHz bands. These MW links can be used for macro and micro sites backhauling. The main disadvantages of these PTMP links are the need for strict LoS and the limited sector throughput of up to 100 Mbps. Therefore, such backhauling does not meet the peak throughput requirements of LTE and LTE-Advanced where a peak throughput of 300 Mbps with MIMO 4×4 is needed, and even higher with CA.
  - PTMP at 42 GHz and 32 GHz with sector throughput of above 1 Gbps. These PTMP technologies are currently being developed to serve mainly small cell backhauling with very high capacity per hub site.

Figure 8.16 summarizes the fixed and wireless broadband access technologies and demonstrates both converge and throughput ranges coverage distance for each technology.

Table 8.3 provides a comparative analysis between the PTMP using OFDM technology and the legacy PTP MW. As demonstrated in the comparison, the OFDM/PTMP solution is more practical and cost efficient for small cell backhauling compared to the legacy PTP MW solution.

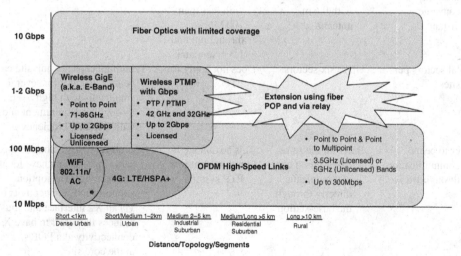

Figure 8.16 Broadband access technologies summaries.

Table 8.3 Comparison between OFDM/PTMP and legacy MW PTP

Specifications	PTP/PTMP	Legacy PTP MW	Remarks
Supported Bands	3.5 GHz, 5 GHz	7, 13, 18, 26, 38 GHz	MW with high frequency band means significantly shorter hop lengths
Duplexing mode	Time division duplex (TDD)	FDD	A TDD system allows the flexibility to configure the entire bandwidth in a single DL direction. This means efficient BW utilization. MW as it is FDD is symmetric in terms of UL/DL throughput. Thus much lower spectral efficiency compared to TDD systems
Modulation from BPSK 1/2 up to 64 QAM	64 QAM	Up to 256QAM	For MW, the maximum sustained propagation with 256 QAM modulation is ~2 km for a a 0.6 m antenna using 13 GHz band
Maximum Throughput	up to 300 Mbps in 40 MHz and 600 Mbps in 2×40 MHz	Maximum of 1 Gbps with 2×56 MHz and using the highest modulation	Antenna size of OFDM is smaller than the MW antenna at same throughput and OFDM at lower frequency bands be extended to 50 km
Antenna and ODU	Integrated	ODU is separate from IDU	–
Maximum converge distance	50 km at 3.5 GHz with 0.6 m antenna	20 km with 13 GHz with less than 100 Mbps throughput and 1.8 m antenna	–
Total sectors per sites	Up to six sectors	Does not apply for PTP	Then high capacity hub site can be used to cater for several small cells with high throughput requirement and with trunking efficiency
Peer to peer communication through BS	Peers on the same base-station can communicate directly through the base-station	This advantage is not present in a PTP system	Therefore, CPEs belong to the same hub site can have local connectivity. This option supports eNB connectivity via an X2 interface and thus reduces the effort to have X2 connectivity at a POP site or at the core site

Table 8.3 (*continued*)

Specifications	PTP/PTMP	Legacy PTP MW	Remarks
System capability (LoS, OLoS (obstructed line of sight), and NLoS (non-line of sight))	All supported	LoS only	Then clear LoS and strict alignment is needed with MW while this is not the case with OFDM PtMP at lower bands
Number of CPE per hub site	10–50 CPEs	1	So, 50 small cell can be backhauled using one hub site
Support IEEE 802.3af POE protocol	Yes	No	MW usually need DC rectifier which is not practical with small cell deployment
MIMO with cross-polarization	Yes with adaptive MIMO for capacity and coverage scenarios	Only XPIC where switch between 1 + 1 and 2 + 0 is not possible after installation	PTMP can use different polarization to reuse same frequency in back to back sectors.

Table 8.4 illustrates a comparison between three PTMP technologies. As shown in the comparison, the MW/PTMP lacks the peak throughput of the OFDM/PTMP. Furthermore, the OFDM/PTMP is more practical in terms of easy installation, light product, support of native IP, and POE (power over ethernet). The MW/PTMP is the first developed PTMP and it has been successfully deployed with 2G and 3G. However, after the LTE evolution, it became difficult to use it, considering the LTE peak throughput requirement. The PTMP at higher frequency bands such as 32 GHz and 42 GHz represents a good opportunity for small cell backhaling thanks to higher capacity per sector and spectrum availability. Also, the spectrum fees at such higher bands is lower than other MW bands. The only limitation is the link distance and the sensitivity to weather conditions specially in regions with heavy rain.

The wireless backhauling solution for small cells can typically include the following capabilities:

- Availability: from 99.99 to 99.999, redundancy may be added for higher reliability.
- Capacity: 300 Mbps to 1 Gbps for one sector, for trunking efficiency.
- Security: AES (advanced encryption standard) 256 encryption available for secure applications when required.
- Triple play services should be tested and deployed on the proposed solution.
- Multi-topology: connects single (PTP) or multiple (PTMP) sites.
- Bandwidth allocation: flexible upload/download configurations such as TDD mode technologies.

- Easy deployment: multiple antenna sizes and implementation options for esthetically sensitive sites.
- High packet size support: support of high MTU (maximum transmission unit) sizes (giant packets).
- Low latency: with less than 1 ms per link.
- Very low jitter (less than tens of picoseconds) and low packet loss (less than 1% packet drops).

8.4.2   Key Features for HetNet Deployment

The HetNet is introduced as an evolution path for the radio access network (RAN), where limited spectrum resources do not meet the data hungry application and services. Therefore, it mainly targets the introduction of small cells with multi-technology and multi-band in order to overcome the limitation in the resources.

On the other hand, the small cell is low power nodes installed within the macro layout to boost the capacity or/and to enhance the coverage. The LTE standard adopts several bands

Table 8.4   Comparison between PTMP technologies

Product specifications	OFDM PTP/PTMP at lower bands	PTMP MW at 26 GHz/28 GHz	PTMP at 32 GHz/42 GHz
Supported bands	3.5 and 5 GHz	26 and 28 GHz	42 or 32 GHz
Technology	OFDM	ATM over the air and currently is evolving to native IP	Native IP with FDM
Application	3G/LTE/SME/small cell backhauling	2G/3G/small cell backhauling	3G/LTE/SME/small cell backhauling
Channel BW	40 MHz	28 MHz	Up to 1 GHz
Duplexing mode	TDD	FDD	TDD
Interfaces	Native IP	E1/ATM/IP	Native IP
Peak sector throughput	300 Mbps (600 Mbps with 2 × 40MHz)	Maximum of 150 Mbps	Up to 2 Gbps
CPE antenna and ODU	Integrated or separated	Separated	Integrated
Sector coverage distance	Up to 50 km with 3.5 GHz	5 km with 26 GHz	3 km with 42 GHz/32 GHz
System capability (LOS, OLOS, and NLOS)	All supported	LOS	LOS
Support of POE (power over ethernet)	Yes	No	Yes
MIMO with cross polarization	Supported	Not supported	Not supported

for small cells, both in FDD and TDD modes. In a typical heterogeneous deployment, the mobile operators are expected to have at least two bands for LTE and may use the TDD as an offloading layer where the spectrum is usually cheaper than the FDD option.

The HetNet guarantees the best user experience and efficient use of the available resources, with the following key concepts:

- Seamless handover across different layers and different bands.
- Efficient and dynamic utilization of available network resources and available spectrum.
- Deployment of necessary smart offloading for best user experience and maximum utilization of the resources.
- Interference cancellation techniques, such as eICIC to improve user experience and enhance the spectral efficiency.
- Adopting the SON to deliver an operational competitive network with optimized OPEX (operational expenditure).

The interference and traffic management techniques are key factors for the overall Het-Net performance. Therefore, re-design and re-evaluation is required for the topics like: inter/intra-RAT mobility optimization, load balancing between RATs, WiFi offloading, inter/intra-band mobility and load balancing, dynamic ICIC, enhanced ICIC, and CoMP.

8.5 Inter-Cell Interference Coordination (ICIC)

In LTE systems, the users occupy physical RBs in both downlink (OFDMA orthogonal frequency division multiple access) and the uplink (SC-FDMA, single-carrier frequency division multiple access) in a cell. The eNB ensures orthogonality in the frequency domain and, therefore, the intra-cell interference is typically low. However, the inter-cell interference can be relatively high, because all neighboring cells can provide services over the entire system band (i.e., the frequency re-use factor is 1). As a result, the inter-cell interference is severe for cell-edge users, in particular.

To increase the cell-edge users' throughput and enhance the coverage area, the inter-cell interference must be mitigated. ICIC is one technique that applies power control and eNB scheduling methods to combat the inter-cell interference. ICIC divides the entire system band into several frequency bands and utilizes them differently, especially at the overlapping regions between the neighboring cells. Hence, the users at the cell edge are preferentially scheduled in the portion of the bands that can effectively mitigate the inter-cell interference.

ICIC, in concept, works as a "cell-breathing" technique, for which the overall band of a cell is adjusted when determining the allocation of different edge-bands to the neighboring cells. The edge-band adjustment takes into account the inter-cell interference and the cell load. In doing so, the eNB can differentiate users within a cell by their signal strength, into near and far cell users. Accordingly, the eNB scheduler assigns different portions of the RBs, and can also provide different power control values for each identified type of UE.

To utilize this technique efficiently, the neighboring cells need to exchange information about the interference and load levels over X2 interfaces. Once any of the edge-band on a cell is being adjusted, the cell needs to notify its neighboring cells of the changes in order for the neighbors to be able to further adjust their own edge-bands. Figure 8.17 illustrates the main concepts of ICIC for UEs in cell-edge and near-cell conditions.

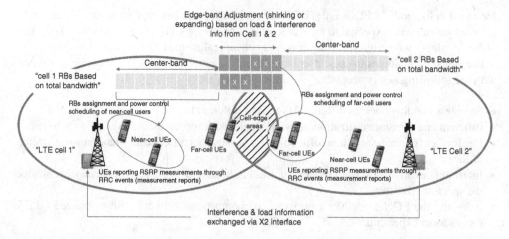

Figure 8.17   ICIC concepts and techniques.

As shown in the figure, the neighboring cells shrink or expand their edge-bands to serve users in different conditions. The coordination between the cells is applied via an X2 interface. Additionally, the adjustment of the edge-bands is done in different methods described next.

The downlink inter-cell interference is mainly caused by eNB power at the cell edge while the uplink inter-cell interference is mainly caused by the transmit power of the UEs at the cell edge. Because the cell-edge users incur interference in both downlink and uplink, ICIC concepts apply to the downlink and uplink scheduling. Hence, the edge-band is adjusted for the downlink PDSCH and/or uplink PUSCH physical channels, as shown in the illustration of Figure 8.17. There are two main types of ICIC: static ICIC, and semi-static ICIC.

In static ICIC, the system statically assigns non-overlapping frequency bands for cell-edge UEs of adjacent cells to cancel the frequency interference. The simple implementation of this approach is by using frequency re-use 3, or what is called fractional frequency re-use. The limitation in this approach is the degradation in the throughput as the users on the cell edge would not be able to use the entire bandwidth in order to mitigate the inter-cell interference. In addition, the users near to the cell center can be possibly limited to certain power levels to avoid interference with the cell-edge users. In addition, the static ICIC can be deployed for both the downlink and uplink traffic.

On the other hand, in the semi-static ICIC technique, the system uses the power control and the eNB scheduler to dynamically assign the frequency resources for cell-edge users. This in turn reduces the inter-cell interference and improves the throughput of the cell-edge users. The frequency resource assignment depends on ICIC messages from neighbor cells. These messages are used to select neighbor cells for ICIC based on the previous message and the reported RSRP, in addition to the overall cell loading. The semi-static ICIC is more efficient in scenarios where the loading on the neighbor cells is unbalanced.

ICIC deployment is very effective in cases where a high number of eNBs are deployed to provide continuous LTE coverage. A typical example is an urban area with low inter-site distance. In addition, ICIC is beneficial when the cells' overlapping areas is large and the traffic utilization becomes high (i.e., when RB utilization is high), causing an increasing interference in such areas. Therefore, these conditions fit within the HetNet deployment where eNBs are densely deployed.

ICIC concepts have been introduced in 3GPP Release 8 and 9. Several enhancements are applied in Release 10, referred to as enhanced inter-cell interference coordination. eICIC utilizes concepts such as almost blanket subframes (ABS), where it can mute certain subframes of one-layer cells to reduce interference to the other layer. This becomes important for HetNet, particularly where multiple layers are deployed with different types of cells.

8.6 Coordinated Multi-Point Transmission and Reception

CoMP transmission and reception is considered for LTE-Advanced Release 11 as a tool to improve the coverage of high data rates, the cell-edge throughput, and also to increase system throughput [6]. In LTE, the uplink signals can be received by the serving cell and neighbor cells at the cell edge or in cell overlapping conditions. LTE does not support the soft handover concept of 3G, where a UE can have multiple cells in its active set. Soft handover in 3G typically achieves macro diversity gains where signals on the uplink from a user are combined, improving the performance at the cell edge. LTE therefore provides a similar concept with enhanced performance by applying CoMP techniques, to improve the cell-edge throughput. Unlike 3G system, with LTE the CoMP can be applied in the DL and UL.

The possible CoMP scenarios include [6]:

- Inter- and intra-site CoMP in homogeneous macro networks.
- Coordination between a cell(s) and the distributed RRHs (remote radio heads) connected to the cell(s): negligible latency is assumed over the interface between a cell(s) and the RRHs connected to the cell(s). The RRHs may or may not form separate cells from the cell to which they are connected.
- Coordination between different cell layers and within a cell layer in HetNet: coordination is performed between a macro cell(s) and small cells in the coverage of the macro cell(s). The small cells may be non-uniformly distributed in the coverage of a macro cell(s).

The following scenarios are evaluated in next section for DL and UL CoMP in [6]:

1. **Scenario 1**: Homogeneous network with intra-site CoMP, as illustrated in Figure 8.18.
2. **Scenario 2**: Homogeneous network with high Tx power RRHs, as illustrated in Figure 8.19. A reference CoMP coordination cell layout for Scenario 2 is illustrated in Figure 8.20.
3. **Scenario 3/4**: HetNet with low power RRHs within the macrocell coverage where the transmission/reception points created by the RRHs have different cell IDs (Scenario 3) or the same cell IDs (Scenario 4) as the macro cell, as illustrated in Figure 8.21.

8.6.1 DL CoMP Categories

The DL CoMP scheme may be categorized into one of the following categories.

1. **Joint processing (JP)**: Data for a UE is available at more than one point in the CoMP cooperating set (definition follows) for a time-frequency resource

(a) **Joint transmission (JT)**: Simultaneous data transmission from multiple points (part of or entire CoMP cooperating set) to a single UE or multiple UEs in a time–frequency resource

(b) **Dynamic point selection (DPS)/muting**: Data transmission from one point (within the CoMP cooperating set) in a time–frequency resource. The transmitting/muting point may change from one subframe to another, including varying over the RB pairs within a subframe. Data are available simultaneously at multiple points. This includes dynamic cell selection (DCS)

(c) **Joint DPS and JT**: DPS may be combined with JT, in which case multiple points can be selected for data transmission in the time–frequency resource.

2. **Coordinated scheduling/coordinated beamforming (CS/CB)**: Data for a UE are only available at, and transmitted from one point in the CoMP cooperating set (DL data transmission is done from that point) for a time–frequency resource but user scheduling/beamforming decisions are made with coordination among points corresponding to the CoMP cooperating set. The transmitting points are chosen semi-statically via semi-static point selection (SSPS). In the SSPS, the transmission to a specific UE is from one point at a time. The transmitting point may only change in a semi-static manner.

3. **Hybrid JP and CS/CB mode**: Data for a UE may be available only in a subset of points in the CoMP cooperating set for a time–frequency resource but user scheduling/beamforming decisions are made with coordination among points corresponding to the CoMP cooperating set. For example, some points in the cooperating set may transmit data to the target UE according to JP while other points in the cooperating set may perform CS/CB.

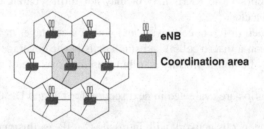

Figure 8.18   Scenario 1 – Homogeneous network with intra-site CoMP. (Source: [6] 3GPP. Reproduced with permission of ETSI.)

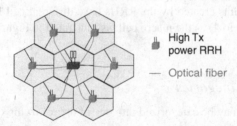

Figure 8.19   Scenario 2 – Homogeneous network with high Tx power RRHs. (Source: [6] 3GPP. Reproduced with permission of ETSI.)

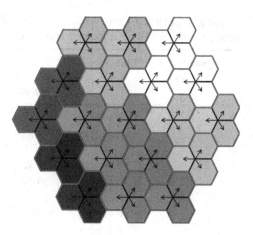

Figure 8.20 Reference CoMP Coordination Cell Layout for Scenario 2. (Source: [6] 3GPP. Reproduced with permission of ETSI.)

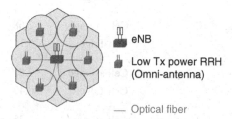

Figure 8.21 Scenario 3/4 – Network with low power RRHs within the macrocell coverage. (Source: [6] 3GPP. Reproduced with permission of ETSI.)

8.6.2 UL CoMP Categories

The UL CoMP scheme may be categorized into one of the following categories.

1. **Joint reception (JR)**: PUSCH transmitted by the UE is received jointly at multiple points (part of or entire CoMP cooperating set) at a time, for example, to improve the received signal quality. This similar to soft handover in UMTS.
2. **CS/CB**: user scheduling and precoding selection decisions are made with coordination among points corresponding to the CoMP cooperating set. Data are intended for one point only.

8.6.3 Performance Evaluation of CoMP

Performance evaluation for different CoMP scenarios is conducted in [6]. Table 8.5 shows the relative gains of CoMP schemes versus single-cell schemes in Scenario 1 FDD DL with closely-spaced cross-polarized antenna and ULA (uniform linear array) deployment scenarios and assuming a full buffer traffic model for the 3GPP case 1 channel model. Table 8.5 also demonstrates the relative gain of CoMP JR schemes versus single-cell schemes in Scenario

Table 8.5 Relative performance gain of DL/UL CoMP in Scenario 1 with full buffer cross-polarized and ULA deployments with FDD/3GPP case 1 channel model [6]

Test scenario	Number of sources	Location	Average gain of cross-polarized antenna (%)	Average gain of ULA (%)
DL CS/CB SU-MIMO 2×2 vs. SU-MIMO 2×2	1	Cell average	3.67	5.15
		Cell-Edge	9.63	11.64
DL CS/CB SU-MIMO 4×2 vs. SU-MIMO 4×2	2	Cell average	7.72	1.89
		Cell-Edge	29.23	13.66
Dl CS/CB MU-MIMO 2×2 vs. MU-MIMO 2×2	2	Cell average	2.23	5.26
		Cell-Edge	4.35	9.17
DL CS/CB MU-MIMO 4×2 vs. MU-MIMO 4×2	4	Cell average	1.00	3.11
		Cell-Edge	4.11	4.10
DL JT MU-MIMO 2×2 vs. MU-MIMO 2×2	7	Cell average	2.68	12.68
		Cell-Edge	26.13	36.68
DL JT MU-MIMO 4×2 vs. MU-MIMO 4×2	7	Cell average	2.81	2.12
		Cell-Edge	20.42	20.63
UL JR SU-MIMO 1×2 vs. SU-MIMO 1×2	1	Cell average	22.25	12.15
		Cell-Edge	41.19	22.00
UL JR MU-MIMO 1×8 vs. MU-MIMO 1×8	1	Cell average	20.20	5.08
		Cell-Edge	25.00	23.08

(Source: [6] 3GPP. Reproduced with permission of ETSI.)

Table 8.6 Relative performance gain of DL/UL CoMP in Scenarios 3 and 4 with full buffer traffic model for uniform UE distribution [6]

FDD full buffer	CoMP JP scn3/4 gains		CoMP CS/CB scn3/4 gains	
	Macro cell area average (%)	5% worst user (%)	Macro cell area average (%)	5% worst user (%)
DL relative gain versus HetNet without eICIC	3.0	24.1	5.1	24.8
DL relative gain versus HetNet with eICIC	3.3	52.8	2.7	19.7
UL relative gain versus HetNet without eICIC	13.5	39.7	N/A	N/A

(Source: [6] 3GPP. Reproduced with permission of ETSI.)

1 FDD uplink with closely-spaced cross-polarized antenna and ULA (uniform linear array) deployment scenarios and assuming a full buffer traffic model for the 3GPP case 1 channel model. With DL scenarios, the average of the cell average throughput improvement over all scenarios is 4%, while the average of cell-edge throughput improvement of all scenarios is 16%. With UL scenarios, the average of the cell average throughput improvement over all scenarios is 15% while the average of cell-edge throughput improvement of all scenarios is 28%. Therefore, DL and UL CoMP will bring major gain to users on the cell edge. Moreover, the UL CoMP offers significant gains compared to DL CoMP.

Table 8.6 provides the relative performance gains of downlink and uplink CoMP in scenarios 3 and 4 with full buffer traffic model for the uniform UE distribution case. We can observe from Table 8.6 that the UL CoMP offers significant average cell throughput gain compared to the DL CoMP.

According to the above performance results, the CoMP offers major performance gain in HetNet deployment with all scenarios. It is worth mentioning that the vendors' roadmaps indicate that the UL CoMP will be deployed before the DL CoMP.

8.7   Self-Organizing, Self-Optimizing Networks (SON)

An increase in the number of network elements in the radio network leads to complex network maintenance and high costs. To address this problem, the LTE system requires that the E-UTRAN supports self-configuration and self-optimization processes, SON.

The initial effort of SON has been provided by the NGMN (Next Generation Mobile Network). A group of operators created the NGMN alliance with the objective to provide business requirements to the new technologies being developed by standardization forums. NGMN's first efforts included high-level requirements for self-optimization network strategies [7]. NGMN's recommendations on SON requirements are done in such a way that they are high-level and also align with the 3GPP features and enhancements. NGMN's efforts stand as an initial guidance to the standards development and, in some cases, complement existing standard functionality. Several of NGMN's SON use cases have already been introduced in the 3GPP standards.

Other organizations contributing to SON developments are LSTI (LTE SAE Trial Initiative) and Socrates (Self-Optimization and Self-Configuration in Wireless Networks). LSTI provides 3GPP with the development of the SON test descriptions and test plans. On the other hand, Socrates is a project linked to NGMN and 3GPP that describes and sets the technical requirements for the performance and functioning of the SON algorithms, in addition to SON solution demonstration and verification.

3GPP initiated the standardization effort of self-optimizing and self-organizing capabilities for LTE starting in Release 8 and Release 9. The standards provide several techniques of network intelligence and automation management features in order to organize and optimize the network elements of the existing and added base-stations, to freely adapt with the surrounding cells and other underlying RATs deployed heterogeneously or homogeneously. The 3GPP effort has advanced in Release 10 with additional coverage and capacity add-ons which allow inter-RAT operation, ICIC, and minimization of operational expenses through minimization of drive testing.

SON features can be categorized into four main components, as shown in Table 8.7. Among these features, the seemingly appealing ones for current LTE deployments are in the

Table 8.7 Basic 3GPP SON features

SON feature	Basic functionality
Self-configuration	• eNB automatic discovery • Plug and play installation • Self-test and report
Self-planning	• Derivation of initial network parameters • Minimize radio network planning • Auto eNB configuration planning • Auto-discovery of environments
Self-optimization	• Parameter optimization with commercial terminal assistance • Reduce driver test • Improve network quality and performance
Self-maintenance, self-healing	• Automatic problem detection • Automatic problem mitigation/solving • Real time performance management • Automatic inventory management

self-optimization category. The self-optimization category typically consists of the following functions:

1. Automatic neighbor relation (ANR)
2. Mobility robust optimization (MRO)
3. Mobility load balancing (MLB).

8.7.1 Automatic Neighbor Relation (ANR)

ANR is a self-optimization function that maintains the efficiency of neighbor cell lists and neighbor relation tables between cells of different topologies or technologies. This function increases handover success rates and improves network performance. In addition, ANR typically operates with a minimal manual intervention, which reduces the costs of network planning and optimization.

For LTE, ANR optimizes the network performance in the cases of missing neighboring cells for handover, physical cell identifier (PCI) conflicts, and abnormal neighboring cell coverage. Based on neighbor relations, ANR is classified into intra-RAT ANR and inter-RAT ANR. These functions are expected to handle the neighbor relations within the same RAT (i.e., LTE intra and inter-frequency) as well as inter-RAT between LTE and UMTS, for example.

ANR functions in most scenarios are left up to each vendor's implementation. However, some standardized procedures are related to the way the UE makes the measurements and reports them. The ANR capabilities of a UE refer to the ability of the UE to read the ECGI (E-UTRAN cell global identifier) of neighboring cells. According to [8], the feature group indicators (FGIs) bit string contained in the UE capability. [3] The ANR function in the serving eNB may therefore request the UE to retrieve the ECGI. Then, the serving eNB can set up

---

[3] Refer to Table 1.8 in Chapter 1 for ECGI, and refer to Chapter 3 for a complete description of FGI bits and an example of the UE's FGI reporting.

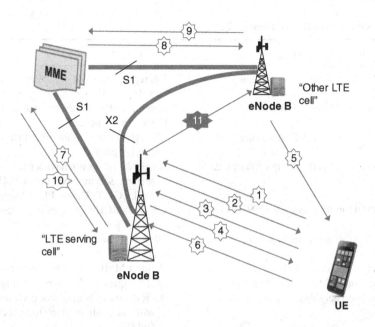

Figure 8.22   UE positioning architecture applicable to E-UTRAN.

the related X2 interfaces automatically to create a neighbor relation table. This ANR function provides the ability to detect the missing neighbors and then creates the neighbor relations for enhancing handover performance.

For example, consider Figure 8.22 of the ANR procedure for missing neighbor relations. The procedures from the figure are described in Table 8.8.

In step 3 in the illustrated procedures, the UE does not report the neighbor PCIs belonging to blacklist eNBs. For example, if the neighbor relation causes cross-cell coverage and leads to unstable handovers, known cells can be added as blacklisted. This procedure may require manual configurations of some cells that are part of the neighbor relation tables in each eNB's ANR database.

Additionally, based on the vendor's implementation, UE reporting for ECGI may not be taken into account to establish ANR functions without UE-assisted measurements. This avoids the case where a UE does not support the ANR functions part of the signaled FGI bits (a case with early deployed LTE devices, for example). However, this may need manual intervention to validate the sites planning and coverage.

8.7.2   Mobility Robust Optimization (MRO)

MRO targets minimizing handover failures, call drops, and improving handover delays. MRO achieves these requirements typically by automatic optimization of handover-related parameter settings. MRO's mobility and parameter optimization can therefore directly improve the user experience of data rate and service interruption in mobility conditions, and in turn, it enhances the network resources.

Table 8.8   Steps description of Figure 8.22

Step	Interface: direction	Description
1	LTE_U-u: from UE to serving eNB	UE reports its ANR capability with FGI Bits: 5, 16, 17, 18, 19
2	LTE_U-u: from serving eNB to UE	Serving eNB delivers configuration for UE measurements
3	LTE_U-u: from UE to serving eNB	UE reports the neighbor cell PCI to the serving eNB
4	LTE_U-u: from serving eNB to UE	If neighbor does not have a relation with serving, serving eNB instructs UE to read the ECGI, TAC, and PLMN neighbor cell
5	LTE_U-u: from neighbor eNB to UE	UE reads BCH messages to retrieve neighbor cell ECGI
6	LTE_U-u: from UE to serving eNB	UE reports neighbor ECGI to the serving eNB
7	S1: from serving eNB to MME	Request MME for X2 transport configuration
8	S1: from MME to neighbor eNB	MME requests X2 transport configuration
9	S1: from neighbor eNB to MME	ANR database in neighbor eNB responds with X2 configuration (logical connection and IP)
10	S1: from MME to serving eNB	Response delivered to serving eNB ANR database
11	X2: Between serving and neighbor eNBs	ANR database on both eNBs set up neighbor cell tables and create relations

MRO functions usually work in close coordination with other ANR functions, in particular, neighbor relations. Several handover failure scenarios can occur in mobility, including early handover, later handover or, to some extent, ping-pong handovers. Figure 8.23 describes the early and delayed handover and how MRO functions interact to resolve such failure scenarios.

From these examples, the way MRO identifies the optimization is by detecting the radio link failure (RLF) and at which cell the RRC re-establishment occurs, following the RLF.

The early handover failure occurs in scenarios where the UE experiences an RLF on a cell *during* the handover. The MRO detects an early handover when the RRC re-establishment attempt is sent by the UE to the original cell (source cell) indicating a suboptimal HO (handover) decision caused by parameters, or an overshooting cell. Another case could be where the UE experiences an RLF on a cell *after* the handover is successfully completed. The MRO detects an early handover depending on the time spent on the new cell until the RLF occurs, and the RRC re-establishment attempt is sent by the UE to the original cell indicating a suboptimal HO decision caused by parameters, or an overshooting cell.

The delayed handover failure occurs in scenarios where the UE experiences a RLF on a cell *during* or *after* the handover. The MRO detects a delayed handover when the RRC re-establishment attempt is sent by the UE to any other cell (not the original source cell) indicating a suboptimal HO decision caused by parameters, or an overshooting cell.

In both cases, the MRO database as well as a manual handover parameter optimization aims at moving the handover into the normal region. MRO, hence, fine-tunes the handover process

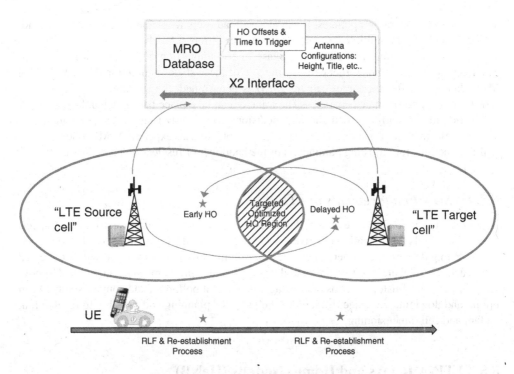

Figure 8.23   An example of MRO function.

based on a long-running evaluation of KPIs (key performance indicators) and specific detections in eNBs which is influenced by operator policies. After a RLF, the UE sends the RRC message (re-establishment) indicating the RLF information to the cell to which it camps after the failure. The MRO database analyzes the parameters and KPIs and then forwards the results in a handover report over the X2 to the cell (or other cells) that needs to rectify the parameters based on handover policies, including the neighbor cell relations defined in the ANR.

MRO eventually needs to interact with handover parameters defined over the air for event- and periodic-based measurements. One improved aspect of LTE is that 3GPP provides speed-dependent handover measurements scaling. 3GPP allows scaling of the handover parameters based on the number of handover measurements performed by the UE. It hence detects "medium or high" speed users and scales their handover parameters accordingly. In particular, it scales the handover time-to-trigger parameters to configurable levels, depending on the speed detection, as a function of the number of handovers performed within a pre-configured time-window.

8.7.3   Mobility Load Balancing (MLB)

The MLB database intelligently coordinates load distribution among intra or inter -RAT cells to maximize network resource usage, increase the access success rate, and improve user experience. To achieve these targets, the MLB checks the load status of the cells, exchanges cell load information, and spreads users from busy cells to cells with more available resources, according to the operator's policies. It can also offload users from one LTE carrier to another if LTE is deployed in multiple carriers in the same or different bands.

MLB can work at the user admission phase or for congestion avoidance during the call. The load in the source and target cells during any of these stages is monitored by a historical MLB database. After the target cell for MLB is determined, the serving cell selects some UEs to transfer between cells/carriers. The transfer can take place based on handovers or cell reselection load reshuffling methods. The decision of load reshuffling takes QoS (quality of service) information into account to ensure an acceptable user experience. MLB hence takes real-time actions for adjusting parameters of the handover or reselection procedures.

8.7.4 SON Enhancements in LTE-A

In addition to the features previously described for SON, prior to Release 10, LTE-A specification also adds minimization of drive tests (MDTs) into SON functions. MDT is designed for collecting data from commercial devices for optimization purposes [9]. The data collected from each UE are then used to monitor and detect coverage problems in the network. The coverage issues can be listed in terms of coverage holes, pilot pollution, and comparison between uplink and downlink coverage. It therefore helps in RF planning and possibly to re-plan link budget and cell dimensioning.

8.8 LTE-A Relays and Home eNodeBs (HeNB)

A relay node (RN) is a low-powered eNB connected to the remaining part of the network. Relays are effective for extending coverage; they also increase the user's data rate in limited-coverage indoor locations, for example. With relaying, the UE communicates to an RN that is connected to a donor cell (DeNB, donor eNodeB) using LTE radio access [10]. In addition to supporting the RN, the donor cell serves the UEs that are directly connected to it. Figure 8.24 illustrates the relaying E-UTRAN architecture.

The RN supports the eNB functionality where it terminates the radio protocols of the E-UTRA radio interface, and the S1 and X2 interfaces. RN also supports a subset of the UE functionality, for example, PHY, MAC, RLC, PDCP, RRC, and NAS functionality, in order to wirelessly connect to the DeNB [10].

The RN connects to the DeNB via the Un interface using the same radio protocols and procedures as a UE connecting to an eNB. The S1 and X2 user and control plane packets are mapped to radio bearers over the Un interface. The RRC layer of the Un interface has functionality to configure and reconfigure an RN subframe through the RN reconfiguration procedure, for transmissions between an RN and a DeNB. The PDCP layer of the Un interface has functionality to provide integrity protection for the user plane.

Handover to or from RN is also supported via DeNB. Based on the availability of the X2 interface between the DeNB and a neighbor eNB, the RN initiates either S1 or X2 handover for the UE. The S1/X2 handover request is sent by the RN to the DeNB, which reads the target cell ID from the message, finds the target node corresponding to the target cell ID, and forwards the message toward the target node to complete the handover.

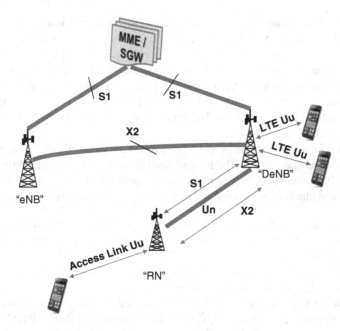

Figure 8.24   Relaying Architecture.

8.9   UE Positioning and Location-Based Services in LTE

8.9.1   LBS Overview

3GPP Release 9 provides a framework for defining the UE location (so-called UE positioning) in order to support a variety of location-based services (LBSs). In Release 9, the positioning reference signals (PRSs) have been introduced to facilitate the determination of the position of the UE, referred to as a UE-assisted positioning technique. A UE-assisted positioning technique involves the following:

- The UE makes some radio signal measurements and
- The network determines the UE location (e.g., latitude and longitude) by processing the measurements reported by the UE. This is conducted by a separate system that processes the reported radio information to identify the UE location.

The PRS are transmitted on antenna port 6 and are sent in a configurable number of consecutive subframes of up to five subframes. The E-UTRAN configures the PRS bandwidth (e.g., a certain number of RBs) and the periodicity of the PRS (e.g., one PRS occurrence every 160 subframes). Within a subframe containing the PRS, they are transmitted on more subcarriers and more OFDM symbols when compared to the regular cell-specific reference signals being sent on an antenna. The utilization of more time–frequency resources within a subframe by the PRS can improve the quality of the UE measurements compared to the use of only the basic cell-specific reference signals.

A pseudo-random sequence is sent on the PRS, and this sequence is a function of numerous factors such as PCI (physical layer cell identity), slot number, OFDM symbol number, and the cyclic prefix. The UE observes the PRS from different neighboring cells and makes certain measurements. Examples of such measurements include OTDOA (observed time difference of arrival) measurements such as RSTD (reference signal time difference). RSTD is the relative timing difference between a neighbor cell and the reference cell. The LBS system processes these OTDOA measurements from the UE (implementation-specific algorithms) to estimate the UE location [11].

In order for OTDOA to function properly, it is mandatory that the eNBs are synchronized to GPS accurate time (i.e., 50 ns). The relative transmission timing of the eNBs must also be known by the LBS system. The timing must also be known by the UEs to facilitate the search for the reference symbols. The accurate relative timing information may be obtained by:

- **Asynchronous mode** – Each eNB is connected to a local GPS receiver to time stamp its own transmissions. The eNB only measures its own time and it does not use GPS as a frame synchronization source.
- **Local synchronized mode** – Each eNB is connected to a local GPS receiver and uses GPS to align the eNB frames in the eUTRAN network.
- **Remote synchronized mode** – The eNB is connected to a remote clock synchronization source via a network time protocol (NTP) or a precise time protocol (PTP). The synchronization signals are transmitted over the transmission network to all eNB.

The uncertainty of the position measurement is network-implementation dependent. The uncertainty of the position information is dependent on the method used, the position of the UE within the coverage area, and the activity of the UE. Several design options of the E-UTRAN system (e.g., size of cell, adaptive antenna technique, path loss estimation, timing accuracy, eNode B surveys) allow the network operator to choose a suitable and cost-effective UE positioning method for their market. The uncertainty may vary between networks as well as from one area within a network to another, based on the clutter type. The uncertainty may be hundreds of meters in some areas and only a few meters in others. In the event that a particular position measurement is provided through a UE-assisted process, the uncertainty may also depend on the capabilities of the UE. In some jurisdictions, there is a regulatory requirement for location service accuracy that is part of an emergency service [12, 13].

A sample regulatory requirement for LBS accuracy is 300 m for 90% percentile and 150 m for 65% percentile. These values are an average accuracy requirement for a network. The accuracy for each clutter will be different due to the inter-site distance. Therefore, with dense urban and urban clutters the accuracy is expected to be better than the values mentioned above, while with suburban and rural, it will be degraded below these values. Figure 8.25 represents a typical CDF (cumulative density function) for the accuracy at different clutters and the average accuracy, which meets the above requirement.

There are many different possible uses for the positioning information. The positioning functions may be used internally by the EPS, by value-added network services, by the UE itself or through the network, and by "third party" services. It may also be used by an emergency service (which may be mandated or "value-added"), but the location service is not exclusively for emergencies.

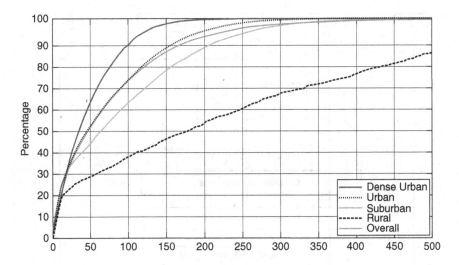

Figure 8.25 CDFs for accuracy for different clutters and overall accuracy.

8.9.2 LTE Positioning Architecture

Figure 8.26 illustrates the EPS architecture for the UE's positioning with E-UTRAN access.

The MME (mobility management entity) receives a request for some location service associated with a particular target UE from another entity (e.g., gateway mobile location center, GMLC). Alternatively, the MME can decide to initiate some location service on behalf of a particular target UE (e.g., for an IMS (Internet protocol multimedia subsystem) emergency call from the UE). The MME then sends a location services request to an E-SMLC (evolved serving mobile location center). The E-SMLC processes the location services request, which may

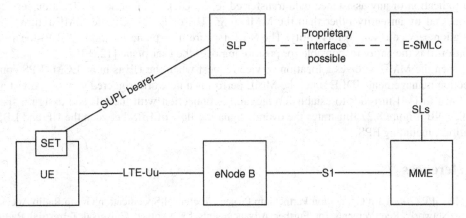

Figure 8.26 UE Positioning Architecture applicable to E-UTRAN. (Source: [12] 3GPP. Reproduced with permission of ETSI.)

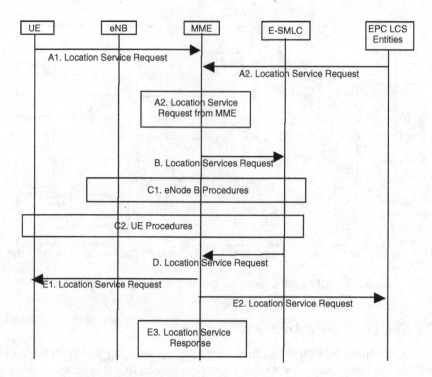

Figure 8.27 Location service signaling flow. (Source: [12] 3GPP. Reproduced with permission of ETSI.)

include transferring assistance data to the target UE to assist with UE-based and/or UE-assisted positioning and/or may include positioning of the target UE. The E-SMLC then returns the result of the location service back to the MME (e.g., a position estimate for the UE and/or an indication of any assistance data transferred to the UE). In the case of a location service requested by an entity other than the MME (e.g., UE or E-SMLC); the MME returns the location service result to this entity. The S-LP (secure user plane location (SULP)-location platform) is the entity responsible for positioning over the user plane [12, 13].

When the MME receives a location service request when the UE is in an ECM (EPS connection management -IDLE state, the MME performs a network triggered service request as defined in [14] in order to establish a signaling connection with the UE and assign a specific eNB. Figure 8.27 illustrates the overall signaling flow of LBS between the UE and LBS entities, including EPS.

References

[1] 3GPP (2009) 3rd Generation Partnership Project; Technical Specification Group Radio Access Network; Requirements for Further Advancements for Evolved Universal Terrestrial Radio Access (E-UTRA), Release 9. TS 36.913 V9.0.0.
[2] 3GPP (2011) 3rd Generation Partnership Project; Technical Specification Group Radio Access Network; Requirements for Further Advancements for Evolved Universal Terrestrial Radio Access (E-UTRA) (LTE-Advanced) (Release 10). TR 36.913 V10.0.0.

[3] 3GPP (2012) 3rd Generation Partnership Project; Technical Specification Group Radio Access Network; Evolved Universal Terrestrial Radio Access (E-UTRA); User Equipment (UE) Radio Access Capabilities. TS 36.306 V10.7.0.

[4] 3GPP (2009) 3rd Generation Partnership Project; Technical Specification Group Radio Access Network; Evolved Universal Terrestrial Radio Access (E-UTRA); Physical Layer Procedures. TS 36.213 V8.8.0.

[5] 3GPP (2012) 3rd Generation Partnership Project; Technical Specification Group Radio Access Network; Evolved Universal Terrestrial Radio Access (E-UTRA); User Equipment (UE) Radio Transmission and Reception (Release 11). TR 36.101 V11.3.0.

[6] 3GPP (2011)Coordinated Multi-point Operation for LTE Physical Layer Aspects. TR36.819 V11.1.0

[7] NGMN Requirement Document (2008) NGMN Recommendations on SON and OAM Requirements, December 2008.

[8] 3GPP (2010, 2011) 3rd Generation Partnership Project; Technical Specification Group Radio Access Network; Evolved Universal Terrestrial Radio Access (E-UTRA); Radio Resource Control (RRC); Protocol Specification (Release 8), (Release 9). TS 36.331 V8.9.0, 2010 and V9.9.0, 2011.

[9] 3GPP (2012) 3rd Generation Partnership Project; Technical Specification Group Radio Access Network; Universal Terrestrial Radio Access (UTRA) and Evolved Universal Terrestrial Radio Access (E-UTRA); Radio Measurement Collection for Minimization of Drive Tests (MDT); Overall Description. TS 37.320 V11.1.0.

[10] 3GPP (2012) 3rd Generation Partnership Project; Technical Specification Group Radio Access Network; Evolved Universal Terrestrial Radio Access (E-UTRA) and Evolved Universal Terrestrial Radio Access Network (E-UTRAN). TS 36.300 V11.3.0.

[11] Kottkamp, M., Rössler, A., Schlienz, J., and Schütz, J. (2011) LTE Release 9 Technology introduction. R&S White Paper.

[12] 3GPP (2012) 3rd Generation Partnership Project; Technical Specification Group Radio Access Network; Evolved Universal Terrestrial Radio Access Network (E-UTRAN); Stage 2 Functional Specification of User Equipment (UE) positioning in E-UTRAN. TS 36.305 V11.1.0.

[13] 3GPP (2012) 3rd Generation Partnership Project; Technical Specification Group Services and System Aspects; Functional Stage 2 Description of Location Services (LCS). TS 23.271 V11.0.0.

[14] 3GPP (2012) 3rd Generation Partnership Project; Technical Specification Group Services and System Aspects; General Packet Radio Service (GPRS) Enhancements for Evolved Universal Terrestrial Radio Access Network (E-UTRAN) Access. TS 23.401 V11.3.0.

Index

Design, Deployment and Performance of 4G-LTE Networks: A Practical Approach, First Edition. Ayman Elnashar, Mohamed A. El-saidny and Mahmoud R. Sherif.
© 2014 John Wiley & Sons, Ltd. Published 2014 by John Wiley & Sons, Ltd.

Printed in the United States
By Bookmasters

Printed in the United States
By Bookmasters